MW01264987

Clinical Aspects of

Functional Foods and Nutraceuticals

Clinical Aspects of

Functional Foods and Nutraceuticals

Edited by

Dilip Ghosh

Debasis Bagchi

Tetsuya Konishi

CRC Press is an imprint of the
Taylor & Francis Group, an **informa** business

CRC Press
Taylor & Francis Group
6000 Broken Sound Parkway NW, Suite 300
Boca Raton, FL 33487-2742

Printed on acid-free paper
Version Date: 20140716

International Standard Book Number-13: 978-1-4665-6910-2 (Hardback)

Visit the Taylor & Francis Web site at
http://www.taylorandfrancis.com

and the CRC Press Web site at
http://www.crcpress.com

Dedication

To my beloved father, the late Mr. Tarak Chandra Bagchi, MSc, AIC.

Contents

Preface

In the past three decades, revolutionary achievements have taken place in the functional food and nutraceutical research, including the introduction of a number of cutting-edge dietary supplements backed by human trials and strong patents. Novel manufacturing technologies including unique extraction process and biodelivery such as nanotechnology and innovative packaging are very important steps for the successful positioning in marketplace. Growing clinical support is another tool for coping up with the stringent regulatory requirements, and finally consumer acceptance and appreciation around the world.

The book covers vast areas of functional food and nutraceuticals from a clinical nutritional perspective. In addition to science underpinning, it also focuses on food innovation, global regulations, problems and challenges, and future direction. It provides an essential overview of the clinical aspects of functional foods and nutraceuticals for key stakeholders, drawing links between areas of knowledge that are often isolated from each other. This form of knowledge integration is essential for practice, especially for policy makers and administrators in government, private, or academic sector.

This book includes six different sections that are explained in detail below.

INTRODUCTION

This section demonstrates the basic understanding on the current scenario in the field of functional foods and nutraceuticals. Four chapters are included in this section. Dr. Yasmine Probst and collaborators from the Smart Foods Centre, University of Wollongong, NSW, Australia, highlighted the progress of food-based dietary guidelines around the globe in Chapter 1. Since "Ayurveda" is an integral and vital component of functional foods and nutraceuticals, Chapter 2 was designed by Dr. Ashok Vaidya and collaborators from Kasturba Health Society, Mumbai, India covering the clinical perspectives of Ayurceuticals and its challenges and opportunities for global health and wellness. Dr. Chunling Wang, a reputed nutrition scientist from COFCO Nutrition and Health Research Institute, Beijing, China, formalized Chapter 3 and Ms Rajshri Roy, accrediting practicing dietitian from the University of Sydney, emphasized about the clinical trials in Chapter 4.

SCIENCE UNDERPINNING HEALTH BENEFITS

The health benefits of functional foods are extensively demonstrated from the pharmacological, toxicological, and clinical standpoints. Dr. Tetsuya Konishi, coeditor of this book and professor in Niigata University of Pharmacy and Applied Life Sciences (NUPALS), Niigata, Japan, discussed about the dual function of food factors in Chapter 5. Dr. Akira Murakami from Kyoto University, Kyoto, Japan, emphasized about the anti-inflammatory and chemopreventive potentials in Chapter 6. Dr. Junji Terao from Tokushima University exhibited the key food factors instrumental to prevent muscle atrophy, especially "sarcopenia," in Chapter 7. Chapter 8 was designed by Dr. Asim K. Duttaroy, University of Oslo School of Medicine, Oslo, Norway on metabolic syndromes and associated concepts. Dr. Kennichi Watanabe from NUPALS, Niigata, Japan, outlined the novel cardioprotection by functional foods in Chapter 9. Chapter 10 was structured by Dr. Akira Kubo of Tokai University, Tokyo, Japan, which discusses about the various diets and their effects on aging.

PROBLEMS AND CHALLENGES TO INDUSTRIES, CONSUMERS, AND POLICY MAKERS

The continuing challenges to the industries, consumers, and policy makers are broadly discussed in this section. Dr. Namrata Taneja from Riddet Institute, New Zealand, designed Chapter 11, which elaborates about probiotics. Chapter 12 on food and drug interaction was highlighted by Dr. Noriaki Yoko, NUPALS, Niigata, Japan. Dr. R.B. Smarta from Interlink Marketing Consultancy, Mumbai, India, formalized Chapter 13, which discusses on the growth of both pharmaceutical and food industries; while Dr. Yuji Naito from Kyoto Prefectural University of Medicine structured Chapter 14 emphasizing on heme oxygenase and its functions.

INNOVATION VERSUS REGULATION

In this section, the controversial issues of innovation versus regulation are discussed in four chapters. In Chapter 15, Dr. Tee E. Siong, the president of Nutrition Society of Malaysia, broadly discussed about functional foods and their effects in Southeast Asia. Dr. Toshio Shimizu, Nagoya Bunri University, Inazawa, Japan, extensively provided the comparative regulatory details in Chapter 16. In Chapter 17, Dr. George Burdock thoroughly discussed challenges associated with presumptive claim(s); while in Chapter 18, the effects of clinical research and clinical trial in health care was demonstrated by a renowned Advocate Kappillil Anilkumar.

FARM TO CLINIC APPROACH/FUNCTIONAL ASPECTS OF FOOD FACTORS

Chapter 19 designed by Drs. Andrea Zangara and Dilip Ghosh, Australia, discusses about herbal medicines. Dr. Shantanu Das of Riddet Institute, New Zealand, highlighted about the innovations in the food industry in Chapter 20, and Chapter 21 was extensively demonstrated by Dr. Bharat B. Aggarwal and his esteemed team from MD Anderson Cancer Center, Texas, extensively demonstrated the properties of Curcumin and its compounds. Chapter 22 formalized by Dr. Chin-Kun Wang, Chung San Medical University, Taiwan, discusses about functional foods use in human trials. Dr. Daniel Roytas from Southern School of Natural Therapies, Fitzroy, VIC, Australia, formalized Chapter 23, which elaborates about pertinent issues related to the safety and quality of Chinese herbal medicine.

FUTURE TRENDS

This section describes where we are and where to go. Dr. V. Prakash, an eminent scientist and the president of Nutrition Society of India, Hyderabad, India, demonstrated the concept of traditional foods and their values for health and wellness in Chapter 24, which was followed by Chapter 25 by Dr. Lynnette Ferguson, the University of Auckland, Auckland, New Zealand, dealing with the concept of food and its supplements of public health. Chapter 26 was formalized by Dr. Dennis Chang, University of Western Sydney, NSW, Australia, which elaborates about ginger and its effects on metabolic syndrome. Dr. Sumita Ghosh, University Technology of Sydney, NSW, Australia, highlighted the importance of local food and its impact on health in Chapter 27 and discussed about the pros and cons of the future of controversial food supplements and its wide acceptance in the community. Chapter 28 demonstrated by Dr. Young Rok Seo, Dongguk University, Seoul, Republic of Korea, deals with differing food preparations and consumptions. Chapter 29 emphasized by Dr. Parikshit Debnath and his esteemed colleagues, SDM College of Ayurveda and Hospital, Hassan, India, discusses about Ayurnutrigenomics.

Health, marketing, and business professionals are continuously struggling to establish the inarticulated needs in conjunction with regulatory hurdles and customer requirements for appropriate nutrition, optimal health, and disease prevention. Although clinical support of any health benefits from foods and nutraceuticals is essential to commercialize the products, an in-depth understanding of the traditional use and benefits is important while accepting such functional foods with medicinal properties. In the organizational context, commitment to clinical research and assimilation of novel technologies is the only way to boost efficiency, productivity, and finally the well-being of humankind.

We, the editors, performed our best to design and implement the best concepts from both Oriental and modern Western world on functional foods and nutraceuticals for our esteemed readers.

Dilip Ghosh
Nutriconnect

Debasis Bagchi
University of Houston

Tetsuya Konishi
Niigata University of Pharmacy and Applied Life Sciences (NUPALS)

Editors

Dilip Ghosh, PhD, FACN, has received his PhD in biomedical science from the University of Calcutta, Kolkata, India. Previously, he held positions in Organon (India) Ltd., a division of Organon International BV and Akzo Nobel, the Netherlands; HortResearch, New Zealand; the US Department of Agriculture (USDA)-Agricultural Research Service (ARS), Human Nutrition Research Center on Aging (HNRCA) at Tufts University, Boston, Massachusetts; the Smart Foods Centre, University of Wollongong, NSW, Australia; and Neptune Bio-Innovation Pty Ltd., NSW, Australia. Dr. Ghosh is a most sought-after international speaker, facilitator, and author. He has been involved for a long time in drug development (both synthetic and natural) and functional food research and development in both academic and industrial domains. He is a fellow of the American College of Nutrition, Clearwater, Florida, and also a member in the editorial board of several journals. Currently, he is the director at Nutriconnect, NSW, Australia; professionally involved with Soho Flordis International, University of Western Sydney, NSW, Australia; and honorary ambassador at Global Harmonization Initiative (GHI), NSW, Australia.

Dr. Ghosh has published more than 60 papers in peer-reviewed journals and numerous articles on food and nutrition in magazines and books. His two recent books, *Biotechnology in Functional Foods and Nutraceuticals* and *Innovation in Healthy and Functional Foods*, have created a strong impact in functional food and nutraceutical domains. He was an associate editor and member of *Toxicology Mechanisms and Methods* (2006–2007) and is a review editor for *Frontiers in Nutrigenomics* (2011 to present), *American Journal of Advanced Food Science and Technology* (2012 to present), *Journal of Obesity and Metabolic Research* (2013 to present), and *Journal of Bioethics* (2013 to present). He can be reached at dilipghosh@nutriconnect.com.au and dghosh@optusnet.com.au.

Debasis Bagchi, PhD, MACN, CNS, MAIChE, received his PhD in medicinal chemistry in 1982. He is a professor in the department of pharmacological and pharmaceutical sciences at the University of Houston, College of Pharmacy, Houston, Texas. Dr. Bagchi is also the chief scientific officer of Cepham Inc., Piscataway, New Jersey. He served as the senior vice president of Research & Development of InterHealth Nutraceuticals Inc., Benicia, California, from 1998 till February 2011, and then as the director of Innovation and Clinical Affairs at Iovate Health Sciences Research Inc., Oakville, Ontario, from February 2011 to June 2013. He received the Master of American College of Nutrition Award in early October 2010. He is currently the chairman of the International Society for Nutraceuticals and Functional Foods (ISNFF), USA, immediate past president of American College of Nutrition, Clearwater, Florida, and past chair of the Nutraceuticals and Functional Foods division of the Institute of Food Technologists (IFT), Chicago, Illinois. He is serving as a distinguished advisor on the Japanese Institute for Health Food Standards, Tokyo, Japan. He is a member of the Study Section and Peer Review Committee of the National Institutes of Health (NIH), Bethesda, Maryland. He has published 299 papers in peer-reviewed journals, 24 books, and numerous patents. He has delivered invited lectures in various national and international scientific conferences, organized workshops, and group discussion sessions. Dr. Bagchi is also a member of the Society of Toxicology, a member of the New York Academy of Sciences, a fellow of the Nutrition Research Academy, and a member of the TCE stakeholder Committee of the Wright Patterson Air Force Base, Fairborn, Ohio. He is the associate editor of the *Journal of Functional Foods* and the *Journal of the American College of Nutrition*, and also serving as an editorial board member of numerous peer-reviewed journals, including *Antioxidants & Redox Signaling, Cancer Letters, Toxicology Mechanisms and Methods, and The Original Internist.*

Dr. Bagchi received funding from various institutions and agencies including the US Air Force Office of Scientific Research, Arlington, Virginia; Nebraska Department of Health & Human Services, Lincoln, Nebraska; Biomedical Research Support Grant from the NIH, Bethesda, Maryland; National Cancer Institute (NCI), Bethesda, Maryland; Health Future Foundation Inc., Quezon City, Philippines; Procter & Gamble Co., Cincinnati, Ohio; and Abbott Laboratories, Abbott Park, Illinois.

Tetsuya Konishi, PhD, was born in 1944 at Shanghai in China and grew up in Akita, mostly in Tokyo, Japan. He graduated from Tokyo College of Pharmacy in 1966 and completed his graduate study in radiopharmacy in 1968. After that, he engaged in research and education as an assistant professor in the physical and analytical chemistry department of Tokyo College of Pharmacy and carried out metabolic studies on lipophilic vitamins using radioisotope tracers. After finishing his PhD in pharmaceutical sciences, he worked with Dr. Lester Packer as a postdoctoral research fellow at the University of California–Berkeley, Berkeley, California, and was also appointed as a research biochemist at the Energy and Environment division of Lawrence Berkeley National Laboratory, Berkeley, California, from 1975 to 1978, where his interest was focused on membrane bioenergetics studies, mainly on bacteriorhodopin, a phototransduction protein in halobacteria. He had his laboratory in the department of biophysics and radiochemistry in Niigata University of Pharmacy and Applied Life Sciences (NUPALS), Niigata, Japan, in 1978 to start his biophysical studies on molecular adaptation of halobacteria to an extreme environment as an associate professor and later as a full professor until 2000. During this period, his works were mainly focused on energy transduction by Na/H antiporter and also on radiation protection in halobacteria. The Na/H antiporter is the first protein found by him in that bacteria, which works as an essential device for energy transduction. Radical biology was an approach for the latter subject. In 2001, he established a functional and analytical food sciences department in the Faculty of Applied Life Sciences and has developed broad spectra of researches on the health beneficial functions of antioxidant natural products and traditional herbal medicines, which include such studies as metabolism and functions of anthocyanins, antioxidant herbal prescriptions in protecting cerebral oxidative damage and aging, and DNA damage checkpoint modulation by food factors. Since 2012 he is a professor emeritus of functional and analytical food sciences and a specific project professor for "The Basic Research on Developing Functional Food for Next Generation" in NUPALS, Niigata, Japan.

His special interest is focused on complementary application of food functions in modern medicine treatment, that is, adjuvant use. He published more than 160 original research papers and reviews in peer-reviewed journals. He has engaged in social activities such as a board member of several academic committees including Niigata International Food Award, Japanese Society of Food Factors, Japanese Society of Flavonoid Research, World Federation of Traditional Chinese Medicine, and so on, and is now the head of Organization for Food and Bio-Research promotion in Niigata. He has also engaged in the editorial board of several science journals such as *Journal of Pharmacological Research*, *Molecular Nutrition & Food Research*, *Cancer Chemotherapy*, *Journal of Biological & Pharmacological Research*, and *Medicine* as well as reviewer of many international journals. He has worked as an examiner of PhD theses at several abroad universities such as India, Hong Kong, and Germany, and as a grant evaluator of Austrian Science Academy. He founded and organized the International Niigata Symposium on Diet and Health (INSDH) in 2008, which is held at Niigata city every year. The INSDH is collaborated with the world-famous Oxygen Club of California (OCC) congress and Linus Pauling Institute (LPI) conference on optimum health and diet, in which he engaged as a board member. He also founded and directed NUPALS Liaison R/D Center from 2007 to 2010. He is directing a five-year consortium project titled "The Basic Research on Development of Functional Foods for Next Generation" at NUPALS supported by the Ministry of Science and Education started from 2010. He is now serving as a visiting professor of Changchun University of Chinese Medicine to develop the international collaborative research on functional food. He can be reached at konishi@nupals.ac.jp or haldkonytk@tlp.ne.jp.

Contributors

Bharat B. Aggarwal
Department of Experimental Therapeutics
The University of Texas: MD Anderson Cancer Center
Houston, Texas

M.A. Alwar
Samskriti Foundation
Mysore, India

Kappillil Anilkumar
The Lawyers' Syndicate
High Court of Kerala
Ernakulam, India

K.I. Anitha
Health & Research Centre
Thiruvananthapuram, India

Somasundaram Arumugam
Department of Clinical Pharmacology
Niigata University of Pharmacy and Applied Life Sciences
Niigata, Japan

Josephine M. Balzac
Burdock Group Consultants
Orlando, Florida

Subhadip Banerjee
Department of Ayurveda Pharmacology
Bengal Institute of Pharmaceutical Sciences
Kalyani, India

Karen Bishop
Faculty of Medical & Health Sciences
The University of Auckland
Auckland, New Zealand

George A. Burdock
Burdock Group Consultants
Orlando, Florida

Dennis Chang
School of Science and Health
University of Western Sydney
Sydney, Australia

Shantanu Das
Goodman Fielder Ltd.
Auckland, New Zealand

Parikshit Debnath
Department of Swasthavritta (Preventive Medicine)
SDM College of Ayurveda and Hospital
Hassan, India

Pratip Kumar Debnath
Department of Philosophy to Science Research
Gananath Sen Institute of Ayurveda and Research
Kolkata, India

Asim K. Duttaroy
Department of Nutrition
University of Oslo
Oslo, Norway

Lynnette Ferguson
Faculty of Medical & Health Sciences
The University of Auckland
Auckland, New Zealand

Dilip Ghosh
Nutriconnect
Sydney, Australia

Sumita Ghosh
School of Built Environment
University of Technology
Sydney, Australia

Derek Haisman
Riddet Institute
Massey University
Palmerston North, New Zealand

Akihito Harusato
Molecular Gastroenterology and Hepatology
Kyoto Prefectural University of Medicine
Kyoto, Japan

Yasuki Higashimura
Molecular Gastroenterology and Hepatology
Kyoto Prefectural University of Medicine
Kyoto, Japan

Myriam Hinojosa
Department of Experimental Therapeutics
The University of Texas: MD Anderson Cancer Center
Houston, Texas

Katsuya Hirasaka
Department of Nutritional Physiology
University of Tokushima Graduate School
Tokushima, Japan

Nishi Karunsinghe
Faculty of Medical & Health Sciences
The University of Auckland
Auckland, New Zealand

Yeo Jin Kim
Department of Life Science
Dongguk University
Seoul, Republic of Korea

Shohei Kohno
Department of Nutritional Physiology
University of Tokushima Graduate School
Tokushima, Japan

Tetsuya Konishi
HALD Food Function Research
and
NUPALS Liaison R/D Promotion Center
Niigata University of Pharmacy and Applied Life Sciences
Niigata, Japan

Akira Kubo
School of Medicine
Tokai University
Tokyo, Japan

M.A. Lakshmithathachar
Samskriti Foundation
Mysore, India

Song Li
COFCO Nutrition and Health Research Institute
Chao Yang District
Beijing, People's Republic of China

Rie Mukai
Department of Food Science
University of Tokushima Graduate School
Tokushima, Japan

Akira Murakami
Division of Food Science and Biotechnology
Kyoto University
Kyoto, Japan

Yuji Naito
Molecular Gastroenterology and Hepatology
Kyoto Prefectural University of Medicine
Kyoto, Japan

Srinivas Nammi
School of Science and Health
University of Western Sydney
Sydney, Australia

Fumi Nihei
Dr. Akira Kubo's Office, Co., Ltd
Chuo-ku, Tokyo

Takeshi Nikawa
Department of Nutritional Physiology
University of Tokushima Graduate School
Tokushima, Japan

Arisa Ochi
Department of Nutritional Physiology
University of Tokushima Graduate School
Tokushima, Japan

Jane O'Shea
School of Medicine
University of Wollongong
Wollongong, Australia

Namyata Pathak
ICMR Advanced Centre of Reverse Pharmacology in Traditional Medicine
Kasturba Health Society
Mumbai, India

V. Prakash
Nutrition Society of India
Indian Council of Medical Research
Hyderabad, India

and

Innovation and Development at JSSMVP
JSS Technical Institutions Campus
Mysore, India

Yasmine Probst
School of Medicine
University of Wollongong
Wollongong, Australia

Rajshri Roy
Clinical Nutrition & Dietetics, School of Public Health
University of Sydney
Sydney, Australia

Daniel Roytas
Department of Nutritional Medicine
Southern School of Natural Therapies
Fitzroy, Australia

Anwesha Sarkar
Riddet Institute
Massey University
Palmerston North, New Zealand

Young Rok Seo
Department of Life Science
Dongguk University
Seoul, Republic of Korea

Hiteshi A. Shah
ICMR Advanced Centre of Reverse Pharmacology in Traditional Medicine
Kasturba Health Society
Mumbai, India

Toshio Shimizu
Department of Health and Human Life
Nagoya Bunri Universty
Nagoya, Japan

Tee E. Siong
TES NutriHealth Strategic Consultancy
Selangor DE, Malaysia

R.B. Smarta
Interlink Marketing Consultancy Pvt. Ltd.
Mumbai, India

Hirohito Sone
Department of Internal Medicine
Division of Hematology, Endocrinology and Metabolism
Niigata University Graduate School of Medical and Dental Sciences
Niigata, Japan

Yu-Ting Sun
School of Science and Health
University of Western Sydney
Sydney, Australia

Kenji Suzuki
Department of Internal Medicine
Division of Gastroenterology and Hepatology
Niigata University Graduate School of Medical and Dental Sciences
Niigata, Japan

Tomohisa Takagi
Molecular Gastroenterology and Hepatology
Kyoto Prefectural University of Medicine
Kyoto, Japan

Amit Taneja
Riddet Institute
Massey University
Palmerston North, New Zealand

Namrata Taneja
Riddet Institute
Massey University
Palmerston North, New Zealand

Junji Terao
Department of Food Science
University of Tokushima Graduate School
Tokushima, Japan

Rajarajan A. Thandavarayan
Department of Clinical Pharmacology
Niigata University of Pharmacy and Applied Life Sciences
Niigata, Japan

Rebecca Thorne
School of Medicine
University of Wollongong
Wollongong, Australia

Ashok Vaidya
ICMR Advanced Centre of Reverse Pharmacology in Traditional Medicine
Kasturba Health Society
Mumbai, India

Chin-Kun Wang
School of Nutrition
Chung Shan Medical University
Taichung, Taiwan

Chunling Wang
COFCO Nutrition and Health Research Institute
Chao Yang District
Beijing, People's Republic of China

Kenichi Watanabe
Department of Clinical Pharmacology
Niigata University of Pharmacy and Applied Life Sciences
Niigata, Japan

Yui Yamashita
Department of Nutritional Physiology
University of Tokushima Graduate School
Tokushima, Japan

Noriaki Yohkoh
Department of Clinical Pharmacy
Niigata University of Pharmacy and Applied Life Sciences
Niigata, Japan

Toshikazu Yoshikawa
Molecular Gastroenterology and Hepatology
Kyoto Prefectural University of Medicine
Kyoto, Japan

Andrea Zangara
Ginsana SA
SFI Group
Bioggio, Switzerland

Section I

Introduction

1 Progress of Food-Based Dietary Guidelines around the Globe

Yasmine Probst, Rebecca Thorne, and Jane O'Shea

CONTENTS

1.1 OVERVIEW

Dietary guidelines have been developed worldwide in an effort to assist populations with healthy food choices. The guidelines may be food or nutrient based and are commonly focused on particular health concerns of the country such as overweight and obesity. There is a large degree of variability in the development of dietary guidelines around the world; some countries have opted for one guideline for all, whereas other countries have developed multiple subsets of guidelines related to particular population groups.

The provision of dietary guidelines of a country is not regulated though they are a recommended approach that has been suggested by the World Health Organization (WHO) as one of the two international organizations supporting such activities. The WHO, alongside the Food and Agriculture Organization of the United Nations (FAO), acknowledge that it has long been known that particular nutrients are needed to sustain human health and well-being, and work to disseminate nutrient information to countries in an effort to encourage the development of national dietary allowances for these nutrients. Such nutrient allowances may directly form dietary guidelines of a nutrient-focused nature; though stemming from such developments, countries may also work toward practical approaches of how to translate the nutrient information into everyday practice, an approach often seeing the development of food-based dietary guidelines.

This chapter provides an overview to some of the existing dietary guidelines available around the globe. Where a country regularly updates its dietary guidelines, a comparison of the most recent sets of guidelines will also be addressed. The focus of the dietary guidelines also varies between countries with some developed directly for the general public and others specifically targeted at health professionals to assist with consistency of education messages utilized within the country. The guidelines reviewed will be those available as the most recent guidelines published in the English language. The guidelines currently undergoing an update have not been included.

1.2 DEVELOPMENT OF DIETARY GUIDELINES

Establishing dietary guidelines has significant clinical and public health impact for a country as recommendations may be set to address the issues related to different age group- and gender-related over- or undernutrition problems. The guidelines are generally developed from the evidence supporting particular selected statements or phrases for the country's dietary guidelines. This evidence may be in the form of consumer consultation, previous food intake surveys, or reviews of the scientific literature. Updates to the guidelines are likely to take place in 10–15 years as new evidence is accumulated or the WHO and the FAO update their position in relation to the nutrients included. Some countries, however, choose to update their guidelines more regularly. The last established report of such recommendations is from the joint FAO and WHO consultation of 1999, leaving many countries to follow the processes outlined by others to ensure that their guidelines remain relevant.

Considerations that must be undertaken by a country before an update of the dietary guidelines occurs within a country include a range of factors such as the following (Beaton 2003):

- Evidence that the guidelines, if followed, would result in dietary patterns with inadequate nutrient requirements
- Evidence that the guides developed from the guidelines require an update or redesign
- Evidence that a time has been reached in which it cannot be concluded whether a single food guide or a collection of food guides is developed to suit distinct groups
- The need to align with emerging or updated dietary reference intake values
- The need to reflect new foods in the marketplace or the new fortification practices
- Transborder health promotion campaigns that may impact on food selection practices
- Emerging international nutrition standards and trade regulations that impact on the existing food supply
- Ineffective current guidelines
- Expectations for a revision by health professionals

Time has also changed the previous approach to dietary guidelines foci of avoidance of nutrient *inadequacy* and rather has now in many instances been teamed with the need to also avoid nutrient *excess*. This shift is being demonstrated by a need for revisions of recommended dietary intake levels of nutrients for particular countries. These intake levels commonly have an upper limit set as a form of risk management (Beaton 2003). Such risk management also sees many countries creating models of exemplar dietary patterns that may result from various combinations of intake that may result from consumer interpretation of the advice given. Such dietary pattern models are generally created based on the existing national food consumption survey data. Therefore, dietary guidelines are not only developed based on the food supply and eating habits within given countries but also underpinned by cultural practice and beliefs within the population or population groups being targeted. Subsections 1.3 through 1.8 provide an overview of selected dietary guidelines across the globe with a focus on those most recently updated.

1.3 DIETARY GUIDELINES OF THE OCEANIA REGION

The dietary guidelines of Australians have recently undergone a 10-year revision, resulting in a significant shift in the detail included within the guidelines though the overall content has remained largely the same (National Health and Medical Research Council 2003, 2013a, 2013b). Endorsed by the National Health and Medical Research Council, the guidelines are developed to target not only the general population but also health professionals and policy makers. The guidelines are updated through an iterative process of expert review and consultation. The focus on variety of food sources remains a standing characteristic while the range of food sources recommended has increased.

The underlying nutrient focus appears via the messages to limit and select specific types of foods from within the food groups. Having safe food practices as well as encouragement of water as the primary beverage of choice remain consistent. The message to support breastfeeding and the need to remain physically active have been maintained, while messages for specific population groups have been added in the latest revision (Table 1.1).

TABLE 1.1
Dietary Guidelines for Australians

2003	2013
	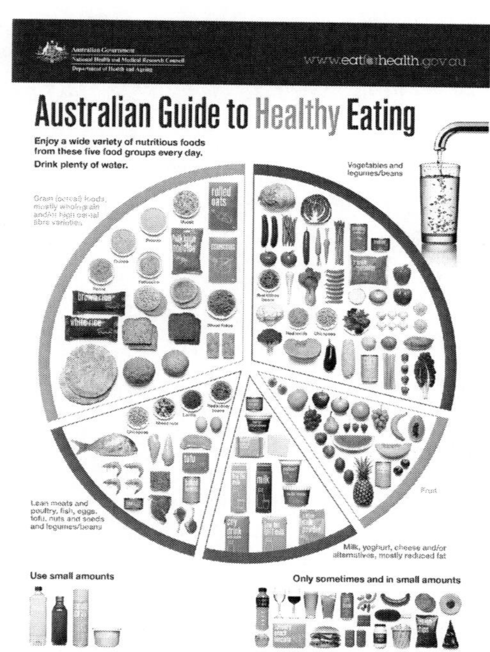
Enjoy a wide variety of nutritious foods: • Eat plenty of vegetables, legumes, and fruits • Eat plenty of cereals (including breads, rice, pasta, and noodles), preferably whole grains • Include lean meat, fish, poultry, and/or their alternatives • Include milk, yogurt, cheeses, and/or their alternatives. Reduced-fat varieties should be chosen, where possible • Drink plenty of water Take care to • Limit saturated fat and moderate total fat intake. • Choose foods low in salt. • Limit your alcohol intake if you choose to drink. • Consume only moderate amounts of sugars and foods containing added sugars.	Eat a wide variety of nutritious foods from the following five groups every day: • Plenty of vegetables, including different types and colors, and legumes/beans • Fruits • Grain (cereal) foods, mostly whole grains, such as breads, cereals, rice, pasta, noodles, polenta, couscous, oats, quinoa, and barley • Lean meat and poultry, fish, eggs, nuts and seeds, and legumes/beans • Milk, yogurt, cheese and/or their alternatives, mostly reduced fat (reduced fat milks are not suitable for children under the age of two years) Drink water. Limit intake of foods and drinks containing saturated and *trans* fats: • Include small amounts of foods that contain unsaturated fats. • Low-fat diets are not suitable for infants.

(*Continued*)

TABLE 1.1
(Continued) Dietary Guidelines for Australians

2003	2013
Prevent weight gain: Be physically active and eat according to your energy needs.	Limit intake of foods and drinks containing added salt: • Read labels to choose lower sodium options among similar foods. • Do not add salt to foods.
Care for your food: Prepare and store it safely.	Limit intake of foods and drinks containing added sugars. In particular, limit sugar-sweetened drinks.
Encourage and support breast-feeding.	If you choose to drink alcohol, limit intake.
	To achieve and maintain a healthy weight you should be physically active and choose amounts of nutritious foods and drinks to meet your energy needs.
	Children and adolescents should eat sufficient nutritious foods to grow and develop normally. They should be physically active every day and their growth should be checked regularly.
	Older people should eat nutritious foods and keep themselves physically active to help maintain muscle strength and a healthy weight.
	Encourage and support breastfeeding.
	Care for your food; prepare and store it safely.

Source: National Health and Medical Research Council, 2003, http://www.nhmrc.gov.au/_files_nhmrc/publications/attachments/n33.pdf; National Health and Medical Research Council, 2013a, http://www.nhmrc.gov.au/_files_nhmrc/publications/attachments/n55a_australian_dietary_guidelines_summary_book_0.pdf; National Health and Medical Research Council, 2013b, http://www.nhmrc.gov.au/_files_nhmrc/publications/attachments/n31.pdf.

The guidelines are not developed in isolation but rather have a closely related food guide that incorporates the key messages. This guide utilizes food portion size and number of servings to translate the guidelines messages into practice-based approaches for the general public/health education.

The dietary guidelines of New Zealand are endorsed by the Ministry of Health with separate guidelines available for specific population groups (Ministry of Health 2009a). Utilizing the same nutrient reference values as Australia, similarities may be seen between the two countries' dietary guidelines as well. The guidelines of New Zealand are fewer in number though also addressing the message of variety, avoidance of particular risk nutrients, suggestions for beverages to avoid and encourage, and the underpinning message of physical activity. Contrary to the Australian guidelines, the New Zealand message for physical activity provides more detail about the level of activity. Similarly, the message relating to the use of salt provides an alternate approach of using iodized salt for consumers. This message has only recently been promoted to the Australian consumers as a result of research into the nutrients' importance in brain development (Table 1.2).

While the guidelines for the general population are now 10 years old, the guidelines for the subgroups of infants and toddlers (Ministry of Health 2008a), children (Ministry of Health 2009b), adolescents, pregnant and breast-feeding women (Ministry of Health 2008b), and older people (Ministry of Health 2010) have more recently been updated.

TABLE 1.2
New Zealand's Food and Nutrition Guidelines for Healthy Adults

Maintain a healthy body weight by eating well and engaging in daily physical activity.[a]

Eat well by including a variety of nutritious foods from each of the following four major food groups each day:

- Eat plenty of vegetables and fruits.
- Eat plenty of breads and cereals, preferably whole grains.
- Have milk and milk products in your diet, preferably reduced or low-fat options.
- Include lean meat, poultry, seafood, eggs, or alternatives.

Prepare foods or choose pre-prepared foods, drinks, and snacks

- With minimal added fat, especially saturated fat.
- That are low in salt; if using salt, choose iodized salt.
- With little added sugar; limit your intake of high-sugar foods.

Drink plenty of liquids each day, especially water.

If you choose to drink alcohol, limit your intake.

Purchase, prepare, cook, and store food to ensure food safety.

Source: Ministry of Health, 2009a, http://www.health.govt.nz/publication/clinical-guidelines-weight-management-new-zealand-adults.

[a] At least 30 minutes of moderate-intensity physical activity on most if not all days of the week, and if possible, do some vigorous exercises for extra health and fitness.

1.4 DIETARY GUIDELINES OF NORTH AMERICA

The dietary guidelines of the United States are updated on a five-yearly basis by a process of scientific consultation. The last two updates of these guidelines are shown in Table 1.3 with the committee already working toward the 2015 update. Between 2005 and 2010, the main difference that is notable is the decreased length in the later set of guidelines (Health.gov 2008; US Department of Agriculture 2012). Although both are published online with supporting detail for each guideline, the 2005 guidelines also coincided with the release of the MyPyramid interactive tool for consumers to build healthy food habits. MyPyramid was developed from extensive consumer consultation in the form of focus groups and Web–TV-based questionnaires to determine the aesthetics of the tool as well as the concepts to be included (Haven et al. 2006). The consumer work ran as a number of phases beginning initially in 2002 and repeated again in 2004 (Britten et al. 2006). Most notably, this research showed that the image of the pyramid was well understood while the detail behind the recommendations was lacking.

The guidelines themselves, however, are broken down into sections. Management of weight was maintained across the two standards of 2005 and 2010, with the section of physical activity from 2005 also being incorporated into this section. The concept of foods to increase were also maintained through the additional nutrients mentioned in the later edition allowed for the incorporation of nutrient sections that were previously separated out on their own. The alcohol, sodium, and potassium sections, and parts of the fat and carbohydrate sections were condensed and formed a new section of food and food components to be reduced, creating a shorter and more direct message overall.

The Canadian guidelines (Health Canada 2007, 2011) are not updated as regularly as those of the United States but did undergo a significant revision in 2007 seeing an expansion of many of the existing guideline messages. The message of variety was incorporated into a collection of messages for each food group that was addressed. These food groups were previously all grouped together as one message. The messages relating to physical activity, alcohol, and caffeine have been removed, while more positive messages encouraging consumption of water for thirst have been used. The wording of the newer guidelines appears to be more positive in its choice of wording.

TABLE 1.3
Dietary Guidelines for Americans

2005[a]

Weight management

- To maintain body weight in a healthy range, expend balance calories from foods and beverages with calories.
- To prevent gradual weight gain over time, make small decreases in food and beverage calories and increase physical activity.
- To lose weight, aim for a slow, steady weight loss by decreasing calorie intake while maintaining an adequate nutrient intake and increasing physical activity.
- For overweight children: Reduce the rate of body weight gain while allowing growth and development. Consult a health-care provider before placing a child on a weight reduction diet.
- For pregnant women: Ensure appropriate weight gain as specified by a health-care provider.
- For breast-feeding women: Moderate weight reduction is safe and does not compromise weight gain of the nursing infant.
- For overweight adults and overweight children with chronic diseases and/or on medication: Consult a health-care provider about weight loss strategies prior to starting a weight reduction program to ensure appropriate management of other health conditions.

Physical activity

- Engage in regular physical activity and reduce sedentary activities to promote health, psychological well-being, and healthy body weight.

2010

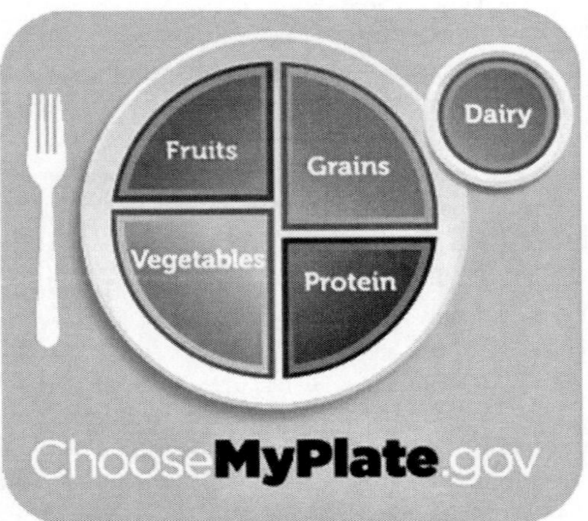

Balancing calories to manage weight

- Prevent and/or reduce overweight and obesity through improved eating and physical activity behaviors.
- Control total calorie intake to manage body weight. For people who are overweight or obese, this will mean consuming fewer calories from foods and beverages.
- Increase physical activity and reduce time spent in sedentary behaviors.
- Maintain appropriate calorie balance during each stage of life—childhood, adolescence, adulthood, pregnancy and breastfeeding, and older age.

Foods and food components to reduce

- Reduce daily sodium intake to <2300 mg and further reduce intake to 1500 mg among persons who are 51 and older and those of any age who are African American or have hypertension, diabetes, or chronic kidney disease. The 1500 mg recommendation applies to about half of the US population, including children, and the majority of adults.
- Consume <10% of calories from saturated fatty acids by replacing them with monounsaturated and polyunsaturated fatty acids.
- Consume <300 mg per day of dietary cholesterol.
- Keep *trans* fatty acid consumption as low as possible by limiting foods that contain synthetic sources of *trans* fats, such as partially hydrogenated oils, and by limiting other solid fats.
- Reduce the intake of calories from solid fats and added sugars.

TABLE 1.3
Dietary Guidelines for Americans

2005[a]	2010
• To reduce the risk of chronic disease in adulthood, engage in at least 30 minutes of moderate-intensity physical activity, above the usual activity, at work or home on most days of the week. • For most people, greater health benefits can be obtained by engaging in physical activity of more vigorous intensity or longer duration. • To help manage body weight and prevent gradual, unhealthy body weight gain in adulthood, engage in approximately 60 minutes of moderate- to vigorous-intensity activity on most days of the week while not exceeding caloric intake requirements. • To sustain weight loss in adulthood, participate in at least 60–90 minutes of daily moderate-intensity physical activity while not exceeding caloric intake requirements. Some people may need to consult with a health-care provider before participating in this level of activity. • Achieve physical fitness by including cardiovascular conditioning, stretching exercises for flexibility, and resistance exercises or calisthenics for muscle strength and endurance. **Food groups to encourage** • Consume a sufficient amount of fruits and vegetables while staying within energy needs. Two cups of fruits and two and a half cups of vegetables per day are recommended for a reference 2000-calorie intake, with higher or lower amounts depending on the calorie level. • Choose a variety of fruits and vegetables each day. In particular, select from all five vegetable subgroups (dark green, orange, legumes, starchy vegetables, and other vegetables) several times a week. • Consume three or more ounce-equivalents of wholegrain products per day, with the rest of the recommended grains coming from enriched or wholegrain products. In general, at least half the grains should come from whole grains. • Consume three cups per day of fat-free or low-fat milk or equivalent milk products. **Fats** • Consume <10% of calories from saturated fatty acids and <300 mg/day of cholesterol, and keep *trans* fatty acid consumption as low as possible. • Keep the total fat intake between 20% and 35% of calories, with most fats coming from sources of polyunsaturated and monounsaturated fatty acids, such as fish, nuts, and vegetable oils.	• Limit the consumption of foods that contain refined grains, especially refined grain foods that contain solid fats, added sugars, and sodium. • If alcohol is consumed, it should be consumed in moderation—up to one drink per day for women and two drinks per day for men—and only by adults of legal drinking age. **Foods and nutrients to increase** Individuals should meet the following recommendations as part of a healthy eating pattern while staying within their calorie needs: • Increase vegetable and fruit intake. • Eat a variety of vegetables, especially dark green, red, and orange vegetables, and beans and peas. • Consume at least half of all grains as whole grains. Increase wholegrain intake by replacing refined grains with whole grains. • Increase intake of fat-free or low-fat milk and milk products, such as milk, yogurt, cheese, or fortified soy beverages. • Choose a variety of protein foods, which include seafood, lean meat and poultry, eggs, beans and peas, soy products, and unsalted nuts and seeds. • Increase the amount and variety of seafood consumed by choosing seafood in place of some meat and poultry. • Replace protein foods that are higher in solid fats with choices that are lower in solid fats and calories and/or are sources of oils. • Use oils to replace solid fats where possible. • Choose foods that provide more potassium, dietary fiber, calcium, and vitamin D, which are nutrients of concern in American diets. These foods include vegetables, fruits, whole grains, and milk and milk products. **Building healthy eating patterns** • Select an eating pattern that meets the nutrient needs over time at an appropriate calorie level. • Account for all foods and beverages consumed and assess how they fit within a total healthy eating pattern. • Follow food safety recommendations when preparing and eating foods to reduce the risk of foodborne illnesses.

(*Continued*)

TABLE 1.3
(Continued) Dietary Guidelines for Americans

2005[a]	2010
• When selecting and preparing meat, poultry, dry beans, and milk or milk products, make choices that are lean, low fat, or fat free. • Limit intake of fats and oils high in saturated and/or *trans* fatty acids, and choose products low in such fats and oils.	
Carbohydrates	
• Choose fiber-rich fruits, vegetables, and whole grains often. • Choose and prepare foods and beverages with little added sugars or caloric sweeteners, such as amounts suggested by the US Department of Agriculture (USDA) Food Guide and the Dietary Approaches to Stop Hypertension (DASH) Eating Plan. • Reduce the incidence of dental caries by practicing good oral hygiene and consuming sugar- and starch-containing foods and beverages less frequently.	
Sodium and Potassium	
• Consume <2300 mg (approximately 1 tsp of salt) of sodium per day. • Choose and prepare foods with little salt. At the same time, consume potassium-rich foods, such as fruits and vegetables.	
Alcoholic beverages	
• Those who choose to drink alcoholic beverages should do so sensibly and in moderation—defined as the consumption of up to one drink per day for women and up to two drinks per day for men. • Alcoholic beverages should not be consumed by some individuals, including those who cannot restrict their alcohol intake, women of childbearing age who may become pregnant, pregnant and lactating women, children and adolescents, individuals taking medications that can interact with alcohol, and those with specific medical conditions. • Alcoholic beverages should be avoided by individuals engaging in activities that require attention, skill, or coordination, such as driving or operating machinery.	

TABLE 1.3
Dietary Guidelines for Americans

2005[a]	2010
Food safety	
To avoid microbial foodborne illness • Clean hands, food contact surfaces, and fruits and vegetables. Meat and poultry should not be washed or rinsed. • Separate raw, cooked, and ready-to-eat foods while shopping, preparing, or storing foods. • Cook foods to a safe temperature to kill microorganisms. • Chill (refrigerate) perishable food promptly and defrost foods properly. • Avoid raw (unpasteurized) milk or any products made from unpasteurized milk, raw or partially cooked eggs or foods containing raw eggs, raw or undercooked meat and poultry, unpasteurized juices, and raw sprouts.	

Source: Health.gov, 2008, http://www.health.gov/dietaryguidelines/dga2005/document/html/executivesummary.htm; United States Department of Agriculture, 2012, http://www.cnpp.usda.gov/DGAs2010-PolicyDocument.htm.

[a] Guidelines for special populations are not shown.

The first official food guide for Canada was produced in 1942 outlining the official food rules. The guide that exists today follows a developmental process similar to that of the United States and Australia by which expert consultation is used and dietary patterns are modeled in conjunction with nutrient reference values so that the recommended food choices reflect that of risk reduction for chronic diseases prevalent within the country (Katamay et al. 2007). During this process, the popularity of particular food choices is identified using previous Federal–Provincial surveys, ensuring that the newly developed messages/statements reflect all of the considerations discussed in Subsection 1.2.

The process of modeling diets that flag nutrient reference values may be achieved through the food supply with ease. This process is followed through to those nutrients that need to be specifically targeted. These nutrients were then mentioned through food suggestions that did not have a significant impact on the total energy intake in the food intake statements (Katamay et al. 2007). Expert consultation and review of scientific evidence then also formed the basis for the development of statements to reduce the risk of chronic disease such as those related to salt, fat, and sugar intake.

The Canadian food guide follows a consumer-friendly format and is developed to guide food choices that are aligned with the food pattern statements. The Health Canada website (www.hc-sc.gc.ca) allows consumers to use "My Food Guide," an interactive guide that not only shows the food intake statements but also informs users of the number of servings of each food group they require based on their gender and user-entered age group (Table 1.4).

The island nations off the coast of North America also have their own developed food-based dietary guidelines. The Caribbean Food and Nutrition Institute, for example, has developed guidelines for St Vincent's and the Grenadines though these are used throughout all the islands of the Caribbean (Pan American Health Organization). The messages are simple and reflect the food intakes of the population. The statements used are closely linked to the food intake target for the region that visually emphasizes the messages as well (Table 1.5).

TABLE 1.4
Canada's Food Guide

1990

- Enjoy a variety of foods.
- Emphasize cereals, breads, other grain products, vegetables, and fruits.
- Choose lower fat dairy products, leaner meats, and foods prepared with little or no fat.
- Achieve and maintain a healthy body weight by enjoying regularly physical activity and healthy eating.
- Limit salt, alcohol, and caffeine.

2007

- Eat at least one dark green and one orange vegetable each day.
- Have vegetables and fruits more often than juice.
- Make at least half of your grain products whole grain each day.
- Drink skim, 1%, or 2% milk each day.
- Have meat alternatives such as beans, lentils, and tofu often.
- Eat at least two Food Guide Servings of fish each week.
- Include a small amount of unsaturated fat each day.
- Satisfy your thirst with water.
- Choose foods lower in fat, salt, and sugar.
- Choose vegetables and fruits prepared with little or no added fat, sugar, or salt.
- Choose grain products that are lower in fat, sugar, or salt.
- Select lower fat milk alternatives.
- Select lean meat and alternatives prepared with little or no added fat or salt.
- Limit foods and beverages high in calories, fat, sugar, or salt.

Source: Health Canada, 2007, http://www.hc-sc.gc.ca/fn-an/food-guide-aliment/context/fg_history-histoire_ga-eng.php; Health Canada, 2011, http://www.hc-sc.gc.ca/fn-an/alt_formats/hpfb-dgpsa/pdf/pubs/res-educat-eng.pdf.

TABLE 1.5
Dietary Guidelines of St Vincent's and the Grenadines (Caribbean)

Eat a variety of foods

- Eat larger amounts of fruits and colored vegetables.
- Eat less fatty, oily, greasy, and barbequed foods.
- Use less salt, salty foods, salty seasonings, and salty snacks.
- Choose to use less sweet foods and drinks.
- Drink more water. It's the healthier choice.
- Drink little or no alcohol.
- Be more physically active. Get moving.

Source: Pan American Health Organization, http://new.paho.org/cfni/index.php?option=com_docman&task=doc_details&gid=22&Itemid=219.

TABLE 1.6
National Dietary Guidelines for Honduras

- Eat healthy foods from all groups.
- Eat fruits and vegetables daily because they have vitamins that prevent disease.
- Eat meat, fish, or viscera, at least twice a week to help grow and strengthen the body.
- Cut back on fried foods and sausages for a healthy heart.
- Cut back on salt, cubes, and soups to prevent high blood pressure.
- Drink at least eight glasses of water a day for the proper functioning of your body.
- Walk at least half an hour a day to stay healthy and stress free.

Source: Institute of Nutrition of Central America and Panama, 2012, http://www.incap.int/index.php/en/2012-04-10-19-01-58/strategic-actions/477-honduras-national-workshop-on-advances-in-updating-the-food-based-dietary-guidelines-for-honduras.

Similarly, the Institute of Nutrition of Central America and Panama (2012) has developed guidelines for Honduras. Development of these guidelines followed a more qualitative approach to that used on the mainland of North America, utilizing an extensive process of workshops and consumer consultation to develop the guideline statements. The workshops held in 2011 and 2012 involved consultation with key organizations including the FAO, World Vision International, and the Ministry of Health. The workshops began by sharing the results of consumer testing to select recommendations that would be feasible to implement within the region. These recommendations formed the messages of the dietary guidelines that were then teamed with an icon that would be accepted on a population level and education material formats as well as a validation methodology for the education materials and an implementation plan for their use (Institute of Nutrition of Central America and Panama 2012) (Table 1.6).

1.5 DIETARY GUIDELINES OF SOUTH AMERICA

The first food distribution program, aimed at the prevention of goiter, was introduced into Brazil in the mid-1950s. Brazil first began the development of its dietary guidelines during the 1990 following the WHO recommendations and generally basing them on those of the United States (Sichieri et al. 2010). The guidelines are supported by the Brazilian Ministry of Health. Although the American guidelines are largely focused on targeting obesity concerns of the population, the greatest public health problem of Brazil is actually iron deficiency, and accordingly, the consumption of foods that are a good source of iron, such as organ, meats, and eggs, is encouraged in the Brazilian guidelines. Unlike other guidelines around the world, the Brazilian guidelines suggest percentage energy targets for macronutrient intakes and group physical activity and food safety as special guidelines at the end of their principal guidelines (Table 1.7).

Reference to processed and genetically modified food, the use of pesticides and hormones, artificial sweeteners, cooking methods, and ready-to-eat meals is included as an appendix to the guidelines (Sichieri et al. 2010). The Brazilian guidelines also refer to complex carbohydrates provided by cereals and roots, though providing only a general mention of wholegrain cereal intake as may be seen in the American guidelines. The Brazilian guidelines generally follow a similar message to other countries, though many statements are highly prescriptive by comparison, while others are open to many different levels of interpretation. Pictorial aids, such as the healthy eating pyramid or the healthy eating plate, have not been adopted for the Brazilian guidelines. However, a number of folders and banners showing the "Ten Steps to Healthy Food" have been produced, including a pocket edition of the guidelines. These are used for the promotion of the guidelines.

TABLE 1.7
Guiding Principles of the Brazilian Dietary Guidelines

Healthy food and meals	• Have at least three meals per day and small snacks between the meals. • Healthy meals are prepared with varied food and include vegetables and animal food. • Exclusive breastfeeding is the better option for a healthy diet during the first six months of life.
Cereals and roots	Complex carbohydrates (starch) from cereals (preferably whole) and roots should provide 45% to 65% of total energy intake. It is recommended to keep the intake of six portions of cereals and roots daily.
Fruits and vegetables	It is recommended to keep the intake of three portions of fruits and three portions of vegetables daily, aiming to provide 9%–12% of total energy intake (for a 2000 kcal diet), weighing around 400 g/day.
Beans	It is recommended to keep the intake of one portion of beans daily; the proportion of rice and beans should be 2:1. Beans should provide 5% of total energy intake.
Milk and dairy, meat, and eggs	It is recommended to keep the intake of three portions of milk and dairy and one portion of lean meat, fish, or eggs. Adults should choose low-fat milk and dairy.
Fat, sugar, and salt	• Oils and fats (preferably vegetable oils, olive oil, and trans-fat-free margarine) should provide between 15% and 30% of total energy intake (one portion/day). • The limit for saturated fat is 10% of total energy intake. • The limit of *trans* fat is 1% of total energy intake (or 2 g/day for a 2000 kcal diet). • Simple sugars should account for a maximum of 10% of total energy intake (one portion/day). • Salt: 5 g/day plus the advice to reduce salty processed food such as seasonings, concentrated broth, sauces, salty snacks, and industrialized soups.
Water	At least 2 l of water daily
Physical activity	Encourage physical activity that is appropriate for the individual's life stage and the maintenance of a healthy weight, at least 30 minutes daily.
Food safety	Prevention and control measures along the productive chain

Source: Sichieri, R. et al., *Cadernos de Saúde Pública,* 26, 2050–2058, 2010.

Other guidelines of South America have been produced in the native language of the country and subsequently are not available in English.

1.6 DIETARY GUIDELINES OF EUROPE

The countries of Europe generally have their own specific guidelines developed to suit the food patterns, availability, and health of the given region. Neither all countries have publicly available guidelines, nor do they produce their guidelines in languages other than those native to the country. A few selected guidelines that are available in the English language have been addressed in the following text.

Bulgaria has guidelines (Ministry of Health National Center of Public Health Protection 2006), which, similar to many other countries addressed in Subsections 1.3 through 1.5, place an emphasis on variety within the diet and consumption of a range of foods from different food groups. The Bulgarian guidelines also focus on regularity of intake as well as the eating environment and the time taken to consume foods. This shows a particular link to the social aspect of food consumption that is not often seen in dietary guidelines. The food group suggestions then follow on with preferential food types from each food group listed and their portion sizes where appropriate. Despite these portions, the use of moderate quantities, for example, is also used, which leave the statements open to interpretation by the user. A collection of messages relating to foods to avoid are also given, and there is an emphasis on food over nutrients in the wording of the statements. Food safety is also included as its own statement, and weight maintenance and physical activity form another statement, similar to other countries (Table 1.8).

The dietary guidelines of Denmark (Council 2009; The Danish Veterinary and Food Administration) are largely open to interpretation with many open statements about leaving portion sizes and frequency, for example, to be interpreted in the users' preferred manner. Only two statements specify the detail for the user, namely, statement for fruit and vegetable consumption and statement for physical activity. The guidelines have a strong food focus with limited reference to nutrients. The guidelines are generally short and do not include the added detail that may be seen for some European guidelines (Table 1.9).

France adopted the use of nine food rules that were linked to their food steps developed in 2002 (van Dooren and Kramer 2012). The steps provide a visual display of foods that should be eaten in the least amounts at the bottom of the staircase and the most at the top of the staircase. The food rules are not quantitative and there is some concern over the amount of sodium suggested in the dietary guidelines. The sodium recommendation is set at 8 g, with the WHO recommendation only suggesting 5 g (Table 1.10).

Greenland's dietary guidelines incorporate the mention of local produce for obtaining the foods eaten (Niclasen and Schnohr 2010). They also begin with their statement about variety as the first guideline for the country simplifying the message down to three words "vary your diet." Following

TABLE 1.8
Dietary Guidelines of Bulgaria

- Eat a nutritious diet with variety of foods. Do eat regularly, take enough time, and enjoy your food in a friendly environment.
- Consume cereals as an important source of energy. Prefer wholegrain bread and other wholegrain products.
- Eat a variety of vegetables and fruits >400 g every day, preferably raw.
- Prefer milk and dairy products with low fat and salt content.
- Choose lean meat and replace meat and meat products often with fish, poultry, or pulses.
- Limit the total fat intake, especially animal fat. Replace animal fats with vegetable oils when cooking.
- Limit the consumption of sugar, sweets, and confectionery. Avoid sugar-containing soft drinks.
- Reduce the intake of salt and salty foods.
- If you drink alcoholic beverages, you should consume moderate quantities.
- Maintain a healthy body weight and be physically active every day.
- Drink plenty of water every day.
- Prepare and store the food in a way to ensure its quality and safety.

Source: Ministry of Health National Center of Public Health Protection, 2006, http://ncphp.government.bg/files/hranene-en.

TABLE 1.9
Dietary Guidelines of Denmark

- Eat six pieces of fruits and vegetables per day.
- Eat fish and fish products several times a week.
- Eat potatoes, rice, pasta, and wholemeal bread every day.
- Limit the intake of sugar, particularly from soft drinks, confectionery, and cakes.
- Eat less fat, particularly fats from meat and dairy products.
- Eat a varied diet and maintain a normal weight.
- Drink water when you are thirsty.
- Engage in physical activity at least 30 minutes a day.

Source: Council, 2009, Food-based dietary guidelines in Europe; The Danish Veterinary and Food Administration.

TABLE 1.10
The French Food Rules

- Fruits and vegetables: at least five servings a day (80–100 g/serving), all forms including freshly pressed juice
- Dairy: three servings a day (three or four for children and adolescents), for example, one yogurt (125 g), quark (100 g), fresh cheese or cottage cheese (60 g), cheese (30 g), or one glass of milk
- Starchy foods at each meal according to appetite: These include bread, rusks, cereals, and legumes. Products with complex carbohydrates and whole grains are preferred.
- Meat, fish, and eggs: once or twice a day. Eat fish at least twice a week (100 g/serving).
- Fat products: limited (includes butter and cream). Vegetable oils, oily fish, and nuts are preferred, as are cooking methods requiring little fat.
- Sweet products: limited
- Salty foods (prepared foods, meats, crackers, and snacks): limited. Not more than 8 g of salt a day
- Water: as much as needed during and between meals. Herbal infusions may be used as an alternative. Tap water is as healthy as mineral water.
- Alcohol: more than two standard glasses for women and three for men (wine, beer, champagne, or liquor), which increases the risk of certain illnesses
- Physical activity: the equivalent of at least 30 minutes of brisk walking per day for adults (at least one hour for children and adolescents)

Source: van Dooren, C. and Kramer, G., 2012, http://www.scp-knowledge.eu/sites/default/files/knowledge/attachments/LiveWell%20for%20Life_2012_Food%20patterns%20and%20dietary%20recommendations%20in%20Spain,%20France%20and%20Sweden.pdf.

this, the statements largely follow a positive tone and include a reference to home cooking and family mealtimes, rarely seen in other guidelines. Some guidelines remain open to interpretation, though the physical activity guideline appears to suggest the greatest amount of physical activity of all guidelines worldwide at one hour per day (Table 1.11).

Iceland has nutritional guidelines for individuals who are two years and older, which have been translated into English (www.landlaeknir.is). They now build their recommendations on the Nordic Nutrition Recommendations of 2004 with the development of new guidelines currently in progress. Similar to other European guidelines, a focus on variety and nutrients is seen in the guideline statements, though the translation may not encompass the correct wording of the messages (Table 1.12).

The Food Safety Authority of Ireland endorses the dietary guidelines of Ireland (Department of Health and the Health Service 2012). Written with a pyramid to support the messages, the guidelines

TABLE 1.11
Guidelines of Greenland

- Vary your diet.
- Eat Greenlandic foods and fish often.
- Eat fruits and vegetables daily.
- Eat brown bread and hulled grains every day.
- Eat potatoes, rice, or pasta often.
- Eat less sugar, candy, chips, and cakes.
- Drink water. Drink syrup and soft drinks only for special occasions.
- Eat often, but not much. Breakfast is the best start on the day.
- Make dinner the family gathering. Eat a homemade hot meal every day.
- Eat fat with care.
- Be physically active at least one hour a day.

Source: Niclasen, B. and Schnohr, C. *Public Health Nutrition,* 13, 1162–1169, 2010.

of Ireland again begin with variety as a key message and also include suggestions about reducing portion sizes and ways of consuming particular food items (Table 1.13).

Rather than beginning with the message of varied intake, Italy's dietary guidelines (Nutrixione 2009) begin with a focus on body weight and physical activity. The statements are all short though

TABLE 1.12
Icelandic Nutritional Guidelines (Icelandic–English)

- Keep a varied diet.
- Fruits and vegetables daily (500 g a day, at least 200 g of vegetables and 200 g of fruits)
- Fish at least two times a week
- Wholegrain and fiber-rich food
- Low-fat (1.5 g or less) milk products with as little sugar as possible
- Oil instead of butter and *trans* fatty acids.
- Reduce salt use (maximum 6 g for women and 7 g for men)
- Fish oil or other form of vitamin D daily (10 mg of vitamin D for six months and older up to 60 years, 15 mg vitamin D for 61 years and older)
- Water is the best drink.
- Keep a healthy bodyweight.

TABLE 1.13
Healthy Eating Guidelines for Ireland

- Enjoy a wide variety of foods from the following five food groups.
  - Bread, rice, potatoes, pasta, and other starchy foods
  - Fruit and vegetables
  - Milk and dairy foods
  - Meat, fish, eggs, beans, and other nondairy sources of protein
  - Foods and drinks high in fat and/or sugar
- Find enjoyable ways to be physically active every day—Balancing your food intake with active living will help protect you against disease and prevent weight gain.
- Keep an eye on your serving sizes—Choose smaller serving sizes and add plenty of vegetables, salad, and fruits.
- Plain wholemeal breads, cereals, potatoes, pasta, and rice provide the best calories for a healthy weight. Base your meals on these simple foods with plenty of vegetables, salad, and fruits.
- Eat plenty of different colored vegetables, salad, and fruits—at least five a day.
- Low-fat milk, yogurt, and cheese are best—Choose milk and yogurt more often than cheese.
- Choose lean meat and poultry; include fish (oily is best); and remember that peas, beans, and lentils are good alternatives.
- Use polyunsaturated and monounsaturated spreads and oils sparingly—Reduced fat spreads are best.
- Grill, bake, steam, or boil food, instead of frying or deep frying.
- Healthy eating can be enjoyed with limited amounts of "other foods" such as biscuits, cakes, savoury snacks, and confectionery. These foods are rich in calories, fat, sugar, and salt so remember—*not* too *much* and *not* too *often*.
- Limit your salt intake.
- Drink plenty of water.
- Everyone should take vitamin D supplement daily. 5 μg per day for those aged 5–50 years and 10 μg per day for those aged 51 years and over.
- All women of childbearing age who are sexually active should take a folic acid supplement (400 μg) every day to help prevent neural tube defects (NTDs) in babies, for example, spina bifida.
- Breast-feeding should be encouraged and supported by everyone in Ireland because it gives babies the very best start in life and helps protect women's health.
- Prepare and store food safely.

Source: Department of Health and the Health Service, 2012, http://www.healthpromotion.ie/hp-files/docs/HPM00796.pdf.

all are open to interpretation. The quality of the fats is referred, unlike other countries that define the types of fat that should be consumed. The wording of the messages seem to be largely angled toward the user with many statements referring to the person reading the messages. The only reference to specific food groups relates to cereals, vegetables, tubers, and fruits with no statements relating to dairy and meat intake (Table 1.14).

The Polish dietary guidelines utilize a combination of statements with both food and nutrient focus (Sekuła et al. 2009). Variety begins the guidelines as the first statement followed by a statement about health in relation to obesity management through physical activity. The food emphasis begins at the third statement in which a recommendation for cereal products as the primary food group intake is suggested followed by the guidelines for the consumption of dairy products with two large glasses of low-fat milk, which is the recommended preference. The use of large for the serving size leaves the guideline open to interpretation by users. Guidance to consume meat and sugars in moderation is again open to interpretation. Fruit and vegetable consumption is suggested daily though absent of quantitative measures to assist with this. Fats are suggested in a negative light with animal fats and those containing cholesterol not recommended. Salt is also worded similarly in terms of avoidance, while alcohol is simply stated to be avoided (Table 1.15).

TABLE 1.14
Italian Food Guidelines

- Watch your weight and be active.
- Eat more cereals, vegetables, tubers, and fruits.
- Eat fat—Choose quality and limit the amount.
- Consume sugars, sweets, and sweet drinks—just the right amount.
- Drink plenty of water every day.
- Consume salt—better if little.
- Consume alcoholic drinks—only if in limited amounts.
- Make varied choices.
- Provide special advice for special people.
- The safety of your food also depends on you.

Source: Nutrixione, 2009, http://www.inran.it/648/linee_guida.html.

TABLE 1.15
Food-Based Dietary Guidelines of Poland

- Take care to eat variety of foods.
- Beware of overweight and obesity, and be physically active.
- Cereal products should be the main source of calories for you.
- Drink at least two large glasses of the low-fat milk. Milk could be substituted for yogurt and kefir, and partly for cheese.
- Eat meat in moderation.
- Eat a lot of vegetables and fruits every day.
- Limit intake of fats, particularly of animal ones and all foods containing cholesterol.
- Be moderate in intake of sugar and sweets.
- Limit salt intake.
- Avoid alcohol.

Source: Sekuła, W. et al., 2009, http://www.fao.org/ag/humannutrition/18852-0c8a9ff3ef1a3948677f067bc6f9275f9.pdf.

Spain updated its dietary guidelines in 2004 with the Mediterranean diet pyramid replacing the previous food pyramid. The new pyramid gives maximum and minimum intake levels for food groups that are also outlined to be consumed every meal, every day, and every week. The pyramid is supported by the Spanish Ministry of Agriculture. Another guideline that was developed in Spain by the Spanish Society for Dietetics and Food Science is the new wheel of foods. This wheel contains different segments representing food groups and quantities of each food group as recommended for a healthy diet. Toward the center, the wheel contains foods that are recommended in lesser quantities, and at the center, physical activity and water represented as basic requirements for healthy living.

Sweden represents its food-based dietary guidelines or Swedish Nutrition Recommendations Objectified using the food circle (Livsmedelsverket) (Barbieri and Lindvall 2005). Neither quantities are shown nor are they mentioned in the guideline statements. The circle suggests that at least three portions of fruits and vegetables, one portion of meat or fish, and one portion of dairy should be consumed every day. The guideline statements do have a nutrient focus with low fat and fatty acid mentioned in Table 1.16.

Table 1.17 provides an outline of the sources of other European food-based dietary guidelines that are primarily available in the native language of the country itself.

TABLE 1.16
Advice Accompanying the Swedish Food Circle

- Fruits and berries (juice may be an alternative)
- Vegetables, including pulses (choose coarser varieties, vary according to season; pulses can sometimes replace meat and fish)
- Potatoes and root vegetables (most people should eat more of these)
- Bread, cereals, pasta, and rice (preferably choose wholegrain alternatives)
- Fats (spread a thin layer on bread and preferably choose low-fat margarine; when cooking, use soft or liquid cooking fat with a good fatty acid composition)
- Milk and cheese (use low-fat cheese and milk products)
- Meat, fish, and eggs (try to choose lean alternatives; eat more fish, including the more fatty species)

TABLE 1.17
Source of Food-Based Dietary Guidelines in Selected European Countries and Language of Publication

Country	Source
Austria	http://www.gesundesleben.at/essen-und-trinken/gesund-essen/ernaehrungspyramide/die-ernaehrungspyramide-baut-auf
Albania	Bejtja, G., Selfo, M., Ceka, N., Cakraj, R., Rama, S., Lakrori, J., Hyska, J., Mihali, E., Molla, L., Shehu, B., Xhindoli, N., and Gjipali, S. Recommendations for a healthy nutrition in Albania, December 2008.
Belgium	http://www.mijnvoedingsplan.be/
Bosnia & Herzegovina	Filipovic Hadziomeragic, A., Vilic Svraka, A., and Jusupovic, F. Vodic o ishrana za odraslu populaciju [Bosnian]. Institute of Public Health of Federation of Bosnia and Herzegovina, 2004.
	Filipovic Hadziomeragic, A., Vilic Svraka, A., and Jusupovic, F. Vodiq o prehrani za odraslu populaciju [Croatian]. Institute of Public Health of Federation of Bosnia and Herzegovina, 2004.
Bulgaria	Petrova, S., Angelova, K., Bajkova, D. et al. Food-based dietary guidelines for adults in Bulgaria. 2006, MH, NCPHP.
Croatia	Antonić Degač, K., Hrabak-Žerjavić, V., Kaić-Rak, A. et al. Dietary guidance for adults, 2002. Croatian National Institute of Public Health, Academy of Medical Sciences of Croatia.

(Continued)

TABLE 1.17
(Continued) Source of Food-Based Dietary Guidelines in Selected European Countries and Language of Publication

Country	Source
Czech Republic	The National Institute of Public Health (http://www.szu.cz/home-original-en)
Estonia	Vaask, S., Liebert, T., Maser, M., Pappel, K., Pitsi, T., Saava, M., Sooba, E., Vihalemm, T., and Villa, I. Estonian nutrition and food recommendations. Estonian Society of Nutritional Science, National Institute for Health Development, 2006, p. 117 (in Estonian).
	www.terviseinfo.ee
Finland	Finnish Nutrition Recommendations, http://www.ravitsemusneuvottelukunta.fi/portal/en/nutrition_recommendations/, 2005 (in Finnish and Swedish)
	Endorsed by the National Nutrition Council (currently under review)
France	www.inpes.sante.fr
	http://www.mangerbouger.fr/
Germany	www.dge.de/modules.php?name=Content&pa=showpage&pid=40
Greece	Supreme Scientific Health Council. Ministry of Health and Welfare.
	http://www.nut.uoa.gr/dietaryENG.html
Hungary	ftp://ftp.fao.org/es/esn/nutrition/dietary_guidelines/hun.pdf
Latvia	www.vm.gov.lv/indew.php?id=198&top=117
Lithuania	Recommendations on a healthy diet. 2005, Kaunas University of Medicine, National Nutrition Centre under the Ministry of Health, Faculty of Medicine of Vilnius University. Endorsed by the Ministry of Health. Published in Lithuanian.
The Netherlands	http://www.voedingscentrum.nl
Poland	Olej rzepakowy-nowy surowiec, nowa prawda (Rapeseed oil—new raw material, new truth). A monograph edited by Krzymanski, J. et al., published in Polish in Warsaw in 2009 and including a chapter contributed by Szostak, W.B., National Food and Nutrition Institute, presenting, inter alia, 10 FBDGs and a pyramid.
	Szostak, W.B. and Cichocka, A. Dieta srodziemnomorska w profilaktyce kardiologicznej. Poradnik dla lekarzy (Mediterranean diet in cardiovascular prevention. A guidebook for physicians).
	Szostak, W.B. and Cichocka, A. Dieta srodziemnomorska w profilaktyce kardiologicznej. Propozycje dla pacjentów (Mediterranean diet in cardiovascular prevention. Guidelines for the patients).
Romania	No FBDG
Serbia	No FBDG
Slovakia	Update of Slovak Inhabitants Nutrition Improvement Programme (approved by the Slovak Government on December 17, 2008); Slovak language, www.uvzsr.sk, Bulletin of the Chief Medical Officer.
Switzerland	Restricted access to members only

Source: Adapted from European Food Information Council, Food-based dietary guidelines in Europe, 2009. Retrieved from http://www.eufic.org/article/en/expid/food-based-dietary-guidelines-in-europe/.

FBDG, food-based dietary guideline.

1.7 DIETARY GUIDELINES OF THE ASIAN REGION

The guidelines of the Arab gulf encompass the countries of Bahrain, Kuwait, Oman, Qatar, Saudi Arabia, and the United Arab Emerates (Musaiger et al. 2012). These countries have a similar socio-economic status and are facing similar nutrition-related issues (Musaiger et al. 2012). The aim is to address both the prevention and control of nutrition-related problems that are occurring as a result of lifestyle-related changes that have occurred across these countries. Not only are lifestyle-related diseases of concern, but nutrient deficiencies also remain a problem. Previously drawing on guidelines of the United States and the United Kingdom, the country of Oman has developed its own guidelines, but it was felt that there was not enough emphasis on the prevention and control of lifestyle-related diseases as those of the Gulf region. They were also prepared for persons or health professionals who educate the public rather than for the general public. They also had a strong emphasis on portion size rather than on food groups. The regional guidelines were based on a process of determining their purpose, characterizing the guidelines, namely, as addressing prevention and control; were based on affordable and available foods; consider local food habits and food safety; were short, clear, and simple; promote healthy lifestyle; and were based on current scientific information. Following this, a review of the literature was undertaken to inform the nutrition situation and food consumption patterns followed by determining the food and nutrition status, addressing which lifestyle patterns are related to which lifestyle-related diseases, and finally formulating draft guidelines and testing of them. This process is very similar to that of Western countries whose food patterns have begun to be followed by the Gulf region (Table 1.18).

An English version of the dietary guidelines of Hong Kong is available online through the government health and medical services (Hong Kong Dietetic Association 2009). The guidelines encourage a healthy balanced diet to avoid disease, improve energy levels, and maintain a healthy weight. The guidelines are similar to those of the Western countries and focus on three main principles of a healthy diet: variety, moderation, and balance. The food pyramid is used to highlight the recommendations. Residents of Hong Kong are encouraged to eat a diet consisting mostly of grains and cereals; eat more fruits and vegetables; include a moderate amount of meat, poultry, fish, eggs, dry beans, and dairy products; and eat less fats, oils, salts, and sweets. Two pieces of fruit and three vegetable servings (80 g) are the recommended daily intake. Special mention is made on Chinese dishes and guidelines for a vegetarian diet (Table 1.19).

Revised dietary guidelines of India were released in May 2011 (Krishnaswamy 2008; Nutrition 2010). This is the first time since they were first published in 1998. The guidelines were compiled by nutrition scientists at the National Institute of Nutrition. Notable changes in the socioeconomic environment necessitated the review as more people start to shift from a diet of traditional foods to the inclusion of more processed foods along with a change in cooking methods and practices. Cereals, pulses, milk, vegetables, and oils remain the staples of the Indian diet (Table 1.20).

TABLE 1.18
Food-Based Dietary Guidelines for Countries of the Arab Gulf

- Eat a variety of different foods every day.
- Eat an adequate amount of fruits and vegetables daily.
- Eat meat, fish, chicken, legumes, and nuts regularly.
- Make sure that your daily diet contains an adequate amount of cereals and their products.
- Consume an adequate amount of milk and dairy products every day.
- Reduce the intake of food rich in fat.
- Reduce the intake of food and drink high in sugar.
- Reduce the use of salt and intake of salty foods.
- Drink adequate amounts of water and other liquids.
- Maintain an appropriate weight for your height.
- Make physical activity a part of your daily routine.
- Do not smoke and reduce the risk of exposure to smoking environments.
- Avoid drinking alcoholic beverages.
- Ensure safety of food eaten.

Source: Musaiger, A.O. et al., *Journal of Nutrition and Metabolism,* 2012, 905303, 2012.

TABLE 1.19
Guide to Healthy Eating of Hong Kong

1. Go for 5-A-Day daily.
   a. Having at least five serves of fruits and vegetables can help lower cancer risks, and smokers may need more. Fruits and vegetables are rich in vitamins and minerals. All vegetables, from the red ones such as tomato and carrot to the green leafy vegetables, are beneficial to our health. Also, they are high in fiber, which is good for bowel movement. It is suggested to have at least two serves of fruits (1 serve = about the size of your fist) and three serves of vegetables (1 serve = half bowl of boiled vegetables) every day.
2. Limit consumption of red meat. Choose a variety of red meat, poultry, fish, nuts, dry beans, and dairy products.
   a. Limit consumption of red meat can help lower bowel cancer risk. In addition, choosing a variety of dry beans and nuts can help lower blood cholesterol levels, thereby reducing the risk of heart disease.
3. Limit alcohol consumption.
   a. Alcohol consumption has been shown to increase the risk of a number of cancers. Thus, it is suggested to consume no more than 1 drink[a] for women and 2 drinks[a] for men.
4. Employ healthy cooking methods.
   a. Choose lean meat. Remove animal skin and trim off visible fats before cooking.
   b. Opt for low-fat dairy, seasoning, and dressing.
   c. Avoid deep-frying.
   d. Use nonstick pan. Minimize oil use.
5. Easy tips for healthy eating
   a. Enjoy a wide variety of foods.
   b. Eat regular meals and moderate portions.
   c. Get your 5-A-Day by choosing different fruits and vegetables every day.
   d. Maintain a healthy weight. Avoid eating too much fat, salts, and sugars.
   e. Always remove animal skin and trim off visible fats.
   f. Avoid deep-fried dishes. Opt for stir-fried, steamed, or soup dishes.
   g. Try using artificial sweetener and skim milk for tea, coffee, and drinks.
   h. Limit spreads such as butter, condensed milk, and peanut butter.
6. Prevent yourself from getting overweight or obese.
   a. Maintain a healthy weight by keeping your BMI within 18.5–22.9.

Source: Hong Kong Dietetic Association, 2009, http://www.hkda.com.hk/index.php?_a=viewDoc&docId=8.
[a] 1 drink = 250 ml beer, 100 ml red wine or white wine, or 25 ml spirit.
BMI, body mass index.

Insufficient and/or imbalanced dietary intakes are the major food concerns, with many children and adults suffering malnutrition, but as affluence improves, overweight and obesity is emerging as a significant problem in urban areas.

Six dietary goals are listed, and sixteen dietary guideline statements are included with examples of how to achieve a balanced diet (Table 1.19). The pyramid diagram is used in line with the guidelines, with recommendations to "consume adequately" grains and cereals; "eat liberally" vegetables and fruits; "eat moderately" fish, egg, seafood, fats, and oils; and to "eat sparingly" the processed and sweet snack foods. Again, the wording is open to interpretation.

The Japanese guidelines updated in 2013 (The Japan Dietetic Association 2013) are one of the few guidelines that refer to the well-being of the person in terms of body and mind rather than the traditional lifestyle-related factors. Reference to a long life is noted as well as interaction with one's family and those involved in food preparation. The "rhythm" of eating or pattern of eating from the first guideline in which daily meals are mentioned flows to the second guideline, which refers to the regularity of the meal patterns and then gives suggestions for particular meal occassions and timing. Meal balance is further tied in, in relation to staple foods and side dishes, and suggestions are made to include a variety of cooking methods and combine meals eaten at home with those eaten outside of

TABLE 1.20
Dietary Guidelines of India

- Eat a variety of foods to ensure a balanced diet.
- Ensure provision of extra food and health care to pregnant and lactating women.
- Promote exclusive breast-feeding for six months and encourage breast-feeding till two years or as long as one can.
- Feed home-based semisolid foods to the infant after six months.
- Ensure adequate and appropriate diets for children and adolescents, in both health and sickness.
- Eat plenty of vegetables and fruits.
- Ensure moderate use of edible oils and animal foods and very less use of ghee, butter, or vanaspati.
- Avoid overeating to prevent overweight and obesity.
- Exercise regularly and be physically active to maintain ideal body weight.
- Restrict salt intake to minimum.
- Ensure the use of safe and clean foods.
- Adopt right precooking processes and appropriate cooking methods.
- Drink plenty of water and take beverages in moderation.
- Minimize the use of processed foods rich in salt, sugar, and fats.

Source: Krishnaswamy, K. *Asia Pacific Journal of Clinical Nutrition,* 17, 66–69, 2008; Nutrition, 2010, http://www.ninindia.org/DietaryguidelinesforIndians-Finaldraft.pdf.

the home. Alcohol consumption, as per other guidelines, is suggested in moderation, leaving it open to the user of the guidelines. In reference to food group-based statements, the Japanese guidelines suggest eating enough grains, which again does not quantify the statement, though reference to local produce is given in relation to the types of grains to be selected. Combining other food groups is suggested with a particular reference to vegetables, fruits, milk products, beans, and fish. The colors are then further addressed in relation to the nutrients as is the consumption of small fish as a source of calcium. Salt intake is quantfied to <10 g daily and closely followed by avoidance of oily and fatty foods. These statements are combined with a reference to check nutrient labels of the chosen food items (Table 1.21).

Being mindful of one's body in relation to body weight, overall health and beauty is included as is reference to carefully chewing the foods consumed and not eating too quickly. These statements further flow onto messages about culture and reiteration of the message to encourage local produce and try new tastes. The guidelines are concluded in relation to food safety and awareness of one's own intake, again bringing the message of family values in by suggesting talk about your diet.

Malaysia follows a similar guidelines structure (Ministry of Health Malaysia 2010; Tee 2011) to other countries worldwide, with variety being the initial focus followed by body weight and physical activity. Endorsed by the Ministry of Health, each of the food groups is subsequenty addressed with very little detail of quantity or frequency. Breast-feeding guidelines are incorporated with six months of exclusive breast-feeding suggested. The final statements of the Malaysian guidelines refer to food safety and label reading again given as short informative statements (Table 1.22).

The dietary guidelines of the Philippines (Tanchoco 2011) begin with the message of variety though, unlike other countries, their second guideline refers to breast-feeding exclusively to six months followed by the introduction of foods while continuing to breast-feed. There is no additional age of the infant given, leaving the statement open to interpretation. The third statement refers to the weight of the children rather than to the weight of the general population, and the statements in relation to specifc food groups then follow. Contrary to other countries, the message about the use of oils is written in a positive light with edible/cooking oil suggested on a daily basis. Reference to iodized salt is made in relation to salt intake though avoidance of salty food is noted in the same statement.

TABLE 1.21
Dietary Guidelines of Japan

Enjoy your meals
- Have delicious and healthy meals that are good for your mind and body.
- Aim to achieve a longer healthy life through your daily meals.
- Enjoy communication at the table with your family and/or other people and participate in the preparation of meals.

Establish a healthy rhythm by keeping regular hours for meals.
- Have breakfast to make a good start to the day.
- Avoid large snacks before bedtime and between meals.
- If you drink alcoholic beverages, do so in moderation.

Eat well-balanced meals with staple food, as well as main and side dishes.
- Make a good combination of various foods.
- Try to cook in various ways.
- Combine wisely homemade meals with eating out, and eating processed and prepared foods.

Eat enough grains such as rice and other cereals.
- Eat grains at every meal to maintain sufficient intake of energy from carbohydrate.
- Make the best use of grains such as rice and other cereals, suited to Japan's climate and soil conditions.

Combine vegetables, fruits, milk products, beans, and fish in your diet.
- Eat enough of vegetables and fruits every day to get vitamins, minerals, and fibers.
- Drink milk and eat green/yellow vegetables, beans, and small fish to get a sufficient amount of calcium intake.

Avoid too much salt and fat.
- Avoid salty foods and reduce the amount of salt intake to <10 g per day.
- Avoid oily and fatty foods, and make a balanced choice of fat from animal, plant, and fish.
- Check nutrition labels in choosing foods and setting menus.

Learn your healthy body weight and balance the calories you eat with physical activity.
- Weigh yourself as soon as you feel like you have gained some weight.
- Have a habit of appropriate physical exercise.
- Good health is essential to beauty. Do not attempt to lose too much weight.
- Chew your food well and do not eat too quickly.

Take advantage of your dietary culture and local food products, while incorporating new and different dishes.
- Enjoy nature's bounty and the changing seasons by using local food products and ingredients in seasons, and by enjoying holiday and special occasion dishes.
- Respect your dietary culture and apply it to daily diet.
- Acquire the knowledge of foods and cooking.
- Be open to trying new foods and dishes.

Reduce leftovers and waste through proper cooking and storage methods.
- Avoid buying and cooking too much food. Try to gauge how much food you need to avoid leftovers.
- Pay attention to "best by" and "consume by" dates on food products.
- Check the contents of your refrigerator and cupboards on a regular basis, and try to create menus that maximize what you have.

Assess your daily eating.
- Set your own health goals and have a habit to assess your diet.
- Think and talk about your diet with your family and friends.
- Learn and practice healthy eating habits at school and at home.
- Promote appreciation of good eating habits from an early stage of life.

Source: The Japan Dietetic Association, 2013, http://www.dietitian.or.jp/english/news/dietary.html.

TABLE 1.22
Malaysian Dietary Guidelines

- Eat a variety of foods within your recommended intake.
- Maintain body weight in a healthy range.
- Be physically active every day.
- Eat adequate amount of rice, other cereal products (preferably whole grain), and tubes.
- Eat plenty of fruits and vegetables every day.
- Consume moderate amounts of fish, meat, poultry, egg, legumes, and nuts.
- Consume adequate amounts of milk and milk products.
- Limit intake of foods high in fats and minimize fats and oils in food preparation.
- Choose and prepare foods with less salt and sauces.
- Consume foods and beverages low in sugar.
- Drink plenty of water daily.
- Practice exclusive breast-feeding from birth until six months and continue to breast-feed until two years of age.
- Consume safe and clean foods and beverages.
- Make an effective use of nutrition information on food labels.

Source: Ministry of Health Malaysia, 2010, http://www2.moh.gov.my/images/gallery/Garispanduan/diet/introduction.pdf; Tee, E.S., *Asia Pacific Journal of Clinical Nutrition,* 20, 455–461, 2011.

TABLE 1.23
Nutrition Guidelines of the Philippines

- Eat a variety of foods every day.
- Breast-feed infants exclusively from birth to six months, and then give appropriate foods while continuing breast-feeding.
- Maintain children's normal growth through proper diet and monitor their growth regularly.
- Consume fish, lean meat, poultry, or dried beans.
- Eat more vegetables, fruits, and root crops.
- Eat foods cooked in edible or cooking oil in your daily meals.
- Consume milk, milk products, and other calcium-rich foods such as small fish and dark leafy vegetables.
- Use iodized salt, but avoid excessive intake of salty food.
- Eat clean and safe food.
- For a healthy lifestyle and good nutrition, exercise regularly, do not smoke, and avoid drinking alcoholic beverages.

Source: Tanchoco, C., *Asia Pacific Journal of Clinical Nutrition,* 20, 462–471, 2011.

A short and sharp message about eating clean and safe food draws in a food safety guideline, and the final statement refers to healthy lifestyle combining physical activity, smoking, and alcohol consumption into one message (Table 1.23).

Thailand follows suit with their message about variety, which is their first statement of the guidelines (Sirichakwal et al. 2011). This same statement also incorporates a message about five food groups and maintenance of proper weight. This weight is not defined. Open statements about the food groups follow with quantification referring only to plenty and sufficient rather than portion amounts. Freuqency of consumption is only defined for dairy produce (specifically milk), while regularity of intake is stated for meats, vegetables, and fruits. Adequacy of grains is the primary focus. Avoidance of sweet and salty foods as well as alcoholic beverages is suggested, and a food safety message worded in terms of eating clean, uncontaminated foods rather than the use of leftover meals as was seen with other countries (Table 1.24).

TABLE 1.24
Thai Dietary Guidelines

- Eat a variety of foods from each of the five food groups daily and maintain proper weight.
- Eat adequate rice or alternative carbohydrate.
- Eat plenty of vegetables and fruits regularly.
- Eat fish, lean meat, eggs, legumes, and pulses regularly.
- Drink sufficient amount of milk every day.
- Take moderate amounts of fat.
- Avoid excessive intake of sweet and salty foods.
- Eat clean and uncontaminated foods.
- Avoid or reduce consumption of alcoholic beverages.

Source: Sirichakwal, P.P. et al., *Asia Pacific Journal of Clinical Nutrition,* 20, 477–483, 2011.

1.8 DIETARY GUIDELINES OF AFRICA

The nutrition guidelines for Namibia (Ministry of Health and Social Services 2000) were developed as part of a broader "Food Security and Nutrition Action Plan" adopted by the National Food Security and Nutrition Council. The plan targeted the issues surrounding food security and malnutrition, as a large portion of the population of Namibia suffers from various forms of malnutrition (National Food Security and Nutrition Council 2000). The FAO (2011) indicates that around 33.9% of the population are undernourished. The guidelines were developed to promote a healthy lifestyle and diet in the general population and are not appropriate for children under two years of age, and special nutrition needs groups such as pregnant or lactating women. The guidelines were developed in conjunction with the "Food Guide" to assist in promoting the consumption of a variety of foods from each of the Namibian four food groups.

Fruit and vegetable consumption is promoted daily due to the common occurrence of micronutrient deficiencies in Namibia (National Food Security and Nutrition Council 2000). Due to the observation of anemia and undernutrition in certain populations, iron-rich and/or protein-rich foods, meat, and beans are promoted to the population with a particular focus on pregnant women and children. Undernutrition is also combated through the promotion of three meals per day. Due to low iodine intakes within the Namibian diet, it is promoted that people use iodized salt to reduce the risk of iodine deficiency disorders, although low salt in general recommended due to its association with increasing blood pressure (National Food Security and Nutrition Council 2000). Food and water safety is promoted to combat the high incidence of diarrhea caused by contaminates in Namibia (National Food Security and Nutrition Council 2000). Overweight and obesity is rising within Namibian population, with around 20% of the population reportedly being overweight or obese (Walker et al. 2001). To combat this issue, the guidelines promote fish as a lean and viable protein source, and it is promoted to achieve and maintain a healthy body weight (Table 1.25).

The most recent dietary guidelines of Nigeria (Abulude 2008) were published in four different languages in 2001. These food-based guidelines were developed as general recommendations for the broader population aimed to maintain good health and consume adequate nutrition. Children are targeted in six different groups up to the age of 18, with breast-feeding being promoted to infants exclusively up to the age of six months. There are general guidelines for adults, but pregnant, breast-feeding women and the elderly are also targeted. In addition to the food based are additional guidelines that promote exercise and healthy lifestyle and warn against consumption of alcohol and use of tobacco (FAO 2009) (Table 1.26).

The South African food-based dietary guidelines (Bourne 2007; Department of Health 2004) were developed over a four-year process through expert consultation and technical review. The guidelines are worded so that they can be based on locally consumed food items due to many

TABLE 1.25
Food Guide for Namibia

- Eat a variety of foods.
- Eat vegetables and fruits every day.
- Eat more fish.
- Eat beans or meat regularly.
- Use wholegrain products.
- Use only iodized salt, but use less salt.
- Eat at least three meals a day.
- Avoid drinking alcohol.
- Consume clean and safe water and food.
- Achieve and maintain a healthy body weight.

Source: Ministry of Health and Social Services, *Food and nutrition guidelines and Namibia:Food Choices for a Healthy Life.* National Food Security and Nutrition Council, Windhoek, Namibia, 2000.

TABLE 1.26
Food-Based Dietary Guidelines for Nigeria

Age Group	Food-Based Dietary Guidelines
Infants (0–6 months)	Start exclusive breast-feeding immediately after birth and continue for six months.
	There should be no bottle-feeding.
Infants (6–12 months)	Continue breast-feeding.
	Introduce complementary feeds made from a variety of cereals, tubers, legumes, fruits, and animal foods, and give with cup and spoon.
Toddlers (12–24 months)	Continue to breast-feed until child is two years.
	Give enriched pap or mashed foods twice daily.
	Give family diet made soft with less pepper and spices.
	Give fruits and vegetables in season.
	Give diet that contains a variety of foods in adequate amounts.
Children (25–60 months)	Add palm oil or vegetable oil to raise the energy level of complementary foods.
	Gradually increase food intake to four to five times daily as baby gets older.
	Provide dark green leafy vegetables, yellow/orange fruits, citrus fruits, cereals, legumes, tubers, and foods of animal origin.
	Limit the consumption of sugary food.
	Continue feeding even when child is ill.
School-aged children (6–11 years)	Give diet that contains a variety of foods in adequate amounts.
	Encourage consumption of good-quality snacks, but limit the consumption of sugary snacks.
Adolescents (12–18 years)	Consume diet containing a variety of foods.
	Most of the energy should be delivered from roots/tubers, legumes, cereals, and vegetables, and less from animal foods.
	An increase in the total food intake is very important at this stage, so is the need to enjoy family meals.
	Snacks, especially pastry, and carbonate drinks should not replace main meals. If you must eat out, make wise food choices.

(*Continued*)

TABLE 1.26
(Continued) Food-Based Dietary Guidelines for Nigeria

Age Group	Food-Based Dietary Guidelines
	Liberal consumption of whatever fruit is in season should be encouraged.
	Females need to eat more iron-containing foods such as meat, fish, poultry, legumes, cereals as well as citrus fruits to enhance the body's use of iron.
Adults (male and female)	Total food intake should take into consideration the level of physical activity.
	Individuals who do manual work need to consume more food than those who do sedentary work.
	Limit the fat intake from animal foods.
	Diet should consist of as wide a variety of foods as possible, for example, cereals, legumes, roots/tubers, fruits, vegetables, fish, lean meat, and local cheese (wara).
	Limit intake of salt, bullion cubes, and sugar.
	Liberal consumption of whatever fruit is in season is encouraged.
Pregnant women	Eat diet that contains a variety of foods in adequate amounts.
	Consume enough food to ensure adequate weight gain.
	Eat more cereals, legumes, fruits, vegetables, dairy products, and animal foods.
	Take iron and folic acid supplements as prescribed.
	Avoid alcohol, addictive substances, and smoking.
	Eat diet that contains a variety of foods in adequate amounts.
Breast-feeding mothers	Eat diet that contains a variety of available food items like cereals, tubers, legumes, meat, fish, milk, fruits, vegetables, etc.
	Consume more foods rich in iron such as liver, fish, beef, etc.
	Eat fruits in season at every meal.
	Consume fluids as needed to quench thirst.
	Avoid alcohol, addictive substances, and smoking.
	Eat diet that contains a variety of available food items such as cereals, tubers, legumes, meat, fish, milk, fruits, and vegetables.
The elderly	Eat diets that are prepared from a variety of available foods, for example, cereals, tubers, fruits, vegetables, etc.
	Increase consumption of fish and fish-based diets.
	Eat more of fruits and vegetables.
	Eat more frequently.

Health Guidelines

Physical activity/exercise	Healthy lifestyles
Physical activity as both short periods of intense exercise or prolonged periods of modest activity on a daily basis generally has beneficial effects. • Children and adolescents should engage in leisure-time exercise. • Adults should undertake some form of exercise as recommended by their doctors.	Some habits and lifestyles, for example, tobacco use and excessive alcohol consumption, have been found to be bad for health. Prolonged indulgence in these lifestyles predisposes to noncommunicable diseases such as cancer, diabetes, heart problems, and hypertension.

TABLE 1.26
Food-Based Dietary Guidelines for Nigeria

Health Guidelines	
Alcohol	**Tobacco**
Too much alcohol consumption can lead to risk of hypertension, liver damage, malnutrition, and various cancers. There is also the problem of alcohol abuse. If you must drink, take alcohol in moderation. • Avoid drinking alcohol when driving a vehicle or operating any machinery.	Tobacco use is associated with lung cancer and other chronic disorders. Smoking during pregnancy can harm the developing baby and can result in low-birth-weight babies. Avoid the use of tobacco in any form.

Source: Abulude, F., *Continental Journal of Medical Research,* 2, 24–27, 2008.

TABLE 1.27
South African Dietary Guidelines

- Enjoy a variety of foods.
- Be active.
- Drink lots of clean, safe water.
- Make starchy foods, the basis of most meals.
- Eat plenty of vegetables and fruits every day.
- Eat dry beans, peas, lentils, and soy regularly.
- Eat chicken, fish, meat, milk, or eggs daily.
- Eat fats sparingly.
- Use salt sparingly.
- Use food and drinks containing sugar sparingly and not between meals.
- If you drink alcohol, drink sensibly.

Source: Bourne, L.T., *Maternal & Child Nutrition,* 3, 227–229, 2007; Department of Health, 2004, ftp://ftp.fao.org/es/esn/nutrition/dietary_guidelines/zaf_eating.pdf.

different cultural practices across the country. The background to the guidelines was that each should be worded to address one simple-and-easy to understand message and then developed visually so that the same message may be understood by those who cannot read. The guidelines were aimed to be user-friendly, to be of a positive tone, to be compatible to the differing cultural eating practices, to be sustainable and environmentally friends, based on foods that are available and affordable for the population, to lead to a selection of foods that may be eaten together, to emphasize the joy of eating, and to be communicated to target populations and address both the under- and overnutrition challenges within the country (Vorster et al. 2001) (Table 1.27).

The many varied dietary guidelines available throughout the globe provide an indication of the complexity of food. Factors relating to intake patterns, culture, food groupings, lifestyle, family and friends, growth and development, and food safety and food labeling have all been incorporated into the many different food-based dietary guidelines addressed in this chapter. These messages become further complicated when the diversity of the food system is layered upon these factors, and the dynamic nature of the food supply and its constantly changing nature is considered in relation to each of the different dietary guidelines. Advances in food technology provides further challenges to the organizations tasked with updating and revising the guidelines though those presented in this chapter are noted as being the most recent at the time of writing.

REFERENCES

Abulude, F. (2008). Dietary guidelines for good health. *Continental Journal of Medical Research, 2*, 24–27.

Barbieri, H.E. and Lindvall, C. (2005). Swedish nutrition recommendations objectified (SNO)—Basis for general advice on food consumption for healthy adults. Retrieved from http://www.slv.se/upload/dokument/rapporter/mat_naring/report_20_2005_sno_eng.pdf.

Beaton, G.H. (2003). Dietary guidelines: Some issues to consider before initiating revisions. *Journal of the American Dietetic Association, 103*(12 Suppl.), 56–59.

Bourne, L.T. (2007). South African paediatric food-based dietary guidelines. *Maternal & Child Nutrition, 3*(4), 227–229.

Britten, P., Haven, J., and Davis, C. (2006). Consumer research for development of educational messages for the MyPyramid Food Guidance System. *Journal of Nutrition Education and Behavior, 38*(6 Suppl.), S108–S123.

Department of Health. (2004). South African guidelines for healthy eating for adults and children over the age of seven years. Retrieved from ftp://ftp.fao.org/es/esn/nutrition/dietary_guidelines/zaf_eating.pdf.

Department of Health and the Health Service. (2012). Your guide to healthy eating using the food pyramid. Retrieved from http://www.healthpromotion.ie/hp-files/docs/HPM00796.pdf.

European Food Information Council. (2009). Food-based dietary guidelines in Europe. Retrieved from http://www.eufic.org/article/en/expid/food-based-dietary-guidelines-in-europe/.

FAO. (2009). Dietary guidelines for Nigeria. Food guidelines by country—FAO Nutrition Information, Communication and Education. Retrieved from http://www.fao.org/ag/humannutrition/nutritioneducation/fbdg/49849/en/nga/.

FAO. (2011). *Food Security Indicators*. Rome, Italy: FAO.

Haven, J., Burns, A., Britten, P., and Davis, C. (2006). Developing the consumer interface for the MyPyramid Food Guidance System. *Journal of Nutrition Education and Behavior, 38*(6 Suppl.), S124–S135.

Health Canada. (2007). Canada's food guides from 1942 to 1992. Retrieved from http://www.hc-sc.gc.ca/fn-an/food-guide-aliment/context/fg_history-histoire_ga-eng.php.

Health Canada. (2011). Eating well with Canada's food guide: A resource for educators and communicators. Retrieved from http://www.hc-sc.gc.ca/fn-an/alt_formats/hpfb-dgpsa/pdf/pubs/res-educat-eng.pdf.

Health.gov. (2008). Dietary guidelines for Americans 2005. Retrieved from http://www.health.gov/dietaryguidelines/dga2005/document/html/executivesummary.htm.

Hong Kong Dietetic Association. (2009). Guide for healthy eating. Retrieved from http://www.hkda.com.hk/index.php?_a=viewDoc&docId=8.

Institute of Nutrition of Central America and Panama. (2012). Honduras: National workshop on advances in updating the food based dietary guidelines for Honduras. Retrieved from http://www.incap.int/index.php/en/2012-04-10-19-01-58/strategic-actions/477-honduras-national-workshop-on-advances-in-updating-the-food-based-dietary-guidelines-for-honduras.

Istituto Nazionale di Ricerca per gli Alimenti e la Nutrixione (2009). Linee Guida. Retrieved from http://www.inran.it/648/linee_guida.html.

Katamay, S.W., Esslinger, K.A., Vigneault, M., Johnston, J.L., Junkins, B.A., Robbins, L.G. et al. (2007). Eating well with Canada's food guide (2007): Development of the food intake pattern. *Nutrition Reviews, 65*(4), 155–166.

Krishnaswamy, K. (2008). Developing and implementing dietary guidelines in India. *Asia Pacific Journal of Clinical Nutrition, 17*(1 Suppl.), 66–69.

Ministry of Food, Agriculture and Fisheries. The Danish Veterinary and Food Administration. The 8 dietary recommendations. Retrieved from http://www.foedevarestyrelsen.dk/english/Nutrition/The_eight_dietary_recommendations/Pages/default.aspx.

Ministry of Health. (2008a). Food and nutrition guidelines for healthy infants and toddlers (aged 0–2): A background paper—Partially revised December 2012. Retrieved from http://www.health.govt.nz/publication/food-and-nutrition-guidelines-healthy-infants-and-toddlers-aged-0-2-background-paper-partially.

Ministry of Health. (2008b). Food and nutrition guidelines for healthy pregnant and breastfeeding women: A background paper. Retrieved from http://www.health.govt.nz/publication/food-and-nutrition-guidelines-healthy-pregnant-and-breastfeeding-women-background-paper.
Ministry of Health. (2009a). Clinical guidelines for weight management in New Zealand adults. Retrieved from http://www.health.govt.nz/publication/clinical-guidelines-weight-management-new-zealand-adults.
Ministry of Health. (2009b). Clinical guidelines for weight management in New Zealand children and young people. Retrieved from http://www.health.govt.nz/publication/clinical-guidelines-weight-management-new-zealand-children-and-young-people.
Ministry of Health. (2010). Food and nutrition guidelines for healthy older people: A background paper. Retrieved from http://www.health.govt.nz/publication/food-and-nutrition-guidelines-healthy-older-people-background-paper.
Ministry of Health Malaysia. (2010). Malaysian dietary guidelines. Retrieved from http://www2.moh.gov.my/images/gallery/Garispanduan/diet/introduction.pdf.
Ministry of Health National Center of Public Health Protection. (2006). Food based dietary guidelines for adults in Bulgaria. Retrieved from http://ncphp.government.bg/files/hranene-en.pdf.
Ministry of Health and Social Services. (2000). *Food and Nutrition Guidelines for Namibia: Food Choices for a Healthy Life.* Windhoek, Namibia: National Food Security and Nutrition Council.
Musaiger, A.O., Takruri, H.R., Hassan, A.S., and Abu-Tarboush, H. (2012). Food-based dietary guidelines for the arab gulf countries. *Journal of Nutrition and Metabolism, 2012*, 905303.
National Food Security and Nutrition Council. (2000). Food and nutrition guidelines for Namibia. Ministry of Health and Social Services, Windhoek, Namibia.
National Health and Medical Research Council. (2003). Australian guide to healthy eating. Retrieved from http://www.nhmrc.gov.au/_files_nhmrc/publications/attachments/n33.pdf.
National Health and Medical Research Council. (2013a). Eat for health—Australian dietary guidelines: Summary. Retrieved from http://www.nhmrc.gov.au/_files_nhmrc/publications/attachments/n55a_australian_dietary_guidelines_summary_book_0.pdf.
National Health and Medical Research Council. (2013b). Food for health—Dietary guidelines for australians. Retrieved from http://www.nhmrc.gov.au/_files_nhmrc/publications/attachments/n31.pdf.
National Institute of Nutrition. (2010). Dietary guidelines for Indians—A manual. Retrieved from http://www.ninindia.org/DietaryguidelinesforIndians-Finaldraft.pdf.
Niclasen, B. and Schnohr, C. (2010). Greenlandic schoolchildren's compliance with national dietary guidelines. *Public Health Nutrition, 13*(8), 1162–1169.
Pan American Health Organization. Food based dietary guidelines for Grenada. Retrieved from http://new.paho.org/cfni/index.php?option=com_docman&task=doc_details&gid=22&Itemid=219.
Sekuła, W., Szostak, W., and Jarosz, M. (2009). Poland's FBDG status, communications & evaluation. *FAO/EUFIC Workshop on Food-Based Dietary Guidelines* (May 18–20, 2009). Retrieved from http://www.fao.org/ag/humannutrition/18852-0c8a9ff3ef1a3948677f067bc6f9275f9.pdf.
Sichieri, R., Chiuve, S.E., Pereira, R.A., Lopes, A.C., and Willett, W.C. (2010). Dietary recommendations: Comparing dietary guidelines from Brazil and the United States. *Cadernos de Saúde Pública, 26*(11), 2050–2058.
Sirichakwal, P.P., Sranacharoenpong, K., and Tontisirin, K. (2011). Food based dietary guidelines (FBDGs) development and promotion in Thailand. *Asia Pacific Journal of Clinical Nutrition, 20*(3), 477–483.
Tanchoco, C. (2011). Food-based dietary guidelines for Filipinos: Retrospects and prospects. *Asia Pacific Journal of Clinical Nutrition, 20*(3), 462–471.
Tee, E.S. (2011). Development and promotion of Malaysian dietary guidelines. *Asia Pacific Journal of Clinical Nutrition, 20*(3), 455–461.
The Japan Dietetic Association. (2013). Japanese health and nutrition information. Retrieved from http://www.dietitian.or.jp/english/news/dietary.html.
US Department of Agriculture. (2012). Dietary guidelines for Americans, 2010. Retrieved from http://www.cnpp.usda.gov/DGAs2010-PolicyDocument.htm.
Van Dooren, C. and Kramer, G. (2012). Food patterns and dietary recommendations in Spain, France and Sweden: Healthy people, healthy planet. LiveWell for Life. Retrieved from http://www.scp-knowledge.eu/sites/default/files/knowledge/attachments/LiveWell%20for%20Life_2012_Food%20patterns%20and%20dietary%20recommendations%20in%20Spain,%20France%20and%20Sweden.pdf.
Vorster, H., Love, P., and Browne, C. (2001). Development of the food-based dietary guidelines for South Africa—The process. *South African Journal of Clinical Nutrition, 14*(3), S3–S6.
Walker, A.R.P., Adam, F., and Walker, B.F. (2001). World pandemic of obesity: The situation in Southern African populations. *Public Health (Nature), 115*(6), 368.

2 Clinical Perspective of Ayurceuticals
Challenges and Opportunities for Global Health and Wellness

Namyata Pathak, Hiteshi A. Shah, and Ashok Vaidya

CONTENTS

2.1 INTRODUCTION

> "The power of traditional food formulations is simply phenomenal. The question is, have we capitalized it fully? What about the health aspects of traditional foods, marketing and consumer preferences? Are we adapting the energy-saving processing equipments? Are we changing ...?"[1]
>
> **V. Prakash**
>
> *International Society for Nutraceuticals, Nutritionals and Naturals*

It has been often observed that when two streams of knowledge intermingle, there are chances of major discoveries and development. However, when systems based on diverse epistemology have to mutually enrich their concepts, practices, and products, a major attitudinal shift and a change in world views of the participants are required. Such a change is much more needed when there is a global disenchantment with modern drugs of chemical origin. The prevalent pharmacophobia is primarily due to reports of severe and sometimes life-threatening adverse reactions to these drugs. The numbers of severe adverse drug reactions and the resultant morbidity and mortality have been alarming.[2] As a consequence, there is a movement of "back to the nature," and an aptitude for "green medicine." The emergence of dietary supplements, nutraceuticals, and functional foods has progressed rapidly over the past few decades. The Dietary Supplement Health and Education Act (DSHEA) of the US Food and Drug Administration (FDA) has been a mixed blessing. It has popularized phytoproducts of other cultures but neglected the very systems from which they were derived. All dietary supplements carry the disclaimer: "This product has not been evaluated by the FDA." However, when adverse reactions such as hepatotoxicity of *Piper methysticum*[3] and cardiotoxicity of *Ephedra sinica*[4] emerge, the dietary supplements are promptly banned and warnings issued.

These occurrences are primarily due to isolation of these products from their matrix in the systems of medicine, disregarding the cautions, dosage, indications, and so on, mentioned in the traditional system. It would only be rational to explore and understand the traditional use of naturals and nutritionals of these systems for avoiding such reactions. Though Ayurveda and traditional Chinese medicine offer a vast potential for innovation in nutraceuticals and functional foods, it is vital that precise guidelines and data substantiating health claims are evolved. In India, the new Food Safety and Standards Authority of India (FSSAI) guidelines do not permit disclaimer like that issued by the FDA.[5] Notwithstanding significant and useful guidelines, bureaucratic interference and delays have thwarted the progress and market growth of nutraceuticals in India.[6]

2.2 AYURVEDA

Ayurveda is an officially recognized system of health care in India, which has been practiced for millenia. In a recent survey of the utilization of traditional system in National Rural Health Mission, it was found that 80%–100% of population in 14 out of 18 states used low health tradition (LHT). Even in the states such as Tamil Nadu, Kerala, and Karnataka, with higher per capita income, nearly 50%–75% usage of LHT was reported.[7] Interestingly, it was found that 55% of allopathic practitioners advised LHT in addition to modern medicine and 75% of the practitioners felt that Ayurveda was not a redundant and suggested way of strengthening it. Two volumes of a comprehensive status report on the system have recently been prepared by Dr. Shailaja Chandra, ex-secretary, Department of Ayurveda, Yoga & Naturopathy, Unani, Siddha and Homoeopathy (AYUSH), Government of India.[8] In view of the long usage and safety of many edible plants with health benefits, Ayurvedic *materia medica* and compendia offer significant opportunities for innovative research and nutraceutical development.[9]

2.2.1 A Nutrition-Centered System

Ayurveda as a holistic system of health, wellness, and longevity was founded upon the *Panchamahabhuta* theory. This theory explains the evolution of microcosm as well as macrocosm. The dynamic exchange of the fundamental elements between the body and its milieu formed the basis of the *Tridosha* theory (*vata*, *pitta*, *kapha*) and their harmony or imbalance. An individual's constitution too is based on the dominance of the basic elements. This was influenced by the nutrition of parents before conception, after conception, and in the early infant nutrition. Minute attention to the compatibility of natural nutrients and the constitution (*prakruti*) of a person forms the strength of Ayurvedic system.

A new science of nutrigenomics is being correlated with *prakruti* and *pathya*. In Ayurveda, nutrients suitable for an individual are called *pathyas*. Lolimbarāja,[10] a physician poet, in the sixteenth century described their role in health and disease in a verse: "If one eats the right kind of food, what is the need of medicine. If one doesn't eat the right kind of food, how will medicine help?" As Ayurvedic nutrition evolved, various ingredients, methods, and processes were incorporated in texts such as *Ksemakutuhalam*, which deals with food-based products.[11] The guiding wisdom of Ayurveda and scientific advances can be fused pragmatically with modern health care. Such a change is seen at the clinical level as functional medicine, integrative Ayurveda, and integrative medicine.

2.2.2 Global Contributions

Ayurveda was globalized millennia back by Indian universities, namely, Nalanda and Takshashila, where students from many countries came to study. *Charaka Samhita*, the major clinical textbook on Ayurveda, was translated into Greek by Ctesias of Cnidus, a contemporary of Hippocrates—the father of medicine. Greek, Arabic, and Chinese travelers to India carried back the knowledge of Ayurveda globally. In the third century BC, Theophrastus, a Greek botanist, included many Ayurvedic plants in his book. The plants from India formed the constituents of European herbal lore. In the sixteenth century, Garcia d'Orta, a Portuguese physician to the governor of Goa, wrote a book *Drugs of Hindustan* that was translated in seven European languages. He had learnt Ayurveda from two contemporary vaidyas. Many of these plants moved into active *materia medica* of Europe.

In a classical textbook of pharmacology, such as by Goodman and Gillman, the roots of several modern drugs have been traced to several Ayurvedic plants.[12] Such discoveries have also occurred in recent times. Hence, it may still be possible to find other beneficial interventions derived from plants. It is interesting to note that the plant-based drug discoveries in the West were founded upon potent biological effects, which were often poisonous in nature, for example, arrow poison. Once the mechanisms were understood, they were used advantageously for new drugs.[13] Unlike some medicinal plants, which have "poisonous" effects, many more have regulating or modulating effects. Plants are suited for their synergistic effects in disease management, preventive effects in high-risk conditions, and health-promoting effects.

2.3 INNOVATIVE RESEARCH APPROACHES IN AYURVEDA DEVELOPED IN INDIA

Recently, there has been a renaissance in research approaches in Ayurveda. India, over the past few decades, has unraveled ways to use Ayurveda's fundamentals for creating scientific products and services. Ayurvedic pharmacoepidemiology and observational therapeutics have emerged as research approaches for developing new products.[12,14] Reverse pharmacology (RP) too has evolved as an innovative way for product development. Pioneered in India (by one of the authors, Vaidya A.B.V.), it is a science of integrating the documented clinical and experiential hits into transdisciplinary exploratory studies and further developing these into drug candidates by experimental and clinical research. Its scope is to understand the mechanisms of action at multiple levels of biological organization and to optimize safety, efficacy, and acceptability of the leads in natural products based on relevant science. RP was initiated in a study of *Rauvolfia serpentina* in hypertension by Kaviraj Gananath Sen. Reserpine, an active molecule of the plant, was used as a tool to understand depression, Parkinson's disease, peptic ulcer, and galactorrhea in terms of the biogenic amines. Innumerable new drugs emerged as a consequence.

The global emergence of transdisciplinary clinical pharmacology, the nature of statistical evidence, and the defined targets of drug action preceded the organized endeavor of RP. The bedside screening of Ayurvedic drugs with a benchside follow-up was evolved by Antarkar, Vaidya, and Joshi in the late 1960s. These efforts in diabetes, Parkinson's disease, and hepatitis were quite fruitful.[15,16]

Recently, two major Government-supported activities have been initiated in India: the Council of Scientific and Industrial Research—New Millennium Indian Technology Leadership Initiative (NMITLI) project and the Indian Council of Medical Research (ICMR) Advanced Centre for Reverse Pharmacology in Traditional Medicine (IACRIT). The former was a nationwide multicentric initiative to develop world-class products inspired from Ayurveda. For this, clinical hits for arthritis, hepatitis, and diabetes were identified based on historical evidence and consensus from contemporary practitioners. However, the IACRIT was a center focused on RP for finding the discipline in areas such as malaria, cancer, sarcopenia, and cognitive decline.[17] Some of the products developed through RP have the possibility of being recognized under novel regulatory categories of Ayurceuticals (under the Food Safety and Standards Act, 2006) and phytopharmaceuticals (under the Drug Controller General of India). Such a potential has led to much industry interest in the RP path (Figure 2.1). With a need to develop Vaidya scientists, who would be Ayurveda doctors trained in basic life sciences, a fellowship program has been initiated at the Institute of Ayurveda and Integrative Medicine by the Department of AYUSH.[18]

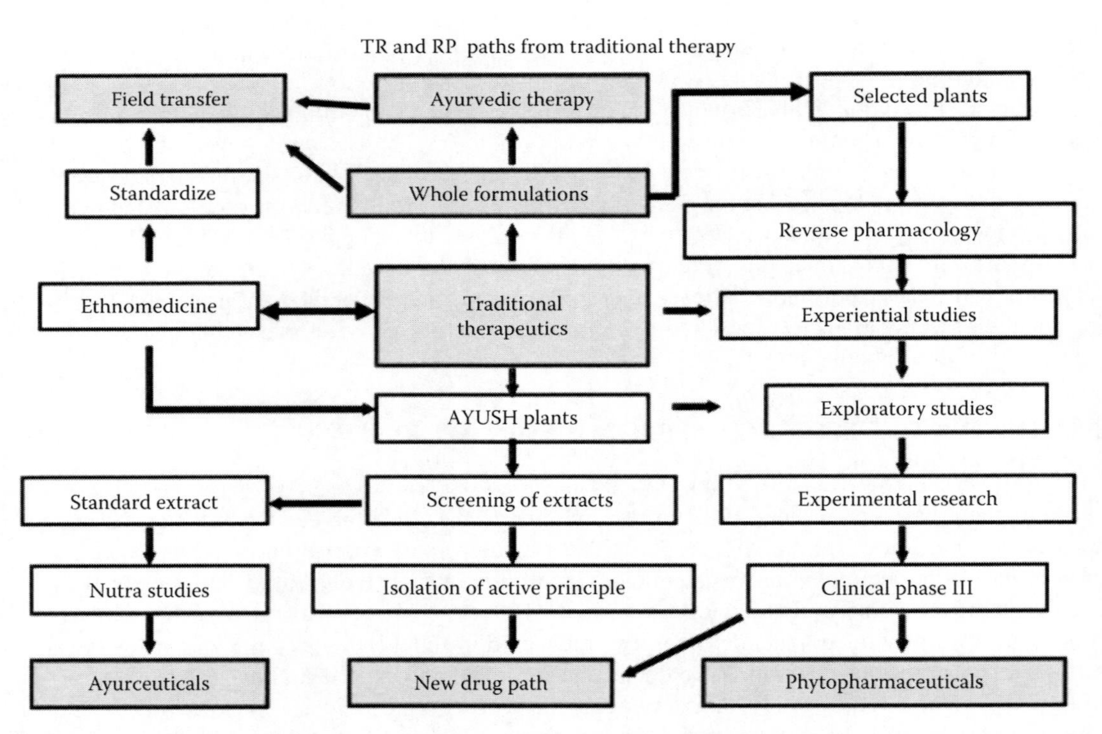

FIGURE 2.1 Development paths in traditional medicine. AYUSH, Department of Ayurveda, Yoga & Naturopathy, Unani, Siddha and Homoeopathy; RP, reverse pharmacology; TR, translational research. (Adapted from Vaidya, A.B., *Proceedings of ICMR Thrust Symposium on Translational Research and Reverse Pharmacology*, Mumbai, IACRIT, Kasturba Health Society, May 31, 2012, p. 124.)

2.4 AYURCEUTICALS

2.4.1 Terminology and Definitions

Ayurceuticals are culture-specific food supplements from Ayurveda, Unani, and Siddha systems. Their scope is to provide for some of the unmet needs of the health care, namely, care of the elderly, metabolic and degenerative diseases, promotion of health and cognitive development of children, and complementary therapy in cancer, allergy, and immune disorders.[19] The spectrum covering Ayurceuticals can be fairly wide as the inspiration for the product is from the wide Ayurvedic usage and categories. Specific food and nutraceuticals can be evolved as per the need and convenience of the user.

Similarly, *rasayanas* can be identified for specific systems or for the whole body. Ayurveda has an in-depth knowledge of taste, properties, and actions of foods *vis-á-vis prakruti*. Ayurceuticals can also be evolved for the reversal of cardiovascular risks or as a complementary to disease management.[20] In Ayurveda, the gastrointestinal (GI) tract is described as "*Mahashotas*, the great stream or path." Its dysfunction leads to several cardiovascular risks. The current developments in gut biota and microbiome open up new opportunities for products such as prebiotics and probiotics, which influence the gut flora.

2.4.2 *Raison D'Être*—Transcultural Challenges

Over the past five decades, botanical product development has been pursued predominantly through phytochemical standardization and screening with biological models of diseases, as defined by contemporary medicine. While this is integral to the process of Ayurceutical development, what strikingly seems amiss is the use of Ayurvedic properties and determinants of clinical response. Such reasoning in the process of product development can enhance their diverse forms, deliveries, and desirable effects, as well as their safety.

A major Ayurceutical example is *Mucuna pruriens*, a natural source of levodopa, which was found to be beneficial in Parkinson's disease in 1978. A further cascade of clinical studies ensued, demonstrating efficacy and certain advantages over the synthetic levodopa supplementation—the mainstay of Parkinson's disease treatment. Consequent global attention to *M. pruriens* has led to products in the markets containing 3%–60% of levodopa (Table 2.1). In Ayurvedic

TABLE 2.1
Some Available Products with *Mucuna pruriens*

Company	Levodopa (%)	Dose Recommendation	Cost ($)
Abhinav	Not mentioned	Two tablets of 300 mg twice a day with a glass of water along with meals	Not mentioned
Himalaya	Not mentioned	One capsule twice a day after meals	71.85/60 capsules
Natural Remedies	10	2 g/day in divided doses	Not mentioned
	20	1 g/day in divided doses	
Sami Labs	10, 15	–	–
Zandopa	Not mentioned	6.525 g in a flavored base; dose as prescribed by physician In half a glass of water (~100 ml) suspend prescribed dose powder, stir, and drink immediately	13.95/175 g
Global Supplements	50	One serving = 1 g, no dose recommendation	99.95/120 servings
Herbal Powers	60	100 mg, one to two times a day, 30 min before meals	35/120 capsules
Ray Sahelin	15	One capsule of 200 mg, in the morning, a few times a week or as directed by your health-care provider	14.95/60 capsules

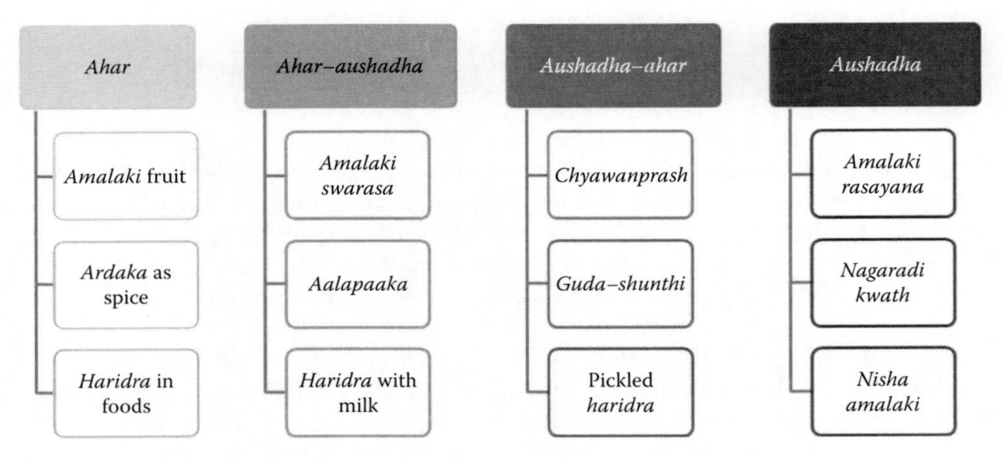

FIGURE 2.2 Food–medicine spectrum in Ayurveda.

formulations, the variations in the composition, processing, dosing, and price are sizable. The strain of *Mucuna* spp., the method of extraction, and the extractive values are mostly not mentioned. Clinical evidence cited is often for a different interventional agent than the one marketed. The dosage forms are not similar to the classical literature or as in the common practice of vaidyas. In addition, in the case of the elderly, the Ayurveda uses *Mucuna* as a jam-like *chyawanprash* or uses it with milk. In these cases, the products are likely to be designed differently for specific health benefits.

Safety should be paramount in using medicinal plants such as *Mucuna*, which are considered as drugs in Ayurveda. It is obvious that the safety index of plants such as *Phyllanthus emblica* and *Curcuma longa*, which are used in diet, would be more than those such as *M. pruriens* used for specific medical and health needs alone. For example, in 1989, when a wild variety of *M. pruriens* was consumed as food during a famine in Mozambique, 203 cases of acute toxic psychosis were reported.[21] Also, dihydroxyphenylalanine (DOPA)-containing *Mucuna* powder, if ingested with a nonselective monoamine oxidase inhibitor (MAOI), can potentially result into a crisis. Hence, when extractions are carried out, the analysis of the sample is desirable but rarely done.[22]

2.4.3 *Ahar* and *Aushadhis*

Ayurveda has dwelt at length on the varying potential of diverse ingredients across the plants. For example, while *Emblica officinalis* as a whole could be consumed as a food, the juice of *Amalaki swarasa* is used as a supplement in diabetes mellitus. For healthy aging, it was converted into a formulation popularly known as *chyawanprash*, but the *Amalaki rasayana*—where the processing includes giving *bhavana* of *Amalaki swarasa* to *Amalaki* powder repeatedly seven times—is being used as a popular complementary treatment of cancer.[23] Similarly, ginger is used as a kitchen spice, but a sweet preparation is used to nourish and strengthen while increasing the *agni*. *Guda shunthi* is used in cold climate, although a special method of consuming the combination is also used to treat odema. Figure 2.2 shows the food–medicine spectrum.

2.5 CLINICAL PERSPECTIVES ON PRODUCTS

Several products from Ayurveda have been used popularly and widely for decades. Here, we share some products that have helped fill certain therapeutic gaps. The basic premise for development of these products had been clinical observations of vaidyas and application of

TABLE 2.2
Examples of Leading Nutra Brands from Ayurveda

Ayurveda Product Brand	Company	Indication	Channel
Liv 52	Himalaya	Hepatitis and liver dysfunction	Consumers and doctors
M2-Tone	Charak	Menopause and menstrual irregularities	Vaidyas and gynecologists
Pancharishta	Zandu	Dyspepsia and flatulence	Vaidyas and consumers
Hajmola	Dabur	Dyspepsia and indigestion	Consumers
Dr. Mom's cough syrup	J.B. Chemicals	Cough	Consumers

TABLE 2.3
Examples of Leading Nutra Ingredients from Ayurveda

Nutraceutical Ingredient	Indication	Related Activity Studies	Product Examples
Aloe vera	Hepatitis	Hegazy et al.[24]	Kumari Asava, Kumaripak
Karela (*Momordica charantia*)	Diabetes	Tsai et al.[25]	Karela (Himalaya), Karela chips
Chyawanprash	Immunity	Parle et al.[23]	Swamla (Dhootpapeshwar), Rajwadi chyawanprash, chyawanprash granules, sugar-free chyawanprash (Dabur)
Brahmi (*Centella asiatica*)	Memory and cognition	Stough et al.[26]	Brahmi prash, Saraswatarishta, Brahmi Ghrita, Brahmi Taila
Vasa (*Adhatoda vasica*)	Cough	Gandhi et al.[27]	Vasavaleha, Vasarishta, Vasakasav, Vasapanak, Vasachandanadi taila, Vasaghritam, Vasaputapak, Gulkand of flowers

technology to create quality products. These are accessed mostly through over-the-counter channels. Recommendations from Ayurvedic physicians and general physicians are often taken. On occasions, conventional specialists do recommend them. The efforts to bring more science to the products are essential for them to be mainstreamed. Their popularity in the market place is a reason for a closer look for their global extension. Table 2.2 lists some Ayurvedic brands that are widely used for indications mentioned and Table 2.3 lists several Ayurvedic plants that are the major ingredients for the specified indications.

2.5.1 GI Wellness

Gut health lies at the heart of Ayurvedic approaches. It is *agni*, the digestive fire essential for the well-being of an individual. Diseases are considered to arise out of vitiations of *agni*. Ayurveda has made available terminology for diseases affecting the GI tract. *Ajirna* (indigestion), *amlapitta* (hyperacidity), *atisaar* (diarrhea), *grahani* (malabsorptive disorders), *aamashaya gata vata* (*vata* in the stomach), *pakwashaya gata vata* (*vata* in the colon), and *chhardi* (vomiting) have been described well with their subtypes. In the Orient, the traditional Asian medicine is often resorted to for relief of the GI complaints. For instance, *shankh vati* is a popular formulation for indigestion, flatulence, and pain in the abdomen. There is room for formulations such as *bilvaavaleha*, a jam of fruit of *Aegle marmelos*, for the irritable bowel syndrome. *Shallaki* is useful in inflammatory bowel disease. Long-standing hyperacidity is managed with *sootshekharrasa*. Table 2.4 describes the four plants used for GI disorders.

TABLE 2.4
GI Ayurceuticals

Ingredient	Proposed Nutra Indication	Ayurvedic Properties[28]	References
Ginger (*Zingiber officinale*)	Nausea and heaviness after meals	*Laghu, Ushna, Snigdha*	Leake et al.[29]
Yashtimadhu (*Glycyrrhiza glabra*)	Heart burn and epigastric discomfort	*Guru, Sheeta, Snigdha*	Wittschier et al.,[30] Fukai et al.[31]
Shallaki (*Boswellia serrata*)	Inflammatory bowel disease	*Laghu, Sheeta, Ruksha*	Joos et al.,[32] Gupta et al.[33]
Bilwa (*Aegle marmelos*)	Irregular bowel movement, gaseousness, and irritable bowel syndrome	*Laghu, Ushna, Ruksha*	Behera et al.,[34] Jindal et al.,[35] Dhuley et al.[36]

2.5.2 Hepatobiliary Protection

Fatty infiltration of liver affects one-fourth of the population. The liver, known as *yakruta* in Ayurveda, is considered central for digestion. Liver health is considered essential in Ayurveda *vis-à-vis* immunity, blood purity and quality, cognition, and musculoskeletal function. Medications such as bitter tonics and common prescriptions such as *arogyavardhini* and *kutki-chirata* help to improve digestion through enhanced liver function. Table 2.5 describes the hepatoprotective plants.

2.5.3 Energy and Metabolism

Despite the adequate intake of substrates for conversion to energy, obese patients complain of low energy and feel fatigued. Ayurveda primarily aims to correct *agni*, albeit at the tissue level, called *dhatwagni*. The corrective agents used are different from those used for *agni* at the GI level. *Santarpana* and *apatarpana* are two opposing ideas of anabolic and catabolic treatments, respectively, which enhance the understanding of *agni* at different levels. The Ayurvedic plants and their ingredients work at the protection of pancreatic islet beta cell mass, prevention of glucolipotoxic damage, and insulin sensitization. Table 2.6 lists the three plants and one medicated wine that will enhance energy and metabolism.

TABLE 2.5
Hepatoprotective Ayurceuticals

Ingredient	Proposed Nutra Indication	Ayurvedic Properties	Reference
Kutki (*Picrorhiza kurroa*)	Fatty liver	*Ruksha, Laghu*	Shetty et al.[37]
Bhui amalaki (*Phyllanthus amarus*)	Viral hepatitis	*Laghu, Ruksha*	Thyagarajan et al.[38]
Trikatu (*Zingiber officinale, Piper longum, P. nigrum*)	Liver-related dyspepsia	*Laghu, Ushna, Tikshna*	Johri et al.[39]

TABLE 2.6
Energy/Metabolic Plants

Ingredient	Proposed Nutra Indication	Ayurvedic Properties	Reference
Amalaki (*Phyllanthus emblica*)	Pancreatic salvage	*Guru, Ruksha*	Balu et al.[40]
Mamejava (*Enicostemma littorale*)	Glucotoxicity	*Laghu, Ushna, Ruksha*	Vaidya et al.[41,42]
Methi (*Trigonella foenum-graecum*)	Insulin resistance	*Ushna, Snigdha, Laghu*	Gupta[43]
Drakshasava (classical formulation)	Fatigue	*Snigdha, Ushna, Laghu*	Sharma[44]

TABLE 2.7
Musculoskeletal Ayurceuticals

Ingredient	Proposed Nutra Indication	Ayurvedic Properties	References
Ashwagandha (*Withania somnifera*)	Muscle weakness	*Laghu*, *Snigdha*	Raut et al.[45]
Guggul (*Commiphora wightii*)	Rheumatic pains	*Laghu*, *Tikshna*, *Snigdha*, *Pichchil*, *Sukshma*, *Sara*	Raut et al.[46]; Chauhan et al.[47]
Rasna (*Pluchea lanceolata*)	Low back ache	*Guru*, *Ushna*	Vaidya and Gogte[48]

TABLE 2.8
Cardiovascular Ayurceuticals

Ingredient	Proposed Nutra Indication	Ayurvedic Properties	Reference
Arjun (*Terminalia arjuna*)	Low heart reserve	*Laghu*, *Ruksha*	Dwivedi et al.[49]
Garlic (*Allium sativum*)	Improvement in blood flow	*Snigdha*, *Tikshna*, *Pichchil*, *Guru*, *Sara*	Banerji et al.[50]
Pomegranate (*Punica granatum*)	Antioxidant damage	*Laghu*, *Snigdha*	Kotamballi et al.[51]

2.5.4 Musculoskeletal Health

A focus on muscle health in nutraceuticals was primarily in sports medicine and nutrition. Recently, the mainstream medicine has recognized sarcopenia as a distinct and important problem. *Commiphora wightii*[47] has been explored for hypolipidemic activity, but traditional practitioners use it widely for rheumatic aches and pains. Low back ache has diverse etiology, but its musculoskeletal component can benefit significantly with *Rasna* and its traditional formulations. Table 2.7 shows the three plants of promise.

2.5.5 Cardiovascular Reserves

With global changes in lifestyle and food habits, cardiovascular morbidity and mortality have emerged as formidable challenges. Besides yoga, exercise, and low-fat diet, there is a need for safe plant-based agents to reduce cardiovascular risk factors. Table 2.8 lists the three plants with a high potential for their development as enhancing cardiovascular reserves.

2.5.6 Respiratory Tract

Ayurveda has much to offer for common upper respiratory tract infection. In India, millions of people avail of simple home remedies for common cold and sore throat. However, the major opportunity lies in exploring Ayurvedic plants for enhancing immunity in recurrent respiratory infections and reducing the frequency, duration, and severity of asthmatic attacks. Table 2.9 lists some products with a potential for respiratory Ayurceuticals.

2.5.7 Kidneys and Bladder Disorders

The classical Ayurveda text elaborates conditions such as dysuria, oliguiria, anuria, and calculi. Many Ayurvedic medicinal plants and formulations are recommended for these conditions and are found to be effective. Plants such as *gokshura*[54] and *punarnava*[55,56] are found to be effective in

TABLE 2.9
Respiratory Ayurceuticals

Ingredient	Proposed Nutra Indications	Ayurvedic Properties	Reference
Chyawanprash	Promote respiratory immunity	*Guru, Snigdha, Ushna*	Parle et al.[23]
Banapsha (*Viola odorata*)	Persistent dry and irritating coughs	*Laghu, Ruksha*	Parihar et al.[52]
Sitopaladi (classical preparation—*Mishri, Vanshlochan, Piper longum Linn, Elettaria cardamomum, Cinnamomum zeylanicum*)	Dry and wet coughs	*Laghu, Snigdha, Ushna*	Sharma[44]
Pippalimoola (*P. longum*)	Bronchial asthma	*Laghu, Ushna*	Fernandez et al.[53]

TABLE 2.10
Urinary Ayurceuticals

Ingredient	Proposed Nutra Indication	Ayurvedic Properties	Reference
Chandan (*Pterocarpus santalinus*)	Protection from urinary tract infection	*Laghu, Ruksha*	Gogte[28]
Punarnava (*Boerhavia diffusa*)	Kidney reserves	*Laghu, Ruksha*	Upadhay et al.[55,56]
Gokshura (*Tribulus terrestris*)	Benign prostatic hyperplasia	*Guru, Snigdha*	Joshi et al.[54]
Kulathi (*Dolichos biflorus*)	Kidney stones	*Laghu, Ruksha, Tikshna*	Singh et al.[57]

TABLE 2.11
Psychoneural Ayurceuticals

Ingredient	Proposed Nutra Indication	Ayurvedic Properties	Reference
Shankapushpi (*Convolvulus pluricaulis*)	Memory enhancer	*Snigdha, Pichchil, Guru, Sara*	Sethiya et al.[61]
Pippalimoola (*Piper longum*, root)	Anxiolytic	*Laghu, Snigdha, Tikshna*	Lee et al.[62]
Saraswatarishta (classical formulation)	Promotion of cognition	*Laghu, Snigdha, Ushna*	Mishra[63]

kidney and urinary disorders. However, plants such as *pashanbhed* and *kulathi* are beneficial in preventing kidney stones.[57] Table 2.10 lists some of these plants.

2.5.8 Psychoneural Health

Ayurveda pays special attention to drugs of the central nervous system (CNS).[59–60] Some of these have been investigated with modern scientific methods. Studies have established the beneficial effects of certain bioactive compounds in medicinal plants, such as memory enhancing, relieving anxiety, and promoting cognition.[59–61] Table 2.11 lists psychoneural Ayurceuticals that highlight the benefits of the bioactive compounds.

2.5.9 Reproduction and Sexual Health

The activities of plants and formulations in sexual health and reproductive disorders were given special attention in Ayurveda as *vrishya dravyas*. Significant research has been initiated in this field. A focused RP research approach needs to be brought to the three plants listed in Table 2.12.

TABLE 2.12
Vrishya Dravyas

Ingredient	Proposed Nutra Indication	Ayurvedic Properties	Reference
Shatavari (*Asparagus racemosus*)	Galactogogue	*Guru, Snigdha*	Jetmalani et al.[64]
Kaunch (*Mucuna pruriens*)	Promotion of virility in the elderly (both sexes)	*Guru, Snigdha*	Singh[65]
Ashoka (*Saraca indica*)	Menorrhagia	*Laghu, Ruksha*	Pradhan et al.[66]

TABLE 2.13
Immunostimulant Ayurceuticals

Ingredient	Proposed Nutra Indication	Ayurvedic Properties	Reference
Tulsi (*Ocimum sanctum*)	Prevention of infections	*Laghu, Ruksha*	Vasudevan et al.[67]
Guduchi (*Tinospora cordifolia*)	Immunomodulator	*Laghu, Snigdha*	More et al.[68]
Daruharidra (*Berberis aristata*)	Antimicrobial	*Laghu, Ruksha*	Bhandari et al.[69]
Kiratatikta (*Swertia chirata*)	Postinfection convalescence	*Laghu, Ruksha*	Pharmacopoeia of India[70]
Kalmegh (*Andrographis paniculata*)	Antiviral	*Laghu, Ruksha, Triksna*	Wiart et al.[71]

TABLE 2.14
Ayurcosmetics/Dermatologics

Ingredient	Proposed Nutra Indication	Ayurvedic Properties	Reference
Khadir (*Acacia catechu*)	Acne	*Laghu, Ruksha*	Gogte[28]
Ajmoda (*Carum roxburghianum*)	Skin allergies	*Laghu, Ruksha, Triksna*	Gogte[28]
Gandhak rasayana (classical formulation)	Adjuvant in skin disorders	*Sheeta, Snigdha, Guru*	Yogratnakar[72]
Jabakusum (*Hibiscus rosa-sinensis*)	Hair fall	*Laghu, Ruksha*	Adirajan et al.[73]
Turmeric (*Curcuma longa*)	Skin allergies	*Laghu, Ruksha*	Shishodia et al.[74]

2.5.10 Immunity and Infections

Ayurveda laid great stress on enhancing resistance to infections rather than developing anti-infectives. *Berberis aristata* and *Tinospora cordifolia* have shown remarkable activity as immune stimulants. Tulsi, also known as holy basil, has been used in Ayurveda since centuries. It is a household plant in India and used in ailments such as common cold, headaches, stomach disorders, and malaria. Table 2.13 lists some immunostimulant Ayurceuticals.

2.5.11 Skin, Hair, and Nails

Having realized the negative effects of chemical-based cosmetics, the need for natural products for beautifying skin, hair, and nails is growing. There are standard references, usage, and research on Ayurvedic cosmeceutics for the indications of skin pigmentation, wrinkles, hair fall, freckles, periorbital darkness, acne, skin infections, and so on. Table 2.14 gives a list of ayurcosmetics. There are multi-ingredient hair oils, which are used by millions in India.

TABLE 2.15
***Rasayana* for Aging**

Ingredient	Proposed Nutra Indication	Ayurvedic Properties	Reference
Triphala (classical combination—*Phyllanthus emblica, Terminalia chebula, Terminalia bellerica*)	Antiaging	*Ruksha, Sheeta, Laghu*	Singh et al.[75]
Garlic (*Allium sativum*)	Antiaging	*Snigdha, Tikshna, Pichchil, Guru, Sara*	Khalid[76,77]

2.5.12 Retardation of Aging

The principle and practice of *kayakalpa* (rejuvenation) is unique to Ayurveda. Ayurveda has a great potential to explain reversal of aging and regeneration of tissues. There are anecdotal reports of reversal of aging in Ayurveda literature, with a regimen of specific medications, milk, and a stay in dark shelter. *Triphala* is widely used as *rasayana* by the aged in India. Table 2.15 lists the two *rasayanas* for aging.

2.6 OPPORTUNITIES AND CHALLENGES

Health professionals and consumers seek a clinical rationale to use a product, besides efficacy with statistical significance. The practice of a long traditional use may generate interest in the ingredients. The backing of contemporary science would enhance the value. This is an opportunity awaiting to be unearthed with the help of a collaborative R&D team.

Reverse nutraceutics is proposed as a new guiding path to create nutra/Ayurceuticals, analogous to what RP has become for modern drug development from Ayurveda. Like RP, reverse nutraceutics has much potential of developing global novel products inspired by products already being safely used in the field (Figure 2.3). Although it is depicted linearly, it may have many feedback loops and would be nonlinear on execution.

As the ingredients for most nutra- and Ayurceuticals are already being used in the field, reverse nutraceutics begins with nutra epidemiology. Through systematic surveys, extensive initial information can be sought: extent of use, safety of use, ease of use, popular practices and processing requirements, modifications used globally, incorporation in daily life, distribution channels, availability, and cost-effectiveness of a potential product (ingredient/formulation). It would also include the study of classical texts and historical evidence of use.

Human observational studies provide case reports or detailed anecdotes of individual experiences by a physician or a sensitive patient. They allow knowing the nature of beneficial response and the temporal relation with intake of ingredient. These are essentially well-described clinical phenomena.

There are three well-designed studies: experiential, exploratory, and experimental.

Experiential nutra studies are open-labeled studies, using common target symptoms and routine investigations as assessments, in a sample group as determined by earlier observations. This stage evolves hits to be taken up further.

Exploratory studies (in vitro, in vivo, *and clinical studies*) need to evolve models and targets relevant to clinically documented effects. The need to innovate models cannot be overemphasized. This stage generates leads from the hits.

Experimental studies (*human nutra studies*) are controlled studies *for* safety and efficacy with defined serving size and defined indications. Experiments for putative mechanistic understanding at all levels of biological organization are included. This stage develops the leads into Ayurceutical candidates.

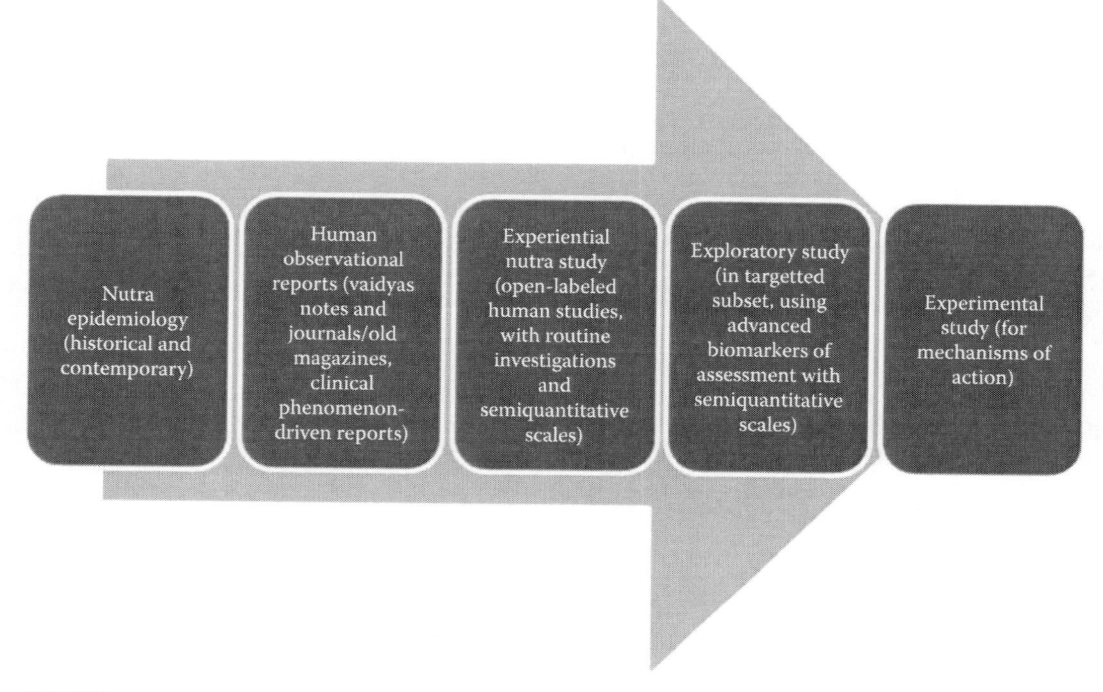

FIGURE 2.3 Reverse nutraceuticals.

The human nutra studies are distinct from drug trials in many ways. The safety of long-term usage in humans provides a comfortable baseline. The range of serving amounts used facilitates dosage. There is also a possibility of a previous field survey of usage (nutra epidemiology), leading to clinically demonstrated effects that guide the development of novel *in vitro* and *in vivo* models.

There is an urgent need of a transdisciplinary team development in Ayurceuticals. The diverse expertises that can be involved are food scientist, food technologist, vaidya partners, nutritionist, botanist, phytochemists, pharmaceutics, pharmacologist, and colleagues. The future of Ayurceutical depends on striving for excellence by such a team.

2.7 CONCLUSIONS

Ayurceuticals are nutraceuticals inspired from the ancient and live system of Ayurveda. The widespread use of Ayurvedic plants makes the potential large. The specific Ayurceutical opportunities, in terms of the plants or formulations, have been listed with appropriate references. It is hoped that reverse ayurceutics from the traditional system of medicine is taken up in earnest within India and globally.

ACKNOWLEDGMENTS

We acknowledge Dr. Mamta Lele for her inputs on Ayurvedic properties of medicinal plants. We thank Sri Dhiru Mehta for the facilities provided at Medical Research Centre (MRC)–Kasturba Health Society, Mumbai, India, and his encouragement for research.

GLOSSARY

***aalapaaka*:** shredded ginger in boiled sugar confectionery
***agni*:** the digestive fire
***ahar*:** diet
***ahar–aushadha*:** dietary products of medicinal value
***amalaki*:** the Indian gooseberry
***aushadha*:** medicine
***aushadha–ahar*:** medicinal products that can be a part of diet
Ayurveda: an Indian traditional medicine
***bhavana*:** a process in which a medicinal substance is impregnated with the juice of another medicinal substance
***chyawanprash*:** a popular jam used as antiaging formulation and for immune enhancement
***dhatwagni*:** moiety of agni that helps in metabolism at the tissue level
***guda–shunthi*:** combination of jaggery and dried ginger
***guru*:** heavy, that which is digested slowly
***haridra*:** Sanskrit name for turmeric
***kapha*:** one of the three doshas (the vital humors)
***laghu*:** light (guna); easy to digest
***nagaradi kwath*:** a decoction containing dried ginger and other herbs
***nisha amalaki*:** combination of turmeric and amla (the Indian gooseberry)
***pathya*:** regimen of diet/lifestyle in a particular disease
***pichchil*:** viscous, sticky
***pitta*:** one of the three doshas (the vital humors)
***prakruti*:** the original nature, character, constitution, or temper of a person
***rasa*:** juice of any medicine; taste perceived by the tongue; mercury; mercurial preparation
***rasayana*:** antiaging phenomenon or formulation
***Ruksha*:** dry
***Sara*:** fluid, liquid (one of the 20 gunas); cathartic, purgative, laxative
***Sheeta*:** cold
Siddha: one branch of traditional medicine that has flourished especially in South India
***Snigdha*:** oily (one of the 20 gunas)
***Sukshma*:** minute, subtle (one of the 20 gunas)
***swarasa*:** plain juice of any medicine
***Tikshna*:** hot, sharp, fast acting
Unani: one branch of traditional medicine
***Ushna*:** hot, heat
***Vaidya*:** a physician of Indian traditional medicine
***vata*:** one of the three doshas (the vital humors)

REFERENCES

1. V. Prakash and Kumar, R. Innovation in functional food industry for health and wellness. In *Innovation in Healthy and Functional Foods*, Dilip, G. and Das, S. (eds.), Boca Raton, FL: CRC Press, 2013.
2. C. Kongkaew, Noyce, P.R., and Darren, M.A. Adverse drug reactions: Hospital admissions associated with adverse drug reactions—A systematic review of prospective observational studies. *Ann Pharmacother* 2008, 42:1017–1025.
3. E. Ernst. A re-evaluation of kava (*Piper methysticum*). *Br J Clin Pharmacol* 2007, 64(4):415–417.
4. C.A. Haller and Benowitz, N.L. Adverse cardiovascular and central nervous system events associated with dietary supplements containing ephedra alkaloids. *N Engl J Med* 2000, 343:1833–1838.
5. Training manual for food safety regulators. Food Safety and Standards Authority of India. http://www.fssai.gov.in/Portals/0/Training_Manual/Volume%20I-%20Intoduction%20to%20Food%20%20and%20Food%20Processing.pdf.

6. J.I. Lewis. Proprietary foods—Balancing diversity with safety. Protein Foods and Nutrition Development Association of India Bulletin, 2012.
7. R. Priya and A.S. Shweta. Status and role of AYUSH and local health traditions under the National Rural Health Mission. New Delhi, India: National Health Systems Resources Centre; Ministry of Health and Family Welfare, 2010, 11.
8. S. Chandra. Status of Indian Medicine and Folk Healing—Parts 1 and 2. 2011, 2013. Available at http://issuu.com/knowledgeforall/docs/ayush_report_partii/3.
9. K.M. Nadkarni and Nadkarni, A.K. *Indian Materia Medica*. Mumbai, India: Popular Book Depot, 1998.
10. L. Lolimbarāja and Saxena, N. *Vaidya Jīvana of Lolimbarāja*, 1st edition. Varanasi, India: Krishnadas Academy, 1978.
11. S. Kṣhema, Ved, D.K., Shankar, D. et al. *Kshemakutuhalam*, 1st edition. Bangalore, India: Indian Institute of Ayurveda & Integrative Medicine (IIAIM), 2009.
12. R.A. Vaidya, Vaidya, A.B., Patwardhan, B. et al. Ayurvedic pharmacoepidemiology: A proposed new discipline. *J Assoc Physicians* 2003, 51:528.
13. B. Patwardhan, Joglekar, V., Pathak, N.Y., and Vaidya, A.B. Vaidya scientists: Catalyzing ayurveda renaissance. *Curr Sci* 2011, 100 (4):476–481.
14. K. Joshi. Insights into ayurvedic biology—A conversation with Professor M.S. Valiathan. *J Ayurveda Integr Med* 2012, 3:226–269.
15. R. Ashwinkumar. *Vaidya Antarkar Memorial Volume*. Mumbai, India: Bharatiya Vidya Bhavan, 2010.
16. J. Sanjeev and Pratima, M. The other Bose: An account of missed opportunities in the history of neurobiology. *Curr Sci* 1997, 97(2):266–269.
17. B. Patwardhan and Mashelkar, R.A. Traditional medicine-inspired approaches to drug discovery: Can Ayurveda show the way forward? *Drug Discov Today* 2009, 14(15):804–811.
18. A.B. Vaidya. An advocacy for Vaidya-Scientist in Ayurveda research. *J Ayurveda Integr Med* 2010, 1(1):6–8.
19. A.B. Vaidya. Reverse pharmacology—A paradigm shift for new drug discovery based on ayurvedic epistemology. In *Ayurveda in Transition*, Muraleedharan, T.S. and Raghava Varier, M.R. (eds.). Kerala, India: Arya Vaidya Sala, 2010, pp. 27–38.
20. N.Y. Pathak. Ayurceuticals: Novel path for the nutra world. *Ingredients South Asia*. Saffron Publication, Chennai. 2013, 38–40.
21. K.P. Gillman. Monoamine oxidase inhibitors (MAOI), dietary restrictions, tyramine, cheese and drug interactions. 2011. http://biopsychiatry.com/maois_diet_full_v2.1.pdf.
22. K.O. Adebowale and O.S. Lawal. Functional properties and retrogradation behaviour of native and chemically modified starch of *Mucuna* beans (*Mucuna pruriens*). *J Sci Food Agr* 2003, 83:1541–1546.
23. B. Parle and Bansal, N. Traditional medicinal formulation, *Chyawanprash*—A review. *Indian J Tradit Knowl* 2006, 5(4):484–488.
24. S.K. Hegazy, El-Bedewy, M., and Yagi, A. Antifibrotic effect of aloe vera in viral infection-induced hepatic periportal fibrosis. *World J Gastroenterol* 2012, 18(17):2026–2034.
25. C.H. Tsai, Chen, E.C., Tsay, H.S., and Huang, C.J. Wild bitter gourd improves metabolic syndrome: A preliminary dietary supplementation trial. *Nutr J* 2012, 11(4):13.
26. C. Stough, Llyod, J., Clarke, J. et al. The chronic effects of an extract of *Bacopa monniera* (Brahmi) on cognitive function in healthy human subjects. *Psychopharmacol* 2001, 156(4):481–484.
27. P.K. Gandhi, Chaudhary, A. and Prajapati, P. Effect of formulations of vasa (avaleha, arishta and ghrita) in the management of tamakashwasa (bronchial ashtma). *Ayurpharm Int J Ayur Alli Sci* 2013, 2:33–40.
28. V.M. Gogte. *Ayurvedic Pharmacology and Therapeutic Uses of Medicinal Plants (Dravyagunavignyan)*. Mumbai, India: Bhavan's SPARC, 2000.
29. I. Leake. Nausea and vomiting: Getting to the root of the antiemetic effects of ginger. *Nat Rev Gastroenterol Hepatol* 2013, 10(5):259.
30. N. Wittschier, Faller, G., and Hensel, A. Aqueous extracts and polysaccharides from liquorice roots (*Glycyrrhiza glabra* L.) inhibit adhesion of *Helicobacter pylori* to human gastric mucosa. *J Ethnopharmacol* 2009, 125(2):218–223.
31. T. Fukai, Marumo, A., Kaitou, K., Kanda, T., Terada, S., and Nomura, T. Anti-*Helicobacter pylori* flavonoids from licorice extract. *Life Sci* 2002, 71(12):1449–1463.
32. S. Joos, Rosemann, T., and Szecsenyi, J. Use of complementary and alternative medicine in Germany—A survey of patients with inflammatory bowel disease. *BMC Complement Altern Med* 2006, 22:6–19.
33. I. Gupta, Parihar A., Malhotra, P. et al. Effects of *Boswellia serrata* gum resin in patients with ulcerative colitis. *Eur J Med Res* 1997, 2(1):37–43.

34. J.P. Behera, Mohanty, B., Ramani, Y.R. et al. Effect of aqueous extract of *Aegle marmelos* unripe fruit on inflammatory bowel disease. *Indian J Pharmacol* 2012, 44(5):614–618.
35. M. Jindal, Kumar, V., Rana, V., and Tiwary, A.K. An insight into the properties of *Aegle marmelos* pectin-chitosan cross-linked films. *Int J Biol Macromol* 2013, 52:77–84.
36. J.N. Dhuley. Investigation on the gastroprotective and antidiarrhoeal properties of *Aegle marmelos* unripe fruit extract. *Hindustan Antibiot Bull* 2003, 45/46(1–4):41–46.
37. S.N. Shetty, Mengi, S., Vaidya, R., and Vaidya, A.D. A study of standardized extracts of *Picrorhiza kurroa* Royle ex Benth in experimental nonalcoholic fatty liver disease. *J Ayurveda Integr Med* 2010, 1(3):203–210.
38. S.P. Thyagarajan, Subramanian, S., Thirunalasundar, T. et al. Effect of *Phyllanthus amarus* on chronic carriers of hepatitis B virus. *Lancet* 1988, 2:764–766.
39. R.K. Johri and Zutshi, U. An ayurvedic formulation "Trikatu" and its constituents. *J Ethnopharmacol* 1992, 37(2):85–91.
40. Sameermahmood, Z., Raji, L., Saravanan, T. et al. Gallic acid protects RINm5F β-cells from glucolipotoxicity by its antiapoptotic and insulin-secretagogue actions. *Phytother Res* 2010, 24(4):632.
41. H. Vaidya, Prajapati, A., Rajani, M. et al. Beneficial effects of swertiamarin on dyslipidemia in streptozotocin-induced type 2 diabetic rats. *Phytother Res* 2012, 26:1259–1261.
42. H. Vaidya, Goyal, R.K., and Cheema, S.K. Anti-diabetic Activity of swertiamarin is due to an active metabolite, gentianine, that upregulates PPAR-γ gene expression in 3T3-L1 cells. *Phytother Res* 2013, 27:624–627.
43. A. Gupta, Gupta, R., and Lal, B. Effect of *Trigonella foenum-graecum* (fenugreek) seeds on glycaemic control and insulin resistance in type 2 diabetes mellitus: A double blind placebo controlled study. *J Assoc Physicians India* 2001, 49:1057–1061.
44. P.V. Sharma. *Çaraka-Samhita: Agniveśa's Treatise Refined and Annotated by Çaraka and Redacted by Dŕdhbala*. Varanasi, India: Chaukhambha Orientalia, 2008.
45. A.A. Raut, Rege, N.N., Tadvi, F.M. et al. Exploratory study to evaluate tolerability, safety and activity of Ashwagandha (*Withania somnifera*) in healthy volunteers. *J Ayurveda Integr Med* 2012, 3(3):111–114.
46. A.A. Raut, Joshi, A.D., Antarkar, D.S. et al. Anti-rheumatic formulation from ayurveda. *Anc Sci Life* 1991, 2:66–69.
47. C.K. Chauhan, Joshi, M.J., and Vaidya, A.B. Growth inhibition of struvite crystals in the presence of herbal extract *Commiphora wightii*. *J Mater Sci Mater Med* 2009, 20(1):85–92.
48. V.M.G. Vaidya. *Ayurvedic Pharmacology and Therapeutic Uses of Medicinal Plants (Dravyagunavignyan)*. Mumbai, India: Bhavan's SPARC, 2000.
49. S. Dwivedi and Jauhari, R. Beneficial effects of *Terminalia arjuna* in coronary artery disease. *Indian Heart J* 1997, 49(5):507–510.
50. K.B. Sanjay and Subir, K.M. Effect of garlic on cardiovascular disorders: A review. *J Nutr* 2002, 1:4.
51. K.N. Chidambara Murthy, Jayaprakasha, G.K, and Singh, R.P. Studies on antioxidant activity of pomegranate (*Punica granatum*) peel extract using in vivo models. *J Agric Food Chem* 2002, 50(17):4791–4795.
52. M. Parihar, Chouhan, A., Harsoliya, M.S. et al. A Review cough and treatments. *Int J Nat Prod Res* 2011, 1(1):9–18.
53. A. Fernandez, Tavares, F., and Athavale, V.B. Asthma in children: A clinical controlled study of *Piper longum* in asthma. *J Paediatr Clin India* 1980, 15:45–47.
54. V.S. Joshi, Parekh, B.B., Joshi, M.J., and Vaidya, A.B. Herbal extracts of *Tribulus terrestris* and *Bergenia ligulata* inhibit growth of calcium oxalate monohydrate crystals in vitro. *J Crystal Growth* 2005, 275:1403–1408.
55. L. Upadhay, Tripathi, K., and Kulkarni, K. A study of prostane in the treatment of benign prostatic hyperplasia. *Phytother Res* 2001, 15(5):411–415.
56. A. Narayana and Subhose, V. Standardization of Ayurvĕdic formulations: A scientific review. *Bull Indian Inst Hist Med Hyderabad* 2005, 35:21.
57. R.G. Singh, Behur, S.K., and Kumar, R. Litholytic property of kulattha (*Dolichous biflorus*) vs potassium citrate in renal calculus disease: A comparative study. *J Assoc Physicians India* 2010, 58:286–289.
58. A.B. Vaidya. The status and scope of Indian medicinal plants on central nervous system. *Indian J Pharmacol* 1997, 29(5):340–343.
59. K. Dhuri, Vaidya, V., Vaidya, A.D.B., and Parikh, K.M. Stress and ayurveda: Selye-mehta dialogue. *J Assoc Phys Ind* 2000, 48:428.
60. V. Ravindranath. *Ayurvedic and Allopathic Medicine and Mental Health: Proceedings of Indo-US Workshop on Traditional Medicine and Mental Health*. Mumbai, India: Bharatiya Vidya Bhavan, 2003.
61. N.K. Sethiya, Nahata, A., Mishra, S.H. et al. An update on Shankhapushpi, a cognition boosting ayurvedic medicine. *J Chin Integrative Medicine* 2009, 7(11):1001–1022.

62. S.A. Lee, Han, X.H., Lee, C. et al. Methylpiperate derivatives from *Piper longum* and their inhibition of monoamine oxidase. *Arch Pharm Res* 2008, 31(6):679–683.
63. M. Siddhi Nandan. *Bhaishajya Ratnavali*. Varanasi, India: Chaukamba Orientalia, 2011.
64. M.H. Jetmalani, Sabins, P.B., and Gaitonde, B.B. A study on the pharmacology of various extracts of Shatavari-*Asparagus racemosus* (Willd). *J Res Ind Med* 1967, 2(1):1–9.
65. R.H. Singh. Rasāyanā and vajikarana (promotive therapy). In *History of Medicine in India*, Sharma P.V. (ed.). New Delhi, India: Indian National Science Academy, 1992, pp. 242–254.
66. P. Pradhan, Joseph, L., Gupta, V. et al. *Saraca asoca* (Ashoka): A review. *J Chem Pharm Res* 2009, 1(1):62–71.
67. D.M.V. Geetha, Kedlaya, R., Deepa, S., and Ballai, M. Activity of *Ocimum sanctum* (the traditional Indian medicinal plant) against the enteric pathogens. *Indian J Med Sci* 2001, 55:434–438.
68. P. More and Pai, K. Immunomodulatory effects of *Tinospora cordifolia* (Guduchi) on macrophage activation. *Biol Med* 2011, 3(2):134–140.
69. D.K. Bhandari, Nath, G., Ray, A.B. et al. Antimicrobial activity of crude extracts from *Berberis asiatica* stem bark. *Pharm Biol* 2000, 38(4):254–257.
70. Ministry of Health. *Pharmacopeia of India*. 2nd edition. New Delhi, India: Government of India Press, 1966.
71. C. Wiart, Kumar, K., Yusof, M.Y. et al. Antiviral properties of ent-labdene diterpenes of *Andrographis paniculata* Nees, inhibitors of herpes simplex virus type 1. *Phytother Res* 2005, 19(12):1069–1070.
72. D.B. Borkar. *Sarth Yogratnakar* (Vol. 2). Pune, India: Sri Gajanan Book Depot, Reprint 1984.
73. N. Adirajan, Ravikumar, T., Shanmugasundaram, N., and Babu, M. In vivo and in vitro evaluation of hair growth potential of *Hibiscus rosa-sinensis* Linn. *J Ethanpharm* 2003, 88:235–239.
74. S. Shishodia, Sethi, G., and Aggarwal, B.B. Curcumin: Getting back to the roots. *Ann N Y Acad Sci* 2005, 1056:206–217.
75. R.H. Singh, Narsimhamurthy, K., and Singh, G. Neuronutrient impact of Ayurvedic Rasayana therapy. *Biogerontology* 2008, 9:369–374.
76. R. Khalid. Garlic and aging: New Insight into an old remedy. *Ageing Res Rev* 2003, 2(1):39–56.
77. H.P. Sarma. *The Kasyapa Samhita: Vrddhajivakiya Tantra*. New Delhi, India: Chaukhambha Sanskrit Sansthan, 2006.

3 Functional Foods and Nutraceuticals

Potential Role in Human Health

Chunling Wang and Song Li

CONTENTS

3.1 INTRODUCTION

The discovery, development, and marketing of functional foods, nutraceuticals, and related products are currently the fastest growing segments of the food industry. This trend is driven by several factors, such as the increasing cost of pharmaceuticals and other health-care costs, the increasing interest of consumers in their own health, the increasing elderly population, the highly competitive nature of the food market with small profit margins, the advances in biotechnology and "omics," and the current consumer perception that "natural is good." Scientific evidence supports the idea that some foods and food ingredients have positive effects on human health and well-being [1] beyond the provision of basic nutritional requirements. Therefore, the use of food or food products to promote health and prevent or cure disease is well established.

As a pivotal part of the research and development of "healthful food," "functional foods and nutraceuticals" have been studied worldwide over the past 30 years [2]. Although we cannot definitively conclude that specific functional foods and/or nutraceuticals can prevent and/or treat diseases in most cases, various animal or clinical studies have revealed that these products have potential roles in human health. The potential role of functional foods and nutraceuticals in human health is reviewed in this chapter.

3.2 AN OVERVIEW OF FUNCTIONAL FOODS AND NUTRACEUTICALS

3.2.1 Definition of Functional Foods and Nutraceuticals

The consumption of food as medicine, or "dietetic therapy," has long been a strong belief in Chinese culture over thousand years. But the modern term "functional foods" was first introduced in Japan in the 1980s [3], and the Japanese interest in this field has also brought awareness for the need of functional foods to Europe and the United States. Although the term functional food has already been defined several times, there is not yet an official or commonly accepted definition for this food group.

As the original region for functional foods, Japan is the only country that recognizes functional foods as a distinct category in its food regulatory system, and the Japanese functional food market is now one of the most advanced in the world. In Japan, functional foods are known as foods with specific uses for health. They are composed of functional ingredients that affect the structure and/or function of the body and are used to maintain or regulate specific health conditions, such as gastrointestinal health, blood pressure levels, and blood cholesterol levels [4].

In other countries, there is no clear boundary between conventional foods and functional foods. One view is that any food is indeed functional because it provides nutrients and has a physiological effect. Therefore, functional food should be considered a marketing term for a food for which attraction lies in its health benefits and the way that the product is perceived.

Various academic institutions and national authorities have tried to define functional foods [1,5–7]. The simplest definition is "foods that provide health benefits beyond basic nutrition," which was approved by the International Food Information Council (IFIC). The expert report by the Institute of Food Technologists (IFT) defined functional foods as "foods and food components that provide a health benefit beyond basic nutrition. These substances provide essential nutrients often beyond quantities necessary for normal maintenance, growth and development, and/or other biologically active components that impart health benefits or desirable physiological effects" [8]. The International Life Sciences Institute (ILSI) Europe and European Commission's Concerted Action on Functional Food Science in Europe (FuFoSE) defined functional foods as follows: "a food product can only be considered functional if, together with the basic nutritional impact, it has beneficial effects on one or more function of the human organism thus either improving the general physical condition or/and decreasing the risk of the evolution of diseases" [1,3,9–11]. FuFoSE also developed a working definition of functional foods as foods that are "satisfactorily demonstrated to affect beneficially one or more target functions in the body, beyond adequate nutrition effects, in a way that is relevant to either an improved state of health and well-being and/or reduction of risk of disease" [1,9,11].

In 1989, Dr. Stephen first coined the term "nutraceuticals," which is derived from "nutrition" and "pharmaceutical," which are both key contributors to human wellness [12]. There is no absolute disparity between foods and drugs in terms of their functionally; the distinction has to be made case by case. According to Stephen, nutraceuticals can be defined as "a food (or part of a food) that provides medical or health benefits, including the prevention and/or treatment of a disease" [13]. As for functional foods, many definitions of nutraceuticals have been developed in the past two decades. The simplest definition of nutraceutical is "foods for specified health use (FOSHU)" as set forth by the Nutritional Improvement Law Enforcement Regulations of Japan [14]. Lachance defined nutraceuticals as naturally occurring bioactive compounds that have health benefits [15]. Health Canada officially defined nutraceuticals as "a product isolated or purified from food, generally sold in medicinal forms not usually associated with food and demonstrated to have a physiological benefit or provide protection against chronic disease" [11]. The range of nutraceuticals was expanded from food plants to include bioactive compounds from nonfood sources. Nutraceuticals can be presented in one form or in a combination of several forms, which is the most evident difference between nutraceuticals and pharmaceuticals, which are drugs or synthesized or purified pure chemicals.

A universally accepted definition of functional foods and nutraceuticals is lacking. Some experts insist that functional foods and nutraceuticals are two related but different concepts, but these two terms cannot be distinguished in most cases. For instance, Dr. Ekta defined nutraceutical as "functional food that aids in the prevention and/or treatment of disease(s) and/or disorder(s) other than anemia" [12]. A functional food for one consumer can act as a nutraceutical for another consumer based on this definition.

In conclusion, many definitions currently exist for the terms functional food and nutraceutical, and they are widely used in the marketplace at the same time. Functional foods and nutraceuticals can be natural foods or foods that have been modified to have a functional influence on the health and well-being of the consumer through the addition, removal, or modification of specific components. The terms functional foods and nutraceuticals should not be used to imply that there are good foods or bad foods but rather that all foods can be incorporated into a healthful varied diet. Regardless of whether a food is considered to be functional, it must always be safe for its intended use.

3.2.2 Classification of Functional Foods and Nutraceuticals

Functional foods and nutraceuticals are classified in various ways [16]. Because functional foods and nutraceuticals target specific health fields or populations, nutraceuticals are often categorized based on the targeted population or their health benefits (disease prevention) (Table 3.1).

Functional foods and nutraceuticals can also be categorized based on contents or food types (Table 3.2): (1) nutrients, including substances with certain physiological function, such as vitamins,

TABLE 3.1
Classification of Functional Foods and Nutraceuticals Based on Health Benefits

Health Fields	Examples of Nutraceuticals That Have Been Applied
Pregnancy	PUFAs, certain amino acids, folic acid, iron, zinc, iodine
Child growth/early development and growth	Cognition, sensory: PUFAs, iron, zinc, iodine
	Growth and body composition: essential amino acids, unsaturated fatty acids
	Skeletal development: calcium, vitamins D and K_2
	Gastrointestinal health: prebiotics, probiotics
	Immune function: vitamins A and D, antioxidant vitamins, trace elements, L-arginine, nucleotides, probiotics, prebiotics, neutral and acidic oligosaccharides
Gastrointestinal health	Probiotics, prebiotics, synbiotics (mixtures of pre- and probiotics), other non- and poorly digestible carbohydrates
Mental state and performance	Glucose and sugar-derived products, B vitamins, *n*-3 PUFAs, *S*-adenosylmethionine, phytochemicals, some plant extracts
Physical performance	Caffeine, specific amino acids, creatine, and carnitine
Cancer prevention	Colon cancer: prebiotics, dietary fiber, calcium, selenium, folate, low-fat dairy, phytochemicals (e.g., carotenoids, curcumin, polyphenolic compounds), some plant extracts
	Breast cancer: folate, vitamin D, isoflavones, α-linolenic acid, phytochemicals (e.g., lignan, resveratrol), some plant extracts (e.g., green tea, pomegranate)
	Prostate cancer: folate, selenium, isoflavones, some plant extracts (e.g., green and black tea)
Alzheimer's disease prevention	Antioxidant vitamins, DHA, phytochemicals, some plant extracts

(*Continued*)

TABLE 3.1
(Continued) Classification of Functional Foods and Nutraceuticals Based on Health Benefits

Health Fields	Examples of Nutraceuticals That Have Been Applied
Energy balance/body weight management/obesity prevention	Dietary fiber, polyols and other poorly digestible carbohydrates, chitosan, conjugated linoleic acid, diglycerides, medium-chain triglycerides, green tea, caffeine, calcium, capsaicin, phytochemicals, some plant extracts, fat and sugar replacements, foods with low glycemic index or glycemic response
Diabetes prevention	Soluble dietary fiber, lipoic acid, dietary fiber, some plant extracts (e.g., cinnamon, coriander, garlic, turmeric), chromium
Cardiovascular disease prevention	Vitamins, minerals, PUFAs, dietary fiber, phytochemicals (e.g., flavonols, flavones, flavanones, flavan-3-ols, isoflavones, anthocyanins, proanthocyanidins), some plant extracts
Bone protection/ musculoskeletal disease prevention	Vitamins (D, K, and C), minerals (e.g., calcium, manganese, copper, zinc) glucosamine, chondroitin, collagen hydrolysate, methylsulfonylmethane, *S*-adenosylmethionine, soybean unsaponifiables, soy protein, conjugated linoleic acid, fructooligosaccharides, inulin
Against oxidative stress	Combination of vitamins C and E, carotenoids and polyphenols including flavonoids, thousands of phytochemicals, and plant-derived products

DHA, docosahexaenoic acid; PUFAs, polyunsaturated fatty acids.

TABLE 3.2
Classification of Functional Foods and Nutraceuticals Based on Contents

Category	Examples of Nutraceuticals That Have Been Applied
Nutrients	Vitamins, minerals, fatty acids, proteins, amino acids, peptides, dietary fibers, functional carbohydrates, certain phytochemicals extracted from plants
Herbs/botanical extracts	Ginseng, garlic, onion, *Ginkgo biloba*, pomegranate, strawberry, grape, mushroom, tomato, chia, cinnamon, broccoli, valerian, chamomile, echinacea, ginger, licorice, St John's wort
Functional diet	Red wine, functional dairy foods, functional drinks, functional eggs, minimally refined grains

minerals, fatty acids, amino acids, and certain ingredients from plant-based substances; (2) herbal or edible plant substances, including herbs or botanical extracts and/or concentrates, such as garlic, ginseng, ginkgo, and St John's wort; and (3) functional diet, which is a mixed substance containing ingredients intended to add functional components to the diet. The ingredients may contain vitamins, minerals, amino acids, enzymes, botanicals, or other dietary supplements. All functional foods or nutraceuticals can be supplied to consumers in different dosage forms (e.g., powders, tablets, liquids, capsules, extracts, and concentrates).

The American Dietetic Association (ADA) has classified functional foods into four groups [11]:

1. *Conventional foods.* These foods are unmodified whole foods or conventional foods such as cereal products, dairy products, fruits, and vegetables that represent the simplest form of a functional food. For example, strawberries, tomatoes, grapes, and broccoli are considered to be functional foods because they are rich in bioactive components, such as anthocyanin, capsaicinoids, and quercetin.
2. *Modified foods.* Functional foods and nutraceuticals also include foods that have been modified through fortification, enrichment, or enhancement. These foods include calcium-fortified fruit juice, plant sterol esters or plant stanol-fortified margarine, phytochemicals

or plant extract-enriched snacks, and beverages containing energy-promoting ingredients (e.g., caffeine, ginseng, guarana, or taurine). Low-fat milk, gluten-free cereals, and non-*trans* fat oil also belong to the group of modified functional foods.
3. *Medical foods.* These foods are usually formulated to be consumed under the supervision of a physician and are intended for the specific dietary management of a disease. Examples of medical foods include the "diabetic dietary formula," among others.
4. *Foods for specific dietary use.* These foods include "infant foods," "weight loss foods," and so on.

Functional foods and nutraceuticals may also be classified based on their sources: plants (e.g., glucan, ascorbic acid, quercetin, luteolin, cellulose, lutein, pectin, tocopherol, allicin, lycopene, zeaxanthin, lignin, carotene, monounsaturated fatty acids [MUFAs], and some minerals), animals (e.g., conjugated linoleic acid, eicosapentaenoic acid [EPA], docosahexaenoic acid [DHA], sphingolipids, choline, lecithin, calcium, coenzyme Q10, creatine, and some minerals), and microbial groups (e.g., *Saccharomyces boulardii*, *Bifidobacterium bifidum*, *Lactobacillus acidophilus*, and *Streptococcus salivarius*). In addition, nonfood sources of nutraceuticals have been sourced by the development of modern fermentation methods. For example, amino acids and their derivatives have been produced by bacteria grown in fermentation systems [17]; EPA can now be produced by bacteria by importing the appropriate DNA through recombination methods [18].

Over the past 10 years, the demand for foods that improve or benefit health has increased dramatically around the world, along with the rising cost of health care, cost of living, and the desire for a higher quality of life. In this regard, functional foods and nutraceuticals offer a powerful and convenient tool that promises specific health benefits related to various food components. All foods are functional, in that foods provide energy and nutrients that are necessary for survival; however, functional foods and nutraceuticals in use today convey health benefits that extend far beyond mere survival.

3.3 FUNCTIONAL FOODS AND NUTRACEUTICALS IN HEALTH PROMOTION AND DISEASE PREVENTION

Foods are currently intended to not only satisfy hunger and to provide necessary nutrients for humans but also prevent chronic diseases and improve the physical and mental well-being of the consumers. Epidemiological and clinical studies have revealed a tight relationship between diet and health status [19]. From a clinical perspective, health benefits have been most studied in several aspects, including cancer, age-related diseases, diabetes, cardiovascular disease, immune deficiency, and mental health. For instance, populations that consume a large proportion of vegetables and fruits or those that have a higher intake of whole grains have a lower risk of certain types of cancers and cardiovascular disease [20,21].

In this section, we introduce the potential health benefits of functional foods and nutraceuticals through two topics that address the identification of functional foods and nutraceuticals and their potential roles.

3.3.1 Identification of Functional Foods and Nutraceuticals

3.3.1.1 Bioactive Compounds and Phytochemicals

Thousands of scientific studies, including several projects funded by scientific authorities, have led to the identification and understanding of the mechanisms of biologically active components in foods, which may improve health and possibly reduce the risk of disease while enhancing the overall well-being. Almost all ingredients, ranging from macronutrients (e.g., proteins, fats, and carbohydrates) to micronutrients (e.g., vitamins and minerals) have specific physiological functions, and some are consumed as functional foods and nutraceuticals (Table 3.3).

TABLE 3.3
Typical Ingredients Used as Functional Foods and Nutraceuticals

Ingredient Category	Examples	Reported Health Benefits
Amino acids, peptides, protein and their derivatives	L-Carnitine	Weight loss, Alzheimer's disease prevention and treatment, improvement in blood cell count
	Taurine	Improvement in mental performance and heart function
	Casein phosphopeptides	Enhanced mineral solubility and absorption, immunomodulatory activity
	Collagen	Promotion of skin health, pain relief, and improvement of joint function in patients with osteoarthritis, improvement of symptoms of rheumatoid arthritis
	Lactoferrin	Support of the immune system
Lipids and fatty acids	Lecithin	Prevention of cognitive impairment and dementia, decreased cholesterol
	Phosphatidylserine	Treatment of age-related cognitive impairment, Alzheimer's disease prevention and treatment, improvement in mental health
	DHA	Maintenance of normal brain function, Alzheimer's disease prevention and treatment, maintenance of normal vision, treatment of depression, decrease in the risk of age-related macular degeneration
	CLA	Obesity prevention, colorectal cancer prevention
	α-Linolenic acid	Maintenance of normal blood cholesterol levels
	γ-Linolenic acid	Cardiovascular disease prevention
Carbohydrates, fiber, prebiotics	Many types of dietary fiber	Blood cholesterol reduction, obesity prevention, prevention of cardiovascular disease and diabetes, colorectal cancer prevention
	Fructooligosaccharide	Improvement in gastrointestinal function
	Pectins	Blood cholesterol reduction, reduction of blood glucose after meals, prostate cancer prevention, improvement in gastrointestinal function
	Lactulose	Improvement in intestinal health
	L-Arabinose	Inhibition of sucrose absorption, prevention of obesity and diabetes, improvement in gastrointestinal function
	β-Glucan	Contribution to the maintenance of normal cholesterol levels, immune function enhancement
Minerals	Calcium	Contribution to normal blood clotting, contribution to normal muscle and bone function, maintenance of normal teeth
	Iron	Anemia prevention and treatment, immunomodulation
	Iodine	Contribution to normal thyroid, cognitive and nervous system function
	Magnesium	Constipation treatment, reduction of fatigue, contribution to normal cognitive function, contribution to normal bone and tooth function
	Zinc	Reduction in the duration and severity of diarrhea, contribution to normal cognitive function, immunomodulatory activity
	Selenium	Blood cholesterol reduction, immunomodulatory activity, prevention of colorectal and gastric cancer
Vitamins	Vitamin A	Breast cancer prevention, contribution to iron metabolism, contribution to the maintenance of normal vision, immunomodulatory activity
	Vitamin B_{12}	Hyperhomocysteinemia treatment, contribution to the maintenance of normal cognitive function, Alzheimer's disease prevention and treatment, reduction of fatigue, immunomodulatory activity
	Vitamin C	Increased iron absorption, contribution to normal collagen formation, contribution to the maintenance of normal cognitive function, reduction of blood pressure, Alzheimer's disease prevention and treatment, cancer prevention

TABLE 3.3
Typical Ingredients Used as Functional Foods and Nutraceuticals

Ingredient Category	Examples	Reported Health Benefits
	Vitamin D	Osteomalacia treatment, rickets prevention and treatment, osteoporosis prevention and treatment, cancer prevention, obesity prevention, cardiovascular disease and diabetes prevention
	Vitamin E	Protection of cells from oxidative stress, decreased risk of age-related macular degeneration, Alzheimer's disease prevention and treatment
	Folic acid	Prevention of neural tube birth defects, contribution to normal homocysteine metabolism, reduction of fatigue, prevention of certain cancers (colorectal and breast), Alzheimer's disease prevention and treatment

CLA, conjugated linoleic acid; DHA, docosahexaenoic acid.

As the number of studies on nutrients has continued to increase, phytochemicals have become more popular in recent years. The term "phytochemicals" refers to a wide variety of chemical compounds that occur naturally in plants. Many bioactive ingredients and phytochemicals have been identified from food of plant origin as well as animal-based products. Some of these compounds protect plants against insects or have other biological functions. Some have antioxidant or hormone-like activity in plants and humans who eat the plants. Because the consumption of vegetables and fruits is associated with a reduced risk of certain diseases, researchers have examined numerous phytochemicals in seeking specific components that may account for the beneficial health effects [22–24]. Examples of phytochemicals include phenolics (e.g., flavonoids, phenolic acids, and phenols), nitrogen-containing compounds, alkaloids, and terpenoids (Table 3.4). The most commonly studied health benefits of phytochemicals are the prevention of cardiovascular disease and cancer as well as immunological effects, which are associated with their high antioxidant activity. However, some specific phytochemicals may have other physiological functions beyond their antioxidant activity. Curcumin and anthocyanins are two well-known bioactive compounds that are ideal examples to illustrate the potential health benefits of phytochemicals.

Curcumin is an active component of turmeric. Its endogenous antioxidant defense mechanisms have been investigated for many years, and it has been suggested to have an anti-inflammatory activity. Biochemical analyses and clinical studies have demonstrated that curcumin promotes human health in various ways including regulation of lipid metabolism as well as through anti-inflammatory and antioxidant properties. Appropriate intake of curcumin has a potential role in efforts to decrease the incidence of obesity and its associated risk factors [25,26]. Curcumin has also been shown to facilitate diabetes prevention through glycemic control, which further supports its role in cardiovascular disorders [27,28]. The ability of curcumin to delay the onset of cancer has also been the topic of extensive research for many years, especially for colorectal and skin cancers [29,30]. Abundant literature is devoted to the mechanisms by which curcumin may mediate the functions discussed in Subsections 3.2.2 and 3.3.1.1. Additionally, it has been shown that oral curcumin can relieve symptoms of dyspepsia [31], osteoarthritis [32], Alzheimer's disease (AD) [33], and rheumatoid arthritis [34]. In conclusion, although the research on curcumin as a functional food and nutraceutical is in its infancy, recently published studies suggest that curcumin is beneficial to human health [35].

In contrast to curcumin, anthocyanins are more widely recognized as nutraceuticals and are a group of water-soluble vacuolar pigments that may appear red, purple, or blue depending on the pH value of the microenvironment. Anthocyanins are flavonoids that exist in many different fruits and vegetables. To date, more than 600 anthocyanins have been identified in natural foods [36]. Human consumption of anthocyanins is among the highest of all flavonoids, and the

TABLE 3.4
Typical Phytochemicals Used as Functional Foods and Nutraceuticals

Category	Class	Subclass	Examples
Phenolics	Flavonoids	Anthocyanins	Cyanidin
		Flavanols	Theaflavin
		Flavonols	Quercetin
		Dihydroflavonols	Taxifolin
		Flavones	Apigenin
		Isoflavonoids	Genistein
		Flavanones	Naringenin
		Dihydrochalcones	Phloretin
	Phenolic acid	Hydroxybenzoic acids	Gallic acid
		Hydroxycinnamic acids	Ferulic acid
	Lignans		Pinoresinol
	Coumarins		Coumarin
		Coumestans	Coumestrol
		Furanocoumarins	Psoralen
	Phenols	Alkylphenols	4-Ethylguaiacol
			5-Heptadecylresorcinol
		Methoxyphenols	Guaiacol
	Phenylpropanoids	Benzodioxoles	Apiole
		Curcuminoids	Curcumin
		Hydroxyphenylpropenes	Eugenol
	Quinines	Benzoquinones	Maesanin
		Naphthoquinones	Phylloquinone
		Anthraquinones	Rubiacardone A
	Stilbenoids		Resveratrol
	Xanthones		Mangostin
Terpenoids	Monoterpenoids		Limonene
		Phenolic terpenes	Thymol
	Sesquiterpenoids		Farnesol
	Diterpenoids		Cafestol
	Triterpenoids	Phenolic terpenes	Vitamin E
		Saponins	Ursolic acid
		Phytosterols	Campesterol
	Tetraterpenoids	Carotenoids	β-Carotene
Alkaloids	Pyridine alkaloids		Trigoneline
	Betalain alkaloids	Betacyanins	Betanin
		Betaxanthins	Indicaxanthin
	Indole alkaloids	Ergolines	Ergine
		Yohimbans	Reserpine
		Tryptolines or β-carbolines	Harman
	Indolizidine alkaloids		Swaisonine
	Pyrrolidine alkaloids		Nicotine
	Quinoline alkaloids		Quinine
	Isoquinoline alkaloids		Berberine
		Morphinans	Morphine
	Steroidal alkaloids		Solanidine
		Saponins	Salanine
	Tropane alkaloids		Atropine

TABLE 3.4
Typical Phytochemicals Used as Functional Foods and Nutraceuticals

Category	Class	Subclass	Examples
Nitrogen-containing compounds	Amines	Benzylamines	Capsaicin
		Phenylethylamines	Ephedrine
		Tryptamines	Psilocybin
	Cyanogenic glycoside		Amygdalin
	Glucosinolates	Aliphatic glucosinolates	Sulforaphane
		Aromatic glucosinolates	Glucobrassicin
	Purines	Xanthenes	Caffeine
	Miscellaneous nitrogen-containing compounds	Indole alcohols	Indole-3-carbinol
Organic acid lipids and carbohydrates	Ascorbic acid, tartaric acid, fructose, sorbitol		

toxicity of dietary anthocyanins is considered to be extremely low. The bioactivity of bioavailable anthocyanins has been a focus of research [37]. Based on many *in vitro* studies, animal models, and human studies, it has been suggested that anthocyanins possess anti-inflammatory and anticarcinogenic activity [38,39] and aid in the prevention of cardiovascular disease [40,41] and neurodegenerative disorders [42,43] as well as the control of obesity and diabetes [44,45]. All of these effects are more or less associated with the antioxidant properties of anthocyanins, but enzymatic inhibition and other pathways may also be relevant.

For metabolic disease prevention, mechanistic studies support the beneficial effects of anthocyanins on the established biomarkers of cardiovascular disease risk, such as nitric oxide (NO), inflammation, and endothelial dysfunction [46]; anthocyanins also have antidiabetic properties. Published data suggest that anthocyanins may reduce blood glucose by improving insulin resistance, protecting β cells, increasing the secretion of insulin, and reducing the digestion of sugars in the small intestine [47,48]. The metabolism, absorption, and bioavailability of anthocyanins as nutraceuticals have been examined over the past decade; however, further study is required to determine the anthocyanins that are required to achieve "optimal" human health.

Numerous studies have been conducted on phytochemicals, and databases on food phytochemicals and their health-promoting effects have been established [49]. These databases greatly contribute research in this field. In addition, more clinical trials are necessary to demonstrate the health benefits of these nutraceuticals.

3.3.1.2 Plants and Plant Parts

Curcumin and anthocyanins can be consumed in different forms or through conventional food carriers. For example, as a natural antioxidant, anthocyanins are abundant in many plant foods, such as strawberry and pomegranate [50]. It is not possible to measure the advantages and disadvantages of the two different anthocyanin supplementation methods, that is, anthocyanin supplements and anthocyanin-fortified foods or certain fruits and vegetables including extracts, concentrates, and juices. Consumers absorb more anthocyanins from supplements, but more bioactive compounds are obtained from the pomegranate itself or its related products such as extracts, concentrates, and juice. Thus, the pomegranate (and its products) can be considered a nutraceutical that contains several beneficial ingredients. Similarly, several nutraceutical substances are found in higher concentrations in specific foods or food families (Table 3.5).

Pomegranate (*Punica granatum*) and its products are currently being widely promoted to consumers as an effective healthy food that is capable of health promotion and disease prevention [51,52]. Potential functions of pomegranate as reported by the literature include its use as an antioxidant [53],

TABLE 3.5
Good Sources of Nutraceuticals

Nutraceuticals	Related Foods
Allyl sulfur compounds	Onions, garlic
Capsaicinoids	Pepper fruit
Catechins	Teas, berries
Curcumin	Turmeric
EPA and DHA	Fish oils
Isoflavones	Soybeans and other legumes, apios
Indoles	Broccoli, cabbage, cauliflower, kale
Inulin and fructooligosacchrides	Onions, whole grain products
Lycopene	Tomatoes
Quercetin	Broccoli, red grapes, citrus fruits, onions

DHA, docosahexaenoic acid; EPA, eicosapentaenoic acid.

an antibacterial, and an anti-inflammatory [54]. The high antioxidant activity of this nutraceutical and other plant-derived products is the basis for many potential benefits. Health benefits of the pomegranate include those related to cardiovascular health [55–58], diabetes prevention [45], protection against age-related disease (e.g., AD) [59], and anticancer properties [60,61]. Some studies have revealed that the health benefits of pomegranate are more abundant than those of anthocyanins due to the diverse bioactive components in the pomegranate. All pomegranate-derived products have their own bioactive compounds. For example, pomegranate leaf extract contains tannins (e.g., punicalin, pedunculata, gallagic acid, and ellagic acid) and flavonoids; the main components of pomegranate flower extracts are polyphenols (e.g., gallic and ellagic acids) and triterpenes (e.g., oleanolic, ursolic, maslinic, and asiatic acids); pomegranate peel contains phenolics, flavonoids, ellagitannins, proanthocyanidin, and complex polysaccharides; and pomegranate juice, which is the most popular pomegranate-derived product, is a rich source of phenolics, flavonoids, polyphenols, tannins, anthocyanins, and antioxidants (e.g., vitamins C and E, coenzyme Q-10) [51].

The inhibition of cancer by pomegranate or pomegranate-derived products has been studied for colon [38,62], breast [63,64], prostate [65], lung [66], and cervical cancers [67]. It has been shown that the daily consumption of pomegranate juice significantly delayed increases in prostate-specific antigen (PSA) [68,69]. Furthermore, several cell culture and animal studies have also indicated an inhibitory effect of pomegranate on prostate cancer generation and development [70]. Although the function of pomegranate as a nutraceutical needs to be further elucidated by clinical studies, the health benefits of these phytonutrient-rich products have been accepted by scientists and consumers to some extent.

Water-soluble tomato concentrate (WSTC) is another typical case of a vegetable- or fruit-derived product used as a nutraceutical. The DSM Company has generated a series of WSTC products called Fruitflow. Fruitflow is the first natural, scientifically substantiated substance that contributes to healthy blood flow that has been approved by the European Food Safety Authority (EFSA) [71]. As reported in at least eight clinical trials, the consumption of WSTC helps to maintain healthy platelet aggregation and improve blood flow [72,73]. This effect can last 12–18 hours. Increased platelet aggregation is a major risk factor in cardiovascular health, and a series of studies demonstrated that WSTC functions by maintaining the smooth shape of platelets to prevent aggregation inside blood vessels. WSTC does not disrupt the normal blood clotting process following injury.

WSTC functions similarly to aspirin and even can be considered a satisfactory natural alternative to low-dose aspirin that prevents thrombus formation. WSTC acts on multiple aspects of the initial clotting mechanism in a reversible manner, whereas aspirin works on only one specific pathway, that is, the inactivation of up to 95% of platelet COX, which results in a blockade of arachidonate production [74]. It is now recognized that 20%–30% of individuals may experience a so-called aspirin-resistance syndrome, which manifests as a disruption in blood clotting. In this regard, WSTC acts as a nutraceutical that offers an advantage over synthetic drugs. Although all WSTC constituents occur naturally in tomatoes, the effective dose (3 g/day) of WSTC is comparable to levels derived from 2.5 tomatoes. In sum, consumption of the tomato-derived nutraceutical WSTC (3 g/day) could benefit public health by helping to maintain platelets in a reversible state and reducing the risk of thrombotic events mediated by platelet activation.

Typical plant extracts that have been used as nutraceuticals are listed in Table 3.6.

3.3.1.3 Conventional Foods That Can Be Used as Functional Foods

Even though phytochemicals and herbal products in the form of capsules, powder, softgels, and gelcaps are currently popular, daily food consumption is still the most convenient source of nutraceuticals. On the one hand, phytochemicals and herb-derived products can be fortified into conventional

TABLE 3.6
Examples of Plant-Based Products as Functional Foods and Nutraceuticals

Plant-Based Products	Bioactive Compounds	Potential Health Benefits
Green tea	Catechins, caffeine, theanine, flavonols, flavan-3-ols	Obesity prevention, cancer prevention (prostate, ovarian, breast, lung), Alzheimer's disease prevention
Strawberry	Vitamins A, C, and E; folate; anthocyanins; flavonols; flovonols; ellagitannins	Cardiovascular disease prevention, colon cancer prevention
Grape	Anthocyanins, flavan-3-ols, prodelphinidins, procyanidins, flavones, resveratrol	Diabetes and cardiovascular disease (hypertension) prevention, maintenance of age-related brain function, immunomodulatory activity, promotion of oral health
Ginseng	Ginsenosides, phenolics	Immunomodulatory activity, diabetes prevention, breast cancer prevention and treatment, Alzheimer's disease prevention and reversion, aphrodisiac effects
Garlic	γ-Glutamylcysteine-derived sulfur-containing compounds (diallyl trisulfide, vinyldithins, ajoenes, etc.)	Hypertension and cardiovascular disease prevention, cancer prevention and treatment (prostate, colorectal, breast, gastric)
Ginkgo biloba	Flavonols (quercetin, kaempferol, isrohamnetin), terpene lactones (ginkgolides and bilobalide)	Alzheimer's disease prevention and reversion, contributes to the maintenance of cognitive function, diabetes prevention
Wolfberry	Vitamin C, phenolic acid (ferulic acid), flavonoids (anthocyanins), polysaccharide, zeaxanthin	Immunomodulatory activity, reduction of cholesterol, cardiovascular and diabetes prevention
Dates (jujube)	Vitamin C, selenium, flavonoids (anthocyanins), cinnamic acid derivatives, coumarins, tocopherols, carotenoids, polysaccharide	Mental health maintenance, cancer prevention, cardiovascular disease prevention, immunomodulatory activity, anemia prevention
Chia (*Salvia hispanica*)	High-qualified protein content, ALA, dietary fiber, vitamins, minerals	Cardiovascular disease prevention, blood glucose regulation, cancer prevention
St John's wort	Hypericin, hyperforins, terpenes, catechin, tannins, proanthocyanidins	Antidepressant for mild-to-moderate depression

ALA, α-linolenic acid.

foods; on the other hand, some foods are now consumed as functional foods or nutraceuticals, such as cereal-derived products, dairy products, drinks, soy products, and even eggs and meat [75].

Epidemiological studies have clearly shown that wholegrain cereals can protect against obesity, diabetes, cardiovascular disease, and cancers [76–78]. The specific effects of wholegrain structure and high fiber content, together with the antioxidant and anticarcinogenic properties of numerous bioactive compounds, are well-recognized mechanisms underlying this protection [79,80]. Nutraceuticals can also be served in drinks, such as cholesterol-lowering drinks [containing a combination of *n*-3 polyunsaturated fatty acids (PUFAs) and soy], "bone health" drinks (containing calcium, vitamin D, and inulin), "eye health" drinks (containing lutein), and "vitamin-rich" drinks, which are popular in some regions. Additionally, *n*-3 fatty acids and mineral-enriched eggs are available in the market as functional foods [75].

Milk and dairy products have been associated with health benefits for many years due to their rich content of specific proteins, bioactive peptides, vitamins, minerals (highly absorbable calcium), and conjugated linoleic acid. Increasing recent findings have revealed the health benefits of milk and dairy products through an array of bioactivities such as modulating digestive and gastrointestinal functions, controlling probiotic microbial growth and immunoregulation, accelerating bone growth, and maintaining bone health [81]. Recently, studies have been conducted to determine new health benefits of milk and dairy foods [82], such as maintenance of cardiovascular health [83], prevention of obesity and diabetes (low-fat milk) [84], and antitumor activity (especially colorectal, bladder, and breast cancers) [85]. For example, numerous cohort studies demonstrated that higher intake of dairy foods, particularly low-fat milk, is associated with a decreased risk of colorectal cancer [86]. This function of dairy foods may not solely result from their calcium content; other potential chemopreventive components are present, which include vitamin D, butyric acid, sphingolipids, conjugated linoleic acid, and probiotic bacteria in fermented dairy products, such as yogurt [86]. However, cancer prevention is also affected by the human genetic background; therefore, this function of dairy foods can vary in different consumers.

As discussed in this subsection, dairy products can be considered nutraceuticals without any fortification; furthermore, almost all types of nutrients can be fortified in milk. Various fortified dairy products are consumed by target populations, including elderly individuals, infants, youth, and pregnant women. Each population exhibits their own health needs, and fortified dairy foods can provide more nutraceuticals than the dairy product itself (Table 3.7). For instance, vitamin A and lutein are the most commonly fortified ingredients for infant visual development; choline, DHA, arachidonate, and taurine are also fortified for infant brain development.

Functional foods and nutraceuticals may include single nutrients, dietary supplements, herbal products, and processed foods, such as cereals, soups, and beverages. A combination of several functional foods and nutraceuticals is termed a "functional diet," which has become popular in the market. For example, the peak bone mass at the end of adolescence can be increased by a functional diet, which is expected to be of great importance to reduce the risk of osteoporosis in later life. The combined effects of calcium and prebiotic fructans, together with other constituents of growing bone, such as proteins, phosphorus, magnesium, and vitamins D and K, offer many possibilities for the development of functional foods and nutraceuticals. However, regardless of the type of anthocyanin supplementation, for example, pomegranate-derived products or fortified milk, people increasingly consume functional foods and nutraceuticals in their daily life, intentionally or not. Additionally, thousands of functional foods and nutraceuticals have appeared on the market, taking various forms from capsules to well-known traditional foods, and play an important role in our diet.

3.3.2 Potential Roles of Functional Foods and Nutraceuticals in Human Health

Functional foods and nutraceuticals may improve the general condition of the body, decrease the risk of some diseases, and could even be used to help cure some illnesses. Functional foods and

TABLE 3.7
Nutraceuticals Used in Dairy Products

Target Population	Nutraceuticals
Infant	Vision: vitamin A, lutein, zeaxanthin
	Physical performance: calcium, vitamin D, iron, lactalbumin
	Cognitive development: choline, taurine, DHA, arachidonic acid (AA)
	Immunoactivity: selenium, β-carotene, prebiotics, α-lactalbumin
Youth	Anemia prevention: iron, vitamin C
	Skeletal development: vitamin D, calcium, proteins with high bioaccessibility and bioavailability
	Cognitive development: choline, taurine, DHA, lecithin
	Micronutrient deficiency prevention: zinc, iodine, vitamin A
The elderly	See Table 3.8
Pregnant women	Capable of pregnant: iron, vitamin C, folate
	Early phase: folate and vitamins
	Middle and late phase: vitamins, minerals (calcium, iron, zinc), DHA, probiotics, choline, proteins with high bioaccessibility and bioavailability
Women	"High-pressure" working women: vitamins, PUFA, lecithin, taurine, γ-aminobutyric acid, phytochemicals, and plant extracts
	Skin health and appearance: vitamins, collagen, dietary fiber, prebiotics, probiotics, conjugated-linoleic acid, L-caritine, phytochemicals, and plant extracts (e.g., dates and wolfberry)
	Menopause regulation: calcium, vitamin D, soy proteins, soy isoflavones, high-bioaccessibility and high-bioavailability proteins, phytochemicals, and plant extracts

DHA, docosahexaenoic acid; PUFA, polyunsaturated fatty acid.

nutraceuticals can improve life, decrease disease risk, and promote good health, thereby augmenting the function of the traditional diet.

3.3.2.1 Maintain Good Health and Function

Functional foods and nutraceuticals can be consumed on a daily basis to help maintain physical functions, such as vision, cognitive function, and muscle and bone health.

Products for gastrointestinal health are widely consumed by the public in the United States, Europe, and Asia. Such products can be considered the most scientifically approved type of nutraceutical that can improve daily life. The gut is an obvious target for the development of functional foods and nutraceuticals because it acts as an interface between the diet and all other body functions. The most common nutraceuticals used for intestinal health are pro- and prebiotics, which improve human gastrointestinal function mainly through microbiotic regulation [87,88]. Probiotics are defined as live microorganisms that, when consumed in proper numbers, confer a health benefit, mainly on the gastrointestinal health of the host [89]. The most studied and widely employed probiotic bacteria are lactic acid bacteria [90] and bifidobacteria [91]. Dairy foods and fruit juice are two novel, appropriate media for fortification with probiotic cultures. For example, various species of lactobacilli and bifidobacteria combined (or not) with *Streptococcus thermophilus* are the main bacteria used as probiotics in yogurts or fermented dairy products. Major health benefits demonstrated in humans include the reduction of the incidence or severity of gastrointestinal infection and the alleviation of lactose intolerance. Because probiotics are only transiently present in the intestinal tract and do not become part of the host's gut microflora, regular consumption is necessary to maintain favorable effects.

Prebiotics are nondigestible substances that improve human health by stimulating the growth of bacteria in the colon [92]. The key criterion for an ingredient to be classified as a prebiotic is that it must not be hydrolyzed or absorbed in the upper part of the gastrointestinal tract so that it reaches the colon in significant amounts. Additionally, an effective prebiotic must be a selective substrate for beneficial bacteria that are stimulated to grow. Among thousands of prebiotics, the most popular substances are resistant starch [93], inulin [94,95], polydextrose [96], lactulose [97], fructooligosaccharide [94], and isomalto-oligosaccharide [98]. These prebiotics regulate the growth and metabolism of probiotics, and optimize the intestinal microenvironment. Mixtures of probiotics and prebiotics, which modify gut flora and its metabolism by increasing the survival of health-promoting bacteria, are described as synbiotics [99]. Probiotics, prebiotics, and synbiotics promote good health through the modification of gut microflora and gastrointestinal modulation, and these nutraceuticals are well accepted by consumers all over the world.

3.3.2.2 Reduce Risk of Disease

"Biotics" enhance the function of the gastrointestinal system, which can be considered the maintenance and improvement of normal physiological function. In addition to their function in modifying the gut microflora, many other health benefits of pro-, pre-, and synbiotics are being investigated, including activation of the immune system [100,101], increased absorption of minerals, increased production of short-chain fatty acids, and inhibition of lesions that are precursors of adenomas and carcinomas [102,103]. Thus, these compounds may decrease disease risk, particularly in degenerative problems in aging.

Aging is defined as a normal decline in survival with advanced age and is often associated with a higher risk of diseases or disorders such as a reduced digestive function, immune dysregulation and inflammatory disorders, cardiovascular disease, type 2 diabetes, osteoporosis, Parkinson's, disease, and AD [104]. Thus, functional foods and nutraceuticals are important for middle-aged and the elderly. Research on natural dietary compounds using animal models and clinical trials has provided a new strategy for antiaging, especially in the prevention of degenerative diseases or age-related diseases such as AD (Table 3.8).

AD is characterized by the progressive loss of short-term memory and other cognitive dysfunctions. The constantly increasing number of individuals affected by neurodegenerative diseases is one of the major social and medical challenges of the twenty-first century [105]. However, there are still no effective pharmaceuticals to prevent, halt, or reverse AD [106,107]. Therefore, the prevention of cognitive decline by functional foods or nutraceuticals is currently a vastly expanding area of research [108]. Numerous clinical studies have demonstrated that traditional nutrients, including vitamins (B group, C, and E) [109,110], choline [111], and fatty acids (DHA) [112–114] may be effective; however, not all clinical trials have demonstrated positive results. Lecithin [115], phosphatidylserine [116,117], and L-carnitine [118,119] have also been used in AD prevention and treatment.

Many phytochemicals and/or plant extracts, such as flavonoids [120], curcumin [121], resveratrol [122], anthocyanin [53], astaxanthin [123], berry fruit extracts (e.g., strawberry, cherry, and pomegranate) [124], ginseng extracts [125], peppermint extracts [126], cinnamon extracts [127], *Ginkgo biloba* extracts [128], and cocoa extracts [129], have potential effects in improving memory, learning, and cognition in AD. Although the preventive value of the consumption of certain nutraceuticals in AD is not firmly established, the effect of each nutraceutical has been partially confirmed by research on mechanisms and clinical studies. For example, in numerous experiments, flavonoids were shown to not only possess excellent antioxidant activity but also attenuate the accumulation of AD-promoting metabolites, such as Aβ, and suppress inflammation by interacting with proinflammatory enzymes [130]. Most data that are considered to provide evidence of neuroprotection by flavonoids are also derived from studies of complex mixtures of compounds with high flavonoid contents, such as green tea and other herbal medicines. However, there is evidence that the role of green tea in AD prevention is not solely mediated by flavonoids, but other ingredients, including catechins (in which the active component is epigallocatechin-3-gallete [EGCG]) [131], L-theanine, caffeine [132], and other bioactive compounds, also exert these effects.

Consumers prefer nutraceuticals instead of medicine to reduce health problems and prevent disease prevention. Various studies have suggested that some functional foods or nutraceuticals are

TABLE 3.8
Nutraceuticals and Functional Foods for Age-Related Diseases

Disease	Nutraceuticals
Cardiovascular diseases	Calcium, vitamins C and D, *n*-3 PUFAs, fish oil, olives, oats, α-linolenic acid, dietary fiber, fruit phenolics, plant sterols, plant stanols, soy proteins, lactotripeptides, lycopene, tomato extracts, cocoa extracts, nuts and nut extracts, berry extracts (e.g., grape), chia, flaxseed, green tea, low-fat dairy foods
Diabetes	Chromium, vitamins B_3 and D, *n*-3 PUFAs, stearidonic acid, α-lipoic acid, *Agaricus* mushrooms, olives, blond psyllium, guar gum, caffeine, coffee, flaxseed, ginseng, oats, prickly pear cactus extracts, glucomannan, berry extracts, aloe extracts, black tea, green tea, stevia extracts, cassia cinnamon, cocoa extracts, chia, diacylglycerol
Immunodeficiency	Selenium, vitamins, β-glucan, polysaccharides, prebiotics, probiotics, ginseng extracts, wolfberry extracts, grape extracts, phenolic-rich plant extracts
Muscle and skeleton degeneration	Calcium (combined with copper, zinc, magnesium, and other appropriate minerals), vitamins D and K, ipriflavone, fish oil, collagen hydrolysate, glucosamine, chondroitin, *S*-adenosylmethionine, phenolics, avocado extracts, soybean extracts, ginger extracts, curcumin combined with lecithin
Malnutrition and anemia	Vitamins B group and C, inulin
Cognitive decline (Alzheimer's disease)	Vitamins (B group, C, and E), DHA, phosphatidylserine, choline, lecithin, L-carnitine, flavonoids, curcumin, resveratrol, sterol, anthocyanin, astaxanthin, caffeine, green tea extracts, *Ginkgo biloba* extracts, berry extracts, ginseng, cocoa extracts, cinnamon, peppermint extracts
Gastrointestinal-related disease	Probiotics, prebiotics, dietary fiber, phenolics, phenolic-rich plant extracts, casein phosphopeptides
Cancer	Colorectal cancer: calcium, folate, low-fat dairy foods, conjugated linoleic acid, flavonoids, phenolic acids, terpenoids, isothiocyanates, curcumin, coffee, garlic extracts, turmeric extracts, berry extracts, dietary fiber, probiotics, prebiotics
	Prostate cancer: strontium, selenium, calcium, EPA, melatonin, α-linolenic acid, green tea, lentinan, lycopene, isoflavones, pomegranate extracts, broccoli extracts, flaxseed, garlic extracts, tomato extracts

DHA, docosahexaenoic acid; EPA, eicosapentaenoic acid; PUFAs, polyunsaturated fatty acids.

effective in disease prevention or reducing the risk of certain diseases. However, it is still too early to draw scientific conclusions based on current research results, and we cannot yet translate the evidence of numerous scientific studies into definite public health recommendations.

3.3.2.3 Improve Health-Related Quality of Life

In addition to maintaining good health and reducing the risk of disease, functional foods and nutraceuticals have also been reported to increase the quality of life. For example, gluten-free grain [133], lactose-free dairy products, and optimal mental state-improving foods can help to resolve health-related problems, gluten allergy, lactose intolerance, and mental health.

The concept of "mental health" or "cognitive function" has garnered considerable interest in recent years because mental health directly impacts the quality of life. Various aspects of mental health have been evaluated, including cognition, alertness, attention, brain function, insight, intelligence quotient (IQ), judgment, learning, memory, mental state, neuropsychology, perception, psychomotor effects, thinking, information processing, and responsiveness. Collective efforts have shown that some functional foods and nutraceuticals can potentially promote optimal mental performance and improve the quality of life with respect to mood and vitality, reaction to stress, vigilance and attention, and even cognitive ability.

Meals high in carbohydrates have been reported to exert beneficial effects on mental performance, including improvements in feelings of sleepiness and calmness and faster information processing. Sweet foods, such as foods with high sucrose content, may relieve distress in young infants and may reduce pain perception in members of the general population [134]. Although it can be viewed or applied as a comfort food that reduces stress and anxiety, a high-carbohydrate diet is not usually recommended because of the risk of obesity.

Abundant studies have been conducted to discover functional ingredients that improve mental health, and some ingredients, phytochemicals, antioxidants, and micronutrients have been proven to be effective by *in vivo* studies and clinical trials. Increasing evidence suggests that proper dietary intake of the n-3 long-chain PUFAs, EPA and DHA [135], contributes not only to preventing various physical illnesses but also to improving mental health. *Ginkgo biloba* and ginseng extracts increase mental energy and improve mood, as confirmed by clinical studies [136–139]. Tyrosine and tryptophan may help in recovery from jet lag, but only a limited amount of scientific evidence supports this effect [140]. γ-aminobutyric acid, taurine, 5-hydroxytryptamine, and some micronutrients (e.g., vitamins [B group, C, and D], lithium, and magnesium) also have been used as nutraceuticals for the maintenance and promotion of mental health. Evidence is currently available from epidemiological, clinical, and interventional studies that investigated nutraceuticals and their function in promoting mental health; however, the limited results from different studies are inconsistent. The inadequate understanding of the biochemical mechanisms underlying mental regulation in humans has hindered progress in discovering and confirming functional ingredients. Nevertheless, it has been ascertained that the relationship between diet and mental health is substantial, and nutraceuticals and functional foods enhance the quality of life by improving mental health and cognitive function.

In sum, functional foods and nutraceuticals have been demonstrated to be effective in maintaining normal function, helping to prevent disease or reduce the risk of disease, and improving the quality of life; these effects could be significant to some extent. However, we should always be aware that functional foods and nutraceuticals are a part of the whole diet. Thus, although functional foods and nutraceuticals provide health benefits, we should also consider the overall dietary pattern and lifestyle.

3.4 FUTURE RESEARCH AND DEVELOPMENT OF FUNCTIONAL FOODS AND NUTRACEUTICALS

New reports are published almost daily on the role of existing food ingredients and advances in identifying bioactive compounds and their health benefits. However, more research is needed to translate basic research into consumer-relevant products. As shown in Figure 3.1, different processes are available for the discovery, design, development, and marketing for a new nutraceutical, and these processes can be divided into two parts: research and marketing.

The image of the potential health benefits of functional foods and nutraceuticals has been tarnished in recent years by numerous published reports that failed to demonstrate the claimed effects. To solve this problem, well-designed scientific research (research process) and effective consumer education programs (marketing process) are necessary [141].

3.4.1 Qualified Studies Are Needed in the Development of Functional Foods and Nutraceuticals

The safety of functional foods and nutraceuticals is tested against a "gold standard" and is regulated strictly by authoritative institutions in each country. Therefore, the key challenges in researching functional foods and nutraceuticals involve determining their function and efficacy. We must realize that many consumers are confused about which products are truly beneficial.

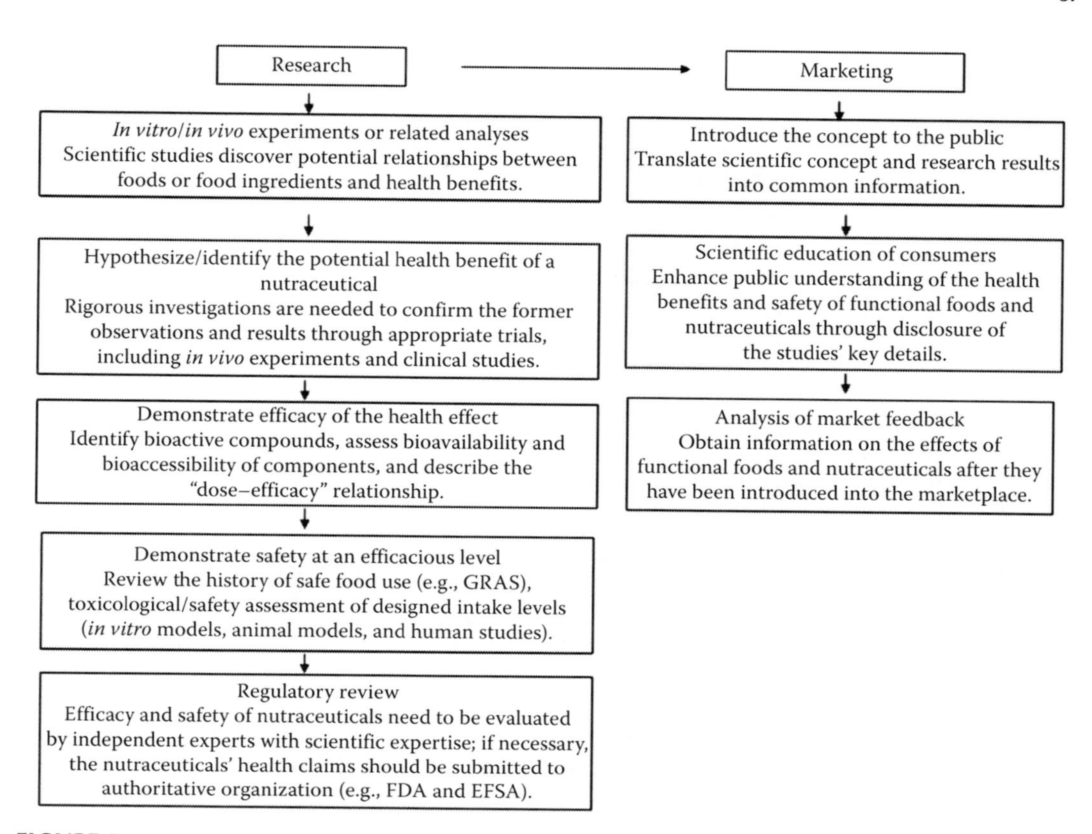

FIGURE 3.1 Steps to bring a nutraceutical to market. GRAS, generally recognized as safe.

In the pharmaceutical industry, candidate drug compounds must undergo a range of *in vitro* and animal studies that assess their efficacy, toxicity, absorption, metabolism, and excretion, and they proceed through three phases of clinical trials before they are first launched. By contrast, only a few functional foods and nutraceuticals even approach this level of study, and there is no standard requirement for clinical trials of these products. However, the strength of evidence is based on scientific reviews (narrative and systematic reviews) of all studies and is weighted by the numbers and quality of studies. The type of research used to determine the level of evidence can range from "soft" to "hard" science and can include *in vitro*, animal, and human studies. Human studies can be further divided into ecological studies, cross-sectional studies, case–control studies, cohort studies, and randomized clinical trials (preferably double-blind, placebo-controlled trials). A meta-analysis is conducted when smaller studies are pooled together to increase the statistical power. A systematic review or meta-analysis cannot be better than the original research data. Ultimately, the weight of evidence obtained through literature reviews should be based on a sufficient number of randomized clinical studies.

From a clinical perspective, the direct measurement of improvements in health and well-being and/or the reduction of disease risk is often difficult to compare to the outcome of a cure by drugs, that is, the state of health and well-being does not always lend itself to quantifiable measurement. For example, AD and cancer are ongoing processes and the time frame for the development of these diseases is long, which makes it difficult to monitor their development under the conditions of a controlled study. Such studies may also involve ethical issues. Thus, although any particular risk factor may present a suitable target for functional foods and nutraceuticals, diseases are the end result of complex biological processes. Therefore, markers that can be used to monitor health state and disease development needed to be identified. For example, elevated serum homocysteine is associated with heart disease, stroke, and AD [142–144]. There is evidence that reducing homocysteine levels in healthy people can prevent the development of certain diseases, and the efficacy of nutraceuticals for specific diseases can thus be partially evaluated by their ability to reduce homocysteine levels. The relevance of selected biomarkers and the incidence of certain diseases must be judged by authoritative organizations on a global scale.

Oat β-glucan was the first substance whose health claims were substantiated by the Food and Drug Administration (FDA) for labeling purposes, following the evaluation of numerous animal and clinical studies that demonstrated a hypocholesterolemic effect [145]. However, most nutraceuticals have few or no appropriate studies to demonstrate their health benefits. Well-designed studies are needed to test the efficacy of a specific nutraceutical and then to communicate these benefits to consumers.

3.4.2 Modern Technologies Are Needed in the Development of Functional Foods and Nutraceuticals

As more functional foods and nutraceuticals have been developed and identified, modern technology has been used to advance their accessibility and availability and to suit the consumer's needs and preferences. Technology can help to achieve this goal in three ways: (1) by creating new functional food components based on traditional materials or in new raw materials; (2) by maximizing the presence of functional food components that already exist in foods by modifying their function or increasing their bioavailability; and (3) by providing the means to monitor the effectiveness of functional foods and nutraceuticals.

Scientists have to face one significant problem that clear cause-and-effect relationships are not always easily determined in human physiology. Differences among population subgroups further complicate the identification of clear cause-and-effect relationships. As we move into the era of "omics," the research technologies have developed rapidly, which has accelerated the development and application of functional foods and nutraceuticals in "individualized nutrition." Nutrigenomics, proteomics, and metabolomics are three new disciplines that will contribute to the rapid development of functional foods and nutraceuticals [146,147]; furthermore, bioinformatics will integrate all data from multiple, disparate cases. It is possible to understand the effects of nutrients and foods at the molecular level in the body and the variable effects of dietary components on each individual.

Nutrigenomics evaluates changes in gene and protein expression and metabolic pathways that occur in response to changes in dietary components. Where such changes occur, markers can be identified and used to demonstrate the effect of bioactive food components on health, thus aiding in the development of functional foods and nutraceuticals that will contribute to health and well-being [148–150].

Proteomics can be viewed as an experimental approach to explain the information contained in genomic sequences in terms of the structure, function, and control of biological processes and pathways. Such technology identifies large number of proteins in humans and elucidates their interactions and bioactivities, which can help us to understand the relationship between nutraceuticals and specific signaling pathways [151–153].

Metabolomics could offer a source of novel biomarkers for nutritional studies. Metabolomics-derived data could be used to more accurately define the molecules in foods that are responsible for changes in metabolic profiles. Thus, metabolomics measures the real outcome of potential changes suggested by genomics and proteomics. This technique enables rapid screening for nutritional status, disease state, food or drug function, and efficacy. Now, this emerging investigative approach is being widely used to assess the efficacy and safety of pharmaceutical agents as well as functional foods and nutraceuticals [154–156].

Food and nutritional science has extended from identifying and correcting nutritional deficiencies to designing foods that promote optimal health and reduce the risks of diseases. A synergy between developments in functional food and nutraceutical science and "omics" may, in the future, result in a situation in which it is possible for individuals to make truly informed choices about the foods that provide the best opportunities for health, well-being, and reduced risks of diseases.

3.4.3 Appropriate Consumer Communication Is Needed in the Development of Functional Foods and Nutraceuticals

Functional foods and nutraceuticals may be available in the market, but for consumers to recognize and understand the benefits of functional foods and nutraceuticals, effective communication is needed. Food labeling is a simple and direct way for consumers and the public to learn about functional foods and nutraceuticals. Although the laws and regulations for functional food labeling vary in different countries or regions, the fundamental principles of food labeling, especially regarding the labeling of health benefits, are similar. The claim should be evaluated by an authoritative institute in each country; for instance, the project "Process for the Assessment of Scientific Support for Claims on Foods (PASSCLAIM)" was conducted to assess the scientific support for health-related claims for foods and food components in the European Union. Similar programs have also been implemented in the United States, Canada, Australia, and Japan by their respective authorities to ensure that the information is scientifically correct and to prevent the presentation of misleading information on food labels. Usually, in most countries, three types of food labeling are used to communicate health-related claims: (nutrient) content claims, health benefit claims, and disease prevention claims. Regulatory details will be discussed in Chapters 15 through 18 of this book.

Consumer acceptance is considered to be the key contribution to successful business, but robust scientific research itself does not make a product successful in the market. What scientists know cannot easily be translated into what common consumers know about a product. The inventor of new products would also hope that the consumers put the nutraceuticals to good use and maximize their intended health benefits. The delivery of scientific information of potential health benefits of functional foods and nutraceuticals, which are obtained from high-quality studies, to the consumers is also a challenging step in the development of functional foods and nutraceuticals. Strategic scientific communication is critical to a company and even the entire food industry.

Despite the rapid development of functional foods and nutraceuticals, various challenges must be overcome before they are well accepted by consumers and the marketplace. The promotion of functional foods and nutraceuticals requires joint efforts from all stakeholders, including those in commerce, industry, agriculture, science, education, government, media, and consumers per se.

3.5 CONCLUSIONS

Recently, research on functional foods and nutraceuticals and how they impact disease prevention and health promotion has garnered much attention. The combination of public desires, advances in nutrition and food technology, and evidence-based science linking functional foods and nutraceuticals to human health has created an unprecedented opportunity to address public health issues through functional foods and nutraceuticals. Scientists are responsible to invent solid evidence-based products, and all stakeholders in areas from academia to government have a responsibility to interpret scientific findings and to translate these findings into practical guidance for consumers and the public.

A growing body of scientific evidence has revealed that functional foods and nutraceuticals have the potential to contribute positively to long-term health and well-being and to reduce the risks of diseases. We must realize that the exploitation of functional foods and nutraceuticals presents its own challenges. The widespread use of functional foods and nutraceuticals relies on strong research and development, along with a successful consumer communication campaign, to extend the benefits of these products to more people.

REFERENCES

1. Howlett, J. 2008. *Functional Foods: From Science to Health and Claims*. Brussels, Belgium: ILSI Europe.
2. Ghosh, D., Das, S., Bagchi, D., and Smarta, R.B. 2012. *Innovation in Healthy and Functional Foods*. Boca Raton, FL: CRC Press.
3. Ozen, A.E., Pons, A., and Tur, J.A. 2012. Worldwide consumption of functional foods: A systematic review. *Nutrition Reviews* 70: 472–81.
4. Arai, S., Morinaga, Y., Yoshikawa, T., Ichiishi, E., Kiso, Y., Yamazaki, M., Morotomi, M., Shimizu, M., Kuwata, T., and Kaminogawa, S. 2002. Recent trends in functional food science and the industry in Japan. *Bioscience, Biotechnology, and Biochemistry* 66: 2017–29.
5. Doyon, M. and Labrecque, J. 2008. Functional foods: A conceptual definition. *British Food Journal* 110: 1133–49.
6. Roberfroid, M.B. 2002. Global view on functional foods: European perspectives. *The British Journal of Nutrition* 88(Suppl 2): S133–8.
7. Arai, S. 2002. Global view on functional foods: Asian perspectives. *The British Journal of Nutrition* 88(Suppl 2): S139–43.
8. Hurley, B.F., Hanson, E.D., and Sheaff, A.K. 2011. Strength training as a countermeasure to aging muscle and chronic disease. *Sports Medicine* 41: 289–306.
9. The British Journal of Nutrition. 1999. Scientific concepts of functional foods in Europe. Consensus document. *The British Journal of Nutrition* 81(Suppl 1): S1–27.
10. Roberfroid, M.B. 2000. A European consensus of scientific concepts of functional foods. *Nutrition* 16: 689–91.
11. Hasler, C.M. and Brown, A.C. 2009. Position of the American Dietetic Association: Functional foods. *Journal of American Dietetic Association* 109: 735–46.
12. Kalra, E.K. 2003. Nutraceutical—Definition and introduction. *AAPS PharmSci* 5(3): E25.
13. DeFelice, S.L. 1993. A comparison of the U.S., European and Japanese nutraceutical health and medical claim rules. *Regulatory Affairs* 5: 163–7.
14. Gupta, S., Chauhan, D., Mehla, K., Sood, P., and Nair, A. 2010. An overview of nutraceuticals: Current scenario. *Journal of Basic and Clinical Pharmacy* 1(2): 55–62.
15. Lachance, P.A. 2003. Nutraceuticals & Food Nutrification. *The Wall Street Transcript* 159: 132–7.
16. Dureja, H., Kaushik, D., and Kumar, V. 2003. Developments in nutraceuticals. *Indian Journal of Pharmacology* 35: 363–72.
17. Leuchtenberger, W., Huthmacher, K., and Drauz, K. 2005. Biotechnological production of amino acids and derivatives: Current status and prospects. *Applied Microbiology and Biotechnology* 69: 1–8.
18. Bajpai, P. and Bajpai, P.K. 1993. Eicosapentaenoic acid (EPA) production from microorganisms: A review. *Journal of Biotechnology* 30: 161–83.
19. Martirosyan, D.M. 2012. *Functional Food Ingredients and Nutraceuticals in Chronic Disease*. CreateSpace Independent Publishing Platform. Richardson, TX.
20. Hu, F.B. 2003. Plant-based foods and prevention of cardiovascular disease: An overview. *The American Journal of Clinical Nutrition* 78: S544–51.
21. Williams, P.G. 2012. Evaluation of the evidence between consumption of refined grains and health outcomes. *Nutrition Reviews* 70: 80–99.
22. Vasanthi, H.R., ShriShriMal, N., and Das, D.K. 2012. Phytochemicals from plants to combat cardiovascular disease. *Current Medicinal Chemistry* 19: 2242–51.
23. Singh, M., Singh, P., and Shukla, Y. 2012. New strategies in cancer chemoprevention by phytochemicals. *Frontiers in Bioscience* 4: 426–52.
24. McGhie, T.K. and Rowan, D.D. 2012. Metabolomics for measuring phytochemicals, and assessing human and animal responses to phytochemicals, in food science. *Molecular Nutrition & Food Research* 56: 147–58.
25. Shehzad, A., Ha, T., Subhan, F., and Lee, Y.S. 2011. New mechanisms and the anti-inflammatory role of curcumin in obesity and obesity-related metabolic diseases. *European Journal of Nutrition* 50: 151–61.
26. Alappat, L. and Awad, A.B. 2010. Curcumin and obesity: Evidence and mechanisms. *Nutrition Reviews* 68: 729–38.
27. Wongcharoen, W. and Phrommintikul, A. 2009. The protective role of curcumin in cardiovascular diseases. *International Journal of Cardiology* 133: 145–51.

28. Aggarwal, B.B. and Harikumar, K.B. 2009. Potential therapeutic effects of curcumin, the anti-inflammatory agent, against neurodegenerative, cardiovascular, pulmonary, metabolic, autoimmune and neoplastic diseases. *The International Journal of Biochemistry & Cell Biology* 41: 40–59.
29. Lopez-Lazaro, M. 2008. Anticancer and carcinogenic properties of curcumin: Considerations for its clinical development as a cancer chemopreventive and chemotherapeutic agent. *Molecular Nutrition & Food Research* 52(Suppl 1): S103–27.
30. Steward, W.P. and Gescher, A.J. 2008. Curcumin in cancer management: Recent results of analogue design and clinical studies and desirable future research. *Molecular Nutrition & Food Research* 52: 1005–9.
31. Thamlikitkul, V., Bunyapraphatsara, N., Dechatiwongse, T., Theerapong, S., Chantrakul, C., Thanaveerasuwan, T., Nimitnon, S. et al. 1989. Randomized double blind study of *Curcuma domestica* Val. for dyspepsia. *Journal of the Medical Association of Thailand = Chotmaihet Thangphaet* 72: 613–20.
32. Kuptniratsaikul, V., Thanakhumtorn, S., Chinswangwatanakul, P., Wattanamongkonsil, L., and Thamlikitkul, V. 2009. Efficacy and safety of *Curcuma domestica* extracts in patients with knee osteoarthritis. *Journal of Alternative and Complementary Medicine* 15: 891–7.
33. Baum, L., Lam, C.W., Cheung, S.K., Kwok, T., Lui, V., Tsoh, J., Lam, L. et al. 2008. Six-month randomized, placebo-controlled, double-blind, pilot clinical trial of curcumin in patients with Alzheimer disease. *Journal of Clinical Psychopharmacology* 28: 110–13.
34. Deodhar, S.D., Sethi, R., and Srimal, R.C. 1980. Preliminary study on antirheumatic activity of curcumin (diferuloyl methane). *The Indian Journal of Medical Research* 71: 632–4.
35. Esatbeyoglu, T., Huebbe, P., Ernst, I.M., Chin, D., Wagner, A.E., and Rimbach, G. 2012. Curcumin—From molecule to biological function. *Angewandte Chemie* 51: 5308–32.
36. He, J. and Giusti, M.M. 2010. Anthocyanins: Natural colorants with health-promoting properties. *Annual Review of Food Science and Technology* 1: 163–87.
37. McGhie, T.K. and Walton, M.C. 2007. The bioavailability and absorption of anthocyanins: Towards a better understanding. *Molecular Nutrition & Food Research* 51: 702–13.
38. Kocic, B., Filipovic, S., Nikolic, M., and Petrovic, B. 2011. Effects of anthocyanins and anthocyanin-rich extracts on the risk for cancers of the gastrointestinal tract. *Journal of the Balkan Union of Oncology* 16: 602–8.
39. Wang, L.S. and Stoner, G.D. 2008. Anthocyanins and their role in cancer prevention. *Cancer Letters* 269: 281–90.
40. Wallace, T.C. 2011. Anthocyanins in cardiovascular disease. *Advances in Nutrition* 2: 1–7.
41. Mazza, G.J. 2007. Anthocyanins and heart health. *Annali Dell'Istituto Superiore di Sanità* 43: 369–74.
42. Kelsey, N., Hulick, W., Winter, A., Ross, E., and Linseman, D. 2011. Neuroprotective effects of anthocyanins on apoptosis induced by mitochondrial oxidative stress. *Nutritional Neuroscience* 14: 249–59.
43. Shih, P.H., Wu, C.H., Yeh, C.T., and Yen, G.C. 2011. Protective effects of anthocyanins against amyloid beta-peptide-induced damage in neuro-2A cells. *Journal of Agricultural and Food Chemistry* 59: 1683–9.
44. Kwon, S.H., Ahn, I.S., Kim, S.O., Kong, C.S., Chung, H.Y., Do, M.S., and Park, K.Y. 2007. Anti-obesity and hypolipidemic effects of black soybean anthocyanins. *Journal of Medicinal Food* 10: 552–6.
45. Ghosh, D. and Konishi, T. 2007. Anthocyanins and anthocyanin-rich extracts: Role in diabetes and eye function. *Asia Pacific Journal of Clinical Nutrition* 16: 200–8.
46. Hidalgo, M., Martin-Santamaria, S., Recio, I., Sanchez-Moreno, C., de Pascual-Teresa, B., Rimbach, G., and de Pascual-Teresa, S. 2012. Potential anti-inflammatory, anti-adhesive, anti/estrogenic, and angiotensin-converting enzyme inhibitory activities of anthocyanins and their gut metabolites. *Genes & Nutrition* 7: 295–306.
47. Zhang, B., Kang, M., Xie, Q., Xu, B., Sun, C., Chen, K., and Wu, Y. 2011. Anthocyanins from Chinese bayberry extract protect beta cells from oxidative stress-mediated injury via HO-1 upregulation. *Journal of Agricultural and Food Chemistry* 59: 537–45.
48. Sancho, R.A.S., and Pastore, M.G. 2012. Evaluation of the effects of anthocyanins in type 2 diabetes. *Food Research International* 46: 378–86.
49. Scalbert, A., Andres-Lacueva, C., Arita, M., Kroon, P., Manach, C., Urpi-Sarda, M., and Wishart, D. 2011. Databases on food phytochemicals and their health-promoting effects. *Journal of Agricultural and Food Chemistry* 59: 4331–48.
50. Tsuda, T. 2012. Dietary anthocyanin-rich plants: Biochemical basis and recent progress in health benefits studies. *Molecular Nutrition & Food Research* 56: 159–70.

51. Johanningsmeier, S.D. and Harris, G.K. 2011. Pomegranate as a functional food and nutraceutical source. *Annual Review of Food Science and Technology* 2: 181–201.
52. Faria, A. and Calhau, C. 2011. The bioactivity of pomegranate: Impact on health and disease. *Critical Reviews in Food Science and Nutrition* 51: 626–34.
53. Zafra-Stone, S., Yasmin, T., Bagchi, M., Chatterjee, A., Vinson, J.A., and Bagchi, D. 2007. Berry anthocyanins as novel antioxidants in human health and disease prevention. *Molecular Nutrition & Food Research* 51: 675–83.
54. Lansky, E.P. and Newman, R.A. 2007. *Punica granatum* (pomegranate) and its potential for prevention and treatment of inflammation and cancer. *Journal of Ethnopharmacology* 109: 177–206.
55. Al-Muammar, M.N. and Khan, F. 2012. Obesity: The preventive role of the pomegranate (*Punica granatum*). *Nutrition* 28: 595–604.
56. Basu, A. and Penugonda, K. 2009. Pomegranate juice: A heart-healthy fruit juice. *Nutrition Reviews* 67: 49–56.
57. Haber, S.L., Joy, J.K. and Largent, R. 2011. Antioxidant and antiatherogenic effects of pomegranate. *American Journal of Health-System Pharmacy* 68: 1302–5.
58. Stowe, C.B. 2011. The effects of pomegranate juice consumption on blood pressure and cardiovascular health. *Complementary Therapies in Clinical Practice* 17: 113–15.
59. Hartman, R.E., Shah, A., Fagan, A.M., Schwetye, K.E., Parsadanian, M., Schulman, R.N., Finn, M.B., and Holtzman, D.M. 2006. Pomegranate juice decreases amyloid load and improves behavior in a mouse model of Alzheimer's disease. *Neurobiology of Disease* 24: 506–15.
60. Adhami, V.M., Khan, N., and Mukhtar, H. 2009. Cancer chemoprevention by pomegranate: Laboratory and clinical evidence. *Nutrition and Cancer* 61: 811–15.
61. Syed, D.N., Afaq, F., and Mukhtar, H. 2007. Pomegranate derived products for cancer chemoprevention. *Seminars in Cancer Biology* 17: 377–85.
62. Khan, S.A. 2009. The role of pomegranate (*Punica granatum* L.) in colon cancer. *Pakistan Journal of Pharmaceutical Sciences* 22: 346–8.
63. Sturgeon, S.R. and Ronnenberg, A.G. 2010. Pomegranate and breast cancer: Possible mechanisms of prevention. *Nutrition Reviews* 68: 122–8.
64. Khan, G.N., Gorin, M.A., Rosenthal, D., Pan, Q., Bao, L.W., Wu, Z.F., Newman, R.A. et al. 2009. Pomegranate fruit extract impairs invasion and motility in human breast cancer. *Integrative Cancer Therapies* 8: 242–53.
65. Bell, C. and Hawthorne, S. 2008. Ellagic acid, pomegranate and prostate cancer—A mini review. *The Journal of Pharmacy and Pharmacology* 60: 139–44.
66. Khan, N., Afaq, F., Kweon, M.H., Kim, K., and Mukhtar, H. 2007. Oral consumption of pomegranate fruit extract inhibits growth and progression of primary lung tumors in mice. *Cancer Research* 67: 3475–82.
67. McDougall, G.J., Ross, H.A., Ikeji, M., and Stewart, D. 2008. Berry extracts exert different antiproliferative effects against cervical and colon cancer cells grown in vitro. *Journal of Agricultural and Food Chemistry* 56: 3016–23.
68. Pantuck, A.J., Zomorodian, N., and Belldegrun, A.S. 2006. Phase-II study of pomegranate juice for men with prostate cancer and increasing PSA. *Current Urology Reports* 7: 7.
69. Paller, C.J., Ye, X., Wozniak, P.J., Gillespie, B.K., Sieber, P.R., Greengold, R.H., Stockton, B.R. et al. 2012. A randomized phase II study of pomegranate extract for men with rising PSA following initial therapy for localized prostate cancer. *Prostate Cancer and Prostatic Diseases*. doi:10.1038/pcan.2012.20.
70. Malik, A. and Mukhtar, H. 2006. Prostate cancer prevention through pomegranate fruit. *Cell Cycle* 5: 371–3.
71. EFSA Report. 2010. Scientific opinion on the modification of the authorisation of a health claim related to water-soluble tomato concentrate and helps to maintain a healthy blood flow and benefits circulation pursuant to Article 13(5) of Regulation (EC) No 1924/2006 following a request in accordance with Article 19 of the Regulation (EC) No 1924/2006. *EFSA Journal* 8: 1689.
72. O'Kennedy, N., Crosbie, L., van Lieshout, M., Broom, J.I., Webb, D.J., and Duttaroy, A.K. 2006. Effects of antiplatelet components of tomato extract on platelet function in vitro and ex vivo: A time-course cannulation study in healthy humans. *The American Journal of Clinical Nutrition* 84: 570–9.
73. O'Kennedy, N., Crosbie, L., Whelan, S., Luther, V., Horgan, G., Broom, J.I., Webb, D.J., and Duttaroy, A.K. 2006. Effects of tomato extract on platelet function: A double-blinded crossover study in healthy humans. *The American Journal of Clinical Nutrition* 84: 561–9.
74. Dutta-Roy, A.K., Crosbie, L., and Gordon, M.J. 2001. Effects of tomato extract on human platelet aggregation in vitro. *Platelets* 12: 218–27.

75. Siro, I., Kapolna, E., Kapolna, B., and Lugasi, A. 2008. Functional food. Product development, marketing and consumer acceptance—A review. *Appetite* 51: 456–67.
76. Ye, E.Q., Chacko, S.A., Chou, E.L., Kugizaki, M., and Liu, S. 2012. Greater whole-grain intake is associated with lower risk of type 2 diabetes, cardiovascular disease, and weight gain. *The Journal of Nutrition* 142: 1304–13.
77. Borneo, R. and Leon, A.E. 2012. Whole grain cereals: Functional components and health benefits. *Food & Function* 3: 110–19.
78. Giacco, R., Della Pepa, G., Luongo, D., and Riccardi, G. 2011. Whole grain intake in relation to body weight: From epidemiological evidence to clinical trials. *Nutrition, Metabolism, and Cardiovascular Diseases* 21: 901–8.
79. Okarter, N. and Liu, R.H. 2010. Health benefits of whole grain phytochemicals. *Critical Reviews in Food Science and Nutrition* 50: 193–208.
80. Fardet, A. 2010. New hypotheses for the health-protective mechanisms of whole-grain cereals: What is beyond fibre? *Nutrition Research Reviews* 23: 65–134.
81. Kliem, K.E. and Givens, D.I. 2011. Dairy products in the food chain: Their impact on health. *Annual Review of Food Science and Technology* 2: 21–36.
82. Alvarez-Leon, E.E., Roman-Vinas, B., and Serra-Majem, L. 2006. Dairy products and health: A review of the epidemiological evidence. *The British Journal of Nutrition* 96(Suppl 1): S94–9.
83. Gibson, R.A., Makrides, M., Smithers, L.G., Voevodin, M., and Sinclair, A.J. 2009. The effect of dairy foods on CHD: A systematic review of prospective cohort studies. *The British Journal of Nutrition* 102: 1267–75.
84. van Meijl, L.E., Vrolix, R., and Mensink, R.P. 2008. Dairy product consumption and the metabolic syndrome. *Nutrition Research Reviews* 21: 148–57.
85. Chagas, C.E., Rogero, M.M., and Martini, L.A. 2012. Evaluating the links between intake of milk/dairy products and cancer. *Nutrition Reviews* 70: 294–300.
86. Pufulete, M. 2008. Intake of dairy products and risk of colorectal neoplasia. *Nutrition Research Reviews* 21: 56–67.
87. Quigley, E.M. 2012. Prebiotics and probiotics: Their role in the management of gastrointestinal disorders in adults. *Nutrition in Clinical Practice* 27: 195–200.
88. Whelan, K. 2011. Probiotics and prebiotics in the management of irritable bowel syndrome: A review of recent clinical trials and systematic reviews. *Current Opinion in Clinical Nutrition and Metabolic Care* 14: 581–7.
89. Mullin, G.E. 2012. Probiotics and digestive disease. *Nutrition in Clinical Practice* 27: 300–2.
90. Clarke, G., Cryan, J.F., Dinan, T.G., and Quigley, E.M. 2012. Review article: Probiotics for the treatment of irritable bowel syndrome—Focus on lactic acid bacteria. *Alimentary Pharmacology & Therapeutics* 35: 403–13.
91. Acharya, M.R. and Shah, R.K. 2002. Selection of human isolates of bifidobacteria for their use as probiotics. *Applied Biochemistry and Biotechnology* 102/103: 81–98.
92. Charalampopoulos, D. and Rastall, R.A. 2012. Prebiotics in foods. *Current Opinion in Biotechnology* 23: 187–91.
93. Higgins, J.A. 2004. Resistant starch: Metabolic effects and potential health benefits. *Journal of AOAC International* 87: 761–8.
94. Meyer, D. and Stasse-Wolthuis, M. 2009. The bifidogenic effect of inulin and oligofructose and its consequences for gut health. *European Journal of Clinical Nutrition* 63: 1277–89.
95. Roberfroid, M.B. 2007. Inulin-type fructans: Functional food ingredients. *The Journal of Nutrition* 137: 2493S–502S.
96. Raninen, K., Lappi, J., Mykkanen, H., and Poutanen, K. 2011. Dietary fiber type reflects physiological functionality: Comparison of grain fiber, inulin, and polydextrose. *Nutrition Reviews* 69: 9–21.
97. Clausen, M.R. and Mortensen, P.B. 1997. Lactulose, disaccharides and colonic flora. Clinical consequences. *Drugs* 53: 930–42.
98. Yen, C.H., Tseng, Y.H., Kuo, Y.W., Lee, M.C., and Chen, H.L. 2011. Long-term supplementation of isomalto-oligosaccharides improved colonic microflora profile, bowel function, and blood cholesterol levels in constipated elderly people—A placebo-controlled, diet-controlled trial. *Nutrition* 27: 445–50.
99. Gourbeyre, P., Denery, S., and Bodinier, M. 2011. Probiotics, prebiotics, and synbiotics: Impact on the gut immune system and allergic reactions. *Journal of Leukocyte Biology* 89: 685–95.
100. Yan, F. and Polk, D.B. 2011. Probiotics and immune health. *Current Opinion in Gastroenterology* 27: 496–501.
101. Lomax, A.R. and Calder, P.C. 2009. Prebiotics, immune function, infection and inflammation: A review of the evidence. *The British Journal of Nutrition* 101: 633–58.

102. Pool-Zobel, B., van Loo, J., Rowland, I., and Roberfroid, M.B. 2002. Experimental evidences on the potential of prebiotic fructans to reduce the risk of colon cancer. *The British Journal of Nutrition* 87(Suppl 2): S273–81.
103. Brady, L.J., Gallaher, D.D., and Busta, F.F. 2000. The role of probiotic cultures in the prevention of colon cancer. *The Journal of Nutrition* 130: S41014.
104. Elmadfa, I. and Meyer, A.L. 2008. Body composition, changing physiological functions and nutrient requirements of the elderly. *Annals of Nutrition & Metabolism* 52(Suppl 1): 2–5.
105. Selkoe, D.J. 2012. Preventing Alzheimer's disease. *Science* 337: 1488–92.
106. Levy, K., Lanctot, K.L., Farber, S.B., Li, A., and Herrmann, N. 2012. Does pharmacological treatment of neuropsychiatric symptoms in Alzheimer's disease relieve caregiver burden? *Drugs & Aging* 29: 167–79.
107. Ness, S., Rafii, M., Aisen, P., Krams, M., Silverman, W., and Manji, H. 2012. Down's syndrome and Alzheimer's disease: Towards secondary prevention. *Nature Reviews Drug Discovery* 11: 655–6.
108. Camfield, D.A., Owen, L., Scholey, A.B., Pipingas, A., and Stough C. 2011. Dairy constituents and neurocognitive health in ageing. *The British Journal of Nutrition* 106: 159–74.
109. von Arnim, C.A., Gola, U., and Biesalski, H.K. 2010. More than the sum of its parts? Nutrition in Alzheimer's disease. *Nutrition* 26: 694–700.
110. Smith, A.D. and Refsum, H. 2009. Vitamin B-12 and cognition in the elderly. *The American Journal of Clinical Nutrition* 89: S707–11.
111. Zeisel, S.H. and Blusztajn, J.K. 1994. Choline and human nutrition. *Annual Review of Nutrition* 14: 269–96.
112. Jia, X., McNeill, G., and Avenell, A. 2008. Does taking vitamin, mineral and fatty acid supplements prevent cognitive decline? A systematic review of randomized controlled trials. *Journal of Human Nutrition and Dietetics* 21: 317–36.
113. Cole, G.M. and Frautschy, S.A. 2010. DHA may prevent age-related dementia. *The Journal of Nutrition* 140: 869–74.
114. Milte, C.M., Sinn, N., and Howe, P.R. 2009. Polyunsaturated fatty acid status in attention deficit hyperactivity disorder, depression, and Alzheimer's disease: Towards an omega-3 index for mental health? *Nutrition Reviews* 67: 573–90.
115. Mozzi, R., Buratta, S., and Goracci, G. 2003. Metabolism and functions of phosphatidylserine in mammalian brain. *Neurochemical Research* 28: 195–214.
116. Heiss, W.D., Kessler, J., Mielke, R., Szelies, B., and Herholz, K. 1994. Long-term effects of phosphatidylserine, pyritinol, and cognitive training in Alzheimer's disease. A neuropsychological, EEG, and PET investigation. *Dementia* 5: 88–98.
117. Kato-Kataoka, A., Sakai, M., Ebina, R., Nonaka, C., Asano, T., and Miyamori, T. 2010. Soybean-derived phosphatidylserine improves memory function of the elderly Japanese subjects with memory complaints. *Journal of Clinical Biochemistry and Nutrition* 47: 246–55.
118. Barrett, A.M. 2001. A 1-year controlled trial of acetyl-L-carnitine in early-onset AD. *Neurology* 56: 425.
119. Montgomery, S.A, Thal, L.J., and Amrein, R. 2003. Meta-analysis of double blind randomized controlled clinical trials of acetyl-L-carnitine versus placebo in the treatment of mild cognitive impairment and mild Alzheimer's disease. *International Clinical Psychopharmacology* 18: 61–71.
120. Spencer, J.P. 2010. The impact of fruit flavonoids on memory and cognition. *The British Journal of Nutrition* 104(Suppl 3): S40–7.
121. Hamaguchi, T., Ono, K., and Yamada, M. 2010. REVIEW: Curcumin and Alzheimer's disease. *CNS Neuroscience & Therapeutics* 16: 285–97.
122. Anekonda, T.S. 2006. Resveratrol—A boon for treating Alzheimer's disease? *Brain Research Reviews* 52: 316–26.
123. Hussein, G., Sankawa, U., Goto, H., Matsumoto, K., and Watanabe, H. 2006. Astaxanthin, a carotenoid with potential in human health and nutrition. *Journal of Natural Products* 69: 443–9.
124. Shukitt-Hale, B., Lau, F.C., and Joseph, J.A. 2008. Berry fruit supplementation and the aging brain. *Journal of Agricultural and Food Chemistry* 56: 636–41.
125. Geng, J., Dong, J., Ni, H., Lee, M.S., Wu, T., Jiang, K., Wang, G., Zhou, A.L., and Malouf, R. 2010. Ginseng for cognition. *The Cochrane Database of Systematic Reviews*. doi:10.1002/14651858.CD007769.pub2.
126. Moss, M., Hewitt, S., Moss, L., and Wesnes, K. 2008. Modulation of cognitive performance and mood by aromas of peppermint and ylang-ylang. *The International Journal of Neuroscience* 118: 59–77.
127. Frydman-Marom, A., Levin, A., Farfara, D., Benromano, T., Scherzer-Attali, R., Peled, S., Vassar, R. et al. 2011. Orally administrated cinnamon extract reduces beta-amyloid oligomerization and corrects cognitive impairment in Alzheimer's disease animal models. *PLoS One* 6: e16564.

128. Herrschaft, H., Nacu, A., Likhachev, S., Sholomov, I., Hoerr, R., and Schlaefke, S. 2012. *Ginkgo biloba* extract EGb 761(R) in dementia with neuropsychiatric features: A randomised, placebo-controlled trial to confirm the efficacy and safety of a daily dose of 240 mg. *Journal of Psychiatric Research* 46: 716–23.
129. Cho, E.S., Jang, Y.J., Kang, N.J., Hwang, M.K., Kim, Y.T., Lee, K.W., and Lee, H.J. 2009. Cocoa procyanidins attenuate 4-hydroxynonenal-induced apoptosis of PC12 cells by directly inhibiting mitogen-activated protein kinase kinase 4 activity. *Free Radical Biology & Medicine* 46: 1319–27.
130. Spencer, J.P. 2008. Flavonoids: Modulators of brain function? *The British Journal of Nutrition* 99(E Suppl 1): ES60–77.
131. Assuncao, M., Santos-Marques, M.J., Carvalho, F., and Andrade, J.P. 2010. Green tea averts age-dependent decline of hippocampal signaling systems related to antioxidant defenses and survival. *Free Radical Biology & Medicine* 48: 831–8.
132. Bryan, J. 2008. Psychological effects of dietary components of tea: Caffeine and L-theanine. *Nutrition Reviews* 66: 82–90.
133. Fric, P., Gabrovska, D., and Nevoral, J. 2011. Celiac disease, gluten-free diet, and oats. *Nutrition Reviews* 69: 107–15.
134. Markus, C.R. 2007. Effects of carbohydrates on brain tryptophan availability and stress performance. *Biological Psychology* 76: 83–90.
135. Richardson, A.J. 2008. *n*-3 Fatty acids and mood: The devil is in the detail. *The British Journal of Nutrition* 99: 221–3.
136. Woelk, H., Arnoldt, K.H., Kieser, M., and Hoerr, R. 2007. *Ginkgo biloba* special extract EGb 761 in generalized anxiety disorder and adjustment disorder with anxious mood: A randomized, double-blind, placebo-controlled trial. *Journal of Psychiatric Research* 41: 472–80.
137. Lieberman, H.R. 2001. The effects of ginseng, ephedrine, and caffeine on cognitive performance, mood and energy. *Nutrition Reviews* 59: 91–102.
138. Kennedy, D.O., Scholey, A.B., and Wesnes, K.A. 2002. Modulation of cognition and mood following administration of single doses of *Ginkgo biloba*, ginseng, and a ginkgo/ginseng combination to healthy young adults. *Physiology & Behavior* 75: 739–51.
139. Kennedy, D.O. and Scholey, A.B. 2003. Ginseng: Potential for the enhancement of cognitive performance and mood. *Pharmacology, Biochemistry and Behavior* 75: 687–700.
140. Garau, C., Aparicio, S., Rial, R.V., Nicolau, M.C., and Esteban, S. 2006. Age related changes in the activity-rest circadian rhythms and c-fos expression of ring doves with aging. Effects of tryptophan intake. *Experimental Gerontology* 41: 430–8.
141. Clydesdale, F. 2005. Functional foods: Opportunities and challenges, IFT Expert Report. Chicago, IL.
142. Selhub, J., Bagley, L.C., Miller, J., and Rosenberg, I.H. 2000. B vitamins, homocysteine, and neurocognitive function in the elderly. *The American Journal of Clinical Nutrition* 71: 614S–20S.
143. Hankey, G.J. and Eikelboom, J.W. 2001. Homocysteine and stroke. *Current Opinion in Neurology* 14: 95–102.
144. Brattstrom, L. and Wilcken, D.E. 2000. Homocysteine and cardiovascular disease: Cause or effect? *The American Journal of Clinical Nutrition* 72: 315–23.
145. Othman, R.A., Moghadasian, M.H., and Jones, P.J. 2011. Cholesterol-lowering effects of oat beta-glucan. *Nutrition Reviews* 69: 299–309.
146. Trujillo, E., Davis, C., and Milner, J. 2006. Nutrigenomics, proteomics, metabolomics, and the practice of dietetics. *Journal of the American Dietetic Association* 106: 403–13.
147. Bagchi, D., Lau, F.C., and Ghosh, D.K. 2010. *Biotechnology in Functional Foods and Nutraceuticals.* Boca Raton, FL: CRC Press.
148. German, J.B., Zivkovic, A.M., Dallas, D.C., and Smilowitz, J.T. 2011. Nutrigenomics and personalized diets: What will they mean for food? *Annual Review of Food Science and Technology* 2: 97–123.
149. Ferguson, L.R. 2009. Nutrigenomics approaches to functional foods. *Journal of the American Dietetic Association* 109: 452–8.
150. Afman, L. and Muller, M. 2006. Nutrigenomics: From molecular nutrition to prevention of disease. *Journal of the American Dietetic Association* 106: 569–76.
151. Moore, J.B. and Weeks, M.E. 2011. Proteomics and systems biology: Current and future applications in the nutritional sciences. *Advances in Nutrition* 2: 355–64.
152. de Roos, B. and McArdle, H.J. 2008. Proteomics as a tool for the modelling of biological processes and biomarker development in nutrition research. *The British Journal of Nutrition* 99(Suppl 3): S66–71.
153. Schweigert, F.J. 2007. Nutritional proteomics: Methods and concepts for research in nutritional science. *Annals of Nutrition & Metabolism* 51: 99–107.

154. Jones, D.P., Park, Y., and Ziegler, T.R. 2012. Nutritional metabolomics: Progress in addressing complexity in diet and health. *Annals of Nutrition & Metabolism* 32: 183–202.
155. Primrose, S., Draper, J., Elsom, R., Kirkpatrick, V., Mathers, J.C., Seal, C., Beckmann, M. et al. 2011. Metabolomics and human nutrition. *The British Journal of Nutrition* 105: 1277–83.
156. Gibney, M.J., Walsh, M., Brennan, L., Roche, H.M., German, B., and van Ommen, B. 2005. Metabolomics in human nutrition: Opportunities and challenges. *The American Journal of Clinical Nutrition* 82: 497–503.

4 Clinical Trial Barriers in Functional Foods and Nutrition

Rajshri Roy

CONTENTS

4.1 FUNCTIONAL FOODS

Functional foods (Table 4.1) are defined as foods and food components that provide health benefits beyond the basic nutrition such as conventional foods, enriched or enhanced foods, and dietary supplements. These substances provide essential nutrients beyond quantities necessary for normal maintenance, growth, development, and other biologically active components that impart health benefits or desirable physiological effects (Institute of Food Technologies 2005).

4.2 NUTRIENTS

Nutrients are defined as vitamins, minerals, and essential fatty acids for which recommended intakes have been established for the population. Other components include phytonutrients or bioactives that have a scientifically established body of evidence for a plausible mechanism, but they do not have a recommended intake (Institute of Food Technologies 2005).

4.3 CLINICAL TRIALS IN FUNCTIONAL FOODS AND NUTRITION

The goal of a clinical trial is to obtain an objective and scientific data from scientific test conducted on humans. It provides the most objective and practical evaluation of product efficacy on humans.

The scientific goals of a clinical study usually include determination and demonstration of efficacy and safety, understanding and validation of the mechanism of action, and understanding for whom

TABLE 4.1
Definitions and Examples of Functional Foods and Nutraceuticals

A functional food

- "is similar in appearance to, or maybe, a conventional food, is consumed as part of a usual diet, and is demonstrated to have physiological benefits and/or reduce the risk of chronic disease beyond basic nutritional functions."[a]
- is "a food or beverage that imparts a physiological benefit that enhances overall health, helps prevents or treat a disease/condition, or improves physical or mental performance via an added functional ingredient, processing modification, or biotechnology."[a]

Examples include tomatoes that are rich in lycopene, spreads containing plant sterols, and eggs enriched in omega-3 fatty acids.

A nutraceutical

- "is a product isolated or purified from foods that is generally sold in medicinal forms not usually associated with food. A nutraceutical is demonstrated to have a physiological benefit or provide protection against chronic disease."[a]

Examples include capsules containing bioflavonoids or gamma-linoleic acid.

Source: Jones, P.J., *Can Med Assoc J* 166, 1555–1563, 2002.
[a] Schema for assessing strength and consistency of scientific evidence leading to significant scientific agreement.

the product works best (Udani 2012). The regulatory goals of a clinical study include establishing substantiation for your proprietary product, determining that the substantiation for your finished product specifically stems from an active ingredient within your formulation, and protecting your company from legal challenges (Udani 2012). The consumer (marketing) benefits of a clinical trial include organizing press-worthy events (scientific poster presentations, peer-reviewed publications) and providing experts with specific documentation to reference in interviews regarding your product (Udani 2012).

Before conducting a clinical trial on a novel and/or functional food, the following prerequisite materials are required:

- Information on the origin and nature of the functional food
- Contents of the functional component
- Processing methods
- Scientific evidence of the safety and efficacy
- Specification and analysis of the final food product
- Validation for analysis of functional food components
- Data on the stability and purity of the food
- Rationale behind dosage amount and duration

Functional food and nutrition research includes the development of cellular and human model systems to study the mechanisms of action of nutrients and/or functional food components—namely, dietary lipids, antioxidants, and plant phytochemicals on atherosclerosis, inflammation, insulin resistance, and obesity—at the molecular, cellular, and whole-body level, as well as the influence of diet–genotype interactions. The researchers use molecular biology and human clinical trial techniques to examine the association between diet, genotype, and chronic disease, and gain an understanding of its molecular basis.

4.4 ROLE OF RESEARCH

Functional foods that are currently on the market represent a small fraction of the possible products. Claims linking the consumption of functional foods or food ingredients with health outcomes require sound scientific evidence and significant scientific agreement. Figure 4.1 differentiates

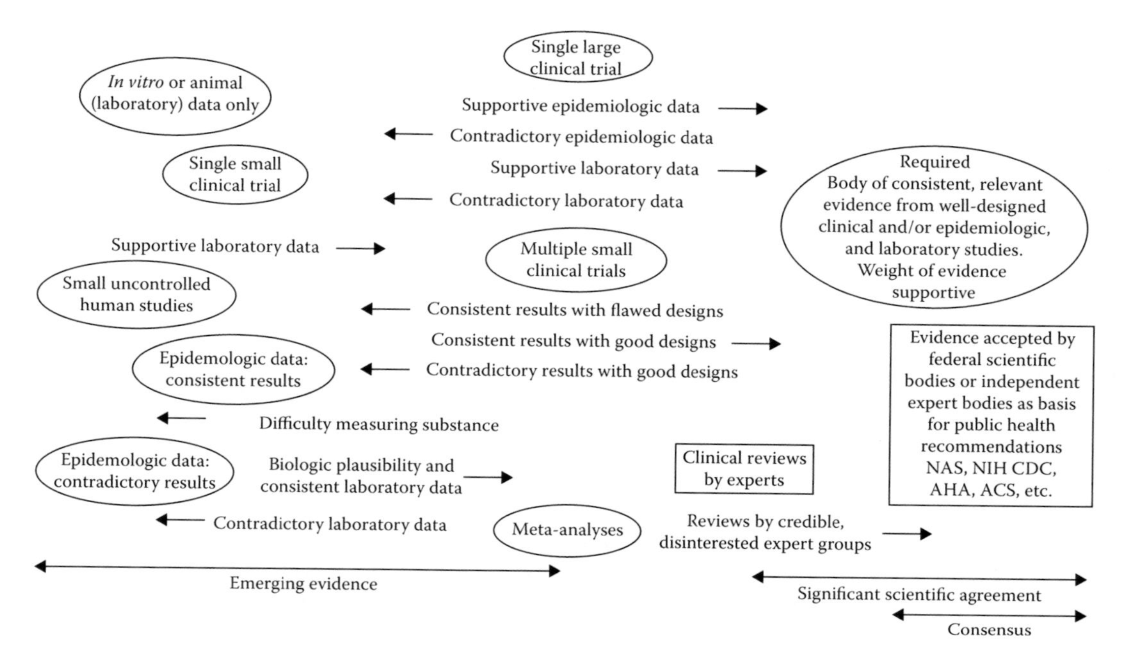

FIGURE 4.1 The FDA's schematic of significant scientific agreement released in December 22, 1999, guidance document. ACS, American Cancer Society; AHA, American Heart Association; CDC, Centers for Disease Control and Prevention; NAS, National Academy of Sciences; NIH, National Institutes of Health. (US Food and Drug Administration, *Guidance for Industry: Significant Scientific Agreement in the Review of Health Claims for Conventional Foods and Dietary Supplements*. US Food and Drug Administration, Washington, DC, 1999.)

"emerging evidence" on the bottom left such as animal, *in vitro*, and uncontrolled human studies from "consensus" on the bottom right, and includes the evidence that has been accepted by federal scientific bodies responsible for public health recommendations regarding functional foods and nutrition. Thus, the evidence for a functional food–health relationship strengthens as we move from left to right in the schematics. Claims about the health benefits of functional foods should be based on sound scientific evidence, but too often only so-called "emerging evidence" is the basis for marketing some functional foods or their components.

Some areas of research that are vital to the development of functional foods are discussed in Subsections 4.4.1 through 4.4.3.

4.4.1 Nutrients and Functional Foods

The basic and applied clinical research must focus on establishing an understanding of the mechanisms of action for known nutrients and novel functional foods. The key aspects include their dose–response relationship, the clinical trial outcomes, and the individual variations in response (Institute of Food Technologies 2005). In nutrition research, there are diverse health effects that are either known or suspected for many nutrients such as selenium, vitamins B and E, and carotenoids. Clinical nutrition journals regularly publish the studies exploring the role of known nutrients in health. For example, Ames et al. (2002) summarize an extensive review of specific nutrient effects on various enzymatic processes in the human body.

The potential health benefits of novel functional food components represent a similar frontier in nutrition research. Examples of such clinical research in the area of functional foods include the potential roles in health for phytochemicals, including sulfur compounds in cruciferous vegetables, polyphenols in teas, and flavonoids in wine, blueberries, and pomegranate (Guhr and LaChance 1997).

The higher intake of whole grains and multiple serves of fruits and vegetables, as repeatedly shown by epidemiological studies, is associated with better health and lower incidence of chronic diseases. The beneficial effects are not only a result of nutrients but also caused by the diverse roles for functional food components in blocking, reversing, or interfering in molecular level processes, which, if left unchecked, would lead to various chronic diseases (Guhr and LaChance 1997).

Ongoing research in the area of nutrition must identify functional food components and determine their mechanisms of action and effects on health, and this knowledge must then be confirmed through well-designed preclinical and clinical trials.

4.4.2 New and Existing Biomarkers

Identifying specific cause–effect relationships between dietary components and health is challenging in the area of functional food research and, in some cases, controversial due to the complexity of human biology and physiology (Institute of Food Technologies 2005).

Biomarkers and their relationship to health status are often identified through observational studies or correlations. At best, correlated factors may suggest a complex, multifactorial relationship between diet and health, and may be supported by scientific theory that appears credible; at worst, the correlations are the result of another unrelated factor and have no basis in fact (Institute of Food Technologies 2005).

Scientists need to identify additional biomarkers that signal changes in health status and then determine the meaning of changes in these biomarkers relative to a defined health condition. In addition, with exposure to specific food components, we need to assess intake, bioavailability, and utilization of potential functional food components. The effects of diet on biomarkers and the entire human body must be validated through prospective clinical trials (Institute of Food Technologies 2005).

4.4.3 Food Vehicles for Functional Food Components

Clinical research needs to be conducted to identify and tailor foods for delivery of novel functional food components. The research in this area is required to provide a stable environment for the functional food components. It increases the knowledge of the interactions between the functional food components, other ingredients in the vehicle matrix, and the human body. The research can be used to maximize the health benefit of functional food components as well as its bioavailability (Lutter and Dewey 2003).

In addition, human clinical trials can also highlight the desirable sensory/organoleptic characteristics of the novel functional foods (Lutter and Dewey 2003).

Research will help scientists better understand each of these issues in functional food development (Lutter and Dewey 2003).

4.5 RESEARCH CHALLENGES FOR FUNCTIONAL FOODS

Functional foods face some key research challenges in which there is a need to identify new functional food ingredients and to gain consumer acceptance of such products.

The development and marketing of functional foods requires significant research efforts, which include identifying functional compounds and assessing their physiological effects; developing a suitable food matrix; and taking into account bioavailability and potential changes during processing and food preparation, consumer education, and clinical trials on product efficacy in order to gain approval for health-enhancing marketing claims (Kotilainen et al. 2006). It is a multistage process that requires input from commercial, academic, and regulatory interests, with a critical need to achieve acceptance by the consumers (Jones and Jew 2007).

Randomized clinical trial data are capable of providing strong experimental evidence to establish causal relationships between functional food components, health and disease/disease risk. However, clinical studies must be well designed in order to optimize the quality of the data they provide (Jones and Jew 2007).

Well-designed and well-conducted clinical trials exist as the gold standard for testing the efficacy and safety of functional foods and their components. Crossover and parallel designs each possess advantages and disadvantages; however, from certain perspectives, it can be argued that crossover designs are more cost-effective. Trials can strictly control background diet or allow for a free-living diet. Strictly controlled metabolic diets are the best choice for testing the efficacy of dietary ingredients; however, a free-living diet with an unsupervised dietary intake can generate data that are more readily generalized and applicable to general populations. Well-conducted clinical trials should carefully define subject inclusion criteria, appropriate control and blinding, food and/or ingredient dosage, as well as frequency and timing of treatment (Institute of Food Technologies 2005).

Moreover, investigators should pay attention to the food matrix that is used to deliver the bioactive ingredient as well as its stability. The trial duration should be short enough to be cost-effective and long enough to ensure efficacy. In addition, the duration for optimal subject compliance is a factor to be considered. Trials should use a validated surrogate as the primary endpoint as these are more practical than clinical endpoints. The safety profile of a product should also be taken into consideration. Use of suitable statistical tests is also necessary in order to appropriately analyze the data and to generate valid conclusions (Institute of Food Technologies 2005).

Finally, trials should have an adequate sample size, not just to ensure statistical significance but also to assess the biological significance of the functional food or its ingredient. Given that health claim legislators view the quality of clinical studies as paramount when considering their inclusion in health claim substantiation dossiers, careful consideration of these factors as well as adherence to Good Clinical Practice (GCP) guidelines are crucial in order to evaluate the validity of relationships between food and health, and food and disease/disease risk for health claim substantiation.

In particular, there is a need to identify potential functional ingredients that could provide benefits in terms of health and well-being; identify individual biological responses that our body has to functional foods; define the bioavailability of functional food ingredients; develop appropriate biomarkers for a wider range of functional endpoints; and develop the potential utility of nutrigenomics, bioinformatics, proteomics, metabolomics, and nanotechnology in the development of functional foods (European Commission 2010).

Additionally, the market needs to anticipate demand for personalized nutrition and the potential role of functional foods, and to ensure stability of functional food ingredients during manufacturing and passage through the gastrointestinal (GI) tract to reach the target organ intact (European Commission 2010).

The nutritionists and the dietitians need to establish Dietary Reference Intakes (DRIs) for a wider range of nutrients to enable commercial exploitation of more functional components.

4.6 BARRIERS IN RESEARCH

Continued presence in the functional foods market requires scientific evidence for product effectiveness. Even though certain foods may have been used for a long time for health enhancement purposes, the definitive scientific support for claims as a functional product is often lacking. This necessary research requires time and financial and human resources, especially for products into the export and home markets as local regulations become stricter. Research investments are also needed for discovering new functional foods or ingredients in local sources (European Commission 2010).

4.6.1 Types of Studies

Both observational studies and experimental trials can provide useful data for identifying diet–disease relationships. Two experimental designs are conventionally used: parallel and crossover. Each of these designs possesses advantages and disadvantages (AbuMweis 2010). For certain functional ingredients, selection of an appropriate control arm is straightforward, whereas for others, it is challenging. Studies should be short enough to optimize subject compliance, be cost-effective, and avoid high subject drop-out rates, and long enough to ensure biological efficacy (AbuMweis 2010). The dose, frequency, and diurnal timing of intake of the active food ingredients all need to be chosen carefully.

Randomized clinical trials testing the efficacy of functional foods may use both validated and emerging surrogate endpoints, and should employ suitable statistical tests for data analysis. Paying attention to all these factors is crucial to the design of quality clinical trials that reliably evaluate the validity of food–health relationship (AbuMweis 2010). Accordingly, clinical studies that incorporate the optimal design elements discussed will yield robust results that are appropriate for the substantiation of health claims on functional foods (AbuMweis 2010).

For novel functional foods, randomized clinical trials are the best type of research because such compounds can be tested and retested successfully. Phytosterols, probiotics, or botanical supplements, for example, fit into this category.

Several clinical trials on phytosterols have shown several times that daily consumption of 1.5–3.0 g of phytosterols/phytostanols can reduce the total cholesterol levels by 8%–17%, representing a significant reduction in the risk of cardiovascular disease. Studies looking at how these results transfer into real free-living populations have backed up the clinical trials, showing a stabilization of cholesterol levels in certain populations.

These clinical trials follow the drug model, however; we supplement the diet with one or two nutrients each at a single dose for a set period of time. Time-constrained randomized trials cannot capture a lifetime of consumption with respect to health and chronic disease in humans.

Randomized clinical trials work by randomly assigning a group of volunteers to receive an active compound, whether a drug or nutrient(s), or a nonactive compound in an inactive form of the active compound or a placebo. Observational studies, as the name suggests, observe a population and relate dietary intakes of foods and nutrients and their effects.

We need to consider science as a whole and not blinker ourselves with results of one big clinical trial on a functional food or nutrient, regardless of how expensive the study was and what universities are involved in the study.

Despite a vast body of observational studies that have linked an increasing dietary intake of antioxidants from fruits and vegetables to reduce the risks of a range of diseases including cancer, cardiovascular disease, and diabetes; however, when such antioxidants have been extracted and put into supplements, the results according to randomized clinical trials do not produce the same benefits.

A Women's Health Initiative (WHI) clinical trial followed up 18,176 postmenopausal women taking calcium (1000 mg) and vitamin D (400IU) supplements. A similar population ($N = 18{,}106$) was given a placebo. The subjects were followed up for about 7 years and the researchers reported that the supplements had no effect on colorectal cancer incidence. None of the women in the study had cancer at the start of the study. However colorectal cancer has a long latency period of 10–20 years (LaCroix et al. 2009) and therefore, the clinical trial posed a barrier as we cannot expect to see any effect in this study.

Unlike pharmaceuticals, nutrients and functional foods often work in synergy with one another and exert effects on multiple body tissues.

Additionally, many randomized clinical trials look at the effect of nutrients on diseased populations, and negative and null results from these trials should not be generalized for the entire population.

There are no viable alternatives available to the barriers of clinical trials. Human trials with randomization are the ideal way to limit bias, and controlling specifically with a placebo places more value on these studies.

4.6.2 Stakeholders and Resources

Additionally, cooperation among a diverse group of stakeholders—including research sponsors (industry, academia, government, nonprofit organizations, and food companies), clinical investigators, customers, payers, physicians, and regulators—is necessary in conducting a functional food clinical trial. Each stakeholder offers a different set of tools to support the essential components of a clinical trial (English 2010). These resources form the infrastructure that currently supports clinical research in nutrition and functional foods. Time, money, personnel, materials, support systems (informatics as well as manpower), and a clear plan for completing the necessary steps in a trial are all part of the clinical research infrastructure (English 2010). Significant time, energy, and money are spent on bringing the disparate resources for each trial together.

4.6.3 Healthy Study Subjects

For natural products such as dietary supplements and functional foods, studies should target nondiseased populations and nondrug endpoints. Statistically, however, there are challenges in performing studies in a nondiseased, healthy population. In healthy volunteers, the magnitude of a physiological change due to an intervention is likely to be smaller than it would be in diseased subjects; as a result of a smaller effect, the likelihood of demonstrating a statistically significant difference is diminished (Udani 2012).

On one hand, the functional foods industry is being asked to perform clinical trials to demonstrate the effects supporting health claims (Udani 2012). On the other hand, if a study involves a diseased population or measures endpoints that are only relevant to drug products, it automatically brands the study product as a drug (Udani 2012).

4.6.4 Sponsor's Goals versus Regulatory Targets

One of the challenges or barriers for clinical research in this area has been designing clinical trials that meet a sponsor's goals while keeping in mind a moving regulatory target.

While studies for functional foods may not include diseased populations, they can include an at-risk population. The rationale for identifying an at-risk population rests in understanding that a subject who is not diseased but who has pathophysiology representing a point on the journey toward the diagnosis of a disease can most likely benefit from the functional food as a dietary supplement, and therefore is the ideal subject for a clinical trial on functional food (Udani 2012).

It is also important to apply a standardized stressor, that is, to overwhelm the subject's physiology to provide the opportunity for the functional food component to potentially have a greater impact and demonstrate a statistically and/or clinically significant distinction when compared to placebo (Udani 2012).

There is also a need to measure the objective and functionally relevant endpoints that have meaning to scientists (by being clinically meaningful and familiar in medical settings), regulators (by staying within the framework established), and subjects (by being understandable and relevant to their health) (Udani 2012).

4.6.5 Cost of Clinical Research

Clinical research in nutrition and functional foods is expensive. Most food companies that manufacture and sell nutraceuticals and functional foods cannot afford to perform clinical trials on their products. However, the medical industry has set up a system in which they pay a minimum of $12,000 for each test subject in one of their studies just to the institution that finds the subject (DiMasi 2003). Private research institutions and universities recruit hundreds to thousands of subjects for each clinical study. A clinical study will cost a drug company several millions of dollars; however, the investment on a drug clinical trial is incurred by the sale of the drugs. By contrast, a nutraceutical or a food company usually lacks the benefit of definite profitable sales. The sales for a particular product may be in the millions, but because the area of nutrition is so competitive, the margins for manufacturers are lean.

4.7 CONCLUSIONS

There is a need to expand clinical research on traditional nutrients, other bioactive food components, and the intersection of genomics and molecular nutrition. Continued basic and applied nutritional clinical trials must further explore the roles and mechanisms of action for traditional nutrients in humans. In addition to traditional nutrients, other bioactive food components with the ability to improve health must be identified and their efficacy proven. The intersection of genomics and molecular nutrition presents opportunities for more definitively understanding diet and health in individuals and population groups, with the potential for personalized diets for optimal health.

There is also a need to expand clinical research on biomarkers and physiological endpoints. Additional biomarkers that signal changes in health status are urgently needed, and the meaning of changes in these biomarkers must be clearly demonstrated in human clinical trials relative to a defined health condition. Clinical research is also needed to expand the validated biomarkers of health status including assessing how genes and gene products relate to disease risk.

REFERENCES

AbuMweis, S.S., Jew, S., and Jones, P.J. (2010). Optimizing clinical trial design for assessing the efficacy of functional foods. *Nutr Rev* 68: 485–499.

Ames, B., Elson-Schwab, I., and Silver, E. 2002. High-dose vitamin therapy stimulates variant enzymes with decreased coenzyme binding affinity (increased Km): Relevance to genetic disease and polymorphisms. *Am J Clin Nutr* 75: 616–658.

DiMasi, J.A., Hansen, R.W. and Grabowski, H.G. (2003). The price of innovation: New estimates of drug development costs. *J Health Econ* 22(2): 151–185.

English, R., Lebovitz, Y., and Giffin, R. (2010). Transforming clinical research in the United States: Challenges and opportunities. Workshop summary, Forum on Drug Discovery, Development, and Translation, Board on Health Sciences Policy, Institute of Medicine of the National Sciences, Retrieved December 2, 2012, from http://www.nap.edu/openbook.php?record_id=12900&page=R1.

European Commission (2010). Functional foods. European Research Area, Food, Agriculture and Fisheries. Retrieved December 2, 2012, from ftp://ftp.cordis.europa.eu/pub/fp7/kbbe/docs/functional-foods_en.pdf.

Guhr, G. and LaChance, P.A. (1997). Role of phytochemicals in chronic disease prevention. In: P.A. Lachance, editor, *Nutraceuticals: Designer Foods III—Garlic, Soy and Licorice*. Trumbull, CT: Food & Nutrition Press, pp. 311–364.

Institute of Food Technologies (2005). Functional foods: Opportunities and challenges. IFT expert panel report. Retrieved December 2, 2012, from http://www.ift.org/Knowledge-Center/Read-IFT-Publications/Science-Reports/Expert-Reports/~/media/Knowledge%20Center/Science%20Reports/Expert%20Reports/Functional%20Foods/Functionalfoods_expertreport_full.pdf.

Jones, P.J. (2002). Clinical nutrition: Functional foods—More than just nutrition. *Can Med Assoc J* 166(12): 1555–1563.

Jones, P.J. and Jew, S. (2007). Functional food development: Concept to reality. *Trends Food Sci Technol* 18: 387–390.

Kotilainen, L., Rajalahti, R., Ragasa, C., and Pehu, E. (2006). Health enhancing foods: Opportunities for strengthening the sector in developing countries. Agriculture and Rural Development Discussion Paper 30, Retreived January 19, 2013, from http://siteresources.worldbank.org/INTARD/Resources/Health_Enhacing_Foods_ARD_DP_30_final.pdf.

LaCroix, A.Z. et al. (2009). Calcium plus vitamin D supplementation and mortality in postmenopausal women: The Women's Health Initiative calcium–vitamin D randomized controlled trial. *J Gerontol A Biol Sci Med Sci* 64A(5): 559–567.

Lutter, C.K. and Dewey, K.G. (2003). Proposed nutrient composition for fortified complementary foods. *J Nutr* 133: 3011S–3020S.

Udani, J. (2012). Clinical trials: Will the wrong trial design make your dietary supplement or functional food an unregistered drug? Medicus Research. Retrieved December 2, 2012, from http://www.nutritionaloutlook.com/article/designing-clinical-trials-4-9622.

US Food and Drug Administration (1999). *Guidance for Industry: Significant Scientific Agreement in the Review of Health Claims for Conventional Foods and Dietary Supplements*. Washington, DC: US Food and Drug Administration.

Section II

Science Underpinning Health Benefits

5 Dual Function of Food Factors as Pharmacological Molecules

Tetsuya Konishi

CONTENTS

5.1 INTRODUCTION

Food and health are strictly related. This has been recognized from ancient times in both the Oriental and Western worlds. The word "nutrition" originated from the Latin *nutrire*, which means "nourish with milk." In ancient times, when food supplies were insufficient, deficiencies in certain nutritional substances caused physiological disorders and diseases [1,2], and thus the role of foods was mainly discussed in the context of nutrition. The search for the essential substances to nourish the body eventually led to establishment of the nutritional sciences [3]. In the developed countries today, the increased incidence of complex diseases such as metabolic syndromes and cancer has become a social problem caused by the excessive intake of calories and imbalanced diets [4,5]. Thus, the role of diet in maintaining and promoting health has attracted considerable attention [6].

5.2 FOOD AS MEDICINE

In ancient times, many field plants were used as medicines, and the concept of food as medicine was accepted worldwide, especially in Oriental countries. It is now well recognized that foods contain a variety of components that are not essential for maintaining life, and thus are not nutrients but do have beneficial health functions such as modulation of physiological reactions or cancer prevention. For example, an estimated 20%–30% reduction in the risk of developing type 2 diabetic complications may be realized by diets high in whole-grain fibers and phytochemicals [7]. The beneficial role of these dietary components, now termed food factors, is well understood and has been discussed in relation to their role in the promotion of human health [8].

The "food functions" are currently classified into three categories: nutritional function, sensory function, and pharmacological or physiological function [9]. The "nutritional function" is the primary and essential function of food. A nutrient is a substance necessary for maintaining cellular architecture and sustaining biochemical activities. The second function is called "sensory function" and is necessary for the promotion of the enjoyable intake of food as necessary nutrition for the body. For this purpose, foods are colored and flavored, and have specific textures and tastes. In addition to these basic functions of food, a "pharmacological or physiological function" has been defined as the third function of food, and extensive studies of various food factors are currently under way, especially for the prevention of cancer and the metabolic syndromes [10,11].

Historically, humans have been aware of the medicinal functions of natural resources, especially of field plants, and many medicines have been developed from plants. Indeed, approximately 70% of anticancer drugs used in modern medicine have been isolated from plants or have been chemically modified from plant extracts [12]. As the importance of preventative medicine has become more widely recognized, these same research efforts have been accelerated to discover certain active ingredients in dietary resources that have been traditionally considered beneficial for health, such as cha (*Camellia sinensis*). An extensive database has been compiled on their physiological functions [13]. Thus, the old dictum "Food as medicine" is now gaining increasing credibility through growing scientific evidence. Based on accumulated knowledge, new definitions of foods or dietary supplements have been proposed, for example, functional foods, nutraceuticals, and pharmafoods, which target not only nutrition but also prevention or even amelioration of diseases [14–18]. "Food function" is now regulated in Japan in the form of Food for Specified Health Use (FOSHU) [19]. So-called functional foods, including dietary supplements, are now manufactured and marketed in the form of enriched active ingredients in foods such as beverages and also in the form of capsules or tablets that contain certain enriched active ingredients.

5.3 FUNCTIONAL PROPERTIES OF FOOD

Even though the pharmacological or physiological activity was defined as the third function of foods [9], there is some debate as to whether the food function is the same as that of medicine. It is believed that food is characterized by its safety and multiplicity or indirectness of function. Food is generally a mixture of many nutritional and nonnutritional ingredients. Thus, the effect of food is due to complex synergistic effects resulting from multiple active components in the mixture. It is also said that the food function is sometimes indirect. For example, fungi modulate certain physiological reactions indirectly by normalizing conditions against pathological attacks such as infection or carcinogenesis. This is believed to occur mainly through its modulatory effects on the immune system. Such a mode of function is termed biological response modifier (BRM) [20]. This effect may also be due to the properties of food, which is a mixture of many functional compounds. Therefore, food may have multiple targets and multiple functions. To better understand this basic character of food function, researchers are currently using the systems biology approach in which several physiological reactions are simultaneously analyzed in an effort to understand the cooperative networking of physiological reactions [21]. Since biological systems are dynamic, this approach is quite reasonable in attempting to understand not only the homeostatic nature of organisms but also the food function.

5.4 FOOD FACTOR AS A PHARMACOLOGICAL MOLECULE

However, medicines are usually synthetic or otherwise highly purified single compounds having a high affinity to a specific target, which leads to the desired pharmacological effect.

When a specific ingredient is isolated from food resources and administered to animals or humans, such a molecule may behave as a synthetic medicine even though the biological activity is not as strong as the synthetic version. This indicates that the only difference may be in the binding affinity to the specific receptor mediating the pharmacological function.

Indeed, several examples show significant differences in effective concentrations between medicines and food factors. Angiotensin-converting enzyme (ACE) is the enzyme that catalyzes angiotensin I to II. As angiotensin II is a vasoconstrictor, ACE became a drug target to manage hypertension. Several ACE-inhibiting drugs such as captopril have been developed and used for treating hypertension and preventing heart disease [22]. Similarly, there has been a search for food ingredients and natural products having ACE inhibitory activity due to the interest in controlling blood pressure by diet [23]. Flavonoids are a class of food ingredients having such activity.

However, the effective concentration of flavonoids is far higher than that of captopril, such that the half maximal effective concentration (EC50) of quercetin-3-glucronide, one of the most effective flavonoids, is almost 1000 times higher than that of captopril [24].

This difference in their effective concentration, that is, the binding affinity to the receptor, may explain in part the characteristic property of the function of food factors such that food ingredients themselves have a rather broad spectrum of function. For example, flavonoids [25] or other polyphenols such as curcumin [26] have a variety of functions such as antioxidant activity, antihypertension, anticancer, antidepressive, and endothelial protection. At the concentrations necessary for a typical pharmacological effect, they may have an opportunity to interact with other targets having even lower affinity of interaction than the primary target, and thus other functions may be possible. Since the binding to the receptor is a necessary step for the pharmacological effect, the multiple functional properties of food may be explained simply by the ligand–receptor interaction theory used for drug action.

5.5 ANTIOXIDATIVE FOOD FACTOR IS MULTIFUNCTIONAL

Antioxidant properties may be an alternative explanation for the multifunctional roles of food factors, since most of them are antioxidant molecules [27]. Antioxidants are molecules having a high reactivity to oxidants or free radicals, preventing oxidative stress induced by, for example, oxidative chain reactions that occur in lipid peroxidation. However, the resulting oxidized form of the antioxidant has a more-or-less oxidant nature. Moreover, the rate of formation of the resultant oxidant is dependent on the oxidation level of the physiological microenvironment of the antioxidant. When the antioxidant molecule has certain specific pharmacological effects, providing that it has certain cellular targets available, there may be a reduced opportunity for the interaction to occur with the receptor due to lower levels of the original active molecule. Therefore, the pharmacological activity of certain antioxidant food ingredients will vary depending on the oxidative stress levels of the organism [28], as was shown in the differential effects of dihydrolipoic acid in diabetic and nondiabetic kidneys [29]. Further, if the oxidatively transformed antioxidant ingredient has some other pharmacological targets, an oxidative stress-dependent alternative function may be possible. Thus, the antioxidant nature of food factors could be another cause of the multiple functionality of food factors. Indeed, the neurotransmitter catecholamine shows strong antioxidant activity because the catechol moiety is highly reactive toward reactive oxygen species (ROS), but it also has significant toxicity toward neuronal cells due to the formation of toxic oxidized compounds such as dopanochrome [30,31]. Enhanced cell toxicity of antioxidant molecules has also been found in other antioxidant food factors.

5.6 STRONG INDIRECT INTERACTION AS AN ALTERNATIVE PATHWAY FOR FOOD FACTOR ACTION

In addition to these mechanisms, we need to discuss alternative examples of food factor action. Usually, almost all food factors are inert toward cellular activity, at least in terms of acute effects, and thus food factors are generally considered to be safe. However, it is also evident that food factors have certain physiological functions, especially after prolonged periods of ingestion or under specific physiological or pathological conditions. For example, several food factors such as squalene [32] or quercetin [33] show an adjuvant effect in anticancer therapies, in which the efficacy of certain anticancer drugs is enhanced in the presence of food factors even though the food factors themselves do not show significant toxicity toward cancer cells. Our studies on Schisandrin B (Sch B), which modulates the DNA damage checkpoint [34,35], provide some evidence for an underlying mechanism of the way food factors may function as adjuvants. The DNA damage checkpoint function is one of the so-called DNA damage response mechanisms in the cell [36]. When cellular DNA is damaged by physiological or artificial stimuli, the cell cycle is arrested at several stages such as G1, S, and G2/M to inspect for DNA damage. If the damage is slight, the repair process is initialized; however, if the damage is too serious, the cell follows the apoptosis pathway. This mechanism is important for maintaining genomic stability and avoiding undesirable genetic transformations being transmitted to the next generation of cells. However, this leads to acquisition of tolerance against anticancer modalities including anticancer drugs or radiation. We found that Sch B, a lignan isolated from *Schisandra chinensis*, inhibits this checkpoint function of the cell. Therefore, the cell fails to repair the DNA damage induced by these modalities and reduces the resistance of cancer cells to these modalities. An interesting observation is that Sch B is rather inert to cancer cells under normal conditions, becoming active only when cellular DNA is damaged. This is believed to occur because Sch B readily attacks ataxia telangiectasia mutated and Rad 3-related (ATR) kinase that is recruited or activated following the DNA damage and manipulates downstream checkpoint signals including p53, cell cycle checkpoint kinase-1 (Chk1), and structural maintenance of chromosomes protein 1 (SMC 1) [37]. In fact, the inhibitory action of Sch B on kinase activity is rather strong, with a half maximal inhibitory concentration (IC50) of ~7.3 μM. Sch B is thus inert to normal cells, as the enzyme is merely expressed. This mode of action of food factors with latent targets appears to be important as a food function for regulation of homeostatic human physiology, and we may term this type of food factor action as the "strong indirect interaction."

As described in this subsection, the checkpoint plays a crucial role in the fate of cancer cells in terms of tolerance acquisition. Checkpoint modulation has attracted much attention as a new target for anticancer drug development [38], and several checkpoint inhibitors such as Chk 1 inhibitor UCN01 [39] and ATR inhibitor VE821 [40] have been developed, although the clinical application of these synthetic drugs was abandoned because of the high risk of adverse effects. As shown for the ACE-inhibiting activity of food factors and clinical medicines discussed in Subsection 5.2, there is a significant difference in the IC50 values of these synthetic checkpoint inhibitors, that is, the 7 nM of UCN01 and the 420 nM of VE821 are approximately 20–1000 times lower than the 7.3 μM of Sch B. These rather weak ligand–target interactions and antioxidant properties explain the functionality of Sch B as a food factor. In fact, Sch B has a number of functions such as brain oxidative stress prevention [41], cellular glutathione (GSH) modulation [42], anti-inflammatory effect [43], acetylcholine esterase inhibition [44], genotoxicity prevention [45], and ATR inhibition [34]. Sch B is a

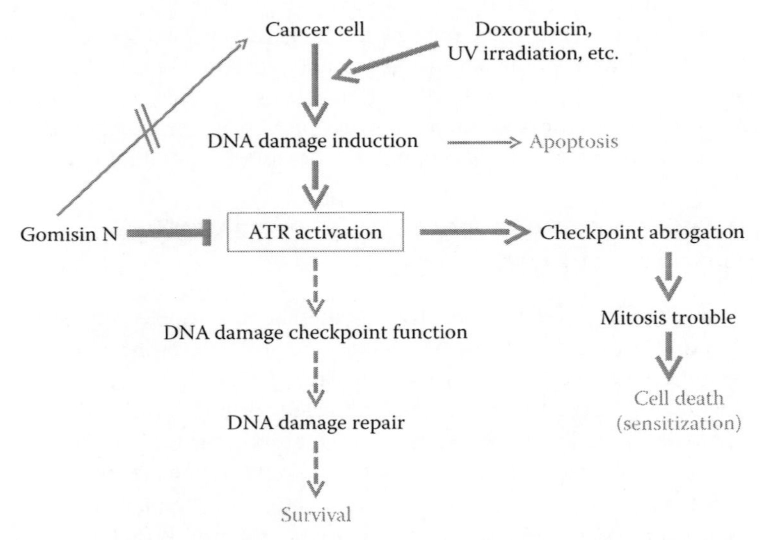

FIGURE 5.1 Mechanism of enhanced cell toxicity of anticancer drug by gomisin N. Gomisin N does not significantly affect cells, and even cancer cells when it is added directly. However, when the cellular DNA is damaged by anticancer drug, ATR is recruited as the target of gomisin N and reacts with rather strong affinity. Resulted abrogation of checkpoint function allows cells to cycle without repairing damaged DNA leading to mitotic catastrophe. This mechanism explains one of the characteristic features of food factor action such that it is pharmacologically active but low or nontoxic to the normal cell.

mixture of two enantiomeric stereoisomers: gomisin N and γ-schisandrin. It is now demonstrated that the ATR inhibitory effect is due to the major component, gomisin N (Figure 5.1) [46].

5.7 CONCLUSION

Food function can be discussed by invoking the same mechanisms used for synthetic single-molecule drugs, that is, the drug–receptor interaction theory. Multifunctional properties of food factors thus can be explained by defining the diverse affinities of food factors for various receptors. Other characteristics contributing to the mode of food factor function include their antioxidant properties and the strong indirect interaction mechanism. Further studies are necessary to fully understand the essential functional differences between food factors and synthetic drugs.

REFERENCES

1. Leger, D. (2008). Scurvy reemergence of nutritional deficiencies. *Can Fam Physician* 54, 1403–1406.
2. Good, R.A., Fernandes, G., Yunis, E.J. et al. (1976). Nutritional deficiency, immunologic function, and disease. *Am J Pathol* 84, 599–614.
3. Carpenter, K.J. (2003). A short history of nutritional science. Part I–IV. *J Nutr* 133, 638–645; 975–984; 3023–3032; 3331–3342.
4. Mottillo, S., Filion, K.B., Genest, J. et al. (2010). The metabolic syndrome and cardiovascular risk: A systematic review and meta-analysis. *J Am Coll Cardiol* 56, 1113–1132.
5. De Heeredia, F.P., Gomez-Martinez, S., and Marcos, A. (2012). Obesity, inflammation and the immune system. *Proc Nutr Soc* 71, 332–338.
6. Kim, K.H. and Park, Y. (2011). Food components with anti-obesity effect. *Annu Rev Food Sci Technol* 2, 237–257.
7. Belobrajdic, D.P. and Bird, A.R. (2013). The potential role of phytochemicals in wholegrain cereals for the prevention of type-2 diabetes. *Nutr J* 12, 62.
8. Shahidi, F., Ho, C.-T., Watanabe, S., and Osawa, T. (2003). *Food Factors in Health Promotion and Disease Prevention.* Washington, DC: American Chemical Society.
9. Pan, M.H., Lai, C.S., and Ho, C.T. (2009). The third function (regulation of physiological function) of food for prevention of lifestyle-related diseases-close linkage to clinical examination. *Rinsho Byori*, 57, 1082–1089.
10. Perera, P.K. and Li, Y. (2012). Functional herbal food ingredients used in type 2 diabetes mellitus. *Pharmacogn Rev* 6, 37–45.
11. Russo, M., Spagnuolo, C., Tedesco, I., and Russo, G.L. (2010). Phytochemicals in cancer prevention and therapy: Truth or dare? *Toxins* 2, 517–551.
12. Newman, D.J. and Gregg, G.M. (2012). Natural products as sources of new drugs over the 30 years from 1981–2010. *J Nat Prod* 75, 311–335.
13. Fujiwara, Y., Kurihara, K., Ida, M. et al. (2011). Metabolomics-driven nutraceutical evaluation of diverse green tea cultivars. *PLoS ONE* 6, e23426.
14. Arai, S., Osawa, T., Ohigashi, H. et al. (2001). A mainstay of functional food science in Japan—History, present status, and future outlook. *Biosci Biotechnol Biochem* 65, 1–13.
15. Chioi, B.D. and Choi, Y.J. (2012). Nutraceutical functionalities of polysaccharides from marine invertebrates. *Adv Food Nutr Res* 65, 11–30.
16. Frisardi, V., Pana, F., Seripa, D. et al. (2010). Nutraceutical properties of Mediterranean diet and cognitive decline: Possible underlying mechanisms. *J Alzheimers Dis* 22, 715–740.
17. Zuchi, C., Ambrosio, G., Luscher, T.F., and Landmesser, U. (2010). Nutraceuticals in cardiovascular prevention: Lessons from studies on endothelial function. *Cardiovasc Ther* 28, 187–201.
18. Rudkowska, I. and Jones, P.J. (2007). Functional foods for the prevention and treatment of cardiovascular diseases: Cholesterol and beyond. *Expert Rev Cardiovasc Ther* 5, 477–490.
19. Uenishi, K. (2010). Nutrition and bone health. Food for specified health use (FOSHU) and bone health. *Clin Calcium* 20, 116–20.
20. Okada, M. and Minamishima, Y. (1987). Protective effect of biological response modifiers on murine cytomegalovirus infections. *Microbial Immunol* 31, 435–447.
21. Hyman, M. (2007). System biology, toxins, obesity, and functional medicine. *Altern Ther Health Med* 13, 5134–5139.

22. Bhuyan, B.J. and Mugesh, G. (2012). Antioxidant activity of peptide-based angiotensin converting enzyme inhibitors. *Org Biomol Chem* 10, 2237–2247.
23. Dong, J., Xu, X., Liang, Y. et al. (2011). Inhibition of angiotensin converting enzyme (CEA) activity by polyphenols from tea (*Camellia sinensis*) and links to processing method. *Food Funct* 2, 310–319.
24. Balasuriya, N. and Rupasinghe, H.P. (2012). Antihypertensive properties of flavonoid-rich apple peel extract. *Food Chem* 135, 2320–2325.
25. Stefek, M. (2010). Natural flavonoids as potential multifunctional agents in prevention of diabetic cataract. *Interdiscip Toxicol* 4, 69–77.
26. Anand, P., Thomas, S.G., Kunnumakkara, A.B. et al. (2008). Biological activities of curcumin and its analogues (congeners) made by man and mother nature. *Biochem Pharmacol* 76, 1590–1611.
27. Jocob, J.K., Tiwari, K., Correa-Betanzo, J. et al. (2012). Biochemical basis for functional ingredient design from fruits. *Annu Rev Food Sci Technol* 3, 79–104.
28. Konishi, T. (2012). Complimentary use of antioxidant dietary factor is promised in cancer treatment. *Chemotherapy* 1, e107.
29. Bhatti, F., Mankhey, R.W., Asico, L. et al. (2005). Mechanisms of antioxidant and prooxidant effects of alpha-lipoic acid in the diabetic and non-diabetic kidney. *Kidney Int* 67, 1371–1380.
30. Bindoli, A., Rigobello, M.P., and Galzigna, L. (1989). Toxicity of aminochromes. *Toxicol Lett*, 48, 3–20.
31. Asanuma, M., Miyazaki, I., and Ogawa, N. (2003). Dopamine- or L-DOPA-induced neurotoxicity: The role of dopamine quinone formation and tyrosinase in a model of Parkinson's disease. *Neurotox Res* 5, 165–176.
32. Narayan, B.H., Tatewaki, N., Girdharan, V. et al. (2010). Modulation of doxorubicin-induced genotoxicity of squalene in Balb/c mice. *Food Funct* 1, 174–179.
33. Spagnuolo, C., Russo, M., Bilotto, S. et al. (2012). Dietary polyphenols in cancer prevention: The example of the flavonoid quercetin in leukemia. *Ann N Y Acad Sci* 1259, 95–103.
34. Nishida, H., Tatewaki, N., Nakajima, Y. et al. (2009). Inhibition of ATR protein kinase activity by Schisandrin B in DNA damage response. *Nucleic Acid Res* 37, 5678–5689.
35. Kawakami, K., Nishida, H., Tatewaki, N. et al. (2011). Persimmon leaf extract inhibits the ATM activity during DNA damage response induced by Doxorubicin in A549 lung adenocarcinoma cells. *Biosci Biotechnol Biochem* 75, 650–655.
36. Shiloh, Y. (2001). ATM and ATR: Networking cellular responses to DNA damage. *Curr Opin Genet Dev* 11, 71–77.
37. Abraham, R.T. (2001). Cell cycle checkpoint signaling through the ATM and ATR kinases. *Genes Dev* 15, 2177–2196.
38. Chen, T., Stephens, P.A., Middleton, F.K., and Curtin, N.J. (2012). Targeting the S and G2 checkpoint to treat cancer. *Drug Discov Today* 17, 194–202.
39. On, K.F., Chen, H.T., Ma, J.P. et al. (2011). Determinants of mitotic catastrophe on abrogation of the G2 DNA damage checkpoint by UCN-01. *Mol Cancer Ther* 10, 784–794.
40. Prevo, R., Fokas, R., Reaper, P.M. et al. (2012). The novel ATR inhibitor VE-821 increases sensitivity of pancreatic cancer cells to radiation and chemotherapy. *Cancer Biol Ther* 13, 1072–1081.
41. Lee, T.H., Jung, C.H., and Lee, D.H. (2012). Neuroprotective effects of Schinsandrin B against transient focal cerebralischemia in Sprague–Dawley rats. *Food Chem Toxicol* 50, 4239–4245.
42. Lam, P.Y. and Ko, K.M. (2011). (-)Schisandrin B ameliorates paraquat-induced oxidative stress by suppressing glutathione depletion and enhancing glutathione recovery in differentiated PC12 cells. *Biofactors* 37, 51–57.
43. Checker, R., Patwardhan, R.S., Sharma, D. et al. (2012). Schisandrin B exhibits anti-inflammatory activity through modulation of the redox-sensitive transcription factors Nrf2 and NFkB. *Free Radic Biol Med* 53, 1421–1430.
44. Giridharan, V.V., Thandavarayan, R.A., Konishi, T. et al. (2011). Prevention of scopolamine-induced memory deficits by Schisandrin B, an antioxidant lignin from *Schisandra chinensis* in mice. *Free Radic Res* 45, 950–958.
45. Giridharan, V.V., Thandavarayan, R.A., Narayan, B.H. et al. (2012). Schisandrin B attenuates cisplatin-induced oxidative stress, genotoxicity and neurotoxicity through modulating NFkB pathway in mice. *Free Radic Res* 46, 40–50.
46. Tatewaki, N., Nishida, H., Yoshida, M. et al. (2013). Differential effect of Schisandrin B stereoisomers on ATR-mediated DNA damage checkpoint signaling. *J Pharmacol Sci* 122, 138–148.

6 Anti-Inflammatory and Chemopreventive Potentials of Citrus Auraptene

Akira Murakami

CONTENTS

6.1 BIOACTIVE COMPOUNDS IN CITRUS FRUITS

Citrus fruits are well known to contain numerous secondary metabolites, which vary in their chemical structures and biological functions (Figure 6.1). These include monoterpenes (D-limonene, etc.), triterpenes (limonoids), flavonoids (nobiletin, hesperidin, etc.), coumarins (auraptene, bergamottin, etc.), and carotenoids (β-carotene, β-cryptoxanthin, etc.). There is also ample evidence that these citrus components have notable physiological functions, such as antioxidative, anti-inflammatory, and anticarcinogenesis activities. For example, D-limonene was reported to be an inhibitor of the oncogenic Ras farnesyltransferase,[1] while limonoids induced apoptosis in human neuroblastoma cells.[2] In addition, hesperidin and naringin, both flavanone-*O*-glycosides, were shown to attenuate experimental colitis[3,4] and prevent tumor promotion, but not initiation, in mouse skin.[5] More strikingly, β-cryptoxanthin has attracted the attention of many researchers ranging from basic scientists to epidemiologists, Nishino et al. the first to suggest that this agent has a more pronounced cancer preventive potential than β-carotene,[6] which was initially believed to play a major role in the chemopreventive activity of green–yellow vegetables and fruits, though the outcome of a clinical trial was puzzling.[7] Some epidemiological surveys have implied that β-cryptoxanthin is beneficial for cancer prevention,[8–10] while rodent experiments that explored the chemopreventive potential of β-cryptoxanthin also provided encouraging findings showing that this unique carotenoid markedly suppressed chemical carcinogenesis in the urinary bladder,[11] lung,[12] tongue,[13] liver,[13] and colon.[13,14] Moreover, polymethoxylated flavonoids, such as nobiletin, have shown pronounced chemopreventive activities in several different animal models of prostate,[15] colon,[15–17] skin,[18] and stomach[19] cancers. As noted later, citrus auraptene[20] has also been reported to show marked physiological functions, especially chemopreventive activities, unique molecular mechanisms, and distinct profiles of metabolism and absorption.

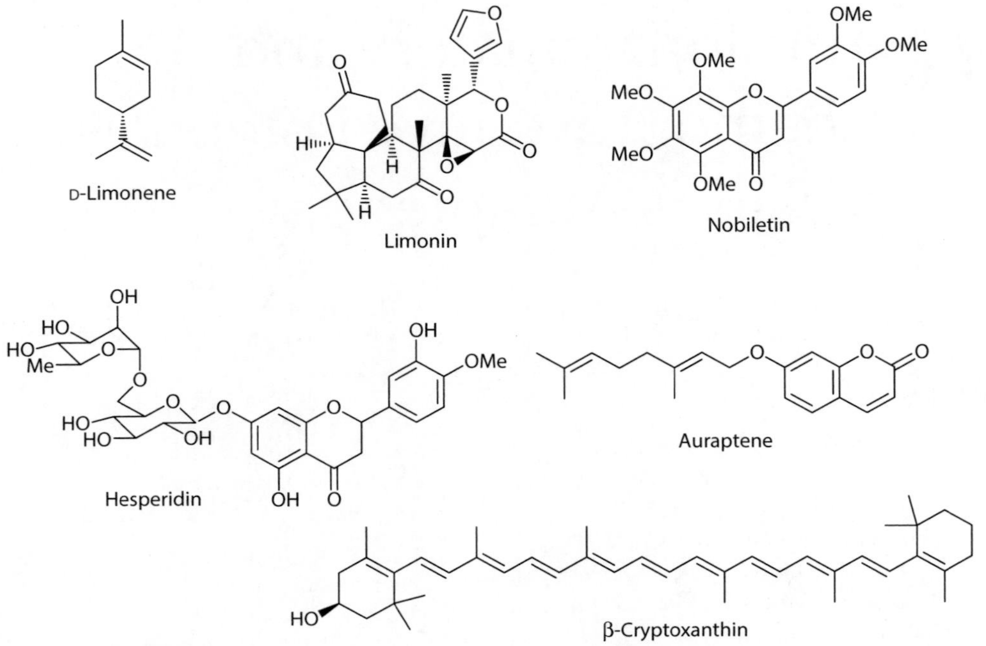

D-Limonene

Limonin

Nobiletin

Hesperidin

Auraptene

β-Cryptoxanthin

FIGURE 6.1 Chemical structures of citrus compounds that have shown notable physiological activities.

6.2 CHEMICAL CHARACTERISTICS OF COUMARINS

Auraptene belongs to a group of coumarin-type of phytochemicals. Coumarins, widely distributed throughout the plant kingdom, have a two-piece ring consisting of a C_6–C_3 structure with an ester moiety. Oxidation of the coumarin ring, for example, by introduction of epoxy and hydroxyl groups, results in a chemical variety by extending the side chains of the alkyl and alkenyl groups. One of the prominent side-chain groups is formed through a tandem connection with an isopentenyl group, which is the chemical unit for terpenoids. The *Citrus* genus has been well described as an abundant source of coumarins with isopentenyl moieties. As described later, coumarins with an isopentenyl moiety have distinct biological activities that are presumably derived from their higher molecular hydrophobicity.

6.2.1 Anti-Inflammatory Activities

Inflammation is a pathophysiological phenomenon seen in numerous diseases, and each human organ has the potential for disease with an inflammatory condition that is essential to its etiology. Of importance, a considerable proportion of chronic inflammatory diseases display an overlap with the onset and development of cancer, such as ulcerative colitis reflux esophagitis, Barrett's esophagus, hepatitis, and gastritis. It is also well documented that infection with microorganisms is closely related to inflammation-derived carcinogenesis.[21] In the process of inflammation, neutrophils and monocytes mature and are recruited, and then infiltrate the inflamed tissue for inducing and producing proinflammatory molecules, such as reactive oxygen and nitrogen species, prostaglandins (PGs), inflammatory cytokines, and chemokines, as well as others. Nitric oxide (NO) is released at high levels by inducible NO synthase (iNOS), along with the formation of stoichiometric amounts of L-citrulline from L-arginine. iNOS-mediated excessive and prolonged NO generation has attracted the attention of researchers on account of the relevance to epithelial carcinogenesis.[22,23] However, there is a large body of evidence showing that cyclooxygenase (COX)-2 expression is involved in the development of certain cancers.[24,25] In contrast to COX-1, COX-2 activity is inducible and its elevation enhances the biosynthesis of PGs, including PGE_2, which is one of the physiologically active and stable PGs produced in the pathways downstream of COX enzymes. PGE_2 is known to stimulate Bcl-2 activity and therefore prevent apoptosis for the development of tumor cell growth.[26] Thus, COX-2-targeted drugs are now considered to be promising for anticancer strategies, though some side effects have recently been reported.[27] However, superoxide anion radical (O_2^-) is a free radical generated through nicotinamide adenine dinucleotide phosphate (NADPH) oxidase, dominantly present in leukocytes, as well as from xanthine oxidase in epithelial cells, and may be subsequently converted into more reactive intermediates such as the hydroxyl radical responsible for DNA mutations.

Auraptene is notably able to attenuate phorbol ester- and endotoxin-induced expressions of both iNOS and COX-2 protein for immune cell inactivation.[28] Interestingly, auraptene was found to not disrupt the levels of iNOS/COX-2 messenger RNA (mRNA) expression, but prevented protein synthesis.[29,30] In addition, this agent suppressed phorbol ester-induced O_2^- generation in differentiated HL-60 cells, which is used as a model for neutrophils.[29,30] It is important to note that auraptene has no potential for free radical scavenging activity, while it suppresses oxidative stress by preventing free radical generation from stimulated leukocytes. In support of these *in vitro* observations, topical application of auraptene significantly suppressed phorbol ester-induced skin inflammation in mice.[29] Interestingly, umbelliferone, an analogous coumarin lacking an isopentenyl moiety, has been found to be virtually inactive in many experimental systems,[29,30] indicating the importance of the presence of a side chain for exhibiting activity.

Meanwhile, Takeda et al. recently presented an interesting observation that feeding of auraptene for seven weeks to Mongolian gerbils suppressed gastric inflammation induced by *Helicobacter pylori* infection.[31] *Helicobacter pylori* is one of the most widespread human pathogens infecting 50% of individuals throughout the world and associated with development of chronic gastritis, as well as gastric and duodenal ulcers.[32] Although two different antibiotics (amoxicillin and clarithromycin) and proton pump inhibitors are widely used in antimicrobial therapy for infected patients, cases of clarithromycin resistance are increasing and treatment failure is an emerging concern, especially in Asian countries.[33,34] Adherent *H. pylori* strains are able to survive in gastric mucosa, colonize at high densities, and also recolonize, while nonadherent strains are readily removed.[35] Such adhesion to gastric epithelial cells is recognized as one of the initial and critical steps in the development of gastritis, as it leads to injection of the definitive virulence factor cytotoxin-associated antigen A.[36]

CD74 was initially characterized for its role in regulating major histocompatibility complex (MHC) class II folding and intracellular sorting, and has been studied in detail in antigen-presenting cells. CD74 plays a crucial role in antigen presentation, as MHC class II processing and regulation do not properly occur in its absence.[37] Interestingly, CD74 was also reported to be a high-affinity binding protein for the multifunctional proinflammatory cytokine, macrophage migration inhibitory factor (MIF).[38–42] This cytokine binds to the extracellular domain of CD74, while its interaction with CD44 is required for activating proliferative and proinflammatory signaling molecules, such as extracellular signal-regulated kinases 1 and 2 (ERK1/2) (Figure 6.2).[43] Recently, Beswick et al. suggested that urease β binds to CD74 in gastric epithelial cells and induces nuclear factor-kappa B activation, thereby stimulating interleukin (IL)-8 production.[44,45] In addition, cell surface-associated urease binds to MHC class II molecules, including CD74, for inducing apoptosis.[46]

We have recently speculated that suppression of cell surface CD74 would be a novel strategy for *H. pylori* regulation and food phytochemicals capable of downregulating CD74 expression may serve as a novel type of anti-*H. pylori* agent. As a result, we established a novel cell-based enzyme-linked immunosorbent assay (ELISA) system for evaluating CD74 expression in NCI-N87 gastric carcinoma cells and screened a total of 25 phytochemicals for their CD74 suppressive activities.[47] Interestingly, several citrus compounds, including bergamottin, nobiletin, and auraptene, were found to be highly suppressive of CD74 protein expression.[47]

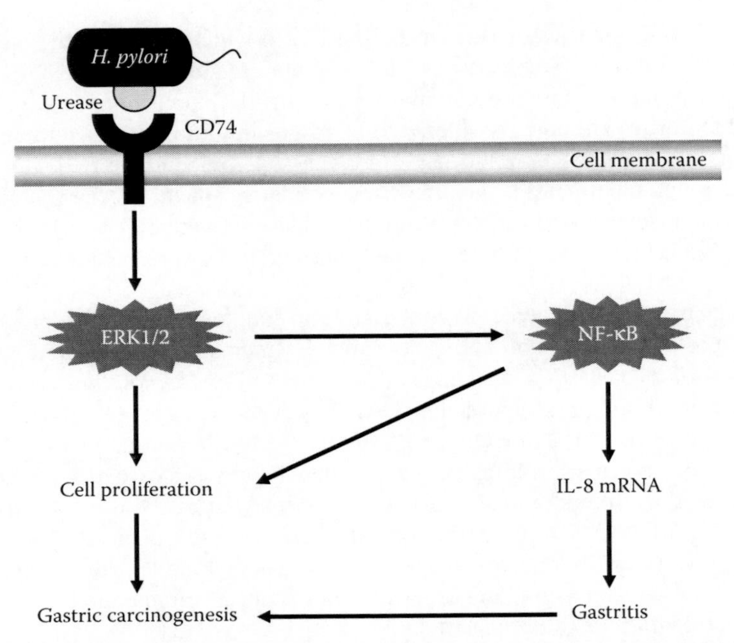

FIGURE 6.2 Role of CD74 of gastric cells in their attachment to *Helicobacter pylori*. ERK1/2, extracellular signal-regulated kinases 1 and 2.

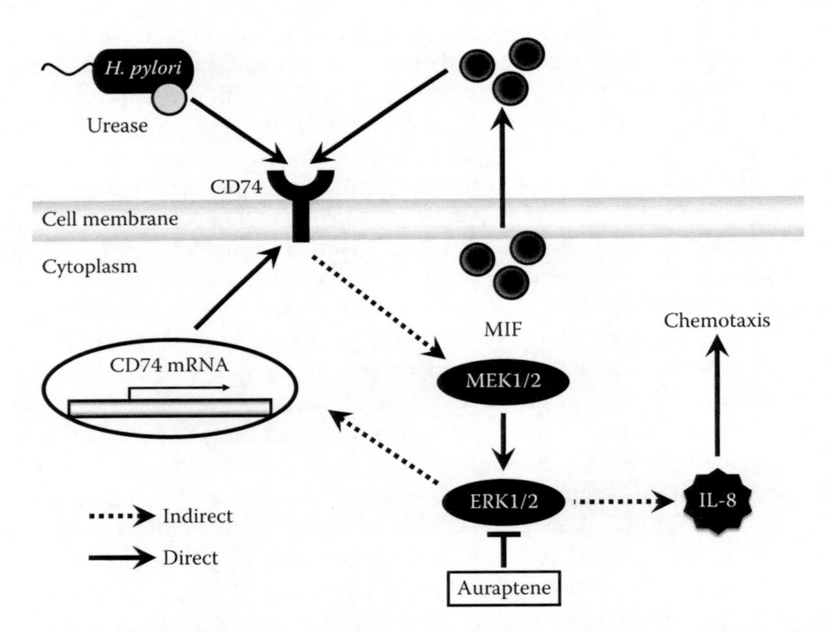

FIGURE 6.3 Molecular mechanisms by which auraptene attenuates CD74 protein expression in NCI-N87 human gastric cancer cells. ERK1/2, extracellular signal-regulated kinases 1 and 2; MEK1/2, mitogen-activated protein kinases 1 and 2; MIF, macrophage migration inhibitory factor.

In addition, auraptene was shown to disrupt serum starvation-induced ERK1/2 activation, and attenuated *H. pylori* adhesion and IL-8 production in a co-culture system using NCI-N87 cells and *H. pylori* (Figure 6.3).[48] Moreover, our recent *in vivo* studies showed that oral administration of auraptene inhibited *H. pylori*-induced expression and/or production of CD74, MIF, IL-1β, and tumor necrosis factor-α in gastric mucosa, together with serum macrophage inhibitory protein 2.[49] Taken together, auraptene can be described as a promising anti-*H. pylori* agent with a unique molecular mechanism.

6.2.2 Chemopreventive Activities

We previously reported that topical applications of auraptene[30] suppressed the tumor promotion stage in two-stage carcinogenesis in mouse skin, while another group noted that colorectal cancer may be one of the best targets for therapy with auraptene.[50] Oral administration of auraptene (100 and 500 ppm in the basal diet) inhibited azoxymethane (AOM)-induced formation of aberrant crypt foci (ACF),[51] one of the histological tumor markers, as well as adenomas and adenocarcinomas[52] in rats. Along the same lines, auraptene and collinin, another isopentenyloxycoumarin, inhibited colitis-related colon carcinogenesis in mice.[53] Furthermore, Hayashi et al. found that dietary auraptene (250 ppm) reduced the numbers of AOM-induced ACF and β-catenin-accumulated crypts in C57BL/KsJ-db/db (db/db) mice with obese and diabetic phenotypes, as well as in wild-type mice.[54] In addition, two-stage carcinogenesis by AOM and dextran sulfate sodium (DSS) was markedly suppressed by auraptene in several independent studies.[53,55,56] However, auraptene (100 and 500 ppm) suppressed 4-nitroquinoline 1-oxide-induced oral carcinogenesis in rats,[57] while dietary supplementation with auraptene inhibited *N,N*-diethylnitrosamine-induced rat hepatocarcinogenesis,[58,59] which was accompanied by decreased β-catenin mutation.[59] Dietary auraptene was also shown to be effective for inhibiting the development of esophageal tumors by *N*-nitrosomethylbenzylamine when given during the initiation as well as postinitiation phases.[60] Moreover, Krishnan et al. found that auraptene (500 ppm) significantly delayed the median time to tumor formation by 39 days compared to an *N*-methyl nitrosourea only group and also observed decreased expression of cyclin D1 in Sprague–Dawley (SD) rats.[61] In addition to the results of chemical carcinogenesis experiments, Tang et al. used transgenic rats that developed adenocarcinomas of the prostate bearing the SV40 T-antigen transgene and reported that a diet containing 500 ppm auraptene decreased those carcinomas.[62] Also, Onuma et al. noted that dietary auraptene reduced not only the number of infiltrating cells but also the expression of iNOS and formation of 8-hydroxy-2′-deoxyguanine, an oxidative biomarker, in inflammatory cells, and therefore suppressed the growth of implanted colon tumor cells in mice.[63] Meanwhile, it is of interest to indicate that combination of all-*trans* retinoic acid with auraptene resulted in a synergistic decrease in formation of squamous cell carcinomas in a xenograft mouse model.[64] In addition, a recent study showed that auraptene suppressed the growth and sphere formation (i.e., surrogate tumors) of FOLFOX-resistant colon cancer cells, which are highly enriched in cancer stem cells.[65]

Tumor progression is characterized as the carcinogenesis stage, in which tumor invasion and metastasis become dominant. Tanaka et al. presented *in vivo* findings showing that dietary supplementation with auraptene led to marked suppression of lung metastasis by melanoma cells in mice.[66] Although the mechanisms of action by which auraptene exhibits antitumor progression activity remain to be fully elucidated, we found that it remarkably inhibited the production of a tumor-invasive and metastatic protein, pro-matrix metalloproteinase (proMMP-7), in HT-29 human colon cancer cells without affecting its mRNA expression level.[67] In addition, our mechanistic studies using small interfering RNA (siRNA) revealed that the phosphorylation levels of 4E-binding protein 1 (4EBP1) at Thr37/46 and Thr70, as well as eukaryotic translation initiation factor 4B (eIF4B) at Ser422, were substantially decreased though auraptene markedly dephosphorylated constitutively activated ERK1/2 for reducing the phosphorylation of eIF4B at Ser422 (Figure 6.4).[67] Furthermore, the agent attenuated the protein, but not mRNA, expression of proliferation- and apoptosis-associated genes, such as c-*myc*, cyclin-dependent kinase 4, and Bcl-xL.[67] Taken together, auraptene targets the translation stop of several tumor progression-related molecules including proMMP-7 by disrupting ERK1/2-mediated phosphorylation of 4EBP1 and eIF4B.[66] In addition, we found that dietary administration of auraptene at 100 ppm ameliorated DSS-induced colitis in mice, which was importantly accompanied with reduced production of MMPs.[68]

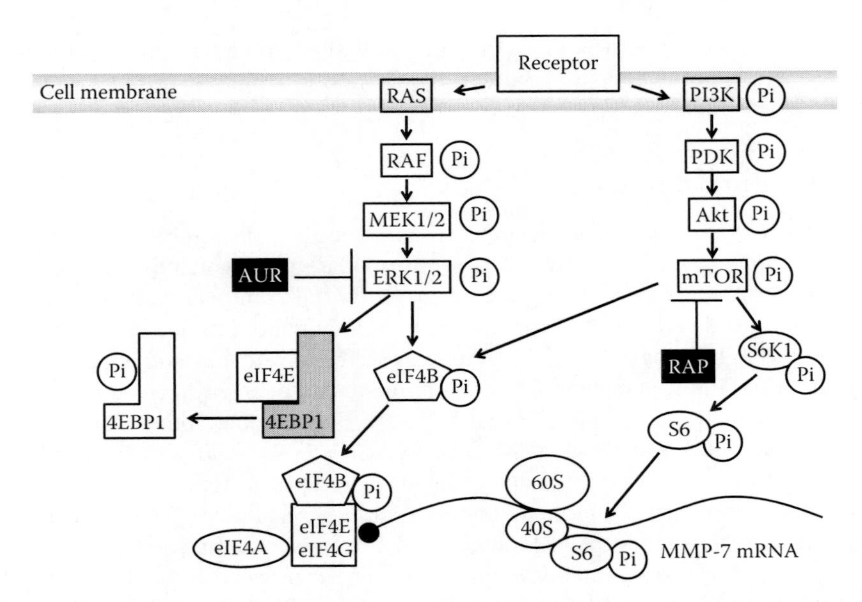

FIGURE 6.4 Molecular mechanisms by which auraptene attenuates proMMP-7 protein synthesis in HT-29 human colon adenocarcinoma cells. AUR, auraptene; ERK1/2, extracellular signal-regulated kinases 1 and 2; MEK1/2, mitogen-activated protein kinases 1 and 2; RAP, rapamycin.

6.2.3 Influence on Drug Metabolism

It is known that exogenous substances are metabolized by drug-metabolizing enzymes, which consist of those in the phase 1 (e.g., cytochrome P450s that add a hydrophilic functional group) and phase 2 (e.g., glutathione [GSH] *S*-transferase [GST], which provides GSH to that metabolized by phase 1 enzymes) groups. For example, most polycyclic aromatic hydrocarbons, called procarcinogens, are biologically inactive in their native structure and activated by a phase 1 enzyme, leading to formation of ultimate carcinogens that can bind to cellular DNA. Alternatively, activated carcinogens are subjected to the reaction by phase 2 enzymes to be inactivated and then excreted from the body. The Keap1/Nrf2 system adaptively functions to protect cells from oxidative and electrophilic damages. In a normal state, the transcription factor Nrf2 is continuously ubiquitinated by the Cul3–Keap1 ubiquitin E3 ligase complex and therefore rapidly subjected to degradation in proteasomes. Electrophilic chemicals and oxidative stresses oxidize the reactive cysteine residues of Keap1 in both direct and indirect manners.[69] This critical step stabilizes Nrf2, thereby inducing robust expressions of a battery of cytoprotective genes. However, in the phase 3 detoxification stage, P-glycoprotein, known as a plasma membrane glycoprotein, confers multidrug resistance to cells by virtue of its ability to exclude cytotoxic drugs in an adenosine triphosphate (ATP)-dependent manner. Previously, auraptene was found to be a selective phase 2 enzyme inducer as demonstrated in several different systems.[51,52,58,59,70] Importantly, Prince et al. reported that phase 2 enzymes were upregulated through the Keap1/Nrf2 system, while they also showed that auraptene-increased hepatic GST activity was significantly decreased in Nrf2(–/–) mice.[71] Thus, selective upregulation of phase 2 enzymes by auraptene is considered to be partially mediated through the Keap1/Nrf2 system for exhibiting its chemopreventive activities.

6.2.4 Bioavailability and Metabolism

Auraptene also has notable characteristics in regard to its metabolism and absorption. We previously reported that 7-ethoxycoumarin permeated the basolateral (portal vein) side of differentiated Caco-2 cells in a time-dependent manner, while auraptene slightly permeated the cells though intracellular accumulation was remarkable.[70] In addition, epoxyauraptene and umbelliferone were detected

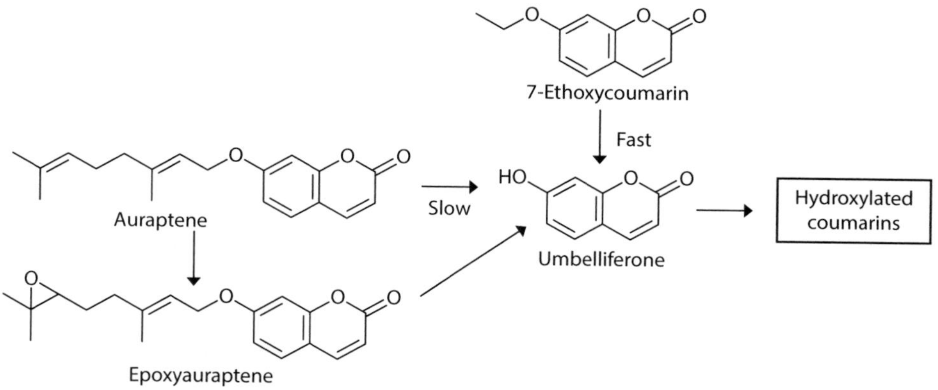

FIGURE 6.5 Metabolism of auraptene and 7-ethoxycoumarin in rat liver S-9 mixture.

when auraptene was treated with S-9 mixture from rat livers (Figure 6.5). In another experiment, 7-ethoxycoumarin was found to be converted to umbelliferone, though its $t^{1/2}$ value of two hours was much shorter than that of auraptene (more than 24 hours).[70] These findings suggest that auraptene, bearing a geranyloxyl side chain, is a relatively metabolism-resistant substrate for cytochrome P450 enzymes, and thus stable in the liver compared to 7-ethoxycoumarin. We also explored the metabolism characteristics of auraptene and 7-ethoxycoumarin in male SD rats,[72] and found that a single gastric intubation of auraptene, but not 7-ethoxycoumarin, led to significant localization in the liver from one to four hours, with comparable contents observed in the gastrointestinal tract. Following 7-ethoxycoumarin administration, treatments of serum and urinary samples with glucuronidase/sulfatase led to greater formation of umbelliferone compared to those with auraptene. Collectively, these results indicate that auraptene has a longer life span than 7-ethoxycoumarin in the body, while its marked chemopreventive effects toward colon carcinogenesis, which have been shown in various animal models, may be associated with its notable localizability in the gastrointestinal tract.

6.3 CONCLUSION

There is a great body of evidence from epidemiological surveys showing the beneficial effects of fresh fruits on human health promotion and disease prevention. However, other approaches directed specifically toward citrus fruits are unfortunately still limited, though some highly encouraging findings obtained in an interesting epidemiological survey were recently reported.[73] Based on accumulated results, there is no doubt that citrus fruits and their constituents are attractive and useful materials from the viewpoints of both basic and applied science, as well as for development of physiologically functional foods. In particular, auraptene has been shown to be a promising chemopreventive phytochemical, as demonstrated by results of experiments with a variety of independent animal models. Furthermore, it should be noted that we have established a method for preparing auraptene in large quantities from citrus peel oil.[74] However, there is scarce information regarding potential side effects of this phytochemical, which is indispensable prior to development for human use. One of the distinct characteristics of citrus fruits, compared to those of other foods, is their variety of active constituents in terms of chemical characteristics and bioactivities. Thus, combination studies using different types of citrus components, including auraptene for enhancing each efficacious characteristic, are warranted.

ACKNOWLEDGMENTS

The author expresses his thanks to Professors Takuji Tanaka (The Tohkai Cytopathology Institute: Cancer Research and Prevention) and Hajime Ohigashi (Fukui Prefectural University) for their long-time collaboration. The study mentioned in this chapter was partly supported by a Grant-in-Aid for Scientific Research (C) (A.M., #23580164).

REFERENCES

1. Gould, M.N., Moore, C.J., Zhang, R. et al. 1994. Limonene chemoprevention of mammary carcinoma induction following direct in situ transfer of v-Ha-ras. *Cancer Research*, 54: 3540–3543.
2. Poulose, S.M., Harris, E.D., Patil, B.S. 2005. Citrus limonoids induce apoptosis in human neuroblastoma cells and have radical scavenging activity. *Journal of Nutrition*, 135: 870–877.
3. Jung, U.J., Lee, M.K., Jeong, K.S. et al. 2004. The hypoglycemic effects of hesperidin and naringin are partly mediated by hepatic glucose-regulating enzymes in C57BL/KsJ-db/db mice. *Journal of Nutrition*, 134: 2499–2503.
4. Crespo, M.E., Gálvez, J., Cruz, T. et al. 1999. Anti-inflammatory activity of diosmin and hesperidin in rat colitis induced by TNBS. *Planta Medica*, 65: 651–653.
5. Berkarda, B., Koyuncu, H., Soybir, G. et al. 1998. Inhibitory effect of hesperidin on tumour initiation and promotion in mouse skin. *Research in Experimental Medicine*, 198: 93–99.
6. Tsushima, M., Maoka, T., Katsuyama, M. et al. 1995. Inhibitory effect of natural carotenoids on Epstein–Barr virus activation activity of a tumor promoter in Raji cells. A screening study for anti-tumor promoters. *Biological & Pharmaceutical Bulletin*, 18: 227–233.
7. Nowak, R. 1994. Cancer prevention. Beta-carotene: Helpful or harmful? *Science*, 264: 500–501.
8. Zhang, J., Dhakal, I., Stone, A. et al. 2007. Plasma carotenoids and prostate cancer: A population-based case-control study in Arkansas. *Nutrition and Cancer*, 59: 46–53.
9. Stram, D.O., Yuan, J.M., Chan, K.K. et al. 2007. Beta-cryptoxanthin and lung cancer in Shanghai, China—An examination of potential confounding with cigarette smoking using urinary cotinine as a biomarker for true tobacco exposure. *Nutrition and Cancer*, 57: 123–129.
10. Yuan, J.M., Stram, D.O., Arakawa, K. et al. 2003. Dietary cryptoxanthin and reduced risk of lung cancer: The Singapore Chinese Health Study. *Cancer Epidemiology, Biomarkers & Prevention*, 12: 890–898.
11. Miyazawa, M., Okuno, Y., Fukuyama, M. et al. 1999. Antimutagenic activity of polymethoxyflavonoids from *Citrus aurantium*. *Journal of Agricultural and Food Chemistry*, 47: 5239–5244.
12. Kohno, H., Taima, M., Sumida, T. et al. 2001. Inhibitory effect of mandarin juice rich in beta-cryptoxanthin and hesperidin on 4-(methylnitrosamino)-1-(3-pyridyl)-1-butanone-induced pulmonary tumorigenesis in mice. *Cancer Letters*, 174: 141–150.
13. Tanaka, T., Tanaka, M., and Kuno, T. 2012. Cancer chemoprevention by citrus pulp and juices containing high amounts of β-cryptoxanthin and hesperidin. *Journal of Biomedicine & Biotechnology*, 2012: 516981.
14. Narisawa, T., Fukaura, Y., Oshima, S. et al. 1999. Chemoprevention by the oxygenated carotenoid beta-cryptoxanthin of *N*-methylnitrosourea-induced colon carcinogenesis in F344 rats. *Japanese Journal of Cancer Research*, 90: 1061–1065.
15. Tang, M.X., Ogawa, K., Asamoto, M. et al. 2011. Effects of nobiletin on PhIP-induced prostate and colon carcinogenesis in F344 rats. *Nutrition and Cancer*, 63: 227–233.
16. Suzuki, R., Kohno, H., Murakami, A. et al. 2004. Citrus nobiletin inhibits azoxymethane-induced large bowel carcinogenesis in rats. *BioFactors*, 22: 111–114.
17. Miyamoto, S., Yasui, Y., Tanaka, T. et al. 2008. Suppressive effects of nobiletin on hyperleptinemia and colitis-related colon carcinogenesis in male ICR mice. *Carcinogenesis*, 29: 1057–1063.
18. Murakami, A., Nakamura, Y., Torikai, K. et al. Inhibitory effect of citrus nobiletin on phorbol ester-induced skin inflammation, oxidative stress, and tumor promotion in mice. *Cancer Research*, 60: 5059–5066.
19. Minagawa, A., Otani, Y., Kubota, T. et al. 2001. The citrus flavonoid, nobiletin, inhibits peritoneal dissemination of human gastric carcinoma in SCID mice. *Japanese Journal of Cancer Research*, 92: 1322–1328.
20. Genovese, S. and Epifano, F. 2011. Auraptene: A natural biologically active compound with multiple targets. *Current Drug Targets*, 12: 381–386.
21. Coussens, L.M. and Werb, Z. 2002. Inflammation and cancer. *Nature*, 420: 860–867.
22. Nathan, C. and Xie, Q.W. 1994. Nitric oxide synthases: Roles, tolls, and controls. *Cell*, 78: 915–918.
23. Ohshima, H. and Bartsch, H. 1994. Chronic infections and inflammatory processes as cancer risk factors: Possible role of nitric oxide in carcinogenesis. *Mutation Research*, 305: 253–264.
24. Taketo, M.M. 1998. Cyclooxygenase-2 inhibitors in tumorigenesis (Part II). *Journal of the National Cancer Institute*, 90: 1609–1620.
25. Taketo, M.M. 1998. Cyclooxygenase-2 inhibitors in tumorigenesis (Part I). *Journal of the National Cancer Institute*, 90: 1529–1536.
26. Fosslien, E. 2000. Molecular pathology of cyclooxygenase-2 in neoplasia. *Annals of Clinical and Laboratory Science*, 30: 3–21.

27. Fujimura, T., Ohta, T., Oyama, K. et al. 2007. Cyclooxygenase-2 (COX-2) in carcinogenesis and selective COX-2 inhibitors for chemoprevention in gastrointestinal cancers. *Journal of Gastrointestinal Cancer*, 38: 78–82.
28. Murakami, A., Shigemori, T., and Ohigashi, H. 2005. Zingiberaceous and citrus constituents, 1′-acetoxychavicol acetate, zerumbone, auraptene, and nobiletin, suppress lipopolysaccharide-induced cyclooxygenase-2 expression in RAW264.7 murine macrophages through different modes of action. *Journal of Nutrition*, 135: 2987S–2992S.
29. Murakami, A., Nakamura, Y., Tanaka, T. et al. 2000. Suppression by citrus auraptene of phorbol ester-and endotoxin-induced inflammatory responses: Role of attenuation of leukocyte activation. *Carcinogenesis*, 21(10): 1843–1850.
30. Murakami, A., Kuki, W., Takahashi, Y. et al. 1997. Auraptene, a citrus coumarin, inhibits 12-O-tetradecanoylphorbol-13-acetate-induced tumor promotion in ICR mouse skin, possibly through suppression of superoxide generation in leukocytes. *Japanese Journal of Cancer Research*, 88: 443–452.
31. Takeda, K., Utsunomiya, H., Kakiuchi, S. et al. 2007. Citrus auraptene reduces *Helicobacter pylori* colonization of glandular stomach lesions in Mongolian gerbils. *Journal of Oleo Science*, 56: 253–260.
32. Parsonnet, J., Vandersteen, D., Goates, J. et al. 1991. *Helicobacter pylori* infection in intestinal- and diffuse-type gastric adenocarcinomas. *Journal of the National Cancer Institute*, 83: 640–643.
33. Debets-Ossenkopp, Y.J., Sparrius, M., Kusters, J.G. et al. 1996. Mechanism of clarithromycin resistance in clinical isolates of *Helicobacter pylori*. *FEMS Microbiology Letters*, 142: 37–42.
34. Maeda, S., Yoshida, H., Matsunaga, H. et al. 2000. Detection of clarithromycin-resistant *Helicobacter pylori* strains by a preferential homoduplex formation assay. *Journal of Clinical Microbiology*, 38: 210–214.
35. Hayashi, S., Sugiyama, T., Asaka, M. et al. 1998. Modification of *Helicobacter pylori* adhesion to human gastric epithelial cells by antiadhesion agents. *Digestive Diseases and Sciences*, 43: 56S–60S.
36. Hatakeyama, M. 2004. Oncogenic mechanisms of the *Helicobacter pylori* CagA protein. *Nature Reviews Cancer*, 4: 688–694.
37. Beswick, E.J. and Reyes, V.E. 2009. CD74 in antigen presentation, inflammation, and cancers of the gastrointestinal tract. *World Journal of Gastroenterology*, 15: 2855–2861.
38. Bacher, M., Metz, C.N., Calandra, T. et al. 1996. An essential regulatory role for macrophage migration inhibitory factor in T-cell activation. *Proceedings of the National Academy of Sciences of the United States of America*, 93: 7849–7854.
39. Calandra, T. and Roger, T. 2003. Macrophage migration inhibitory factor: A regulator of innate immunity. *Nature Reviews Immunology*, 3: 791–800.
40. Roger, T., David, J., Glauser, M.P. et al. 2001. MIF regulates innate immune responses through modulation of Toll-like receptor 4. *Nature*, 414: 920–924.
41. Kleemann, R., Hausser, A., Geiger, G. et al. 2000. Intracellular action of the cytokine MIF to modulate AP-1 activity and the cell cycle through Jab1. *Nature*, 408: 211–216.
42. Hudson, J.D., Shoaibi, M.A., Maestro, R. et al. 1999. A proinflammatory cytokine inhibits p53 tumor suppressor activity. *Journal of Experimental Medicine*, 190: 1375–1382.
43. Shi, X., Leng, L., Wang, T. et al. 2006. CD44 is the signaling component of the macrophage migration inhibitory factor-CD74 receptor complex. *Immunity*, 25: 595–606.
44. Beswick, E.J., Bland, D.A., Suarez, G. et al. 2005. *Helicobacter pylori* binds to CD74 on gastric epithelial cells and stimulates interleukin-8 production. *Infection and Immunity*, 73: 2736–2743.
45. Beswick, E.J., Pinchuk, I.V., Minch, K. et al. 2006. The *Helicobacter pylori* urease B subunit binds to CD74 on gastric epithelial cells and induces NF-kappaB activation and interleukin-8 production. *Infection and Immunity*, 74: 1148–1155.
46. Fan, X., Gunasena, H., Cheng, Z. et al. 2000. *Helicobacter pylori* urease binds to class II MHC on gastric epithelial cells and induces their apoptosis. *Journal of Immunology*, 165: 1918–1924.
47. Sekiguchi, H., Washida, K., and Murakami. A. 2008. Suppressive effects of selected food phytochemicals on CD74 expression in NCI-N87 gastric carcinoma cells. *Journal of Clinical Biochemistry and Nutrition*, 43: 109–117.
48. Sekiguchi, H., Irie, K., and Murakami, A. 2010. Suppression of CD74 expression and *Helicobacter pylori* adhesion by auraptene targeting serum starvation-activated ERK1/2 in NCI-N87 gastric carcinoma cells. *Bioscience, Biotechnology, and Biochemistry*, 74: 1018–1024.
49. Sekiguchi, H., Takabayashi, F., Irie, K. et al. 2012. Auraptene attenuates gastritis via reduction of *Helicobacter pylori* colonization and pro-inflammatory mediator production in C57BL/6 mice. *Journal of Medicinal Food*, 15: 658–663.

50. Epifano, F., Genovese, S., Carlucci, G. et al. 2012. Novel prodrugs for the treatment of colonic diseases based on 5-aminosalicylic acid, 4′-geranyloxyferulic acid, and auraptene: Biological activities and analytical assays. *Current Drug Delivery*, 9: 112–121.
51. Tanaka, T., Kawabata, K., Kakumoto, M. et al. 1997. Citrus auraptene inhibits chemically induced colonic aberrant crypt foci in male F344 rats. *Carcinogenesis*, 18: 2155–2161.
52. Tanaka, T., Kawabata, K., Kakumoto, M. et al. 1998. Citrus auraptene exerts dose-dependent chemopreventive activity in rat large bowel tumorigenesis: The inhibition correlates with suppression of cell proliferation and lipid peroxidation and with induction of phase II drug-metabolizing enzymes. *Cancer Research*, 58: 2550–2556.
53. Kohno, H., Suzuki, R., Curini, M. et al. 2006. Dietary administration with prenyloxycoumarins, auraptene and collinin, inhibits colitis-related colon carcinogenesis in mice. *International Journal of Cancer*, 118: 2936–2942.
54. Hayashi, K., Suzuki, R., Miyamoto, S. et al. 2007. Citrus auraptene suppresses azoxymethane-induced colonic preneoplastic lesions in C57BL/KsJ-db/db mice. *Nutrition and Cancer*, 58: 75–84.
55. Tanaka, T., Yasui, Y., Ishigamori-Suzuki, R. et al. 2008. Citrus compounds inhibit inflammation- and obesity-related colon carcinogenesis in mice. *Nutrition and Cancer*, 60(Suppl 1): 70–80.
56. Tanaka, T., de Azevedo, M.B., Durán, N. et al. 2010. Colorectal cancer chemoprevention by 2 beta-cyclodextrin inclusion compounds of auraptene and 4′-geranyloxyferulic acid. *International Journal of Cancer*, 126: 830–840.
57. Tanaka, T., Kawabata, K., Kakumoto, M. et al. 1998. Chemoprevention of 4-nitroquinoline 1-oxide-induced oral carcinogenesis by citrus auraptene in rats. *Carcinogenesis*, 19: 425–431.
58. Sakata, K., Hara, A., Hirose, Y. et al. 2004. Dietary supplementation of the citrus antioxidant auraptene inhibits *N*,*N*-diethylnitrosamine-induced rat hepatocarcinogenesis. *Oncology*, 66: 244–252.
59. Hara, A., Sakata, K., Yamada, Y. et al. 2005. Suppression of beta-catenin mutation by dietary exposure of auraptene, a citrus antioxidant, in *N*,*N*-diethylnitrosamine-induced hepatocellular carcinomas in rats. *Oncology Report*, 14: 345–351.
60. Kawabata, K., Tanaka, T., Yamamoto, T. et al. 2000. Suppression of *N*-nitrosomethylbenzylamine-induced rat esophageal tumorigenesis by dietary feeding of auraptene. *Journal of Experimental & Clinical Cancer Research*, 19: 45–52.
61. Krishnan, P., Yan, K.J., Windler, D. et al. 2009. Citrus auraptene suppresses cyclin D1 and significantly delays *N*-methyl nitrosourea induced mammary carcinogenesis in female Sprague–Dawley rats. *BMC Cancer*, 9: 259.
62. Tang, M., Ogawa, K., Asamoto, M. et al. 2007. Protective effects of citrus nobiletin and auraptene in transgenic rats developing adenocarcinoma of the prostate (TRAP) and human prostate carcinoma cells. *Cancer Science*, 98: 471–477.
63. Onuma, K., Suenaga, Y., Sakaki, R. et al. 2011. Development of a quantitative bioassay to assess preventive compounds against inflammation-based carcinogenesis. *Nitric Oxide*, 25: 183–194.
64. Kleiner-Hancock, H.E., Shi, R., Remeika, A. et al. 2010. Effects of ATRA combined with citrus and ginger-derived compounds in human SCC xenografts. *BMC Cancer*, 10: 394.
65. Epifano, F., Genovese, S., Miller, R. et al. 2012. Auraptene and its effects on the re-emergence of colon cancer stem cells. *Phytotherapy Research*, 27(5): 784–786.
66. Tanaka, T., Kohno, H., Murakami, M. et al. 2000. Suppressing effects of dietary supplementation of the organoselenium 1,4-phenylenebis(methylene)selenocyanate and the citrus antioxidant auraptene on lung metastasis of melanoma cells in mice. *Cancer Research*, 60: 3713–3716.
67. Kawabata, K., Murakami, A., and Ohigashi, H. 2006. Citrus auraptene targets translation of MMP-7 (matrilysin) via ERK1/2-dependent and mTOR-independent mechanism. *FEBS Letters*, 580: 5288–5294.
68. Kawabata, K., Murakami, A., and Ohigashi, H. 2006. Auraptene decreases the activity of matrix metalloproteinases in dextran sulfate sodium-induced ulcerative colitis in ICR mice. *Bioscience, Biotechnology, and Biochemistry*, 70: 3062–3065.
69. Kobayashi, M. and Yamamoto, M. 2005. Molecular mechanisms activating the Nrf2-Keap1 pathway of antioxidant gene regulation. *Antioxidants & Redox Signaling*, 7: 385–394.
70. Murakami, A., Wada, K., Ueda, N. et al. 2000. In vitro absorption and metabolism of a citrus chemopreventive agent, auraptene, and its modifying effects on xenobiotic enzyme activities in mouse livers. *Nutrition and Cancer*, 36: 191–199.
71. Prince, M., Li, Y., Childers, A. et al. 2009. Comparison of citrus coumarins on carcinogen-detoxifying enzymes in Nrf2 knockout mice. *Toxicology Letters*, 185: 180–186.
72. Kuki, W., Hosotani, K., and Ohigashi, H. 2008. Metabolism and absorption of auraptene (7-geranyloxylcoumarin) in male SD rats: Comparison with 7-ethoxycoumarin. *Nutrition and Cancer*, 60: 368–372.

73. Sugiura, M., Matsumoto, H., Kato, M. et al. 2004. Seasonal changes in the relationship between serum concentration of beta-cryptoxanthin and serum lipid levels. *Journal of Nutritional Science and Vitaminology (Tokyo)*, 50: 410–415.
74. Takahashi, Y., Inaba, N., Kuwahara, S. et al. 2002. Rapid and convenient method for preparing aurapten-enriched product from hassaku peel oil: Implications for cancer-preventive food additives. *Journal of Agricultural and Food Chemistry*, 50: 3193–3196.

7 Food Functions Preventing Muscle Atrophy

Junji Terao, Rie Mukai, Yui Yamashita, Arisa Ochi, Shohei Kohno, Katsuya Hirasaka, and Takeshi Nikawa

CONTENTS

7.1 INTRODUCTION

Skeletal muscle comprises 40%–50% of human body weight. Weight and strength of skeletal muscle decrease with aging. This phenomenon is known as sarcopenia and is assumed to be a major reason for bone fractures caused by accidental falls in bedridden people. The population of bedridden people is approaching nearly two million in Japan, as the nation is now facing the advent of a "super-aged" society earlier and more rapidly than any other country. Sarcopenia is therefore one of the serious problems to be resolved in an aging society. The decline in mechanical loading of the muscles by weightlessness, plaster casts, and bed rest also diminishes the skeletal muscle weight, similarly to sarcopenia. This so-called disuse muscle atrophy and sarcopenia involve a common mechanism underlying the loss of skeletal muscle. Both oxidative stress and disrupted insulin-like growth factor (IGF)-1 signaling are suggested to trigger the loss of muscle protein [1]. Disuse muscle atrophy occurs rapidly over a short period of time and the muscle weight can be recovered by gravitation loading. By contrast, sarcopenia progresses slowly year by year and its recovery appears to be difficult. The differences between sarcopenia and disuse muscle atrophy are shown in Table 7.1.

For example, disuse muscle atrophy mainly involves a decrease in slow twitch muscle fibers, whereas sarcopenia occurs in fast twitch muscle fibers. Understanding of each mechanism is essential for the application of functional food factors for the prevention and treatment of these muscle dysfunctions. In this chapter, we focus on disuse muscle atrophy and its prevention by dietary antioxidants and functional peptides.

7.2 IGF-1 SIGNALING AND MUSCLE ATROPHY

Under normal conditions, synthesis of muscle proteins and their degradation are well balanced, resulting in constant muscle weight [2,3]. IGF-1 plays an essential role in the growth of skeletal muscles. This polypeptide is synthesized in the liver, muscle cells, and osteoblasts, and promotes the growth of muscle and bone. At first, tyrosine kinase activity is elevated by the binding of IGF-1

TABLE 7.1
Differences between Disuse Muscle Atrophy and Sarcopenia

Item	Disuse muscle atrophy	Sarcopenia
Symptom	Acute	Chronic
Causes	Unloading, weightlessness	Aging, inflammation
Degree	Severe	Mild
Recovery	Reversible	Irreversible
Target muscle	Slow twitch fiber	Fast twitch fiber

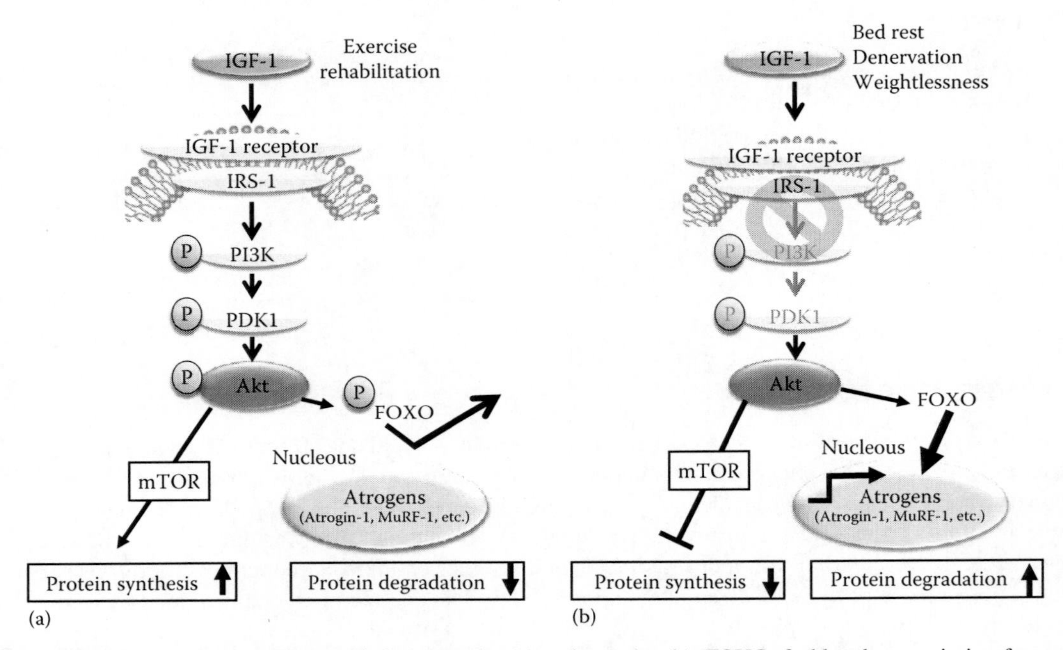

FIGURE 7.1 Mechanism for muscle hypertrophy (a) and atrophy (b). FOXO, forkhead transcription factor; IGF, insulin-like growth factor; IRS-1, insulin receptor substrate-1; mTOR, mammalian target of rapamycin; MuRF-1, muscle RING-finger protein-1; PDK1, phosphoinositide-dependent protein kinase 1; PI3K, phosphatidylinositol 3-kinase.

to its receptor, and then insulin receptor substrate (IRS)-1, phosphatidylinositol 3-kinase, and Akt are phosphorylated successively. Phosphorylated Akt induces protein synthesis through mammalian target of rapamycin (mTOR) signaling (Figure 7.1a). By contrast, protein degradation happens through the ubiquitin–proteasome, calcium-dependent calpain, or cathepsin-derived lysosomal system. Among them, the ubiquitin–proteasome system seems to play a major role in the protein degradation in disuse muscle atrophy [4] (Figure 7.1b).

The unloading state causes IGF-1 resistance in skeletal muscle, resulting in suppression of Akt phosphorylation. This event switches off the IGF-1 signal for protein synthesis. Instead, forkhead transcription factor (FOXO) escapes from phosphorylation, which is also located downstream of Akt phosphorylation, is translocated into the nucleus to induce expression of muscle atrophy-related genes, such as muscle RING-finger protein (MuRF)-1 and muscle atrophy F-box protein (MAFbx)/atrogin-1. MuRF-1 and MAFbx/atrogin-1 are ubiquitin ligases for myosin heavy chain and toroponin-1/calcineurin, respectively. Finally, degradation of muscle protein is provoked by the ubiquitination of these proteins and following breakdown in the proteasomes. Nikawa et al. [5]

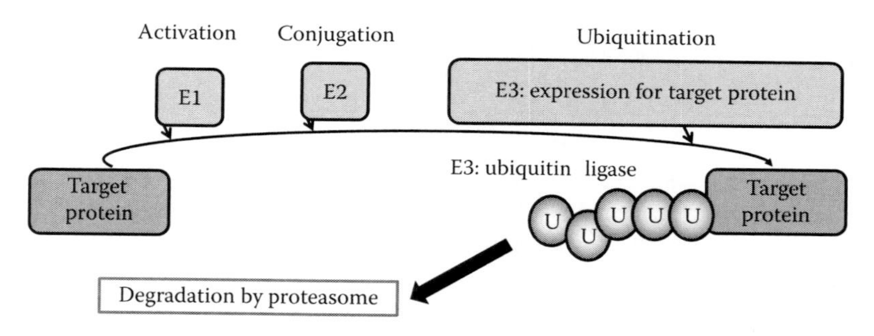

FIGURE 7.2 Mechanism of protein degradation via the ubiquitin–proteasome system in skeletal myotubes.

found in the space shuttle (STS-90) experiment that gene expression of ubiquitin ligase Cbl-b was markedly elevated in the atrophied muscle of rats under conditions of weightlessness. Thereafter, Cbl-b was identified as a negative regulator for IGF-1 signaling during muscle atrophy caused by unloading [6]. Cbl-b specifically binds to IRS-1, resulting in IGF-1 resistance to degradation of IRS-1 and inhibition of the signaling pathway located downstream of IRS-1. Therefore, Cbl-b is apparently one of the targets for functional food factors to prevent disuse muscle atrophy.

7.3 INHIBITOR OF UBIQUITIN LIGASE AND FUNCTIONAL PEPTIDES

The ubiquitin–proteasome system is illustrated in Figure 7.2. Target proteins are ubiquitinated by the enzymes of activation (El), conjugation (E2), and ubiquitination (E3: ubiquitin ligase). Ubiquitinated proteins are selectively hydrolyzed by 26S proteasomes. Cbl-b catalyzes the transfer of ubiquitin molecules from E2 enzyme to IRS-1 selectively.

As discussed in Subsections 7.1 and 7.2, disuse muscle atrophy may be prevented and/or treated by inhibiting ubiquitination activity of Cbl-b. This idea is based on the fact that IRS-1 selectively acts as the target protein of enhanced ubiquitination activity of Cbl-b in an unloading state. Cbl-b recognizes the steric structure of the target protein for its binding and ubiquitination. Nakao et al. [6] demonstrated that a synthetic pentapeptide, DGpYMP, can inhibit ubiquitination of IRS-1 by Cbl-b in a cell-free ubiquitination assay system. This pentapeptide, which is named as Cblin, inhibited IRS-1 degradation in the skeletal muscle of denervated mice. Thus, selective inhibition of ubiquitination of IRS-1 seems to be a powerful tool for the prevention of unloading-associated muscle atrophy. In this sense, functional peptides possessing Cblin-like activity should be screened from foods and foodstuffs. It is well known that a variety of functional peptides have been developed as functional foods in which some peptides for lowering blood pressure have been permitted as Food for Specified Health Uses (FOSHU) [7]. Table 7.2 summarizes the functional peptides derived from food proteins [8]. A soy protein, 11S glycinin, possesses an inhibitory peptide against Cbl-b ubiquitin ligase. Furthermore, Abe et al. [9] recently reported that 11S glycinin diet significantly prevented denervation-induced muscle loss and inhibited the ubiquitination and degradation of IRS-1 in murine muscle. This study predicts that the intake of soy protein is helpful for prevention of disuse muscle atrophy. It is reported that feeding meals containing soy protein after exercise stimulates protein synthesis and translation initiation in the skeletal muscle of male rats [10]. This implies that soybean serves as a powerful foodstuff to prevent disuse muscle atrophy.

7.4 OXIDATIVE STRESS AND ANTIOXIDANTS

Oxidative stress is a critical factor to stimulate the expression of muscle atrophy-related genes in the skeletal muscle. It is known that muscle atrophy induced by unloading stress and plaster cast is accompanied by the generation of reactive oxygen species (ROS) [11]. In fact, thiobarbituric acid reactive substance (TBARS), a specific biomarker of lipid peroxidation, and oxidized glutathione are

TABLE 7.2
Functional Peptides Derived from Food Proteins

Peptide	Source	Physiological Activity
	Opioid Peptide	
β-Casomorphin	Casein	Suppression of intestinal motility
Gluten exorphin A	Gluten	Improvement of learning
Rubiscolin	Rubisco	Improvement of learning
	Vasorelaxing Peptide	
Ovokinin	Ovalbumin	Antihypertension
	Immunostimulating Peptide	
Soypeptide	Glycinin	Immunostimulation, protection from hair loss
	Inhibitor for ACE	
FFVAPFPEVFGK	Casein	Antihypertension
IPP, VPP	Casein	Antihypertension
LKPNM	Fish protein	Antihypertension
	Modulator for Intestinal Absorption	
Casein phosphopeptide	Casein	Improvement of calcium absorption
Soybean HMF	Soybean protein	Hypocholesterolemic effect
	Antimicrobial Peptide	
Lactoferricin	Lactoferrin	Anti-infection
	Antioxidative Peptide	
LLPHH	Soybean protein	Antioxidation

ACE, angiotensin-converting enzyme; HMF, high-molecular-weight fraction.

elevated with the decrease in reduced glutathione in the atrophied muscle of rats after immobilization via osmotic pumps [12]. It is known that mitochondrial dysfunction occurs in the skeletal muscle under unloading conditions [13]. Therefore, ROS that escape from the mitochondrial antioxidant system may induce disuse muscle atrophy. Powers et al. [14] proposed a specific mechanism for the increase in mitochondrial production of ROS induced by skeletal muscle inactivity: mitochondrial uptake of calcium, depressed protein transport into the mitochondria, increased mitochondrial level of fatty acid hydroperoxides, and increased mitochondrial fission. Mitochondrial release of ROS results in activation of the intracellular signaling pathway involving nuclear factor kappa B (NF-κB), activator protein 1 (AP-1), and FOXO transcription factors, which increase both protein degradation and cytotoxicity in the skeletal muscle [15] (Figure 7.3).

It is suggested that ROS stimulate protein catabolism in skeletal muscle by upregulating the ubiquitin conjugation systems [16]. Bhattacharya et al. [17] suggested that fatty acid hydroperoxides released from denervated muscle mitochondria are an important determinant of muscle atrophy.

Activities of antioxidant enzymes involving superoxide dismutase, catalase, and glutathione peroxidase are reduced in muscular atrophy [18]. By contrast, supplementation of the antioxidant nutrient cysteine prevents unloading-induced ubiquitination in association with redox regulation in rat skeletal muscle [19]. Our preliminary study using 3D clinorotation of cultured myotubes, a model of unloading stress, indicated that antioxidative flavonoids, epicatechin, epicatechin gallate, epigallocatechin gallate, and quercetin can attenuate the expression of (MAFbx)/atrogin-1 and MuRF-1 during unloading by attenuating the mitogen-activated protein kinase pathway [20].

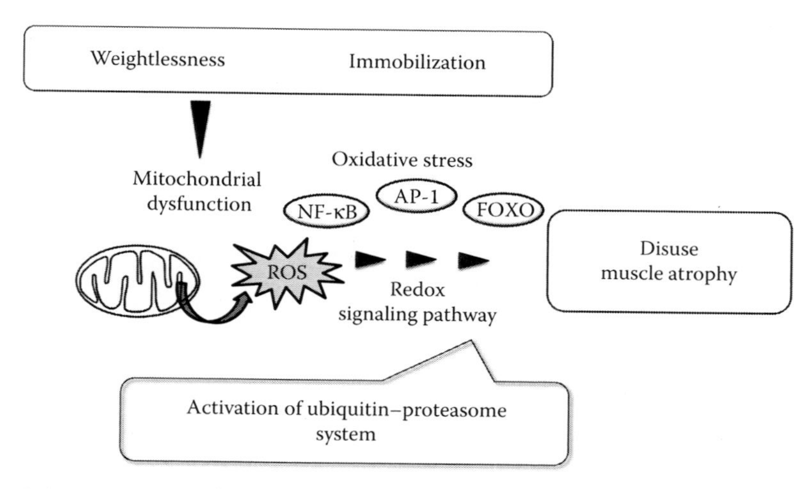

FIGURE 7.3 Relationship between oxidative stress and muscle protein degradation contributing to skeletal muscle dysfunction. AP-1, activator protein-1; FOXO, forkhead transcription factor; ROS, reactive oxygen species.

TABLE 7.3
Effect of Dietary Antioxidants on Muscular Dysfunctions

Antioxidant	Animal	Method	Main Result	Reference
Trolox	C57BL6 mouse	Hindlimb suspension	Inhibition of oxidative stress	[22]
Vitamin E	Wistar rat	Hindlimb suspension	Prevention of muscle weight decrease	[23]
Resveratrol	*mdx* mouse	Animal model of Duchenne muscular dystrophy	Inhibition of oxidative stress	[24]
Resveratrol	Rat	Hindlimb suspension	Inhibition of oxidative stress	[25]
Green tea catechins	BALB/c mice	Hindlimb suspension	Prevention of muscle force decrease	[26]
Epigallocatechin gallate	C57BL6 mouse	Tumor bearing	Suppression of NF-κB signal	[27]
Green tea extract	*mdx* mouse	Animal model of Duchenne muscular dystrophy	Suppression of the necrosis of fast twitch muscle	[28]
Epigallocatechin gallate	*mdx* mouse	Animal model of Duchenne muscular dystrophy	Improvement of muscular dystrophy	[29]
Vitamin C	Wistar rat	Overload the plantaris muscle	Prevention of overload-induced skeletal muscle hypertrophy	[30]

Furthermore, quercetin as well as *N*-acetylcysteine prevented unloading-derived disuse muscle atrophy by attenuating the induction of ubiquitin ligase (MAFbx)/atrogin-1 in tail suspension mice [21]. In this study, both antioxidants suppressed the increase in TBARS in the skeletal muscle, indicating that they prevent disuse muscle atrophy when concentrated in the target muscle. Table 7.3 summarizes studies that investigated the preventive effect of dietary antioxidants on muscle dysfunction.

7.5 PRENYLFLAVONOIDS FOR MUSCLE ATROPHY

Prenylflavonoids are flavonoid derivatives containing one or more isoprene (dimethylallyl) units. Flavanones, flavones, and chalcones are major classes of prenylflavonoids in the plant kingdom. These lipophilic flavonoids are frequently found in Leguminosae, Moraceae, and Asteraceae, and are ingredients of Oriental medicine [31,32]. There are many reports on the biological activities

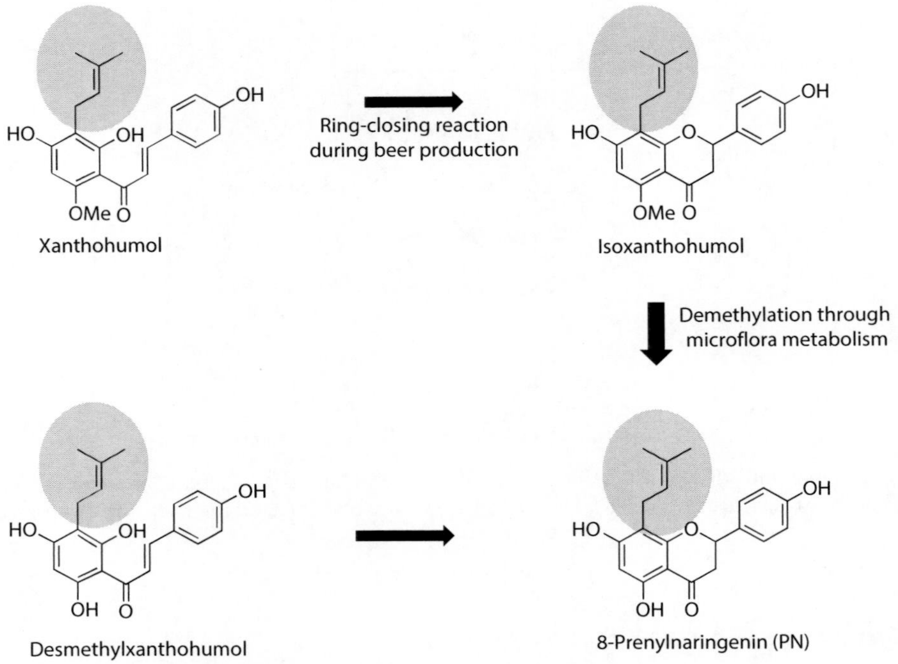

FIGURE 7.4 Production of 8-PN during the process of beer manufacture.

of specific prenylflavonoids, such as inhibition of ABC transporter-dependent drug efflux for 6-prenylchrysin [33], apoptosis induction and antioxidant effects for xanthohumol [34,35], and cell cycle arrest for icaritin [36]. A cell culture experiment demonstrated that melanin biosynthesis inhibitory activity of luteolin was strengthened by prenylation at its 3-position [37]. Introduction of a prenyl group to the basic diphenylpropane of flavonoid structure may enhance the cellular uptake by increasing the lipophilicity, resulting in promotion of its biological effect. Therefore, it seems to be of interest to estimate the effect of prenylation on the biological activity of flavonoids in relation to disuse muscle atrophy.

Estrogenic activity may affect recovery from disuse muscle atrophy by accelerating the Akt-signaling pathway [38]. Kretzschmar et al. [39] found that estrogenicity of naringenin was enhanced by prenylation at its 8-position [40]. This 8-prenylnaringenin (8-PN) is suggested to be a potent phytoestrogen present in hops (*Humulus lupulus*) [40]. 8-PN can be produced from desmethyl-xanthohumol by the process of beer manufacture or formed from isoxanthohumol by demethylation through microflora metabolism [41] (Figure 7.4).

The inhibitory effect of 8-PN and that of nonprenylated narigenin (N) on disuse muscle atrophy were compared by adding them to the diet of denervated mice [42]. Consumption of 8-PN, but not N, prevented the loss of gastrocnemius muscle weight. Consumption of 8-PN inhibited the induction of (MAFbx)/atrogin-1 protein content in accordance with the activation of Akt phosphorylation. It should be noted that 8-PN content in the target muscle was much higher than that of N. Therefore, 8-PN is likely to be effectively concentrated into the skeletal muscle to exert its preventive effect. Antioxidant activity may not be essential for this prenylflavonoid to exert its preventive effect, although modulation of the IGF-1-signaling pathway is definitively involved in the mechanism of action. This *in vivo* study stimulated the screening of the novel biological functions of prenylflavonoids.

7.6 SUMMARY

Oligopeptides derived from soybean glycinin and a prenylflavonoid present in hops possess the potential to exert a preventive effect on disuse muscle atrophy by attenuating the activity and/or induction of ubiquitin ligases responsible for protein degradation in the skeletal muscle. It is likely that daily consumption of such functional food factors is helpful in the prevention of disuse muscle atrophy, although human intervention studies should be performed to obtain evidence for the benefit of application of food factors to bedridden people. Drug treatment may cause serious side effect in patients suffering from muscle dysfunction. Functional food factors are a promising tool in the prevention of disuse muscle atrophy and other muscular disorders.

ACKNOWLEDGMENT

This work was supported by the Program for Promotion of Basic and Applied Researches for Innovations in Bio-Oriented Industry in Japan.

REFERENCES

1. Perrini, S., Laviola, L., Carreira, M.C., Cignarelli, A., Natalicchio, A., and Giorgino, F.J. 2010. The GH/IGF1 axis and signaling pathways in the muscle and bone: Mechanism underlying age-related skeletal muscle wasting and osteoporosis. *J. Endocrinol.* 205:201–210.
2. Sadowski, C.L., Wheeler, T.T., Wang, L.H., and Sadowski, H.B. 2001. GH regulation of IGF-1 and suppressor of cytokine signaling gene expression in C2C12 skeletal muscle cells. *Endocrinology* 142:3890–3900.
3. Goldspink, G., Williams, P., and Simpson, H. 2002. Gene expression in response to muscle stretch. *Clin. Orthop. Relat. Res.* 403:146–152.
4. Sandri, M., Sandri, C., Gilbert, A., Skurk, C., Calabria, E., Picard, A., Walsh, K., Schiaffino, S., Lecker, S.H., and Goldberg, A.L. 2004. Foxo transcription factors induce the atrophy-related ubiquitin ligase atrogin-1 and cause skeletal muscle atrophy. *Cell* 117:399–412.
5. Nikawa, T., Ishidoh, K., Hirasaka, K., Ishihara, I., Ikemoto, M., Kano, M., Kominami, E. et al. 2004. Skeletal muscle gene expression in space-flown rats. *FASEB J.* 18:522–524.
6. Nakao, R., Hirasaka, K., Goto, J., Ishidog, K., Yamada, C., Ohno, A., Okumura, Y. et al. 2009. Ubiquitin ligase Cbl-b is a negative regulator for insulin-like growth factor 1 signaling during muscle atrophy caused by unloading. *Mol. Cell. Biol.* 29:4798–4811.
7. Arai, S., Yasuoka, A., and Abe, K. 2008. Functional food science and food for specified health use policy in Japan: State of art. *Curr. Opin. Lipidol.* 19:69–73.
8. Arai, S., Osawa, T., Ohigashi, H., Yoshikawa, M., Kaminogawa, S., Watanabe, M., Ogawa, T. et al. 2001. A mainstay of functional food science in Japan-history, present status, and future outlook. *Biosci. Biotechnol. Biochem.* 65:1–13.
9. Abe, T., Kohno, S., Yama, T., Ochi, A., Suto, T., Hirasaka, K., Ohno, A. et al. 2013. Soy glycinin contains a functional inhibitory sequence against muscle atrophy-associated ubiquitin ligase Cbl-b. *Int. J. Endocrinol.* doi:10.1155/2013/907565.
10. Anthony, T.G., McDaniel, B.J., Knoll, P., Bunpo, P., Paul, G.L., and McNurlan, M.A. 2007. Feeding meals containing soy or whey protein after exercise stimulates protein synthesis and translation initiation in the skeletal muscle of male rats. *J. Nutr.* 137:357–362.
11. Bar-Shai, M., Carmeli, E., Ljubuncic, P., and Reznick, A.Z. 2008. Exercise and immobilization in aging animals: The involvement of oxidative stress and NF-κB activation. *Free Radic. Biol. Med.* 44:202–214.
12. Kondo, H., Miura, M., and Itokawa, Y. 1993. Antioxidant enzyme systems in skeletal muscle atrophied by immobilization. *Pflügers Arch.* 422:404–406.
13. Bonetto, A., Penna, F., Mascaritoli, M., Minero, V., Fanelli, F.R., Baccino, F.M., and Costelli, P. 2009. Are antioxidants useful for treating skeletal muscle atrophy? *Free Radic. Biol. Med.* 47:906–916.
14. Powers, S.K., Wiggs, M.P., Duarte, J.A., Zergeroglu, A.M., and Demirel, H.A. 2012. Mitochondrial signaling contributes to disuse muscle atrophy. *Am. J. Physiol. Endocrinol. Metab.* 303:E31–E39.
15. Powers, S.K., Smuder, A.J., and Criswell, D.S. 2011. Mechanistic links between oxidative stress and disuse muscle atrophy. *Antioxid. Redox Signal.* 15:2519–2528.

16. Li, Y.P., Chen, Y., Li, A.S., and Reid, M.B. 2009. Hydrogen peroxide stimulates ubiquitin-conjugating activity and expression of genes for specific E2 and E3 proteins in skeletal muscle myotubes. *Am. J. Physiol. Cell Phyisol.* 285:C806–C812.
17. Bhattacharya, A., Muller, F.L., Liu, Y., Sabia, M., Liang, H., Song, W., Jang, Y.C., Ran, Q., and Van Remmen, H. 2009. Denervation induces cytosolic phospholipase A2-mediated fatty acid hydroperoxide generation by muscle mitochondria. *J. Biol. Chem.* 284:46–55.
18. Linke, A., Adamas, V., Schulze, P.C., Erbs, S., Gielen, S., Fiehn, E., Mobius-Winker, S., Schubert, A., Schuler, G., and Hambrecht, R. (2005). Antioxidative effects of exercise training patients with chronic heart failure: Increase in radical scavenger enzyme activity in skeletal muscle. *Circulation* 111:1763–1770.
19. Ikemoto, M., Nikawa, T., Kano, M., Hirasaka, K., Kitano, T., Watanabe, C., Tanaka, R., Yamamoto, T., Kamada, M., and Kishi, K. 2002. Cysteine supplementation prevents unweighting-induced ubiquitination in association with redox regulation in rat skeletal muscle. *Biol. Chem.* 383:715–721.
20. Hemdan, D.I., Hirasaka, K., Nakao, R., Kohno, S., Kagawa, S., Abe, T., Harada-Sukeno, A. et al. 2009. Polyphenols prevent clinorotation-induced expression of atrogenes in mouse C2C12 skeletal myotubes. *J. Med. Invest.* 56:26–32.
21. Mukai, R., Nakano, R., Yamamoto, H., Nikawa, T., Taekda, E., and Terao, J. 2010. Quercetin prevents unloading-derived disused muscle atrophy by attenuating the induction of ubiquitin ligases in tail-suspension mice. *J. Nat. Prod.* 73:1708–1710.
22. Desaphy, J.-F., Pierno, S., Liantonio, A., Giannuzzi, V., Digennaro, C., Dinardo, M.M., Camerino, G.M. et al. 2010. Antioxidant treatment of hindlimb-unloaded mouse counteracts fiber type transition but not atrophy of disused muscles. *Pharmacol. Res.* 61:553–563.
23. Sevais, S., Letexier, D., Favier, R., Duchamp, C., and Desplanches, D. 2007. Prevention of unloading-induced atrophy by vitamin E supplementation: links between oxidative stress and soleus muscle proteolysis. *Free Radic. Biol. Med.* 42:627–635.
24. Hori, Y.S., Kuno, A., Hosoda, R., Tanno, M., Miura, T., Shimamoto, K., and Horio, Y. 2011. Resveratrol ameliorates muscular pathology in the dystrophic max mouse, a model for Dunchenne muscular dystrophy. *J. Pharmacol. Exp. Ther.* 338:784–794.
25. Jackson, J.R., Ryan, M.J., Hao, Y., and Always, S.E. 2010. Mediation of endogenous antioxidant enzymes and apoptotic signaling by resveratrol following muscle disuse in the gastrocnemius muscles of young and old rats. *Am. J. Physiol. Regul. Integr. Comp. Physiol.* 299:R1572–R1581.
26. Ota, N., Soga, S., Haramizu, S., Yokoi, Y., Hase, T., and Murase, T. 2011. Tea catechins prevent contractile dysfunction in unloaded murine soleus muscle: A pilot study. *Nutrition* 27:955–959.
27. Wang, H., Lai, Y.J., Chan, Y.L., Li, T.L., and Wu, C.J. 2011. Epigallocatechin-3-gallate effectively attenuates skeletal muscle atrophy caused by cancer cachexia. *Cancer Lett.* 305:40–49.
28. Buetler, T.M., Renard, M., Offord, E.A., Schneider, H., and Ruegg, U. 2002. Green tea extract decreases muscle necrosis in *mdx* mice and protects against reactive oxygen species. *Am. J. Clin. Nutr.* 75:749–753.
29. Nakae, Y., Dorchies, O.M., Stoward, P.J., Zimmermann, B., Ritter, C., and Ruegg, U. 2012. Quantitative evaluation of the beneficial effects in the *mdx* mouse of epigallocatechin gallate, an antioxidant polyphenol from green tea. *Histochem. Cell Biol.* 137:811–827.
30. Nakanae, Y., Kawada, S., Sasaki, K., Nakazato, K., and Ishii, N. 2013. Vitamin C administration attenuates overload-induced skeletal muscle hypertrophy in rats. *Acta Physiol.* 208:57–65.
31. Barron, D. and Ibrahim, R.K. 1996. Isoprenylated flavonoids—A survey. *Phytochemistry* 43:921–982.
32. Botta, B., Vitali, A., Menendez, P., Misiti, D., and Delle Monache, G. 2005. Prenylated flavonoids: Pharmacology and biotechnology. *Curr. Med. Chem.* 12:713–739.
33. Ahmed-Belkacem, A., Pozza, A., Munoz-Martinez, F., Bates, S.E., Castanys, S., Gamarro, F., Di Pietro, A., and Perez-Victoria, J.M. 2005. Flavonoid structure-activity studies identify 6-prenylchrysin and tectochrisin as potent and spcific inhibitors of breast cancer resistance protein ABCG2. *Cancer Res.* 65:4852–4860.
34. Pan, L., Becker, H., and Gerhauser, C. 2005. Xanthhumol induces apoptosis in cultured 40-16 human colon cancer cells by activation of the death receptor- and mitochondrial pathway. *Mol. Nutr. Food Res.* 49:837–843.
35. Stevens, J.F., Miranda, C.L., Frei, B., and Buhler, D.R. 2003. Inhibition of peroxynitrite-mediated LDL oxidation by prenulated flavonoids: The alpha, beta-unsaturated keto functionality of 2′-hydroxychalcones as a novel antioxidant pharmacophore. *Chem. Res. Toxicol.* 16:1277–1286.
36. Huang, X., Zhu, D., and Lou, Y. 2007. A novel anticancer agent, icaritin, induced cell growth inhibition, G1 arrest and mitochondrial transmembrane potential drop in human prostate carcinoma PC-3 cells. *Eur. J. Pharmacol.* 564:26–36.

37. Arung, E.T., Shimizu, K., Tanaka, H., and Kondo, R. 2010. 3-Prenyl luteolin, a new prenylated flavone with melanin biosynthesis inhibitory activity from wood of *Artocarpus heterophyllus*. *Fitoterapia* 81:640–643.
38. McClung, J.M., Davis, J.M., Wilson, M.A., Goldsmith, E.C., and Carson, J.A. 2006. Estrogen status and skeletal muscle recovery from disuse atrophy. *J. Appl. Physiol.* 100:2012–2023.
39. Kretzschmar, G., Zierau, O., Wober, J., Tischer, S., Metz, P., and Vollmer, G. 2010. Prenylation has a compound specific effect on the estrogenicity of naringenin and genistein. *J. Steroid Biochem. Mol. Biol.* 118:1–6.
40. Milligan, S.R., Kalita, J.C., Pocock, V., Van de Kauter, V., and Stevens, J.F. 2000. The endocrine activities of 8-prenylnaringenin and related hop (*Humulus lupulus* L.) flavonoids. *J. Clinical Endocrinol. Metabolism* 85:4912–4915.
41. Stevens, J.F., Taylor, A.W., Clawson, J.E., and Deinzer, M.L. 1999. Fate of xanthohumol and related prenylflavonoids from hops to beer. *J. Agric. Food Chem.* 47:2421–2428.
42. Mukai, R., Horikawa, H., Fujikura, Y., Kawamura, T., Nemoto, H., Nikawa, T., and Terao, J. 2012. Prevention of disuse muscle atrophy by dietary ingestion of 8-prenylnaringenin in denervated mice. *PLoS One*. 7:e45048.

8 Effects of Nutraceuticals on Metabolic Syndromes

Asim K. Duttaroy

CONTENTS

8.1 INTRODUCTION

Metabolic syndrome (MetS) is a cluster of conditions such as central obesity, hypertension, insulin resistance, low-grade inflammation, fasting glucose, and dyslipidemia that occur together and increase the risk of cardiovascular disease (CVD), stroke, and diabetes [1–3]. A growing number of studies have highlighted the involvement of a chronic low-grade inflammatory state that often accompanies MetS as a major contributing factor both in the development of MetS and its associated pathophysiological consequences. Chronic inflammation in MetS is thought to originate from dysfunction of visceral adipose tissue. In fact, adipose tissue plays a critical role in CVD pathogenesis due to its functions as an endocrine organ producing several factors that modulate inflammation, intermediary metabolism, endothelial function, thrombosis/fibrinolysis, and blood pressure. Figure 8.1 shows the different features of MetS.

The widespread prevalence and deleterious effects of MetS have become a major public health challenge that needs to be addressed pharmacologically and/or changes lifestyle and diet [4,5]. Epidemiological evidence has suggested a protective role for diets such as Mediterranean diet low in saturated fats and rich in fruits and vegetables against the development and progression of CVD. A significant reduction was observed in people with MetS in central obesity with improvements in lipid profile, diastolic blood pressure, and Framingham Risk Scores after nutraceutical supplementation [6,7]. Since no single pathogenic pathway has yet to be identified as a therapeutic target in the MetS, current management addresses the various components of MetS individually by means of both lifestyle modifications (diet, physical activity, etc.) and pharmacological therapy. Nutraceuticals target the pathogenesis of MetS and its complications by favorably modulating biochemical and clinical endpoints [8]. These compounds include antioxidant vitamins, flavonoids, fatty acids, minerals, polyphenols, phytoestrogens, and dietary fibers. Nutraceuticals exert their cardioprotective effects via both antioxidant and nonantioxidant pathways [8–10]. For example, fruits, vegetables, oily seeds, and marine fish oils exert their lipid-lowering effects via inhibition

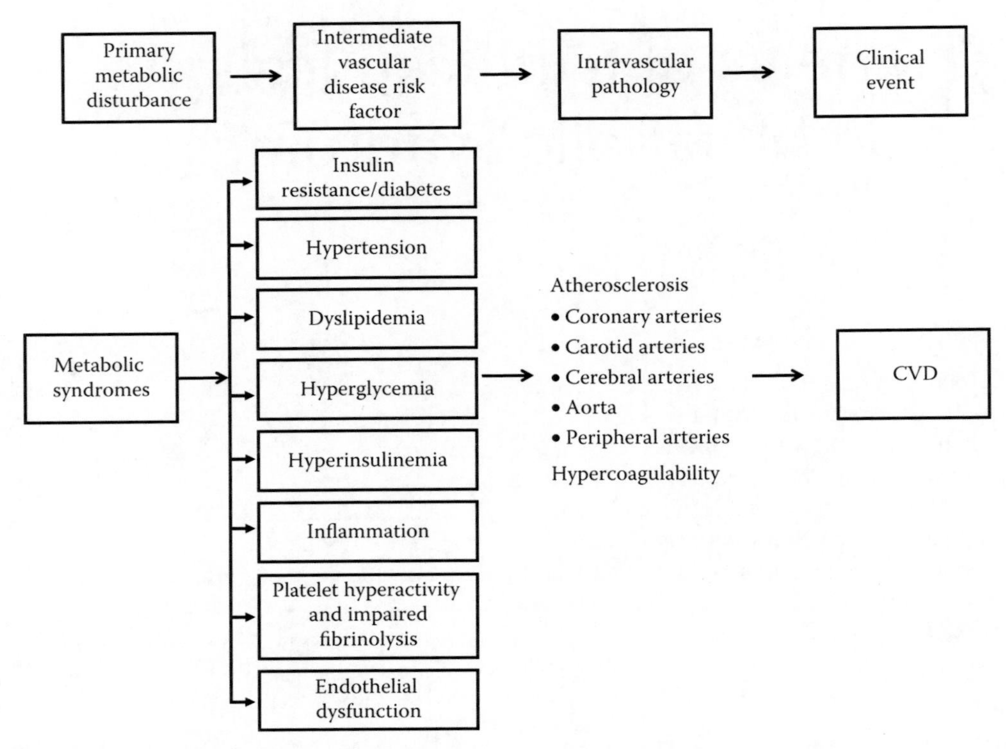

FIGURE 8.1 Multiple factors associated with metabolic syndromes give rise to an increased risk of cardiovascular disease.

of fat absorption and suppression of hepatic cholesterol synthesis. Lifestyle changes including weight control, healthy heart diet, and regular exercises have been proposed as first-line treatment to decrease CVD risks in MetS individuals. In addition, improving insulin resistance and glucose metabolism, controlling blood pressure as well as modulating dyslipidemia can also delay or reverse the progression of CVD in MetS. Several studies suggested that nutraceuticals may also provide prevention and/or treatment to the onset of the MetS and the occurrence of diabetes and CVD. The advances in the knowledge of these nutraceuticals may provide new avenues to develop dietary strategies to prevent the development of MetS. This chapter deals with only the roles of nutraceuticals in MetS (Figure 8.2).

8.2 INSULIN RESISTANCE AND NUTRACEUTICALS

Insulin resistance, defined as an impaired responsiveness of the body to insulin, is the critical factor for the development of MetS [11]. Insulin resistance is an important target for intervention well ahead of the development of diabetes and its cardiovascular complications. Hyperinsulinemia also causes a substantial increase in circulating noradrenaline concentration and reabsorption of sodium accompanied by an increase in blood pressure. The increased production of angiotensin II (AngII) by large insulin-resistant adipocytes results in excess lipids in the form of triglycerides that are stored in ectopic tissues, such as the liver or muscle cells [12]. This ectopic storage of lipids gives rise to insulin resistance and MetS. The high frequency of nutritional imbalances due to an overload in fat and/or refined carbohydrates is also leading causes of obesity, insulin resistance, and MetS [13]. By contrast, some food components may counteract insulin resistance by potentiating effects of insulin [14,15]. Supplementation with low-fat diet, phytosterols, soy protein including isoflavones, hop plant-derived rho iso-alpha acids, and proanthocyanidins improves insulin sensitivity [16,17]. Studies with myo-inositol-containing foods (barley, the fruit *Cucurbita ficifolia*, soy, and buckwheat) demonstrated its insulin-sensitizing effects in MetS, as well as improved polycystic ovary syndrome [18,19]. Grape seed extract (GSE) has various beneficial effects including antihyperlipidemic, insulin sensitivity, and anti-inflammatory. Bioactive compounds in GSE exhibit favorable antidiabetic effects by stimulating glucose uptake in insulin-sensitive cell lines and inhibit intestinal α-glucosidases, pancreatic α-amylase activities, and the process of glycation [20]. The hypoglycemic potential components in bitter melon have been identified as glycosides, saponins, alkaloids, triterpenes, polysaccharides, proteins, and steroids [21]. Several components such as saponins or charantins were isolated from bitter melon and their effects as hypoglycemic chemicals are being studied [22].

8.3 INFLAMMATION AND NUTRACEUTICALS

Chronic inflammation in MetS is originated from dysfunction of adipose tissue. Numerous studies have demonstrated increased circulating biomarkers of inflammation in MetS [23]. The relationship between inflammation and MetS is supported by several studies, as is the relationship between increased visceral fat mass and MetS [24]. A simultaneous increase in proinflammatory cytokines and a decrease in anti-inflammatory cytokines have been observed in MetS [25]. Proinflammatory cytokines induce hypertriglyceridemia and insulin resistance, which are the main components of MetS. Furthermore, low interleukin (IL)-10 production has been associated with high plasma glucose and dyslipidemia that predispose individuals to the MetS and diabetes, while the opposite situation would confer protection. Polyphenols, which are abundant in fruits and vegetables, have shown antioxidant, anti-inflammatory, and hypolipidemic effects [26]. Moreover, polyphenols inhibit the expression and secretion of proinflammatory compounds in several cell lines [27]. Grapes contain numerous polyphenols, such as flavans, anthocyanins, flavonols, and stilbenes (e.g., resveratrol), shown to prevent low-density lipoprotein (LDL) oxidation, oxidative stress, dyslipidemia, and inflammation [26]. These effects may potentially ameliorate MetS [28]. Simultaneous evaluation of the effects of polyphenols on plasma and mRNA expression of inflammation, oxidative stress markers, and peripheral sources of NO are required in MetS. A Mediterranean dietary

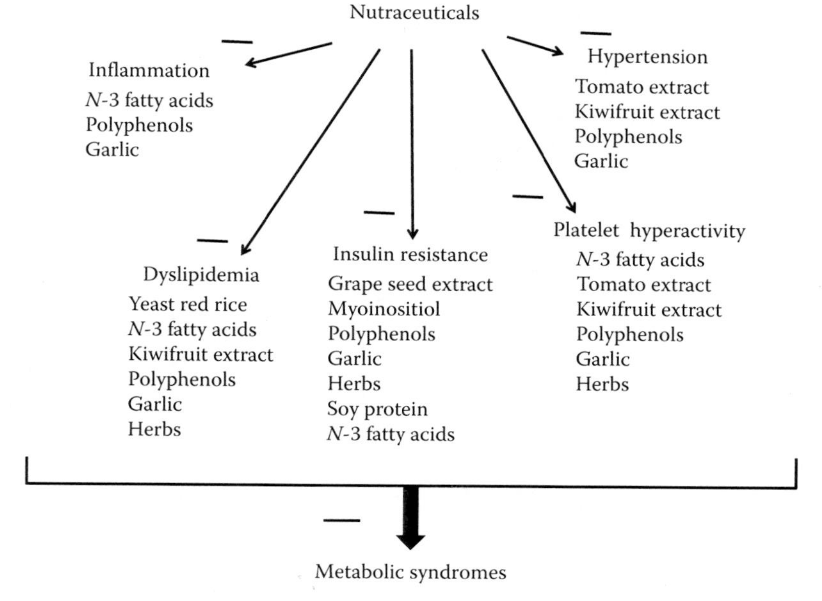

FIGURE 8.2 Targets of nutraceuticals on different pathways of metabolic syndromes. "—" represents, downregulation.

pattern has been associated with a reduced risk of the MetS in several cross-sectional studies. Few studies have confirmed the benefits of following a Mediterranean diet on MetS risk in obese individuals and in those with the MetS. The evidence, albeit limited, for a link between healthy diets based on other diet quality scores and the MetS supports a similar inverse association for the primary and secondary prevention of the MetS. Adhering to healthy diets containing polyphenols such as the Mediterranean diet can reduce inflammation and the MetS.

8.4 HYPERTENSION AND NUTRACEUTICALS

Hypertension leads to endothelial dysfunction as the main pathophysiological basis leading to an increased atherosclerosis and a prothrombotic state in MetS. Hyperinsulinemia also causes an increase in blood pressure. Certain blood pressure-lowering drugs such as mainly thiazides and β-blockers may induce insulin resistance and increase the risk of hyperglycemia [29,30]. Hypertensive patients on β-blocker therapy had a 28% higher risk of diabetes [31]. A meta-analysis of 22 clinical trials showed new onset of diabetes mellitus after antihypertensive therapy with β-blockers [32]. Angiotensin-converting enzyme inhibitors (ACEIs) and AngII receptor blockers (ARBs) have been demonstrated to be the better option to treat hypertensive patients with insulin resistance and/or diabetes [33]. The American Diabetes Association recommends the use of ACEIs or ARBs in hypertensive and diabetic patients. Almost all of the renin–angiotensin–aldosterone system (RAAS) genes (e.g., AGT, REN, ACE, AT1R, AT2R) are expressed in adipose tissue [34–36]. The genes encoding for both ACE and AngII receptor, type 1 or AT_1 receptor expression are significantly upregulated in obese and hypertensive patients, thus increasing the formation of AngII and its effects transmitted by the AT_1 receptor [35]. Obese patients have higher circulating renin, angiotensinogen, aldosterone, and ACE, suggesting that RAAS is significantly induced in MetS [35]. ACE, a carboxyl-terminal dipeptidyl exopeptidase responsible for vasoconstriction, plays a crucial role in the regulation of blood pressure as well as cardiovascular function. ACE converts decapeptide AngI to octapeptide AngII, a potent vasoconstrictor, and inactivates the vasodilator bradykinin and endogenous opioid peptide met-enkephalin. ACE inhibitors play a role in regulating the rennin–angiotensin system (RAS) by preventing the formation of AngII and decreasing the metabolism of bradykinin, leading to systematic dilation of the arteries and veins and a decrease in arterial blood pressure. In addition, inhibiting AngII formation diminishes AngII-mediated aldosterone secretion from the adrenal cortex, leading to a decrease in water and sodium reabsorption, and a reduction in extracellular volume, which further reduces blood pressure.

ACEIs have been characterized from many common food sources, such as milk, fish, soybean, cereal, fruits, and vegetables [37]. They provide advantageous cardiovascular health by inhibiting ACE activity, reducing free radical formation, inducing vasorelaxation, and lowering of blood pressure and lipid levels. These bioactive nutraceuticals are particularly attractive since they have a chronic rather than an acute effect on health, thus they are safe and efficient for preventing hypertension. The combination of policosanol, red yeast rice extract, berberine, folic acid, and coenzyme Q10 may have an antihypertensive effect in MetS [9]. However, human clinical studies are required for definitive conclusions. Recently, we have shown that specially prepared aqueous extracts of kiwifruit and tomato have anti-ACE activity (unpublished data). However, further studies are required on the supplementation of these extracts on hypertensive individuals to determine its bioavailability and expression of the effects on blood pressure.

8.5 PLATELET HYPERACTIVITY AND NUTRACEUTICALS

In the adult human body, billions of blood platelets continuously flow over 1000 m^2 of vascular surface with normally minimal adhesion or aggregation. Upon disruption of the vessel wall or at the sites of activated or damaged endothelium (atherosclerotic plaque), rapid and complex interactions occur among platelets, vascular cells, and the coagulation system [38–42]. Platelet aggregation is fundamental to a wide range of physiological and pathological processes, including the induction of thrombosis and arteriosclerosis [43–46]. Since platelets are involved in the thrombotic event as well as in the initiation and progression of atherosclerotic plaque [47], hyperactive platelets in many conditions such as MetS, diabetes mellitus, obesity, insulin resistance, and smoking, may contribute to the pathogenesis of CVD [41,48–53]. In fact, an increased prothrombotic state induced by platelet hyperactivity is a major risk factor for the development of heart attacks, strokes, and venous thromboembolism [54–57]. The understanding of platelet activation in MetS states has increased dramatically over the past decade. The pathologic states such as hyperlipidemia, hypertension, obesity, insulin resistance, smoking, diabetes, and high-fat diets have been proved intensively associated with platelet hyperreactivity [51,58]. Lowering platelet reactivity is therefore considered to be the corner stone in the prevention and treatment of CVD [55,56]. Now, there is an increasing use of antiplatelet therapy as "blood thinner" to normalize the blood flow. Dietary components have been shown to modify platelet activation and/or hemostasis pathways through a variety of mechanisms [59]. Current antiplatelet treatments are mainly based on inhibition of two important pathways of platelet activation: thromboxane A_2 (TxA_2) mediated (aspirin) and adenosine diphosphate (ADP)–P2Y12 receptor mediated. Data suggest that aspirin resistance and its side effects are concerns in CVD therapeutic as well as preventive treatment [60]. With more and more problems of current antiplatelet agents appearing, people begin to concern conventional antiplatelet strategies only targeting the signaling pathway of platelet activation [61,62]. Therefore, given the serious, unintended adverse health effects of aspirin therapy, this led to the discovery of a broad range of natural compounds, including foods and spices, with demonstrable platelet-inhibiting activity [59]. It is now acknowledged that populations whose diet results in a suppression of platelet activation (e.g., a high sea fish diet or a Mediterranean diet) obtain measurable health benefits in terms of reduction of CVD risk. Dietary components, such as *Allium* species, ginger, *Ginkgo biloba* (*GB*), tomatoes, kiwifruits, flavanols and proanthocyanidins, resveratrol, alcohol, and fatty acids, have been shown to modify platelet activation and/or hemostasis pathways through a variety of mechanisms [59,63,64]. In addition, there are compounds present in fruits and vegetables, the mode of actions of which is quite different from antioxidant activity. Some of these compounds may inhibit phosphodiesterase, cyclooxygenase, or lipase activity that may affect platelet function. A number of potent inhibitors of phospholipase A_2 (PLA_2) have been isolated from natural sources. These include the sesterterpenoids, luffarielolide, and manoalide from marine sponges (*Luffariella* spp.) and the phenanthrene carboxylic acid derivative, aristolochic acid [65]. Among food plants, *Allium* spp. have provided a rich source of compounds that affect arachidonic acid (AA) metabolism [66]. Capsaicin, a component of red and chili peppers (*Capsicum* spp.), strongly inhibits calcium-triggered PLA_2 activity by a mechanism independent of calcium and substrate concentration, but appears to involve a direct interaction with the enzyme protein [67]. Conversely, methylallyltrisulfide from garlic oil, a potent inhibitor of platelet aggregation, promoted the release of labeled AA from rabbit platelets implying the stimulation of PLA_2. Various other natural products have been found to inhibit PLA_2 including retinoids, α-tocopherol, glycyrrhizin, and the sesquiterpenes of feverfew (*Tanacetum parthenium*) [59]. Quercetin, a flavonol commonly encountered in food plants and a potent inhibitor of platelet aggregation, is the only agent of dietary origin thought to directly inhibit PLA C.

Organosulfur compounds from *Allium* species, particularly *Allium cepa* (onion) and *A. sativum* (garlic), have been shown to be potent inhibitors of platelet aggregation [68]. Ginger (*Zingiber officinale*) extracts are also reported to be powerful inhibitors of human platelet aggregation and to inhibit prostanoid formation [69]. The link between inhibition of the AA cascade with curcumin (and demethoxycurcumin and bisdemethoxycurcumin) from turmeric (*Curcuma* spp.) was demonstrated [70]. Resveratrol (3,4′,5-trihydroxystilbene) has antiplatelet activity [71]. The effects of wine phenolics (quercetin, catechin, and epicatechin) and antioxidants (α-tocopherol, hydroquinone, and butylated hydroxytoluene), *trans*-resveratrol, and quercetin demonstrated a dose-dependent inhibition of both thrombin- and ADP-induced platelet aggregation, whereas ethanol inhibited only thrombin-induced aggregation. Epicatechin-rich extracts of green tea, nutmeg oil, and curcumin from turmeric also inhibit platelet aggregation and cyclooxygenase activity of prostaglandin–endoperoxide synthase [59]. *GB* inhibits platelet aggregation and TxA_2 synthesis in the presence of both collagen and ADP, whereas it does not significantly affect AA-induced TxA_2 synthesis [72]. Among all fruits tested *in vitro* for their antiplatelet property, tomatoes and kiwifruits had the highest activity followed by grapefruit, melon, and strawberry, whereas pear and apple had little or no activity [63]. The antiplatelet components (molecular weight [MW] < 1000 Da) in tomatoes and kiwifruits are water soluble and heat stable [63,64]. Nuclear magnetic resonance (NMR) and mass spectroscopy studies indicated that tomatoes contain antiplatelet compounds in addition to adenosine [63]. Unlike aspirin, the tomato-derived compounds inhibit thrombin-induced platelet aggregation.

8.6 DYSLIPIDEMIA AND NUTRACEUTICALS

Dyslipidemia result from changes in the synthesis and catabolism of lipoprotein particles in MetS. They are treated aggressively to avoid CVD. The American Diabetes Association guidelines suggest that all adults with diabetes should be managed to achieve LDL cholesterol < 100 mg/dl employing statins as first-line therapy. Use of statin in diabetic patients demonstrated a sufficient reduction in the relative risk of CVD mortality. Because of the pleiotropic effects of statins, they have been used as a first-line therapy for hypercholesterolemia in hypertensive and diabetic patients. Along with pharmacologic intervention, newer approaches to reduce cholesterol blood levels currently include nutraceuticals. In addition to eating a healthy diet low in saturated fat, with plenty of whole grains, fruits, and vegetables, some specific foods and supplements may help lower cholesterol. Several studies have shown that soluble fiber and beta-glucan lower LDL cholesterol and triglycerides [73,74]. Consuming soy protein helps lower blood cholesterol levels [75]. Plant sterols/stanols reduce cholesterol absorption in the intestinal gut, thereby reducing plasma LDL concentrations. Plant sterols/stanols are abundant in vegetable oils and olive oil, but also in fruits and nuts. Beta-sitosterol (a plant sterol) and policosanol (derived from sugarcane and yams) can lower LDL cholesterol [76]. A meta-analysis showed that intake of 2 g/day of stanols or sterols reduced LDL concentrations by about 10%–11% [77]. In fact, the National Cholesterol Education Program (NCEP) Adult Treatment Panel III (ATP III) has recommended, since 2001, phytosterol-enriched functional foods as part of an optimal dietetic prevention strategy in primary and secondary prevention of CVDs. Recently released guidelines are more critical with food supplementation with phytosterols and have drawn attention to significant safety issues based on large epidemiological studies. The results of the Prospective Cardiovascular Münster (PROCAM) study showed that patients afflicted with myocardial infection (MI) or sudden cardiac death had increased plant sterol concentrations [78]. Upper normal levels of plant sterols were also associated with a threefold increase of risk for coronary events among men in the highest tertile of coronary risk according to the PROCAM algorithm. Similar data are available for the plant sterol campesterol from the Multinational Monitoring of Trends and Determinants in Cardiovascular Disease (MONICA)/Cooperative Health Research in the Region of Augsburg (KORA) study [79]. In this prospective study, campesterol correlated directly with the incidence of infarction. Another natural compound capable of reducing cholesterol levels is red yeast rice (*Monascus purpureus*). This fermented rice contains numerous monacolins that are naturally occurring 5-hydroxy-3-methylglutaryl-coenzyme A (HMG-CoA) reductase inhibitors [80]. A meta-analysis involving 9625 patients in 93 randomized trials with three different commercial preparation of red yeast rice produced a mean reduction in total cholesterol, LDL cholesterol, triglyceride, and a mean rise in high-density lipoprotein (HDL) cholesterol [81]. More recently, a double-blind, multicenter trial has demonstrated that Xuezhikang administration (a commercial red yeast rice preparation) at a dose of 0.6 g twice daily was associated with a reduction nonfatal MI and death from CVD [82]. Studies suggested that the flavonoids quercetin, resveratrol, and catechins help reduce the risk of atherosclerosis by protecting against the damage caused by LDL cholesterol. Garlic has been widely reported to protect against CVD by reducing serum cholesterol concentrations and blood pressure and by inhibiting platelet aggregation.

8.7 MECHANISMS OF ACTION OF NUTRACEUTICALS

Due to the diverse structure, nutraceuticals target the pathogenesis of MetS by modulating various biochemical pathways [83]. Mechanistic studies have shown that specific nutraceuticals regulate insulin signaling through complex nutrigenomic and intracellular signaling networks [83]. Some of these nutraceuticals influence gene expression, transcription, and protein expression and function, preventing metabolic disturbance in cellular systems [84]. Nutraceuticals regulate metabolism, stress resistance, cellular survival, cellular senescence/aging endothelial functions, and circadian rhythms via activating the sirtuin 1 gene [85]. Furthermore, there is an increasing data, indicating that specific nutraceuticals have discrete actions on kinase-mediated intracellular signaling processes. For instance, flavonoids bind to Akt/protein kinase B (PKB), Fyn, Janus kinase 1, mitogen-activated PK4, and phosphoinositide-3-kinase (PI3K) [86]. Adipocytes isolated from insulin-resistant patients exhibit impaired insulin signaling, reduced insulin receptor (IR) substrate 1 (IRS-1) gene and protein expression, and dysfunctional insulin-stimulated PI3K and Akt/PKB signaling [87]. Insulin stimulates the IR tyrosine kinase, leading to the tyrosine phosphorylation of the IRS family of proteins. Activated IRS then displays the binding sites for numerous signaling partners such as PI3K, a key player in insulin function through the activation of the Akt/PKB. When stimulated, Akt/PKB promotes glycogen synthesis via upregulation of the glycogen synthase enzyme that occurs with inhibition of glycogen synthase kinase (GSK)-3. Additionally, insulin activates glucose uptake via a family of glucose transporters (GLUTs). Through negative feedback, PI3K, Akt/PKB, and GSK-3 can result in serine phosphorylation of IRS and subsequent inactivation. Activation of G-protein-coupled receptors can lead to activation of PKC. Nutraceuticals selectively modulated kinases within the insulin signaling cascade and produced favorable insulin-sensitizing effects [88]. An extract from bitter melon (*Momordica charantia*) had the agonist activity for adenosine monophosphate (AMP)-activated PK in 3T3-L1 adipocytes and enhanced glucose disposal in insulin-resistant mice [89]. Excessive stress through inflammatory mediators such as tumor necrosis factor α or through metabolic overflow of lipid from adipose tissue could also impact the insulin signaling cascade. There are undoubtedly a number of different mechanisms by which nutraceuticals from different foods and spices can modulate cellular signaling and metabolic processes. All these suggest that the bioactive substances in food have potential use for safe and effective prevention and management of MetS [90]. The newly recognized roles of specific nutraceuticals as modulators of kinase-dependent intracellular signal transduction and also as activators of enzymes such as sirtuin 1 open the door for robust mechanistic investigations.

8.8 CONCLUSIONS

The relationship between nutraceuticals and MetS has been a major focus of health research for almost a half-century; however, it is not clear whether individual nutraceuticals or their combinations may be responsible for any protective effects. In order for nutraceuticals to be effective against MetS, detailed information is required on the identification of active compounds and respective modes of action, followed by validation in controlled clinical trials with clearly defined endpoints. Moreover, to demonstrate the function *in vivo*, bioactive constituents have to be adsorbed efficiently through the intestine in active form without degradation and exert the activity after reaching the target organ. It is also necessary to further elucidate the underlying mechanisms of physiological activity attributed to these food-derived bioactive constituents. The advances in the knowledge of both the disease processes and healthy dietary components may provide new avenues to develop dietary strategies to prevent MetS.

ACKNOWLEDGMENT

The author is grateful to Throne Holst Foundation, Norway.

REFERENCES

1. Reaven, G.M. 2011. The metabolic syndrome: Time to get off the merry-go-round? *Journal of Internal Medicine*, 269: 127–136.
2. Alberti, K.G., R.H. Eckel, S.M. Grundy, P.Z. Zimmet, J.I. Cleeman, K.A. Donato, J.C. Fruchart, W.P. James, C.M. Loria, and S.C. Smith, Jr. 2009. Harmonizing the metabolic syndrome: A joint interim statement of the International Diabetes Federation Task Force on Epidemiology and Prevention; National Heart, Lung, and Blood Institute; American Heart Association; World Heart Federation; International Atherosclerosis Society; and International Association for the Study of Obesity. *Circulation*, 120: 1640–1645.
3. Gade, W., J. Schmit, M. Collins, and J. Gade. 2010. Beyond obesity: The diagnosis and pathophysiology of metabolic syndrome. *Clinical Laboratory Science: Journal of the American Society for Medical Technology*, 23: 51–61.
4. Esfahani, A., J.M. Wong, A. Mirrahimi, K. Srichaikul, D.J. Jenkins, and C.W. Kendall. 2009. The glycemic index: Physiological significance. *Journal of the American College of Nutrition*, 28: 439S–445S.
5. Georgiou, N.A., J. Garssen, and R.F. Witkamp. 2011. Pharma-nutrition interface: The gap is narrowing. *European Journal of Pharmacology*, 651: 1–8.
6. Izzo, R., G. De Simone, R. Giudice, M. Chinali, V. Trimarco, N. De Luca, and B. Trimarco. 2010. Effects of nutraceuticals on prevalence of metabolic syndrome and on calculated Framingham Risk Score in individuals with dyslipidemia. *Journal of Hypertension*, 28: 1482–1487.
7. Alissa, E.M. and G.A. Ferns. 2012. Functional foods and nutraceuticals in the primary prevention of cardiovascular diseases. *Journal of Nutrition and Metabolism*, 2012: 569486.
8. Davi, G., F. Santilli, and C. Patrono. 2010. Nutraceuticals in diabetes and metabolic syndrome. *Cardiovascular Therapeutics*, 28: 216–226.
9. Rozza, F., G. De Simone, R. Izzo, N. De Luca, and B. Trimarco. 2009. Nutraceuticals for treatment of high blood pressure values in patients with metabolic syndrome. *High Blood Pressure & Cardiovascular Prevention: The Official Journal of the Italian Society of Hypertension*, 16: 177–182.
10. Cabo, J., R. Alonso, and P. Mata. 2012. Omega-3 fatty acids and blood pressure. *British Journal of Nutrition*, 107: S195–S200.
11. Salazar, M.R., H.A. Carbajal, W.G. Espeche, C.A. Dulbecco, M. Aizpurua, A.G. Marillet, R.F. Echeverria, and G.M. Reaven. 2011. Relationships among insulin resistance, obesity, diagnosis of the metabolic syndrome and cardio-metabolic risk. *Diabetes & Vascular Disease Research: Official Journal of the International Society of Diabetes and Vascular Disease*, 8: 109–116.
12. Fuentes, P., M.J. Acuna, M. Cifuentes, and C.V. Rojas. 2010. The anti-adipogenic effect of angiotensin II on human preadipose cells involves ERK1,2 activation and PPARG phosphorylation. *Journal of Endocrinology*, 206: 75–83.
13. Medina-Gomez, G. and A. Vidal-Puig. 2005. Gateway to the metabolic syndrome. *Nature Medicine*, 11: 602, 603.
14. Broadhurst, C.L., M.M. Polansky, and R.A. Anderson. 2000. Insulin-like biological activity of culinary and medicinal plant aqueous extracts in vitro. *Journal of Agricultural and Food Chemistry*, 48: 849–852.
15. Qin, B., K.S. Panickar, and R.A. Anderson. 2010. Cinnamon: Potential role in the prevention of insulin resistance, metabolic syndrome, and type 2 diabetes. *Journal of Diabetes Science and Technology*, 4: 685–693.
16. Lerman, R.H., D.M. Minich, G. Darland, J.J. Lamb, J.L. Chang, A. Hsi, J.S. Bland, and M.L. Tripp. 2010. Subjects with elevated LDL cholesterol and metabolic syndrome benefit from supplementation with soy protein, phytosterols, hops rho iso-alpha acids, and *Acacia nilotica* proanthocyanidins. *Journal of Clinical Lipidology*, 4: 59–68.
17. Lerman, R.H., D.M. Minich, G. Darland, J.J. Lamb, B. Schiltz, J.G. Babish, J.S. Bland, and M.L. Tripp. 2008. Enhancement of a modified Mediterranean-style, low glycemic load diet with specific phytochemicals improves cardiometabolic risk factors in subjects with metabolic syndrome and hypercholesterolemia in a randomized trial. *Nutrition & Metabolism*, 5: 29.
18. Costantino, D., G. Minozzi, E. Minozzi, and C. Guaraldi. 2009. Metabolic and hormonal effects of myo-inositol in women with polycystic ovary syndrome: A double-blind trial. *European Review for Medical and Pharmacological Sciences*, 13: 105–110.
19. Maeba, R., H. Hara, H. Ishikawa, S. Hayashi, N. Yoshimura, J. Kusano, Y. Takeoka et al. 2008. Myo-inositol treatment increases serum plasmalogens and decreases small dense LDL, particularly in hyperlipidemic subjects with metabolic syndrome. *Journal of Nutritional Science and Vitaminology (Tokyo)*, 54: 196–202.

20. Irandoost, P., M. Ebrahimi-Mameghani, and S. Pirouzpanah. 2013. Does grape seed oil improve inflammation and insulin resistance in overweight or obese women? *International Journal of Food Sciences and Nutrition*, 64(6): 706–710.
21. Grover, J.K. and S.P. Yadav. 2004. Pharmacological actions and potential uses of *Momordica charantia*: A review. *Journal of Ethnopharmacology*, 93: 123–132.
22. Raman, A. and C. Lau. 1996. Anti-diabetic properties and phytochemistry of *Momordica charantia* L. (Cucurbitaceae). *Phytomedicine: International Journal of Phytotherapy and Phytopharmacology*, 2: 349–362.
23. Cornier, M.A., D. Dabelea, T.L. Hernandez, R.C. Lindstrom, A.J. Steig, N.R. Stob, R.E. Van Pelt, H. Wang, and R.H. Eckel. 2008. The metabolic syndrome. *Endocrine Reviews*, 29: 777–822.
24. Hotamisligil, G.S. 2006. Inflammation and metabolic disorders. *Nature*, 444: 860–867.
25. Furukawa, S., T. Fujita, M. Shimabukuro, M. Iwaki, Y. Yamada, Y. Nakajima, O. Nakayama, M. Makishima, M. Matsuda, and I. Shimomura. 2004. Increased oxidative stress in obesity and its impact on metabolic syndrome. *Journal of Clinical Investigation*, 114: 1752–1761.
26. Zern, T.L. and M.L. Fernandez. 2005. Cardioprotective effects of dietary polyphenols. *Journal of Nutrition*, 135: 2291–2294.
27. Magrone, T. and E. Jirillo. 2010. Polyphenols from red wine are potent modulators of innate and adaptive immune responsiveness. *Proceedings of the Nutrition Society*, 69: 279–285.
28. Barona, J., C.N. Blesso, C.J. Andersen, Y. Park, J. Lee, and M.L. Fernandez. 2012. Grape consumption increases anti-inflammatory markers and upregulates peripheral nitric oxide synthase in the absence of dyslipidemias in men with metabolic syndrome. *Nutrients*, 4: 1945–1957.
29. Sowers, J.R. 2004. Treatment of hypertension in patients with diabetes. *Archives of Internal Medicine*, 164: 1850–1857.
30. Sowers, J.R. and G.L. Bakris. 2000. Antihypertensive therapy and the risk of type 2 diabetes mellitus. *The New England Journal of Medicine*, 342: 969, 970.
31. Gress, T.W., F.J. Nieto, E. Shahar, M.R. Wofford, and F.L. Brancati. 2000. Hypertension and antihypertensive therapy as risk factors for type 2 diabetes mellitus. Atherosclerosis risk in communities study. *The New England Journal of Medicine*, 342: 905–912.
32. Elliott, W.J. and P.M. Meyer. 2007. Incident diabetes in clinical trials of antihypertensive drugs: A network meta-analysis. *Lancet*, 369: 201–207.
33. The Heart Outcomes Prevention Evaluation (HOPE) Investigators. 2000. Effects of ramipril on cardiovascular and microvascular outcomes in people with diabetes mellitus: Results of the HOPE study and MICRO-HOPE substudy. *Lancet*, 355: 253–259.
34. Boustany, C.M., K. Bharadwaj, A. Daugherty, D.R. Brown, D.C. Randall, and L.A. Cassis. 2004. Activation of the systemic and adipose renin-angiotensin system in rats with diet-induced obesity and hypertension. *American Journal of Physiology—Regulatory, Integrative and Comparative Physiology*, 287: R943–R949.
35. Engeli, S., R. Negrel, and A.M. Sharma. 2000. Physiology and pathophysiology of the adipose tissue renin-angiotensin system. *Hypertension*, 35: 1270–1277.
36. Gorzelniak, K., S. Engeli, J. Janke, F.C. Luft, and A.M. Sharma. 2002. Hormonal regulation of the human adipose-tissue renin-angiotensin system: Relationship to obesity and hypertension. *Hypertension*, 20: 965–973.
37. Huang, W.Y., S.T. Davidge, and J. Wu. 2013. Bioactive natural constituents from food sources-potential use in hypertension prevention and treatment. *Critical Reviews in Food Science and Nutrition*, 53: 615–630.
38. Camera, M., M. Brambilla, L. Facchinetti, P. Canzano, R. Spirito, L. Rossetti, C. Saccu, M.N. Di Minno, and E. Tremoli. 2012. Tissue factor and atherosclerosis: Not only vessel wall-derived TF, but also platelet-associated TF. *Thrombosis Research*, 129: 279–284.
39. Harker, L.A. 1986. Clinical trials evaluating platelet-modifying drugs in patients with atherosclerotic cardiovascular disease and thrombosis. *Circulation*, 73: 206–223.
40. Harker, L.A. and V. Fuster. 1986. Pharmacology of platelet inhibitors. *Journal of the American College of Cardiology*, 8: 21B–32B.
41. Pamukcu, B., G.Y. Lip, V. Snezhitskiy, and E. Shantsila. 2011. The CD40-CD40L system in cardiovascular disease. *Annals of Medicine*, 43: 331–340.
42. Dutta-Roy, A.K., T.K. Ray, and A.K. Sinha. 1986. Prostacyclin stimulation of the activation of blood coagulation factor X by platelets. *Science*, 231: 385–388.
43. Dutta-Roy, A.K. and A.K. Sinha. 1987. Purification and properties of prostaglandin E1/prostacyclin receptor of human blood platelets. *Journal of Biological Chemistry*, 262: 12685–12691.

44. Dutta-Roy, A.K., N.N. Kahn, and A.K. Sinha. 1989. Prostaglandin E1: The endogenous physiological regulator of platelet mediated blood coagulation. *Prostaglandins, Leukotrienes, and Essential Fatty Acids*, 35: 189–195.
45. Hamet, P., H. Sugimoto, F. Umeda, and D.J. Franks. 1983. Platelets and vascular smooth muscle: Abnormalities of phosphodiesterase, aggregation, and cell growth in experimental and human diabetes. *Metabolism*, 32: 124–130.
46. Kroll, M.H. and A.I. Schafer. 1989. Biochemical mechanisms of platelet activation. *Blood*, 74: 1181–1195.
47. Palomo, I., E. Fuentes, T. Padro, and L. Badimon. 2012. Platelets and atherogenesis: Platelet anti-aggregation activity and endothelial protection from tomatoes (*Solanum lycopersicum* L.). *Experimental and Therapeutic Medicine*, 3: 577–584.
48. Shimodaira, M., T. Niwa, K. Nakajima, M. Kobayashi, N. Hanyu, and T. Nakayama. 2013. Correlation between mean platelet volume and fasting plasma glucose levels in prediabetic and normoglycemic individuals. *Cardiovascular Diabetology*, 12: 14.
49. Pamukcu, B., H. Oflaz, I. Onur, A. Cimen, and Y. Nisanci. 2011. Effect of cigarette smoking on platelet aggregation. *Clinical and Applied Thrombosis/Hemostasis: Official Journal of the International Academy of Clinical and Applied Thrombosis/Hemostasis*, 17: E175–E180.
50. Ferroni, P., S. Basili, A. Falco, and G. Davi. 2004. Platelet activation in type 2 diabetes mellitus. *Journal of Thrombosis and Haemostasis*, 2: 1282–1291.
51. Ferroni, P., S. Basili, A. Falco, and G. Davi. 2004. Oxidant stress and platelet activation in hypercholesterolemia. *Antioxidants & Redox Signaling*, 6: 747–756.
52. Huang, Y., Z. Yang, Z. Ye, Q. Li, J. Wen, X. Tao, L. Chen et al. 2012. Lipocalin-2, glucose metabolism and chronic low-grade systemic inflammation in Chinese people. *Cardiovascular Diabetology*, 11: 11.
53. Park, Y. and W.S. Harris. 2009. Dose-dependent effects of n-3 polyunsaturated fatty acids on platelet activation in mildly hypertriglyceridemic subjects. *Journal of Medicinal Food*, 12: 809–813.
54. Diener, H.C., E.B. Ringelstein, R. von Kummer, H. Landgraf, K. Koppenhagen, J. Harenberg, I. Rektor et al. 2006. Prophylaxis of thrombotic and embolic events in acute ischemic stroke with the low-molecular-weight heparin certoparin: Results of the PROTECT trial. *Stroke*, 37: 139–144.
55. Davi, G. and C. Patrono. 2007. Platelet activation and atherothrombosis. *The New England Journal of Medicine*, 357: 2482–2494.
56. Ferroni, P., F. Santilli, F. Guadagni, S. Basili, and G. Davi. 2007. Contribution of platelet-derived CD40 ligand to inflammation, thrombosis and neoangiogenesis. *Current Medicinal Chemistry*, 14: 2170–2180.
57. Ferroni, P., F. Martini, R. D'Alessandro, A. Magnapera, V. Raparelli, A. Scarno, G. Davi, S. Basili, and F. Guadagni. 2008. In vivo platelet activation is responsible for enhanced vascular endothelial growth factor levels in hypertensive patients. *Clinica Chimica Acta*, 388: 33–37.
58. Natarajan, A., A.G. Zaman, and S.M. Marshall. 2008. Platelet hyperactivity in type 2 diabetes: Role of antiplatelet agents. *Diabetes & Vascular Disease Research: Official Journal of the International Society of Diabetes and Vascular Disease*, 5: 138–144.
59. Dutta-Roy, A.K. 2002. Dietary components and human platelet activity. *Platelets*, 13: 67–75.
60. Coccheri, S. 2012. Antiplatelet therapy: Controversial aspects. *Thrombosis Research*, 129: 225–229.
61. Xiang, Y.Z. 2008. Adrenoreceptors, platelet reactivity and clopidogrel resistance. *Thrombosis Haemostasis*, 100: 729, 730.
62. Xiang, Y.Z., Y. Xia, X.M. Gao, H.C. Shang, L.Y. Kang, and B.L. Zhang. 2008. Platelet activation, and antiplatelet targets and agents: Current and novel strategies. *Drugs*, 68: 1647–1664.
63. Dutta-Roy, A.K., L. Crosbie, and M.J. Gordon. 2001. Effects of tomato extract on human platelet aggregation in vitro. *Platelets*, 12: 218–227.
64. Duttaroy, A.K. and A. Jorgensen. 2004. Effects of kiwi fruit consumption on platelet aggregation and plasma lipids in healthy human volunteers. *Platelets*, 15: 287–292.
65. Reynolds, L.J., E.D. Mihelich, and E.A. Dennis. 1991. Inhibition of venom phospholipases A2 by manoalide and manoalogue. Stoichiometry of incorporation. *Journal of Biological Chemistry*, 266: 16512–16517.
66. Moriguchi, T., H. Matsuura, Y. Itakura, H. Katsuki, H. Saito, and N. Nishiyama. 1997. Allixin, a phytoalexin produced by garlic, and its analogues as novel exogenous substances with neurotrophic activity. *Life Sciences*, 61: 1413–1420.
67. Savitha, G., S. Panchanathan, and B.P. Salimath. 1990. Capsaicin inhibits calmodulin-mediated oxidative burst in rat macrophages. *Cell Signal*, 2: 577–585.
68. Srivastava, K.C. 1986. Onion exerts antiaggregatory effects by altering arachidonic acid metabolism in platelets. *Prostaglandins, Leukotrienes, and Medicine*, 24: 43–50.
69. Kiuchi, F., S. Iwakami, M. Shibuya, F. Hanaoka, and U. Sankawa. 1992. Inhibition of prostaglandin and leukotriene biosynthesis by gingerols and diarylheptanoids. *Chemical & Pharmaceutical Bulletin*, 40: 387–391.

70. Srivastava, K.C., A. Bordia, and S.K. Verma. 1995. Curcumin, a major component of food spice turmeric (*Curcuma longa*) inhibits aggregation and alters eicosanoid metabolism in human blood platelets. *Prostaglandins, Leukotrienes, and Essential Fatty Acids*, 52: 223–227.
71. De Lange, D.W., S. Verhoef, G. Gorter, R.J. Kraaijenhagen, A. van de Wiel, and J.W. Akkerman. 2007. Polyphenolic grape extract inhibits platelet activation through PECAM-1: An explanation for the French paradox. *Alcoholism, Clinical and Experimental Research*, 31: 1308–1314.
72. Dutta-Roy, A.K., M.J. Gordon, C. Kelly, K. Hunter, L. Crosbie, T. Knight-Carpentar, and B.C. Williams. 1999. Inhibitory effect of *Ginkgo biloba* extract on human platelet aggregation. *Platelets*, 10: 298–305.
73. Wolever, T.M., A.L. Gibbs, J. Brand-Miller, A.M. Duncan, V. Hart, B. Lamarche, S.M. Tosh, and R. Duss. 2011. Bioactive oat beta-glucan reduces LDL cholesterol in Caucasians and non-Caucasians. *Nutrition Journal*, 10: 130.
74. Knopp, R.H., H.R. Superko, M. Davidson, W. Insull, C.A. Dujovne, P.O. Kwiterovich, J.H. Zavoral, K. Graham, R.R. O'Connor, and D.A. Edelman. 1999. Long-term blood cholesterol-lowering effects of a dietary fiber supplement. *American Journal of Preventive Medicine*, 17: 18–23.
75. Maki, K.C., D.N. Butteiger, T.M. Rains, A. Lawless, M.S. Reeves, C. Schasteen, and E.S. Krul. 2010. Effects of soy protein on lipoprotein lipids and fecal bile acid excretion in men and women with moderate hypercholesterolemia. *Journal of Clinical Lipidology*, 4: 531–542.
76. Myrie, S.B., D. Mymin, B. Triggs-Raine, and P.J. Jones. 2012. Serum lipids, plant sterols, and cholesterol kinetic responses to plant sterol supplementation in phytosterolemia heterozygotes and control individuals. *American Journal of Clinical Nutrition*, 95: 837–844.
77. Talati, R., D.M. Sobieraj, S.S. Makanji, O.J. Phung, and C.I. Coleman. 2010. The comparative efficacy of plant sterols and stanols on serum lipids: A systematic review and meta-analysis. *Journal of the American Dietetic Association*, 110: 719–726.
78. Assmann, G., P. Cullen, J. Erbey, D.R. Ramey, F. Kannenberg, and H. Schulte. 2006. Plasma sitosterol elevations are associated with an increased incidence of coronary events in men: Results of a nested case-control analysis of the Prospective Cardiovascular Munster (PROCAM) study. *Nutrition, Metabolism, and Cardiovascular Diseases*, 16: 13–21.
79. Meisinger, C., H. Loewel, W. Mraz, and W. Koenig. 2005. Prognostic value of apolipoprotein B and A-I in the prediction of myocardial infarction in middle-aged men and women: Results from the MONICA/KORA Augsburg cohort study. *European Heart Journal*, 26: 271–278.
80. Bogsrud, M.P., L. Ose, G. Langslet, I. Ottestad, E.C. Strom, T.A. Hagve, and K. Retterstol. 2010. HypoCol (red yeast rice) lowers plasma cholesterol—A randomized placebo controlled study. *Scandinavian Cardiovascular Journal*, 44: 197–200.
81. Liu, J., J. Zhang, Y. Shi, S. Grimsgaard, T. Alraek, and V. Fonnebo. 2006. Chinese red yeast rice (*Monascus purpureus*) for primary hyperlipidemia: A meta-analysis of randomized controlled trials. *Chinese Medicine*, 1: 4.
82. Lu, Z., W. Kou, B. Du, Y. Wu, S. Zhao, O.A. Brusco, J.M. Morgan, D.M. Capuzzi, and S. Li. 2008. Effect of Xuezhikang, an extract from red yeast Chinese rice, on coronary events in a Chinese population with previous myocardial infarction. *American Journal of Cardiology*, 101: 1689–1693.
83. Minich, D.M. and J.S. Bland. 2008. Dietary management of the metabolic syndrome beyond macronutrients. *Nutrition Reviews*, 66: 429–444.
84. Barnes, S. 2008. Nutritional genomics, polyphenols, diets, and their impact on dietetics. *Journal of the American Dietetic Association*, 108: 1888–1895.
85. Chung, S., H. Yao, S. Caito, J.W. Hwang, G. Arunachalam, and I. Rahman. 2010. Regulation of SIRT1 in cellular functions: Role of polyphenols. *Archives of Biochemistry and Biophysics*, 501: 79–90.
86. Hou, D.X. and T. Kumamoto. 2010. Flavonoids as protein kinase inhibitors for cancer chemoprevention: Direct binding and molecular modeling. *Antioxidants & Redox Signaling*, 13: 691–719.
87. Smith, U. 2002. Impaired ("diabetic") insulin signaling and action occur in fat cells long before glucose intolerance—Is insulin resistance initiated in the adipose tissue? *International Journal of Obesity and Related Metabolic Disorders*, 26: 897–904.
88. Minich, D.M., R.H. Lerman, G. Darland, J.G. Babish, L.M. Pacioretty, J.S. Bland, and M.L. Tripp. 2010. Hop and acacia phytochemicals decreased lipotoxicity in 3T3-L1 adipocytes, db/db mice, and individuals with metabolic syndrome. *Journal of Nutrition and Metabolism*. doi:10.1155/2010/467316.
89. Tan, M.J., J.M. Ye, N. Turner, C. Hohnen-Behrens, C.Q. Ke, C.P. Tang, T. Chen et al. 2008. Antidiabetic activities of triterpenoids isolated from bitter melon associated with activation of the AMPK pathway. *Chemistry & Biology*, 15: 263–273.
90. Dwyer, J.T. 2007. Do functional components in foods have a role in helping to solve current health issues? *Journal of Nutrition*, 137: 2489S–2492S.

9 Cardioprotective Food Function as Underpinning Strategy for Disease Prevention
Dietary Antioxidants in Diabetic Cardiac Complications

Kenichi Watanabe, Somasundaram Arumugam, Rajarajan A. Thandavarayan, Kenji Suzuki, and Hirohito Sone

CONTENTS

9.1 INTRODUCTION

A majority of hospitalizations for complications of diabetes are linked to cardiovascular diseases, and heart failure is the serious comorbidity of diabetes mellitus. Cardiac complications during diabetes mellitus are mostly associated with vascular complications, but there are cases without significant coronary obstruction suggesting the existence of diabetic cardiomyopathy (Hamblin et al. 2007). Using various medicaments for its treatment is the available option, but it will be fascinating to realize the involvement of dietary factors in either treatment or prevention of these diabetic cardiac complications such as diabetic cardiomyopathy. Oxidative stress, which is mediated by several reactive species and free radicals, is implicated in the progression of diabetic complications. It begins during the early stages of the disease and worsens with its progression. As oxidative stress can affect both vascular and metabolic functions, it can mediate diabetic cardiomyopathy via impaired glucose and lipid metabolism in the cardiac tissue (Wiernsperger 2003). There are reports suggesting both physiological and pathological roles for the oxidative stress mediators such as free radicals and reactive species, so targeting oxidative stress is a debate over several years in the treatment of diabetic cardiomyopathy. Treatment with antioxidants has provided both positive and negative results during these pathological conditions. Thus, it will be of interest to know the structure and function of each antioxidant along with its concentration achieved at the site of action as these factors can affect their property *in vivo*. In addition, the selection of a suitable antioxidant for specific conditions has an important value in the treatment. Instead, if modification of the dietary factors could provide some beneficial effects against the progression of diabetic cardiomyopathy, patients will get additional benefit along with their regular therapy. We can discuss various dietary components studied for their effects on diabetic cardiomyopathy due to their antioxidant and other specific properties, which may provide a useful means for picking a suitable dietary element to be either supplemented or shunned from the regular diet of a diabetic patient.

9.2 DIETARY ADJUSTMENTS

Most of the cardiovascular disorders crop up because of improper dietary practice and sedentary lifestyle. As diabetes mellitus is a metabolic disorder with erroneous carbohydrate and lipid metabolism, either aggravation or repression of its complications mostly relies on dietary elements. Thus, modification of daily dietary composition may have some preventive effects of cardiovascular complications associated with diabetes mellitus such as diabetic cardiomyopathy. Dietary recommendation published in the *Journal of the American Dietetic Association* suggests that a proper modification of the daily food intake can prevent cardiovascular disorders. A diet that is low in saturated fatty acids (<7%), *trans*-fatty acids (<1% calories), and dietary cholesterol (<200 mg); rich in *n*-3 fatty acids, eicosapentaenoic acid, and decosahexaenoic acid; fish consumption at least twice a week; ample in total dietary fiber (30 g/day) with emphasis on soluble fiber; unsalted nuts (1 oz) as tolerated and limited by energy needs; other vegetable protein sources such as soy and legumes; skim/low-fat dairy foods and/or other calcium/vitamin D-rich sources; rich in vitamins, minerals, phytochemicals, and antioxidants from multiple servings of fruits and vegetables; low in sodium (<2300 mg/day); rich in vitamin B and fiber from food sources such as whole grains and vegetables; that may include plant sterols and stanols in high-risk individuals; which represents the ideal approach to treat the cardiovascular disordered patients. But physician recommendation has to be considered if needed in patients (Van Horn et al. 2008). There are many dietary recommendations prevailing in different countries, but most of them did not include the recommendations for herbs and spices. Most of the diet which we consume is always with one or the other spices obtained from various herbs. Thus, inclusion of these components in dietary recommendation may also be an important step toward disease prevention with dietary changes (Tapsell et al. 2006). But these recommendations must be made with the scientific support for the disease-preventing effects including antioxidative functions of that particular spice or herb. We will discuss the functions of some of the components of regular diet so that each one can understand its importance in devastating conditions such as diabetic cardiomyopathy.

9.2.1 Amino Acids

Amino acids are the basic components of proteins and are necessary for nucleotide synthesis. They regulate energetic metabolism at both anaerobic metabolism and mitochondrial respiration levels, modulate gene expression, and stimulate protein synthesis in eukaryotic cells. It will be interesting if the amino acids consumed via diet can modulate the cardiovascular disease severity. Amino acid supplementation in the diet has proven value against cardiac complications. It has been reported that supplementation with a mixture of 11 amino acids (high composition of leucine, lysine, isoleucine, and valine) to diabetic rats counteracted the diabetes-induced damage to the myocardial contractility and enzymatic dysfunction, which may be mediated via serine threonine kinase-mediated mammalian target of rapamycin (mTOR) activation and insulin-independent mechanism (Pellegrino et al. 2008).

Similarly, dietary administration of taurine, arginine, and carnitine to diabetic animals resulted in the improvement of various pathological parameters; among them, taurine has attenuated diabetes-mediated changes in cardiac contractile function and ultrastructure without significant changes in blood lipid and glucose levels (Tappia et al. 2011). L-Arginine, an α-amino acid, is a precursor of nitric oxide formed via the action of endothelial nitric oxide synthases. It can activate the transcription factor nuclear factor-erythroid 2 (NF-E2)-related factor 2, thereby improving the antioxidant status of the left ventricular myocardium of diabetic rats (Ramprasath et al. 2012).

9.2.2 Fatty Acids

Consistent reports suggest the involvement of lipotoxicity as a mediator of diabetes-mediated cardiovascular complications including diabetic cardiomyopathy. But most of the studies reported using saturated fatty acids and used a common term as lipids or fatty acids without classifying them based on their functions. When considering Western diet, it is composed mostly of ω-6 fatty acids obtained from plant or animal sources. Reports suggest that ω-6 fatty acids have few negative effects on diabetic heart, whereas ω-3 fatty acids present in fish and marine products can prevent these diabetic cardiac complications by counterbalancing lipotoxicity-mediated apoptotic and necrotic cardiac cell death. Thus, properly adjusting the lipid components based on the type of fat may provide some beneficial effects against diabetic cardiomyopathy mediated by hyperglycemia- or lipotoxicity-induced oxidative stress (Ghosh and Rodrigues 2006).

9.2.3 Vitamins

Abnormalities in the metabolism of essential fatty acids during diabetes mellitus result in increased oxidative stress that is implicated in diabetic cardiomyopathy. Several reports suggest the use of ω-3 polyunsaturated fatty acids along with fat-soluble vitamins for the treatment of these cardiac complications (Ceylan-Isik et al. 2007; Hunkar et al. 2002; Zobali et al. 2002).

Vitamin A or its metabolite retinoic acid is reported to be involved in the cardioprotection against hyperglycemia-induced oxidative stress, cardiomyocytes apoptosis, and activation of renin–angiotensin system via receptors called retinoic acid receptor (RAR) and retinoid X receptor (RXR) (Guleria et al. 2011). Vitamin A supplemented with fish oils such as cod liver oil (Ceylan-Isik et al. 2007; Hunkar et al. 2002) or insulin (Zobali et al. 2002) has been reported to provide protection against diabetic cardiomyopathy.

Vitamin B_1 (thiamine) is an indispensable coenzyme and required at intracellular glucose metabolism. High-dose therapy with vitamin B_1 can decrease the flux through hexosamine biosynthesis pathway and therefore prevent cardiac fibrosis (Kohda et al. 2008). Similarly, benfotiamine (a vitamin B_1 analog) is also a cardioprotective agent in diabetic cardiomyopathy as it prevents cardiac dysfunction in diabetic animals via activation of prosurvival signaling pathways (Katare et al. 2010), and treatment with high dose of benfotiamine rescues cardiomyocyte contractile dysfunction in streptozotocin-induced diabetes mellitus (Ceylan-Isik et al. 2006).

Vitamin C or ascorbic acid has also been reported to improve lipid abnormalities and myocardial performance in diabetes mellitus (Dai and McNeill 1995). In complex with molybdenum, it showed some insulin-mimetic actions and cardioprotective effects in streptozotocin-induced diabetes mellitus (MacDonald et al. 2006). It can improve the oxidant status of the diabetic patients, thereby having a possible cardioprotective function in diabetic cardiomyopathy (Dakhale et al. 2011; Mazloom et al. 2011).

Vitamin E (mostly α-tocopherol) is a potent antioxidant and its presence in the biological membranes represents the major defense system against free radical-mediated lipid peroxidation. Reports are available for both the protective effect and the inefficiency of vitamin E against diabetic complications of cardiovascular system, but the negative results did not consider the oxidative stress conditions of the patients. If administered properly to selected patients, vitamin E can be a good candidate for treating the diabetic cardiac complications such as cardiomyopathy (Hamblin et al. 2007). The protective effects of vitamin E against diabetic cardiomyopathy has been reported by several preclinical studies, suggesting its antioxidative function against apoptosis, lipid peroxidation, protein oxidation, QT interval prolongation, and myocardial contractile dysfunction (Aydemir-Koksoy et al. 2010; Shirpoor et al. 2009).

9.2.4 Minerals

Cardiac dysfunction during type 1 and type 2 diabetes results from multiple parameters including glucotoxicity, lipotoxicity, fibrosis, and mitochondrial uncoupling. Several studies have reported beneficial effects of a therapy with antioxidant agents, including trace elements and other antioxidants, against the cardiovascular system consequences of diabetes (Vassort and Turan 2010). Copper (Cu) is a metal required to be present in the normal diet, and most of the Cu in the physiological system is available as conjugate and almost no free Cu is available in the circulation. Strict homeostasis of Cu ion concentration is required for normal physiological functions, whereas both deficiency and overload can cause cardiovascular dysfunction. Intracellular increase in the levels of Cu in the heart does not strictly affect its function, whereas it can affect the redox system adversely so as to adversely affect the myocardial structure and function. This adverse effect of increased free Cu ion concentration is confirmed by a study using its chelator trientine, where its treatment has increased the urinary Cu output and improved cardiac structure in diabetic rats and humans. In addition, it can reverse heart failure in diabetic rats after chronic treatment. Possible mechanisms reported include regeneration of F-actin and normalization of collagen. Excess loosely bound Cu is thus implicated in the mechanism by which diabetes damages the heart (Cooper et al. 2004).

Magnesium (Mg) is one of the most abundant ions present in living cells, and plasma and intracellular Mg concentrations are tightly regulated by several factors including insulin. A poor intracellular Mg concentration impairs insulin action and causes worsening of insulin resistance in non-insulin-dependent diabetic patients. By contrast, daily Mg administration in diabetic patients restores a more appropriate intracellular Mg concentration and contributes to improve insulin-mediated glucose uptake (Barbagallo et al. 2003). Thus, dietary Mg is also an essential factor that can modulate the development of diabetic cardiovascular complications.

Selenium (Se), a dietary trace mineral essential for humans and animals, exerts its effects mainly through its incorporation into selenoproteins. Adequate Se intake is needed to maximize the activity of selenoproteins, among which glutathione (GSH) peroxidases have been shown to play a critical role in protecting aerobic tissues from oxygen radical-initiated cell injury (Tanguy et al. 2012). It is known that Se compounds can restore some metabolic parameters in experimental diabetes, and the beneficial effects of sodium selenite treatment appears to be the result of the restoration-altered activities of the antioxidant enzymes in diabetic heart tissue (Ulusu and Turan 2005). Combined doses of insulin and Se were effective in the control of blood glucose and correction of altered glucose transporter 4 distribution in diabetic rat hearts (Xu et al. 2010). Compelling biological pathways suggest that Se may lower the onset of type 2 diabetes mellitus, and at dietary levels of intake, individuals with higher toenail Se levels are at lower risk for type 2 diabetes mellitus (Park et al. 2012). The beneficial effects of sodium selenite treatment on the mechanical and electrical activities of the diabetic rat hearts appear to be due to the restoration of the diminished K^+ currents partially related to the restoration of the cell GSH redox cycle (Ayaz et al. 2004). In diabetic rat hearts, alterations of the *discus intercalaris* and nucleus were corrected, and degenerations seen in myofilaments and Z-lines were reversed by Se treatment (Ayaz et al. 2002).

Zinc (Zn) is a vital element in maintaining the normal structure and physiology of cells. The fact that it has an important role in states of cardiovascular diseases has been studied and described by several research groups (Little et al. 2010). Lower consumption of dietary Zn and low serum Zn levels were associated with an increased prevalence of coronary artery disease (CAD) and diabetes and several of their associated risk factors including hypertension, hypertriglyceridemia, and other factors suggestive of mild insulin resistance in urban subjects (Singh et al. 1998). Abstract Zn is one of the essential trace elements and has numerous physiological functions. Zn acts as an antioxidant and also as a part of other antioxidant-related proteins, such as metallothionein and Zn–Cu superoxide dismutase (SOD). Zn deficiency often occurs in patients with diabetes mellitus. Diabetic patients should be monitored and treated for Zn deficiency to avoid the acceleration of diabetes-induced cardiac and renal injury (Li et al. 2013). Zn supplementation of animals and humans has been shown to ameliorate oxidative stress induced by diabetic cardiomyopathy (Kumar et al. 2012).

9.2.5 Dietary Flavonoids

Green tea polyphenols such as catechin derivatives shown to ameliorate mineral metabolism and membrane-bound Ca^{2+}-adenosine triphosphatase (ATPase) and Na^+/K^+-ATPase activity by impeding dyslipidemia, lipid peroxidation, and protein glycation in the heart of streptozotocin-diabetic rats (Babu et al. 2006). Administration of green tea extract initiated six weeks after induction of diabetes improved myocardial collagen changes in diabetic rats. In addition to the antioxidant and chelating effect, the antihyperglycemic effect of green tea may be responsible for this activity (Babu et al. 2007). Administration of tea and tea polyphenols has been reported to prevent or attenuate decreases in antioxidant enzyme activities in a number of animal models of oxidative stress. Although consumption of tea or tea polyphenols by humans frequently results in modest transient increases in the total antioxidant capacity of plasma, recent research suggests that concomitant increases in plasma urate account for much, if not all, of the increased plasma antioxidant capacity (Frei and Higdon 2003).

Icariin is a flavonol and constituent of some Chinese medicinal herbs, which is reported to reduce mitochondrial oxidative stress, thereby preventing diabetic cardiac injury (Bao and Chen 2011). French maritime pine bark extract called Pycnogenol, primarily composed of procyanidins and phenolic acids, corrects diabetic cardiac dysfunction, probably by its metabolic and direct radical scavenging activity without affecting the molecular maladaptations of reactive oxygen species-producing enzymes and cytoskeletal components (Klimas et al. 2010). Breviscapine, a flavonoid isolated from the traditional Chinese medicinal herb *Erigeron breviscapus*, could regulate the expression of protein kinase C, protein phosphatase inhibitor 1, phospholamban, and sarco/endoplasmic reticulum Ca^{2+}-ATPase and have protective effect on diabetic cardiomyopathy (Wang et al. 2010). It also prevented cardiac hypertrophy in diabetic rats by inhibiting protein kinase C-mediated nuclear factor kappaB/c-*fos* signal transduction pathway (Wang et al. 2009).

Hydroxysafflor yellow A, an important constituent of *Carthamus tinctorius* L., significantly improved the cardiac function and downregulated the endothelin system and the reactive oxygen species pathway, resulting in a reversal of the abnormalities of expression of calcium-handling proteins and the cardiac performance in diabetic cardiomyopathy (Cheng et al. 2007). Phytoestrogenic isoflavones daidzein and genistein may reduce glucose toxicity-induced cardiac mechanical dysfunction, and thus possess therapeutic potential against diabetes-associated cardiac defects (Hintz and Ren 2004).

Curcumin is the best-characterized component (1,7-bis[4-hydroxy-3-methoxyphenyl]-1,6-heptadiene-3,5-dione [diferuloylmethane]) found in turmeric (*Curcuma longa*) and is accompanied by demethoxy and bisdemethoxy derivatives. Curcumin has shown surprisingly beneficial effects in experimental studies of acute and chronic diseases characterized by an exaggerated inflammatory reaction as both prevention and treatment (Bengmark 2006). There is a suggestion that curcumin exerts antioxidant and antiangiogenic actions via inhibition of both cellular Ca^{2+} entry and protein kinase C (Balasubramanyam et al. 2003). Curcumin attenuates oxidative stress and activation of mitogen-activated protein kinase pathway so that effective in prevention of diabetic cardiomyopathy (Soetikno et al. 2012). As curcumin is a potent inhibitor of transcriptional coactivator p300, it regulates cardiomyocyte hypertrophy during diabetes mellitus (Feng et al. 2008). Apart from these effects, it also has differential actions on vasoactive factor expression in the diabetic heart, indicating its importance in the maintenance of tissue microenvironment during the development of diabetic cardiomyopathy (Farhangkhoee et al. 2006).

9.2.6 α-Lipoic Acid

α-Lipoic acid (also called as thioctic acid) is a naturally occurring dithiol acid compound synthesized in human tissues. Both oxidized and reduced forms show the antioxidant property and get accumulated in the tissues upon dietary administration. It has been shown to improve the diabetes-mediated and other cardiac complications using various animal models of ischemia–reperfusion, heart failure, and hypertension (Ghibu et al. 2009). It can effectively protect the diabetic myocardium from mitochondria-mediated apoptosis with concomitant increase in mitochondrial Mn-SOD activity and GSH level (Li et al. 2009). Several reports prove the cardioprotective effect of α-lipoic acid using various transgenic animals such as Otsuka Long-Evans Tokushima fatty rats (Lee et al. 2012), Apolipoprotein$^{-/-}$ mice (Yi et al. 2012), and acyl-coenzyme A (CoA) synthase transgenic mice (Lee et al. 2006). It induces reactive oxygen species production mildly so as to create a preconditioning-like effect via activation of extracellular signal-regulated kinases 1 and 2 to protect the myocardium from hyperglycemia-induced excess oxidative stress (Yao et al. 2012). A clinical study used a combination of α-lipoic acid and other antioxidants in chronic diabetic patients and proved to improve their oxidant status (Palacka et al. 2010). In addition, there are few clinical trials reporting the beneficial effect of α-lipoic acid on cardiac autonomic regulation in diabetic patients (Navarese et al. 2011; Ziegler et al. 1997).

9.2.7 Reduced Folic Acid Derivatives

Reduced folates can reconstitute the proper activity of uncoupled endothelial nitric oxide synthase and scavenge peroxynitrite radical—or, more likely, nitrogen dioxide and carbonate radical derived from carbonate-induced degradation of peroxynitrite (McCarty et al. 2009). Thus, it would be beneficial to assess the effect of high-dose reduced folates against the disorders mediated via peroxynitrite-induced oxidative stress including diabetic cardiomyopathy.

9.2.8 Melatonin

Chronically elevated blood glucose induces toxic reactive species causing nitro-oxidative stress, which harms most of the cellular components including lipids, proteins, and DNA leading to severely compromised metabolic activity. Melatonin, a multifunctional indoleamine, has the capability of scavenging oxygen- and nitrogen-based reactants and blocking transcriptional factors that induce proinflammatory cytokines, and contributes to antioxidative, anti-inflammatory, and possibly epigenetic regulatory properties. Additionally, it can ease the effects of insulin resistance via restoration of adipocyte glucose transporter-4 loss (Korkmaz et al. 2012). Study using streptozotocin-induced diabetic rats confirmed the antioxidant potential of melatonin on various organs including the heart. Treatment with 10 mg/kg dose of melatonin has reversed the oxidative stress in the heart, thereby suggesting its importance in the diet for prevention of oxidative stress-mediated cardiovascular complications (Aksoy et al. 2003). In addition, melatonin treatment protected the heart from ischemia–reperfusion injury after diet-induced obesity (Nduhirabandi et al. 2011), where it has prevented the metabolic abnormalities, suggesting that these effects may also be beneficial to the hearts in diabetic condition.

9.2.9 Ginseng

North American ginseng prevents the diabetes-induced retinal and cardiac biochemical and functional changes probably through inhibition of oxidative stress (Sen et al. 2012). Ginsenoside Rb1, a major pharmacological extract of ginseng, exerts cardioprotective effects against myocardial ischemia–reperfusion injury in diabetic rats, which is partly through the activation of phosphatidylinositol 3-kinase/Akt pathway (Wu et al. 2011). It has also prevented peroxidative injury in the myocardium of diabetic rats. The mechanisms of antiperoxidation effect of ginseng might be by lowering the level of fasting blood glucose; decreasing the rate of monosaccharide auto-oxidation and partially protecting the production of free radicals; elevating the activity of enzymatic free radical scavenger in cells, such as SOD; and directly eliminating the superfluous free radicals (Xie et al. 1993).

9.2.10 Garlic and *S*-Allyl Cysteine

Chronic hyperglycemia causes increased protein glycation and the formation of advanced glycation end products, which is the major reason behind the diabetic cardiovascular complications. Aged garlic extract and its major component *S*-allyl cysteine can prevent the formation of advanced glycation end products (Ahmad et al. 2007). Garlic and its components such as *S*-allyl cysteine deserve more attention as possible cost-effective, nontoxic candidates for delaying or preventing complications of diabetes and aging (Ahmad and Ahmed 2006). This property may be effective against the pathogenesis of diabetic cardiomyopathy, and thus addition of garlic in daily diet may provide a better source of antioxidants that can suppress the oxidative stress-mediated cardiovascular complications.

9.3 CONCLUSION

Diabetes mellitus is a devastating metabolic disorder that leads to several complications affecting cardiovascular, nervous, and excretory systems affecting the quality and duration of life. Most of its cardiovascular complications are mediated via hyperglycemia and improper lipid metabolism. These metabolic derangements cause oxidative stress via various reactive oxygen and nitrogen species, which cause tissue injury mediated by several cell signaling pathways. Prevention of oxidative stress or improvement of endogenic antioxidant system is considered to be effective against these cardiovascular complications including diabetic cardiomyopathy. Since diet and lifestyle are important causes for these metabolic disorders, modification of the components of diet can be a good strategy for preventing these complications. To make such changes, we must have good knowledge about each component of diet and their role as antioxidant in preventing these metabolic complications such as diabetic cardiomyopathy. Diabetes mellitus under nontreated conditions can lead to several cardiac complications including diabetic cardiomyopathy. One of the possible reasons behind this condition is the involvement of oxidative stress, and there are various antioxidants available in nature as food components.

REFERENCES

Ahmad, M.S. and Ahmed, N. 2006. Antiglycation properties of aged garlic extract: Possible role in prevention of diabetic complications. *J Nutr* 136(3 Suppl): 796S–799S.

Ahmad, M.S., Pischetsrieder, M., and Ahmed, N. 2007. Aged garlic extract and *S*-allyl cysteine prevent formation of advanced glycation end products. *Eur J Pharmacol* 561(1–3): 32–38.

Aksoy, N., Vural, H., Sabuncu, T., and Aksoy, S. 2003. Effects of melatonin on oxidative-antioxidative status of tissues in streptozotocin-induced diabetic rats. *Cell Biochem Funct* 21(2): 121–125.

Ayaz, M., Can, B., Ozdemir, S., and Turan, B. 2002. Protective effect of selenium treatment on diabetes-induced myocardial structural alterations. *Biol Trace Elem Res* 89(3): 215–226.

Ayaz, M., Ozdemir, S., Ugur, M., Vassort, G., and Turan, B. 2004. Effects of selenium on altered mechanical and electrical cardiac activities of diabetic rat. *Arch Biochem Biophys* 426(1): 83–90.

Aydemir-Koksoy, A., Bilginoglu, A., Sariahmetoglu, M., Schulz, R., and Turan, B. 2010. Antioxidant treatment protects diabetic rats from cardiac dysfunction by preserving contractile protein targets of oxidative stress. *J Nutr Biochem* 21(9): 827–833.

Babu, P.V., Sabitha, K.E., and Shyamaladevi, C.S. 2006. Green tea impedes dyslipidemia, lipid peroxidation, protein glycation and ameliorates Ca^{2+}-ATPase and Na^+/K^+-ATPase activity in the heart of streptozotocin-diabetic rats. *Chem Biol Interact* 162(2): 157–164.

Babu, P.V., Sabitha, K.E., Srinivasan, P., and Shyamaladevi, C.S. 2007. Green tea attenuates diabetes induced Maillard-type fluorescence and collagen cross-linking in the heart of streptozotocin diabetic rats. *Pharmacol Res* 55(5): 433–440.

Balasubramanyam, M., Koteswari, A.A., Kumar, R.S., Monickaraj, S.F., Maheswari, J.U., and Mohan, V. 2003. Curcumin-induced inhibition of cellular reactive oxygen species generation: Novel therapeutic implications. *J Biosci* 28(6): 715–721.

Bao, H. and Chen, L. 2011. Icariin reduces mitochondrial oxidative stress injury in diabetic rat hearts. *Zhongguo Zhong Yao Za Zhi* 36(11): 1503–1507.

Barbagallo, M., Dominguez, L.J., Galioto, A., Ferlisi, A., Cani, C., Malfa, L., Pineo, A., Busardo A., and Paolisso G. 2003. Role of magnesium in insulin action, diabetes and cardio-metabolic syndrome X. *Mol Aspects Med* 24(1–3): 39–52.

Bengmark, S. 2006. Curcumin, an atoxic antioxidant and natural NFkappaB, cyclooxygenase-2, lipooxygenase, and inducible nitric oxide synthase inhibitor: A shield against acute and chronic diseases. *JPEN J Parenter Enteral Nutr* 30(1): 45–51.

Ceylan-Isik, A., Hünkar, T., Aşan E., Kaymaz, F., Ari, N., Söylemezoğlu, T., Renda, N., Soncul, H., Bali, M., and Karasu, C.; (Antioxidants in Diabetes-Induced Complications) The ADIC Study Group. 2007. Cod liver oil supplementation improves cardiovascular and metabolic abnormalities in streptozotocin diabetic rats. *J Pharm Pharmacol* 59(12): 1629–1641.

Ceylan-Isik, A.F., Wu, S., Li, Q., Li, S.Y., and Ren, J. 2006. High-dose benfotiamine rescues cardiomyocyte contractile dysfunction in streptozotocin-induced diabetes mellitus. *J Appl Physiol* 100(1): 150–156.

Cheng, M., Gao, H.Q., Xu, L., Li, B.Y., Zhang, H., and Li, X.H. 2007. Cardioprotective effects of grape seed proanthocyanidins extracts in streptozocin induced diabetic rats. *J Cardiovasc Pharmacol* 50(5): 503–509.

Cooper, G.J., Phillips, A.R., Choong, S.Y., Leonard, B.L., Crossman, D.J., Brunton, D.H., Saafi, L. et al. 2004. Regeneration of the heart in diabetes by selective copper chelation. *Diabetes* 53: 2501–2508.

Dai, S. and McNeill, J.H. 1995. Ascorbic acid supplementation prevents hyperlipidemia and improves myocardial performance in streptozotocin-diabetic rats. *Diabetes Res Clin Pract* 27(1): 11–18.

Dakhale, G.N., Chaudhari, H.V., and Shrivastava, M. 2011. Supplementation of vitamin C reduces blood glucose and improves glycosylated hemoglobin in type 2 diabetes mellitus: A randomized, double-blind study. *Adv Pharmacol Sci* 2011: 195271.

Farhangkhoee, H., Khan, Z.A., Chen, S., and Chakrabarti, S. 2006. Differential effects of curcumin on vasoactive factors in the diabetic rat heart. *Nutr Metab* 3: 27.

Feng, B., Chen, S., Chiu, J., George, B., and Chakrabarti, S. 2008. Regulation of cardiomyocytes hypertrophy in diabetes at the transcriptional level. *Am J Physiol Endocrinol Metab* 294(6): E1119–E1126.

Frei, B. and Higdon, J.V. 2003. Antioxidant activity of tea polyphenols in vivo: Evidence from animal studies. *J Nutr* 133(10): 3275S–3284S.

Ghibu, S., Richard, C., Vergely, C., Zeller, M., Cottin, Y., and Rochette, L. 2009. Antioxidant properties of an endogenous thiol: Alpha-lipoic acid, useful in the prevention of cardiovascular diseases. *J Cardiovasc Pharmacol* 54: 391–398.

Ghosh, S. and Rodrigues B. 2006. Cardiac cell death in early diabetes and its modulation by dietary fatty acids. *Biochim Biophys Acta* 1761(10): 1148–1162.

Guleria, R.S., Choudhary, R., Tanaka, T., Baker, K.M., and Pan, J. 2011. Retinoic acid receptor-mediated signaling protects cardiomyocytes from hyperglycemia induced apoptosis: Role of the renin-angiotensin system. *J Cell Physiol* 226(5): 1292–1307.

Hamblin, M., Smith, H.M., and Hill, M.F. 2007. Dietary supplementation with vitamin E ameliorates cardiac failure in type I diabetic cardiomyopathy by suppressing myocardial generation of 8-iso-prostaglandin F2alpha and oxidized glutathione. *J Card Fail* 13(10): 884–892.

Hintz, K.K. and Ren, J. 2004. Phytoestrogenic isoflavones daidzein and genistein reduce glucose-toxicity-induced cardiac contractile dysfunction in ventricular myocytes. *Endocr Res* 30(2): 215–223.

Hunkar, T., Aktan, F., Ceylan, A., and Karasu, C.; Antioxidants in Diabetes-Induced Complications (ADIC) Study Group. 2002. Effects of cod liver oil on tissue antioxidant pathways in normal and streptozotocin-diabetic rats. *Cell Biochem Funct* 20(4): 297–302.

Katare, R.G., Caporali, A., Oikawa, A., Meloni, M., Emanueli, C., and Madeddu, P. 2010. Vitamin B1 analog benfotiamine prevents diabetes-induced diastolic dysfunction and heart failure through Akt/Pim-1-mediated survival pathway. *Circ Heart Fail* 3(2): 294–305.

Klimas, J., Kmecova, J., Jankyova, S., Yaghi, D., Priesolova, E., Kyselova, Z., Musil, P. et al. 2010. Pycnogenol improves left ventricular function in streptozotocin-induced diabetic cardiomyopathy in rats. *Phytother Res* 24(7): 969–974.

Kohda, Y., Shirakawa, H., Yamane, K., Otsuka, K., Kono, T., Terasaki, F., and Tanaka, T. 2008. Prevention of incipient diabetic cardiomyopathy by high-dose thiamine. *J Toxicol Sci* 33(4): 459–472.

Korkmaz, A., Ma, S., Topal, T., Rosales-Corral, S., Tan, D.X., and Reiter, R.J. 2012. Glucose: A vital toxin and potential utility of melatonin in protecting against the diabetic state. *Mol Cell Endocrinol* 349(2): 128–137.

Kumar, S.D., Vijaya, M., Samy, R.P., Dheen, S.T., Ren, M., Watt, F., Kang, Y.J., Bay, B.H., and Tay, S.S. 2012. Zinc supplementation prevents cardiomyocyte apoptosis and congenital heart defects in embryos of diabetic mice. *Free Radic Biol Med* 53(8): 1595–1606.

Lee, J.E., Yi, C.O., Jeon, B.T., Shin, H.J., Kim, S.K., Jung, T.S., Choi, J.Y., and Roh, G.S. 2012. Alpha-lipoic acid attenuates cardiac fibrosis in Otsuka Long-Evans Tokushima fatty rats. *Cardiovasc Diabetol* 11: 111.

Lee, Y., Naseem, R.H., Park, B.H., Garry, D.J., Richardson, J.A., Schaffer, J.E., and Unger, R.H. 2006. Alpha-lipoic acid prevents lipotoxic cardiomyopathy in acyl CoA-synthase transgenic mice. *Biochem Biophys Res Commun* 344: 446–452.

Li, B., Tan, Y., Sun, W., Fu, Y., Miao, L., and Cai, L. 2013. The role of zinc in the prevention of diabetic cardiomyopathy and nephropathy. *Toxicol Mech Methods* 23(1): 27–33.

Li, C.J., Zhang, Q.M., Li, M.Z., Zhang, J.Y., Yu, P., and Yu, D.M. 2009. Attenuation of myocardial apoptosis by alpha-lipoic acid through suppression of mitochondrial oxidative stress to reduce diabetic cardiomyopathy. *Chin Med J (Engl)* 122: 2580–2586.

Little, P.J., Bhattacharya, R., Moreyra, A.E., and Korichneva, I.L. 2010. Zinc and cardiovascular disease. *Nutrition* 26(11/12): 1050–1057.

MacDonald, K., Bailey, J., MacRory, C., Friis, C., Vogels, C.M., Broderick, T., and Westcott, S.A. 2006. A newly synthesised molybdenum/ascorbic acid complex alleviates some effects of cardiomyopathy in streptozocin-induced diabetic rats. *Drugs R D* 7(1): 33–42.

Mazloom, Z., Hejazi, N., Dabbaghmanesh, M.H., Tabatabaei, H.R., Ahmadi, A., and Ansar, H. 2011. Effect of vitamin C supplementation on postprandial oxidative stress and lipid profile in type 2 diabetic patients. *Pak J Biol Sci* 14(19): 900–904.

McCarty, M.F., Barroso-Aranda, J., and Contreras, F. 2009. High-dose folate and dietary purines promote scavenging of peroxynitrite-derived radicals—Clinical potential in inflammatory disorders. *Med Hypotheses* 73(5): 824–834.

Navarese, E.P., Mollo, R., and Buffon, A. 2011. Effect of alpha lipoic acid on cardiac autonomic dysfunction and platelet reactivity in type 1 diabetes: Rationale and design of the AUTOnomic function and platelet REACTivity trial (AUTO-REACT protocol). *Diabetes Res Clin Pract* 92: 375–329.

Nduhirabandi, F., Du Toit, E.F., Blackhurst, D., Marais, D., and Lochner, A. 2011. Chronic melatonin consumption prevents obesity-related metabolic abnormalities and protects the heart against myocardial ischemia and reperfusion injury in a prediabetic model of diet-induced obesity. *J Pineal Res* 50(2): 171–182.

Palacka, P., Kucharska, J., Murin, J., Dostalova, K., Okkelova, A., Cizova M., Waczulikova, I., Moricova, S., and Gvozdjakova, A. 2010. Complementary therapy in diabetic patients with chronic complications: A pilot study. *Bratisl Lek Listy* 111: 205–211.

Park, K., Rimm, E.B., Siscovick, D.S., Spiegelman, D., Manson, J.E., Morris, J.S., Hu, F.B., and Mozaffarian D. 2012. Toenail selenium and incidence of type 2 diabetes in U.S. men and women. *Diabetes Care* 35(7): 1544–1551.

Pellegrino, M.A., Patrini, C., Pasini, E., Brocca, L., Flati, V., Corsetti, G., and D'Antona, G. 2008. Amino acid supplementation counteracts metabolic and functional damage in the diabetic rat heart. *Am J Cardiol* 101: 49E–56E.

Ramprasath, T., Kumar, P.H., Puhari, S.S.M., Murugan, P.S., Vasudevan, V., and Selvam, G.S. 2012. L-Arginine ameliorates cardiac left ventricular oxidative stress by upregulating eNOS and Nrf2 target genes in alloxan-induced hyperglycemic rats. *Biochem Biophys Res Commun.* doi:10.1016/j.bbrc.2012.10.064.

Sen, S., Chen, S., Wu, Y., Feng, B., Lui, E.K., and Chakrabarti, S. 2012. Preventive effects of North American ginseng (*Panax quinquefolius*) on diabetic retinopathy and cardiomyopathy. *Phytother Res.* doi:10.1002/ptr.4719.

Shirpoor, A., Salami, S., Khadem-Ansari, M.H., Ilkhanizadeh, B., Pakdel, F.G., and Khademvatani, K. 2009. Cardioprotective effect of vitamin E: Rescues of diabetes-induced cardiac malfunction, oxidative stress, and apoptosis in rat. *J Diabetes Complications* 23(5): 310–316.

Singh, R.B., Niaz, M.A., Rastogi, S.S., Bajaj, S., Gaoli, Z., and Shoumin, Z. 1998. Current zinc intake and risk of diabetes and coronary artery disease and factors associated with insulin resistance in rural and urban populations of North India. *J Am Coll Nutr* 17(6): 564–570.

Soetikno, V., Sari, F.R., Sukumaran, V., Lakshmanan, A.P., Mito, S., Harima, M., Thandavarayan, R.A. et al. 2012. Curcumin prevents diabetic cardiomyopathy in streptozotocin-induced diabetic rats: Possible involvement of PKC-MAPK signaling pathway. *Eur J Pharm Sci* 47(3): 604–614.

Tanguy, S., Grauzam, S., de Leiris, J., and Boucher, F. 2012. Impact of dietary selenium intake on cardiac health: Experimental approaches and human studies. *Mol Nutr Food Res* 56(7): 1106–1121.

Tappia, P.S., Thliveris, J., Xu, Y.J., Aroutiounova, N., and Dhalla, N.S. 2011. Effects of amino acid supplementation on myocardial cell damage and cardiac function in diabetes. *Exp Clin Cardiol* 16: e17–e22.

Tapsell, L.C., Hemphill, I., Cobiac, L., Patch, C.S., Sullivan, D.R., Fenech, M., Roodenrys, S. et al. 2006. Health benefits of herbs and spices: The past, the present, the future. *Med J Aust* 185(4 Suppl): S4–S24.

Ulusu, N.N. and Turan, B. 2005. Beneficial effects of selenium on some enzymes of diabetic rat heart. *Biol Trace Elem Res* 103(3): 207–216.

Van Horn, L., McCoin, M., Kris-Etherton, P.M., Burke, F., Carson, J.A., Champagne, C.M., Karmally, W., and Sikand G. 2008. The evidence for dietary prevention and treatment of cardiovascular disease. *J Am Diet Assoc* 108(2): 287–331.

Vassort, G. and Turan B. 2010. Protective role of antioxidants in diabetes-induced cardiac dysfunction. *Cardiovasc Toxicol* 10(2): 73–86.

Wang, M., Zhang, W.B., Zhu, J.H., Fu, G.S., and Zhou, B.Q. 2009. Breviscapine ameliorates hypertrophy of cardiomyocytes induced by high glucose in diabetic rats via the PKC signaling pathway. *Acta Pharmacol Sin* 30(8): 1081–1091.

Wang, M., Zhang, W.B., Zhu, J.H., Fu, G.S., and Zhou, B.Q. 2010. Breviscapine ameliorates cardiac dysfunction and regulates the myocardial Ca(2+)-cycling proteins in streptozotocin-induced diabetic rats. *Acta Diabetol* 47(Suppl 1): 209–218.

Wiernsperger, N.F. 2003. Oxidative stress as a therapeutic target in diabetes: Revisiting the controversy. *Diabetes Metab* 29(6): 579–585.

Wu, Y., Xia, Z.Y., Dou, J., Zhang, L., Xu, J.J., Zhao, B., Lei, S., and Liu, H.M. 2011. Protective effect of ginsenoside Rb1 against myocardial ischemia/reperfusion injury in streptozotocin-induced diabetic rats. *Mol Biol Rep* 38(7): 4327–4335.

Xie, Z.C., Qian, Z.K., and Liu, Z.W. 1993. Effect of ginseng on antiperoxidate injury in myocardium and erythrocytes in streptozocin-induced diabetic rats. *Zhongguo Zhong Xi Yi Jie He Za Zhi* 13(5): 262, 289–290.

Xu, T.J., Yuan, B.X., Zou, Y.M., and Zang, W.J. 2010. The effect of insulin in combination with selenium on blood glucose and GLUT4 expression in the cardiac muscle of streptozotocin-induced diabetic rats. *Fundam Clin Pharmacol* 24(2): 199–204.

Yao, Y., Li, R., Ma, Y., Wang, X., Li, C., Zhang, X., Ma, R., Ding, Z., and Liu L. 2012. α-Lipoic acid increases tolerance of cardiomyoblasts to glucose/glucose oxidase-induced injury via ROS-dependent ERK1/2 activation. *Biochim Biophys Acta* 1823: 920–929.

Yi, X., Xu, L., Hiller, S., Kim, H.S., and Maeda, N. 2012. Reduced alpha-lipoic acid synthase gene expression exacerbates atherosclerosis in diabetic apolipoprotein E-deficient mice. *Atherosclerosis* 223: 137–143.

Ziegler, D., Schatz, H., Conrad, F., Gries, F.A., Ulrich, H., and Reichel, G. 1997. Effects of treatment with the antioxidant alpha-lipoic acid on cardiac autonomic neuropathy in NIDDM patients. A 4-month randomized controlled multicenter trial (DEKAN Study). Deutsche Kardiale Autonome Neuropathie. *Diabetes Care*. 20: 369–373.

Zobali, F., Avci, A., Canbolat, O., and Karasu, C. 2002. Effects of vitamin A and insulin on the antioxidative state of diabetic rat heart: A comparison study with combination treatment. *Cell Biochem Funct* 20(2): 75–80.

10 Clinical Application of Antiaging Food Factors in Functional Foods

Akira Kubo and Fumi Nihei

CONTENTS

10.1 INTRODUCTION

For an effective antiaging diet, it is critical to consider the following components:

- Quantity of food
- Nutrients contained in the diet (including nutrients from foods and supplements)
- Sequence of eating

There are various other issues that affect the effectiveness of an antiaging diet, such as individual's serum vitamin/mineral concentrations and genetic factors. Since it is impossible to cover all these issues, this chapter focuses on some of the major topics that have gained attention in recent years.

Before discussing the details of an antiaging diet (defined as foods and supplements in this chapter), it is important to note that diet alone cannot be a complete strategy—other factors such as physical activity and mental management are also integral. Although these are not covered in this chapter, the neuronal circuit model proposed by Mattson[1]—describing how stimulations from caloric restriction, optimum physical activity, and cognitive challenge are integrated at glutamatergic neuron in the hippocampus—may be useful in clinical practice (Figure 10.1). In terms of caloric restriction and longevity, the results reported in 2012 by the National Institute on Aging (NIA) differed from the previous positive results of the Wisconsin National Primate Research Center (WNPRC) reported in 2009. Mattison et al.[2] explain that composition of diets in the two diets or genetic factors may have led to these differences. Whether these study results can be applied to human, and if so to what extent, is another important issue that needs further investigation.

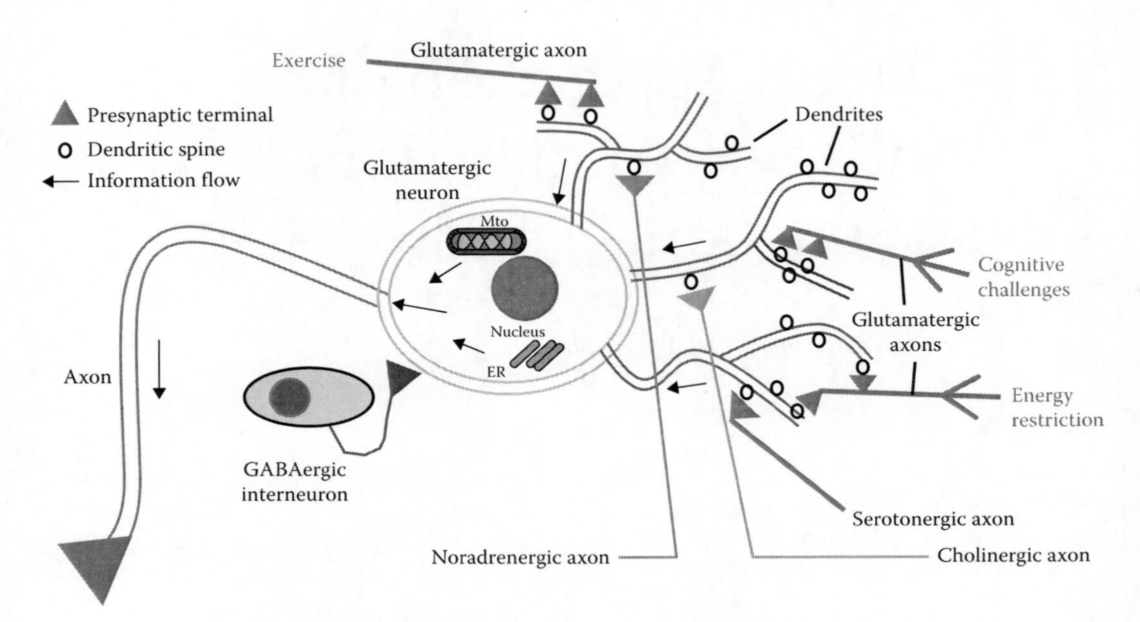

FIGURE 10.1 Neuronal circuit model. ER, endoplasmic reticulum; Mto, mitochondria. (Adapted from Mattson, M.P., *Cell Metab*, 16(6), 706–722, 2012.)

10.2 ANTIAGING DIET PATTERNS

There are various diets considered effective for antiaging, which differ in dietary compositions, such as salt, fat, carbohydrate, or protein. Figure 10.2 lists the nutrient composition and cardioprotective effects of four popular antiaging diets.[3] Regarding the Mediterranean diet, a report by Estruch et al.[4] in 2013 suggested a 30% decrease in the risk of major cardiovascular events. This was a cohort study with a follow-up of 4.8 years, evaluating the effects of Mediterranean diet supplemented with extra-virgin olive oil or with mixed nuts.[4] For vegetarian diets, Huang et al.[5] suggested that vegetarians had reduced risk of ischemic heart disease mortality and cancer incidence based on a systematic review and meta-analysis (Table 10.1).

Contrary to these antiaging diets, meat consumption has been associated with increased mortality. Pan et al.[6] analyzed the data from two cohort studies and estimated that substituting one serving per day of red meat with other foods could reduce the mortality risk by 7%–19%. Although the mechanism is not well understood, eating less meat may be one component of an antiaging diet. Fish and its major dietary component, omega-3 polyunsaturated fatty acids, are promoted in many antiaging diets and have been studied extensively. Wu et al.[7] conducted a systematic review and meta-analysis, suggesting that fish consumption was inversely associated with colorectal cancer incidence.

10.3 ANTIAGING FOOD FACTORS

The antiaging property of fish is thought to be the effect of omega-3 polyunsaturated fatty acids. Omega-3 fatty acids may prevent arteriosclerosis via inhibiting the synthesis of inflammatory mediators (prostaglandins and leukotrienes) from omega-6 fatty acids and producing anti-inflammatory mediators (resolvin and protectin). Regardless of better understanding of the potential mechanisms, results from clinical studies are yet inconclusive. Kwak et al.[8] conducted a meta-analysis of 14 randomized controlled studies and found a 9% reduction in cardiovascular death with omega 3 fatty acid supplementation. This effect, however, disappeared when a study with major methodological problem was excluded.[8] Additionally, in terms of polyunsaturated fatty acids, analysis of

	Nutrients							Evidence for CVD prevention			
	CHO protein SFA MUFA PUFA (cal %)	Cholesterol (mg)	DF (g)	K (mg)	Ca (mg)	Mg (mg)	Na (mg)	Ecologic studies of clinical endpoints	Randomized trials of risk factors	Prospective cohort studies of clinical endpoints	Randomized trials of clinical endpoints
DASH	48–58 15–25 6 13–21 8–10	150	29	4450	1181	473	2190		⬇	⬇	
Mediterranean	39–47 15–18 10–13 10–23 5–7	–	20–29	4589	1028	396	2532	⬇	⬇	⬇	
Vegetarian	–	–	–	–	–	–	–			↓	
Traditional Japanese	79 13 2 2–3 3–4	–	22	2623	315	317	2370	⬇		↓	

⬇⬆ : Consistent evidence from multiple well-conducted studies, with little or no evidence to the contrary.

⇩⇧ : Fairly consistent evidence from several well-conducted studies, but with some perceived shortcomings in the available evidence or some evidence to the contrary that precludes a more definite judgment.

↓↑→ : Some evidence from a relatively limited number of studies, but with relevant evidence to the contrary that raises important questions.

FIGURE 10.2 Nutrient composition and cardioprotective effects of four popular diets. CHO, carbohydrate; CVD, cardiovascular disease; MUFA, monounsaturated fatty acid; PUFA, polyunsaturated fatty acid; SFA, saturated fatty acid. (From Mozaffarian, D. et al., *Circulation* 123, 2870–2891, 2011. With permission.)

TABLE 10.1
Results of Meta-Analyses on the Effects of Vegetarian Diet

Endpoint	Effect Size (95% CI)
All-cause mortality	0.91 (0.66, 1.16)
Ischemic heart disease mortality	0.71 (0.56, 0.87)
Cerebrovascular disease mortality	0.88 (0.70, 1.06)
Cancer incidence	0.82 (0.67, 0.97)

Source: Huang, T. et al., *Ann Nutr Metab* 60, 233–240, 2012.
CI, confidence interval.

the Nurses' Health Study suggested that linoleic acid intake, as well as waist circumference, was inversely associated with telomere length among middle- and older age women.[9] Future studies with carefully selected clinical endpoints are needed to further evaluate the potential benefits of polyunsaturated fatty acids.

For cerebrovascular diseases, Cassidy et al.[10] found that women with high intake of flavanone (one of the major flavonoid subclasses) had a reduced risk of ischemic stroke based on a prospective study with a follow-up of 14 years. Another cohort study in Finland with a median follow-up of 12 years showed a reduced stroke risk for men with high serum lycopene concentration.[11]

Flavonoids, similar to vitamin D, are expected to have pleiotropic effects. For example, anthocyanin, another subclass of flavonoids, has shown to be associated with a lower risk of type 2 diabetes.[12] Lycopene may also have various antiaging effects, besides preventing cerebrovascular disease; a study by Stahl et al.[13] suggested that lycopene, along with β-carotene, may protect the skin from light-induced photo-oxidative processes.

As mentioned in this subsection, the role of vitamin D goes beyond bone health and calcium metabolism, acting on muscles, blood vessels, and immune system. Therefore, the antiaging property of vitamin D and calcium should be considered in a broader perspective, including not only osteoporosis and fracture, but also atherosclerosis and cardiovascular diseases. A caution is needed, however, as the report from the National Institutes of Health–American Association of Retired Persons (AARP) Diet and Health Study in 2013 revealed that men with high supplemental calcium intake had an increased risk of cardiovascular death.[14]

When evaluating a diet or a food factor for its antiaging property, many studies investigate whether it can prevent the incidence of cardiovascular diseases. Rautiainen et al. calculated the total antioxidant capacity, taking into account all antioxidants and their synergistic effects. Their study showed that women with the highest dietary total antioxidant capacity had a 20% lower risk of myocardial infarction.[15]

In 2011, Nadtochiy et al. investigated the Mediterranean diet for its cardioprotective property (part of the antiaging property discussed in this chapter). They identified three components—phenolic compounds (resveratrol, etc.), nitrate/nitrite, and polyunsaturated fatty acids—and revealed their mechanisms of action on transcription factors, inhibiting inflammation and oxidation processes, and promoting autophagy (Figure 10.3).[16]

Resveratrol has been gaining attention for various antiaging effects, including anti-inflammation, antioxidation, platelet aggregation inhibition, protection of vascular endothelial function, and increase in metabolism. Moreover, recent studies suggest its effect on Sirtuin 1 (Sirt 1) and estrogen receptors, activation of 5′-adenosine monophosphate (AMP)-activated protein kinase,[17] and inhibition of cyclooxygenase-1 (COX-1). Although resveratrol appears to be a promising antiaging food factor, many issues still remain, such as the optimum dose and appropriate marker when evaluating its effectiveness.

10.3.1 Vitamins

Among various antiaging food factors, vitamins are one of the factors most extensively studied and are important for clinical use. Figure 10.4 shows the result from an antiaging medical checkup: serum vitamin C and E (α-tocopherol) concentrations stratified by gender, age group, and the use of supplements. The graph in Figure 10.4 shows that for all age group and gender, subjects not taking supplements (bars in the middle) have lower serum vitamin C and E levels than those taking supplements (bars on the left). Furthermore, investigating the association between these serum levels and carotid intima-media thickness (IMT) revealed that there was a trend for inverse association between serum vitamin C and carotid IMT (Figure 10.5).

Recent studies suggest how genetic factors may influence serum vitamin levels and the extent of their health effects. Block et al.[18] suggested that serum vitamin C concentrations may be affected by the genotype of glutathione *S*-transferases (GSTs) and England et al.[19] suggested that the anti-inflammatory effect of vitamin E may depend on variants in the genes encoding tumor necrosis factor (TNF)-α, interleukin (IL)-10, and glutathione *S*-transferase pi 1 (*GSTP1*) gene. These studies suggest the importance of considering individual's genetic factors when interpreting serum vitamin C and E levels.

In 2012, a large randomized controlled trial (Physicians' Health Study II [PHS II]) involving more than 14,000 physicians revealed that daily multivitamin significantly reduced the risk of total cancer by 8%.[20] Anti-inflammatory effect, which is an integral component of antiaging

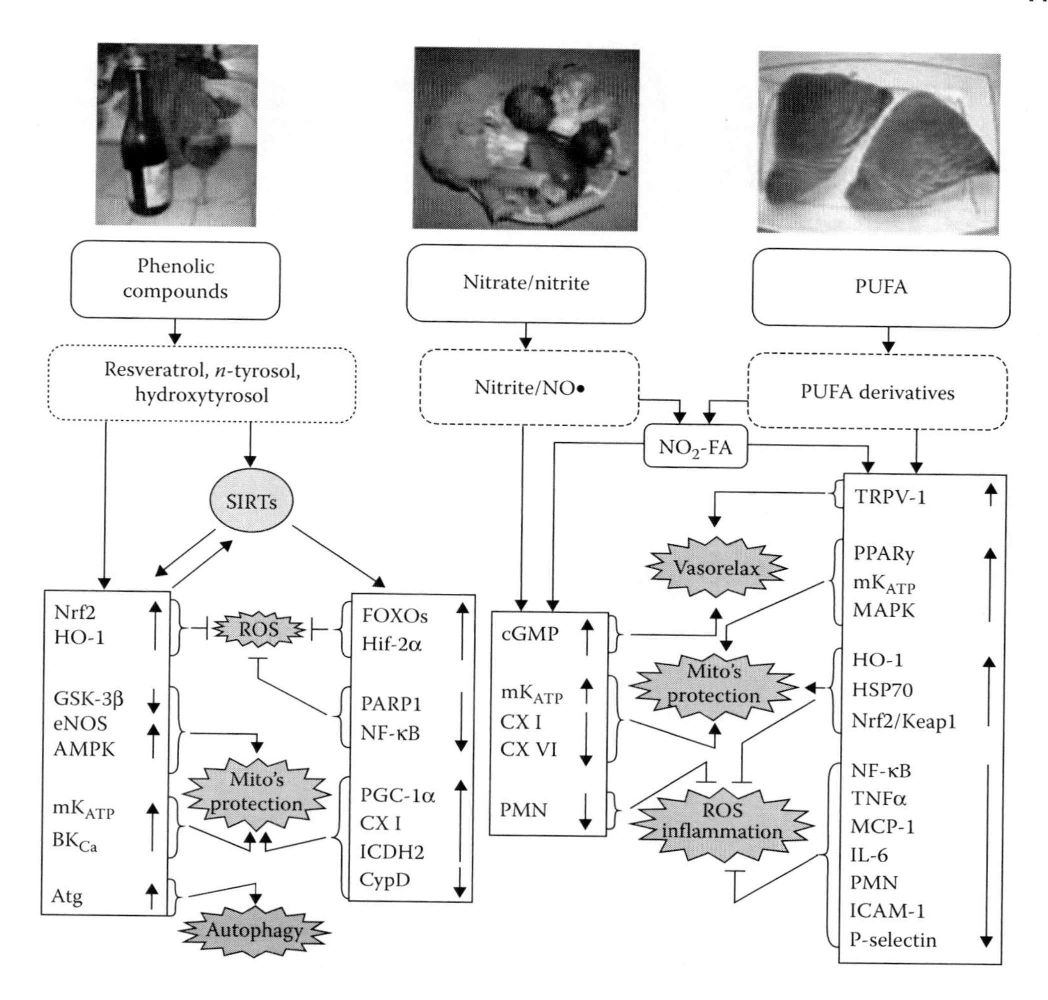

FIGURE 10.3 Mechanism underlying the cardioprotective effects of Mediterranean diet. (Nadtochiy, S.M. et al., *Nutrition* 27, 733–744, 2011. With permission.)

effects, is thought to be one of the mechanisms of this observed effect of cancer prevention. When investigating whether vitamins and other food factors have an antiaging effect, it is important to consider all of the factors, such as dose, serum concentrations, and genetic factors that may influence any of the reactions involved in the mechanism of its antiaging action.

10.4 CONCLUSION

An appropriate dietary approach for antiaging should be considered in the context of the whole lifestyle, including physical exercise and other factors.

Total calories, various nutrients, and timing and sequence of eating may all have some influence on the degree of aging. As a dietary pattern, the Mediterranean diet including fish (omega-3 fatty acids) and olive oil has been extensively studied in clinical trials, and in terms of nutrients, flavanone, vitamin D, and resveratrol are few of the nutrients of increased interests.

Further well-designed studies, especially in terms of appropriate clinical markers to evaluate the effectiveness of these various diets and nutrients, are awaited.

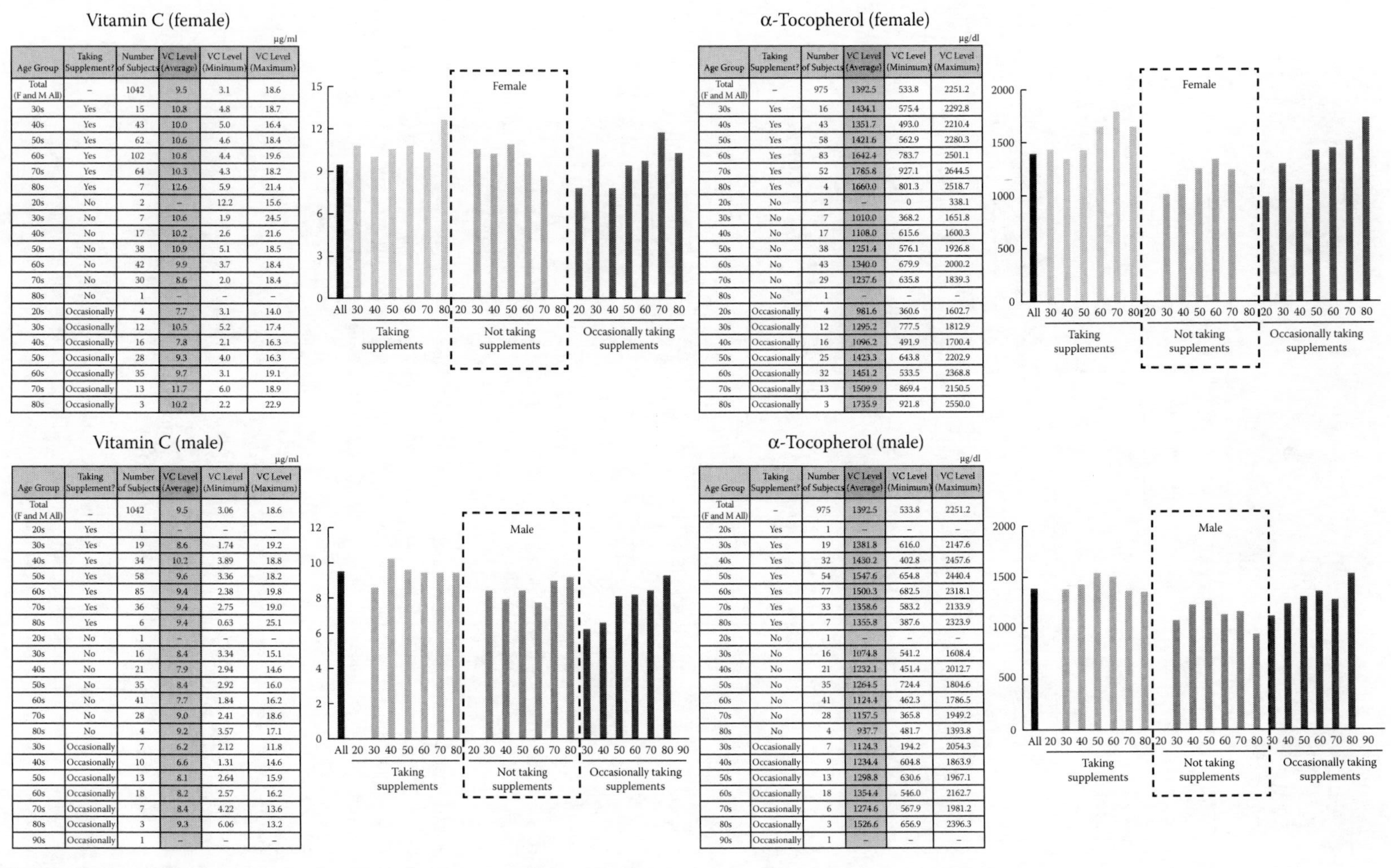

Vitamin C (female)

μg/ml

Age Group	Taking Supplement?	Number of Subjects	VC Level (Average)	VC Level (Minimum)	VC Level (Maximum)
Total (F and M All)	–	1042	9.5	3.1	18.6
30s	Yes	15	10.8	4.8	18.7
40s	Yes	43	10.0	5.0	16.4
50s	Yes	62	10.6	4.6	18.4
60s	Yes	102	10.8	4.4	19.6
70s	Yes	64	10.3	4.3	18.2
80s	Yes	7	12.6	5.9	21.4
20s	No	2	–	12.2	15.6
30s	No	7	10.6	1.9	24.5
40s	No	17	10.2	2.6	21.6
50s	No	38	10.9	5.1	18.5
60s	No	42	9.9	3.7	18.4
70s	No	30	8.6	2.0	18.4
80s	No	1	–	–	–
20s	Occasionally	4	7.7	3.1	14.0
30s	Occasionally	12	10.5	5.2	17.4
40s	Occasionally	16	7.8	2.1	16.3
50s	Occasionally	28	9.3	4.0	16.3
60s	Occasionally	35	9.7	3.1	19.1
70s	Occasionally	13	11.7	6.0	18.9
80s	Occasionally	3	10.2	2.2	22.9

α-Tocopherol (female)

μg/dl

Age Group	Taking Supplement?	Number of Subjects	VC Level (Average)	VC Level (Minimum)	VC Level (Maximum)
Total (F and M All)	–	975	1392.5	533.8	2251.2
30s	Yes	16	1434.1	575.4	2292.8
40s	Yes	43	1351.7	493.0	2210.4
50s	Yes	58	1421.6	562.9	2280.3
60s	Yes	83	1642.4	783.7	2501.1
70s	Yes	52	1785.8	927.1	2644.5
80s	Yes	4	1660.0	801.3	2518.7
20s	No	2	–	0	338.1
30s	No	7	1010.0	368.2	1651.8
40s	No	17	1108.0	615.6	1600.3
50s	No	38	1251.4	576.1	1926.8
60s	No	43	1340.0	679.9	2000.2
70s	No	29	1237.6	635.8	1839.3
80s	No	1	–	–	–
20s	Occasionally	4	981.6	360.6	1602.7
30s	Occasionally	12	1295.2	777.5	1812.9
40s	Occasionally	16	1096.2	491.9	1700.4
50s	Occasionally	25	1423.3	643.8	2202.9
60s	Occasionally	32	1451.2	533.5	2368.8
70s	Occasionally	13	1509.9	869.4	2150.5
80s	Occasionally	3	1735.9	921.8	2550.0

Vitamin C (male)

μg/ml

Age Group	Taking Supplement?	Number of Subjects	VC Level (Average)	VC Level (Minimum)	VC Level (Maximum)
Total (F and M All)	–	1042	9.5	3.06	18.6
20s	Yes	1	–	–	–
30s	Yes	19	8.6	1.74	19.2
40s	Yes	34	10.2	3.89	18.8
50s	Yes	58	9.6	3.36	18.2
60s	Yes	85	9.4	2.38	19.8
70s	Yes	36	9.4	2.75	19.0
80s	Yes	6	9.4	0.63	25.1
20s	No	1	–	–	–
30s	No	16	8.4	3.34	15.1
40s	No	21	7.9	2.94	14.6
50s	No	35	8.4	2.92	16.0
60s	No	41	7.7	1.84	16.2
70s	No	28	9.0	2.41	18.6
80s	No	4	9.2	3.57	17.1
30s	Occasionally	7	6.2	2.12	11.8
40s	Occasionally	10	6.6	1.31	14.6
50s	Occasionally	13	8.1	2.64	15.9
60s	Occasionally	18	8.2	2.57	16.2
70s	Occasionally	7	8.4	4.22	13.6
80s	Occasionally	3	9.3	6.06	13.2
90s	Occasionally	1	–	–	–

α-Tocopherol (male)

μg/dl

Age Group	Taking Supplement?	Number of Subjects	VC Level (Average)	VC Level (Minimum)	VC Level (Maximum)
Total (F and M All)	–	975	1392.5	533.8	2251.2
20s	Yes	1	–	–	–
30s	Yes	19	1381.8	616.0	2147.6
40s	Yes	32	1430.2	402.8	2457.6
50s	Yes	54	1547.6	654.8	2440.4
60s	Yes	77	1500.3	682.5	2318.1
70s	Yes	33	1358.6	583.2	2133.9
80s	Yes	7	1355.8	387.6	2323.9
20s	No	1	–	–	–
30s	No	16	1074.8	541.2	1608.4
40s	No	21	1232.1	451.4	2012.7
50s	No	35	1264.5	724.4	1804.6
60s	No	41	1124.4	462.3	1786.5
70s	No	28	1157.5	365.8	1949.2
80s	No	4	937.7	481.7	1393.8
30s	Occasionally	7	1124.3	194.2	2054.3
40s	Occasionally	9	1234.4	604.8	1863.9
50s	Occasionally	13	1298.8	630.6	1967.1
60s	Occasionally	18	1354.4	546.0	2162.7
70s	Occasionally	6	1274.6	567.9	1981.2
80s	Occasionally	3	1526.6	656.9	2396.3
90s	Occasionally	1	–	–	–

FIGURE 10.4 Serum vitamins C and E (α-tocopherol) concentrations. (Data from Kenko Jumyo Dock, Antiaging checkup.)

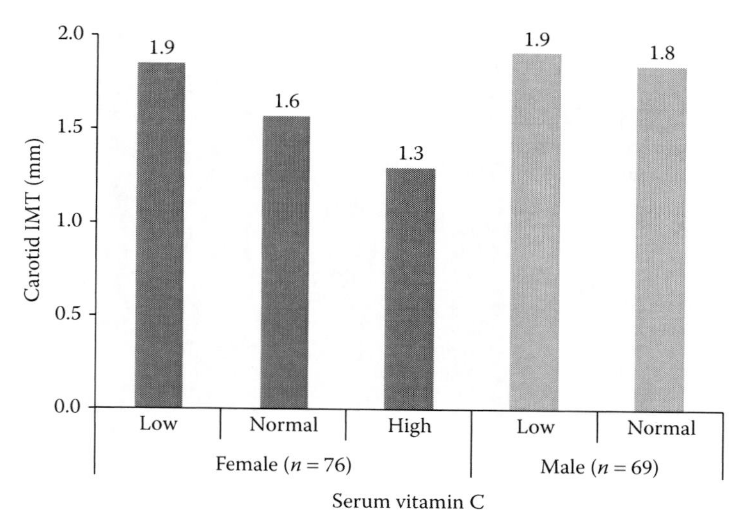

FIGURE 10.5 Association between serum vitamin C concentration and carotid intima-media thickness. (Data from Kenko Jumyo Dock, Antiaging checkup.)

REFERENCES

1. Mattson, M.P. 2012. Energy intake and exercise as determinants of brain health and vulnerability to injury and disease. *Cell Metab* 16(6):706–722.
2. Mattison, J.A. et al. 2012. Impact of caloric restriction on health and survival in rhesus monkeys from the NIA study. *Nature* 489(7415):318–321.
3. Mozaffarian, D. et al. 2011. Components of a cardioprotective diet: New insights. *Circulation* 123(24):2870–2891.
4. Estruch, R. et al. 2013. Primary prevention of cardiovascular disease with a Mediterranean diet. *N Engl J Med* 368:1279–1290.
5. Huang, T. et al. 2012. Cardiovascular disease mortality and cancer incidence in vegetarians: A meta-analysis and systematic review. *Ann Nutr Metab* 60(4):233–240.
6. Pan, A. et al. 2012. Red meat consumption and mortality: Results from 2 prospective cohort studies. *Arch Intern Med* 172(7):555–563.
7. Wu, S. et al. 2012. Fish consumption and colorectal cancer risk in humans: A systematic review and meta-analysis. *Am J Med* 125(6):551–559.e5.
8. Kwak, S.M. et al. 2012. Efficacy of omega-3 fatty acid supplements (eicosapentaenoic acid and docosahexaenoic acid) in the secondary prevention of cardiovascular disease: A meta-analysis of randomized, double-blind, placebo-controlled trials. *Arch Intern Med* 172(9):686–694.
9. Cassidy, A. et al. 2010. Associations between diet, lifestyle factors, and telomere length in women. *Am J Clin Nutr* 91(5):1273–1280.
10. Cassidy, A. et al. 2012. Dietary flavonoids and risk of stroke in women. *Stroke* 43(4):946–951.
11. Karppi, J. et al. 2012. Serum lycopene decreases the risk of stroke in men: A population-based follow-up study. *Neurology* 79(15):1540–1547.
12. Wedick, N.M. et al. 2012. Dietary flavonoid intakes and risk of type 2 diabetes in US men and women. *Am J Clin Nutr* 95(4):925–933.
13. Stahl, W. et al. 2012. β-Carotene and other carotenoids in protection from sunlight. *Am J Clin Nutr* 96(5):1179S–1184S.
14. Xiao, Q. et al. 2013. Dietary and supplemental calcium intake and cardiovascular disease mortality: The National Institutes of Health–AARP Diet and Health Study. *JAMA Intern Med* 4:1–8.
15. Rautiainen, S. et al. 2012. Total antioxidant capacity from diet and risk of myocardial infarction: A prospective cohort of women. *Am J Med* 125(10):974–980.
16. Nadtochiy, S.M. et al. 2011. Mediterranean diet and cardioprotection: The role of nitrite, polyunsaturated fatty acids, and polyphenols. *Nutrition* 27(7/8):733–744.
17. Price, N.L. et al. 2012. SIRT1 is required for AMPK activation and the beneficial effects of resveratrol on mitochondrial function. *Cell Metab* 15(5):675–690.

18. Block, G. et al. 2011. Serum vitamin C and other biomarkers differ by genotype of phase 2 enzyme genes GSTM1 and GSTT1. *Am J Clin Nutr* 94(3):929–937.
19. England, A. et al. 2012. Variants in the genes encoding TNF-α, IL-10, and GSTP1 influence the effect of α-tocopherol on inflammatory cell responses in healthy men. *Am J Clin Nutr* 95(6):1461–1467.
20. Gaziano, J.M. et al. 2012. Multivitamins in the prevention of cancer in men: The Physicians' Health Study II randomized controlled trial. *JAMA* 308(18):1871–1880.

Section III

Problems and Challenges to Industries, Consumers, and Policy Makers

11 Maximizing the Survival of Probiotic Bacteria in Food to Improve Their Potential Health Benefit

Namrata Taneja, Derek Haisman, and Shantanu Das

CONTENTS

11.1 INTRODUCTION

Probiotics are defined by the Food and Agricultural Organization of the United Nations (FAO) and the World Health Organization (WHO) as "live microorganisms, which when administered in adequate amounts, confer a health benefit on the host" (FAO/WHO 2001). Probiotics have been known for improving gut health since the beginning of the nineteenth century. In addition to gut health benefits such as alleviation of constipation and protection against traveler's diarrhea, they have also been documented to exert many other health-promoting effects (Ooi and Liong 2010; Soccol et al. 2010). There is sufficient evidence that the oral administration of lactobacilli and bifidobacteria can restore the normal balance of microbial populations in the gut, but further human trials are required to fully substantiate other specific health claims (Shah 2007). The European Food Safety Authority has so far refused to allow claims of health benefits from the consumption of probiotics, although it is expected that this issue will eventually be resolved (Rijkers et al. 2011).

The probiotic potential of different bacterial strains, even within the same species, differs. Different strains of the same species are always unique and may have differing areas of adherence (site specific), and their specific immunological effects and actions on a healthy or an inflamed mucosal milieu may be distinct from each other (Veld and Havenaar 1991).

Probiotics have been classified as functional food ingredients and are incorporated in an increasing variety of food products. However, ensuring that they are effective in a food is a considerable technological challenge. They must be able to be incorporated into foods without producing off-flavors or undesirable textures, and they should be viable but not growing (Kopp-Hoolihan 2001). They also have to be available at a high concentration, typically 10^6–10^7 cfu g^{-1} of product. To achieve that they need to survive the manufacturing process, preservation and storage conditions, and gastrointestinal stress factors. All of these stress factors need to be considered during the selection of a probiotic strain (Kosin and Rakshit 2006) and have been investigated at the Riddet Institute, Massey University, New Zealand, over a number of years (Poddar et al. 2012).

The most extensively studied and widely used probiotics are strains of lactic acid bacteria (LAB). They are widely used in the food and pharmaceutical industry, and large quantities of their biomass and the end products of their metabolism are produced today (Zacharof et al. 2010). The genus *Lactobacillus* is by far the largest of the genera included in LAB and includes the most acid tolerant of the LAB (Axelsson 1998). These bacteria are thermophilic bacteria, and thus have the advantage of withstanding higher temperatures during the drying process required for their prolonged storage. Their biotechnological importance in the production of cheeses and fermented milk, which all require incubation at a relatively high temperature of 45°C or above, during their manufacturing process, is well known (Kosin and Rakshit 2006).

The most popular LAB used as probiotics are strains of genera *Bifidobacterium* and *Lactobacillus*, which are known to be resistant against gastric acid, bile salts, and pancreatic enzymes, and can adhere to colonic mucosa and readily colonize the intestinal tract. Soccol et al. (2010) listed 24 species of bifidobacteria and lactobacilli with probiotic properties that are commercially available and that include *B. bifidum*, *B. lactis*, *L. acidophilus*, *L. casei*, and *L. plantarum*, among many others. Lactobacilli are almost always preserved for subsequent use by dehydration.

In the production of probiotic bacteria to be incorporated in a food product, there are four main stages:

1. Growth of a suitably concentrated culture broth
2. Harvesting and adaptation of the bacteria for subsequent drying
3. Dehydration
4. Storage and then incorporation in the food product

The conditions in each stage can affect the viability and consequent effectiveness of the bacteria in the product.

11.2 GROWTH

The standard for fermented milks is that at least 10^7 cfu· g^{-1} of probiotic bacteria should be present at the time of consumption (Codex Alimentarius 2003). Thus, the development of concentrated viable starter cultures is required to successfully incorporate these microorganisms into food products in the required numbers (Miao et al. 2008). There are few publications defining the optimal cell concentration for dehydration. Historically, high concentrations of $>1 \times 10^8$ cells ml^{-1} have been selected on the basis that the higher the initial cell concentration, the greater the number of surviving cells subsequently. Costa et al. (2000) found that loss of viability was greater when initial cell concentrations were high, but was better related to the protective medium used. Palmfeldt et al. (2003) also showed a correlation between cell concentration and protective media. However, Linders et al. (1997b) found that the initial cell density directly correlated to the residual activity of normally grown *L. plantarum* after drying.

Lactobacilli are fastidious bacteria with numerous requirements for growth. There have been many studies on optimization of the growth medium to increase cell density and biomass, mainly concerned with the influence of various carbon and nitrogen sources, other growth factors (amino acids and vitamins), and culture conditions, such as temperature, pH, and the level of aeration (e.g., Bevilacqua et al. 2008; Fung et al. 2008; Liew et al. 2005). Liew et al. (2005) found that yeast extract, glucose, vitamins, and pH could significantly affect the growth of *L. rhamnosus*. Fung et al. (2008) showed that meat extract, vegetable extract, and peptone significantly influenced the growth of *L. acidophilus*. Meat extract contains a greater amount of total nitrogen (12% w/w) than yeast extract (10% w/w) and offers a cheap alternative to yeast extract (Polak-Berecka et al. 2010). Regarding the effect of carbohydrates, Liew et al. (2005) showed a slight effect of glucose on the viable cell count of *L. rhamnosus*, and lactose and maltose have been found to increase the biomass production of *L. plantarum* concentration. The effect of lactose was not linear as it required a threshold value of 20 g l^{-1} (Bevilacqua et al. 2008).

Most *Lactobacillus* species grow well in the De Man, Rogosa, and Sharpe (MRS) medium, which contains glucose, yeast and meat extracts, casein peptone, and various salts, but this is rather expensive for mass production (De Man et al. 1960). Consequently, alternative carbon and nitrogen sources have been tested, including whey (Alvarez et al. 2010; Bernardez et al. 2008; Brinques et al. 2010), goat milk whey (Aguirre-Ezkauriatza et al. 2010), and corn steep liquor (Ha et al. 2003), with some success.

LAB cultivations are characterized by a product inhibition kinetic that influences growth rate and productivity (Schiraldi et al. 2003). In a typical batch fermentation process, once the lactic acid concentration increases, the cell growth rate is severely diminished (Aguirre-Ezkauriatza et al. 2010; Alvarez et al. 2010; Yoo et al. 1996), and the upper limit of the concentration of cells is 10^9 cfu ml^{-1}. Even if the pH of the culture is kept at the optimum value of around 6.5 by the addition of alkali solution, the maximum cell concentration of lactobacilli appears to be limited to 10^{10} cfu ml^{-1} (Bernardez et al. 2008; Hayakawa et al. 1990). Avonts et al. (2004) reported that fermentation in MRS media controlled at pH 6.5 resulted in a maximal cell count of above 9 log cfu ml^{-1}, with all the glucose exhausted between 10 and 17 hours of fermentation. There has been a lot of development of fermentation strategies that could control the lactate concentration inside the medium, thus maintaining it below the toxic level, especially because this low-molecular-weight organic acid is a valuable product on the market with a widespread field of applications (Bai et al. 2003; Ding and Tan 2006; Zhang et al. 2010).

The effect of culture medium composition on the cell survival during the subsequent drying process is mainly through its influence on the metabolic flux of the cell and cell viability, which depends strongly on the osmotic pressure of the medium. The cell membrane of LAB is readily permeable to water but presents a more effective barrier to most other solutes. Therefore, when the external concentration of water changes because of increase or decrease in the concentrations of extracellular solutes that are excluded by the membrane, water moves out of or into the cells (Csonka and Hanson 1991). A lowering of the external water activity (hyperosmotic conditions) causes a rapid efflux of cellular water and loss of turgor; ultimately, the cells may plasmolyze, that is, the cytoplasmic membrane may retract from the cell wall. Similarly, upon hypo-osmotic shock, water flows into the cell and increases the cytoplasmic volume and/or turgor pressure. To survive osmotic stress, significant physiological changes have been reported in bacteria, which include the induction of specific stress proteins as well as the accumulation of specific solutes under hyperosmotic conditions and release of these under hypo-osmotic conditions. Such solutes are often referred to as compatible solutes (Hartke et al. 1996; Poolman and Glaasker 1998) because they can be accumulated to high levels by *de novo* synthesis or transport without interfering with vital cellular processes. They include potassium ions, amino acids (e.g., glutamate, proline), amino acid derivatives (peptides, *N*-acetylated amino acids), quaternary amines (e.g., glycine betaine, carnitine), sugars (e.g., sucrose, trehalose), and tetrahydropyrimidines (ectoines). The protective effect of sucrose on the viability of bacteria has been attributed to its action as a compatible solute (Santivarangkna et al. 2008). Glycine betaine and carnitine, present in the yeast and meat extracts in MRS medium (Robert et al. 2000), have proven to be protective during the growth and drying of LAB but are only accumulated by the cell when it is osmotically stressed (Kets and de Bont 1994; Kets et al. 1996; Wood et al. 2001). It is still unclear how these compatible solutes work, but a possible explanation could be their role in altering the physical properties of cell membranes and preventing the aggregation of cellular proteins (Carvalho et al. 2004).

Different strains of LAB can differ in their capability to accumulate glycine betaine as a compatible solute, which is due to some genotypical differences. This accumulation affects their ability to survive drying (Kets et al. 1996). O'Callaghan and Condon (2000) also found different strains of *Lactococcus lactis* can differ in their capability to accumulate glycine betaine as a compatible solute due to the variations in the genotype.

11.3 ADAPTATION

The accumulation of compatible solutes is one example of a preadaptation process that improves the long-term viability of bacteria. This process generally involves applying some sort of stress condition to the cells in or subsequent to the growth phase. In response, the bacteria undergo a metabolic reprogramming that leads to a cellular state of enhanced resistance (Hartke et al. 1996). Typical is osmotic stress induced by sodium chloride (NaCl) to increase the accumulation of compatible solutes as a protection against osmotic shock in subsequent dehydration. The disadvantage is that the stress may also inhibit the growth of the bacteria. Linders et al. (1997b) showed that *L. plantarum* grown under osmotic stress (1 mol l^{-1} NaCl) had lower glucose fermentation activities after fluidized bed (FB) drying or air-convective drying, despite having a higher accumulation of the compatible solutes betaine and carnitine in comparison with cells grown without NaCl. Nag and Das (2012c) found that exposing lactobacilli and bifidobacteria to osmotic stress, followed by fluid bed drying, greatly improved their viability over freeze-dried cultures.

The addition of sucrose to the culture medium of lactobacilli that led to improved stress tolerance can also be considered as an adaptation effect, given that sucrose is not universally utilizable by LAB. *Lactobacillus delbrueckii ssp. bulgaricus* grown in the presence of 20 g l^{-1} sucrose had improved survival during storage following spray drying compared to the control (Silva et al. 2004). In another case, the same species, when grown in the presence of sucrose (11.3 g l^{-1}) had a tenfold increase in cell survival after drying compared to the control (Tymczyszyn et al. 2007). The reason was suggested to be the cell adaptation to the decrease in water activity caused by the sucrose in the growth medium.

Adaptation to osmotic stress is able to exert cross-protection toward heat and other stresses (Desmond et al. 2002; Poolman and Glaasker 1998), but exposure of cells to heat to induce a heat-shock (HS) response has also been investigated for its effect in improving the cell survival during thermal drying. Desmond et al. (2002) found that the probiotic strain *L. paracasei* when prestressed by either heat (52°C for 15 min) or salt (0.3 M for 30 min), survived up to 300-fold better than unadapted control cells during heat stress and 18-fold better during spray drying at outlet temperatures of 100°C–105°C. The lower improvement in cell viability after spray drying of preadapted cells could be due to the dehydration damage caused by the drying process. HS and osmotic shock also resulted in an increased cell survival of *L. rhamnosus* during storage following FB drying (Prasad et al. 2003).

The major problems encountered by cells at high temperature are the denaturation of proteins and their subsequent aggregation, destabilization of macromolecules such as ribosomes and RNA, and damage of cell walls and cell membranes. The HS response has been studied notably in *Escherichia coli* and *Bacillus subtilis*, and entails the synthesis of a number of proteins known as HS proteins. They bind substrate proteins in a transient noncovalent manner, prevent premature folding, promote the attainment of the stable state *in vivo*, and belong to a category of proteins known as "molecular chaperones." Their common names (e.g., DnaK, DnaJ, GrpE, GroES, and GroEL) refer to their genetic family. Among the HS proteins in LAB, chaperone proteins and proteases (Clp, HtrA, and FtsH) have often been identified. The chaperones help to stabilize RNA and repair misfolded proteins, whereas proteases degrade denatured proteins, both leading to an increased heat resistance of cells subjected to HS (Suokko et al. 2008). Prasad et al. (2003) found that the expression of classical HS proteins GroEL and DnaK was upregulated as a result of HS as well as by osmotic shock treatment, ensuring the protection of cellular proteins and other macromolecules from process-induced damage and denaturation. A strain of GroESL-overproducing *L. paracasei*, made by gene transfer, exhibited approximately tenfold better survival after spray drying than control cultures (Corcoran et al. 2006).

11.4 HARVESTING

The growth of bacteria has four distinct phases: lag, log, stationary, and death phases when grown in batch culture. Cells harvested in the stationary phase are well known to have the highest stress tolerance, compared to those harvested at the lag and log phases of a batch culture, as well as those harvested at the stationary phase of a continuous culture (Fu and Chen 2011). Thus, the stationary phase cells are commonly used in different kinds of preservation methods including freeze-drying, spray drying, FB drying, and vacuum drying. Bacteria spend most of their time in the stationary phase, which is most often due to nutrient starvation brought on by bacterial growth. Generally, cell size decreases with increasing time of starvation. Moreover, the overall rate of molecular synthesis is reduced in the stationary phase. Lipid, DNA, and RNA contents are diminished during the transition from growth to nongrowth. Although the rate of total protein synthesis decreases under nutrient limitation conditions, it seems that synthesis of specific "starvation-induced proteins" is important to protect the cells against deprivation. Thus, the stationary phase also provides protection against other environmental stresses. Many studies have shown that starved bacteria are far more resistant to killing doses of heat, hydrogen peroxide, acid pH, UV irradiation, or NaCl than their exponentially growing counterparts. Some of these starvation-induced proteins overlap with proteins synthesized in response to HS, ethanol, and oxidative or osmotic stress (Giard et al. 1996).

The survival response enables the cell to survive desiccation and adverse temperatures. From the comparable cell counts before drying (8.54 and 8.87 log cfu ml^{-1}) of *L. delbrueckii* ssp. *bulgaricus* harvested at log and stationary phases, cell counts after the spray drying of the culture were 5.80 and 7.45 log cfu ml^{-1}, respectively (Teixeira et al. 1995). Spray-dried *L. rhamnosus* cells gave the highest recovery when harvested from stationary-phase cells (31%–50%), whereas early log-phase cells exhibited 14% survival and lag-phase cells were the least robust, with only 2% cell survival (Corcoran et al. 2004). Prasad et al. (2003) reported higher storage stability of the dried *L. rhamnosus* HN001 that was heat shocked after the stationary phase to that of the culture that was heat shocked after the log phase. The HS proteins (DnaK and GroEL) were found to be upregulated after both log- and stationary-phase growth, indicating that there are other stationary phase-related viability factors that can further improve its survival. It is well established that starvation conditions prevail in stationary-phase growth cultures that could confer multiple stress resistance and lead to improved storage stability.

11.5 DEHYDRATION

Dehydration inactivation in lactic acid starter cultures may occur either solely during the drying of cells at their physiological temperatures or in concert with thermal- or cryoinactivation during the drying of cells at an extreme low or high temperature. Since water molecules contribute to the stabilities of proteins, DNA, and lipids as well as confer the structural order upon cells, the removal of water imposes physiological constraints on the cells. When cells are dried to low water content, a number of cellular components will be affected. Damage resulting from the dehydration process can be attributed to two primary causes: (1) changes in the physical state of the lipid membrane (Beney and Gervais 2001) and/or (2) changes in the structure of sensitive proteins (Teixeira et al. 1996). The cytoplasmic membrane is considered to be the most sensitive component and the main factor in the dehydration inactivation process because there is a loss of several intracellular components when the membrane is damaged. Membrane lipid bilayer structures are thermodynamically unstable, and therefore, the lipid membrane is a primary target for dehydration-induced damage. Cell membrane damage was demonstrated by measuring the hydrolyzed DNA from rehydrated *L. plantarum* cells after incubating with DNase (Lievense et al. 1994). The heightened level indicated the leakage of DNA from the cells. Lievense et al. (1994) concluded that cell wall and/or cell membrane damage is an important mechanism of dehydration inactivation of *L. plantarum*, but that thermal inactivation (up to 60°C) occurs by a different mechanism. Minimizing osmotic pressure variation (Gervais and Marechal 1994; Mille et al. 2004) during dehydration should enhance microbial survival because membrane integrity is preserved (Beney and Gervais 2001).

Drying rate and both initial and final water activity and moisture content are expected to influence the inactivation of bacteria by dehydration during FB/vacuum drying (Selmer-Olsen et al. 1999b). The water content of the drying air along with its temperature determines the drying time needed to dehydrate bacteria to a desired moisture level. High or low drying temperatures can induce different inactivation mechanisms on microbial cells. At higher temperatures, in the lethal range for bacteria, heat is more likely to denature the proteins. By contrast, at lower temperatures, dehydration damage becomes more dominant. The rate of drying depends on both the temperature and the moisture content. A study by Lievense et al. (1992) on *L. plantarum* showed that the drying rate had no significant influence on dehydration inactivation during drying up to a temperature of 55°C in an FB dryer. Linders et al. (1996) also found no significant influence of the drying rate on cell viability of *L. plantarum* at 5°C. By contrast, Tymczyszyn et al. (2008) found that at a mild drying temperature of 30°C, slow dehydration caused more dehydration damage to the cytoplasmic membrane than rapid dehydration. The initial water activity and the water content of the matrix containing the microorganism prior to drying influence the heat resistance of microorganisms during the subsequent drying process. Santivarangkna et al. (2006) found that the greatest loss of viability in *L. helveticus* occurred when the water activity fell from 0.85 to 0.6, although at this stage the rate of drying was quite low. Similar results were reported by Selmer-Olsen et al. (1999b) who found the viability of *L. helveticus* dried by an FB dryer (5°C and ~55% relative humidity [RH]) decreased when the water content fell below 0.3–0.4 g H_2O (g dry weight)$^{-1}$. This may be due to physiological damage from large lateral compressive stresses in the plane of the membrane at low water potential.

At high drying rates, the fast diffusion of water out of the cells may promote the formation of the glassy state and improve the stability of starter cultures during storage (Santivarangkna et al. 2008). At low temperatures (low drying rates), crystallization of the cytoplasmic components can occur and, as a consequence, the disruption of cell structures takes place. Other studies on mild-temperature thermal drying, osmotic shock of microbial cells at ambient temperatures and freeze-drying all suggested that a dramatic change in osmotic pressure tends to cause excessive cell death (Friesen et al. 2005; Mille et al. 2004). The actual effect of the drying rate on cell survival, overall, still remains unquantified and sometimes confusing; the response is both species specific and dependent on other factors, such as organism age, history, and support media during drying, and thus needs to be experimentally determined for each drying environment (Fu and Chen 2011).

Low initial water content should be more favorable than high initial water content, since there is less water to evaporate, thus a higher processing efficiency. When cells are dried in FB drier or vacuum drier, the feed is in a solid form such as cell pellets or cell pastes; the decrease of initial water content under these circumstances may improve the stress tolerance of cells. However, there is a risk that the osmotic shock caused by the decrease in the water content prior to drying may still lead to undesirable cell death (Mille et al. 2004).

11.6 PROTECTIVE CARRIERS/DRYING MEDIA

Addition of protective carriers prior to drying is a common means to protect cells during drying and storage. Fu and Chen (2011) grouped the commonly used protective carriers into three categories on the basis of their physical states when mixed with cells and after being dried: carrier in liquid state, carrier in solid state, and carrier used to encapsulate cells. The selection of carrier largely depends on the drying process used, but its effectiveness could be strain specific. For low-temperature drying methods such as FB and vacuum drying, the feed in solid form is preferred (Bayrock and Ingledew

1997; Tymczyszyn et al. 2007) because, first, it saves the cost and, second, it improves the drying efficiency because less water needs to be removed. In this case, the bacteria are mixed directly with the protective carrier, which may also act as a support material. This can impose an osmotic stress upon the bacteria (Mille et al. 2004). By contrast, in high-temperature drying methods such as spray drying, the feed solution is in liquid form. The high water content during spray drying helps keep the temperature of drying material at around the wet-bulb temperature, consequently preventing cells from being overheated by the high inlet temperature of a spray drier. Incorporating a protectant in the liquid feed can stabilize the cellular structure of the bacteria and may also alleviate the osmotic stress during drying (Fu and Chen 2011).

One of the most extensively investigated protectants is trehalose which facilitates microorganisms to survive dehydration. Survival in extreme dessication has been observed in the nature and is known as anhydrobiosis (Crowe et al. 1992). Trehalose can be used as a carrier in both solution and solid forms. Its protective effect can be attributed to the following factors. First, trehalose molecules are able to act as an alternative to water molecules in sustaining the original conformation of the lipid bilayer of cell membrane (Crowe et al. 1992). The barrier properties of cytoplasmic membranes are critically dependent on the fluidity of the lipid bilayers. In the fluid state, the lipids are hydrated and in constant lateral motion, and quickly reseal any imperfections or gaps in the membrane, thus preserving the integrity of the cell. As water is removed, the lipid bilayers become compressed and gradually change to a more solid or gel state, as the van der Waals attraction forces between their hydrocarbon chains increase. During rehydration, the dry lipids undergo a phase transition from the gel back to a liquid crystalline phase and are more prone to discontinuities and leaks in the lipid membrane. Crowe et al. (1992) hypothesized that trehalose can prevent this leakage by substituting for water in the lipid bilayer so that it retains some fluidity during dehydration. As a result, the dry lipids do not undergo a gel/liquid crystal transition during rehydration and do not leak. This mechanism is well discussed by Santivarangkna et al. (2008).

Second, trehalose has a high-glass transition temperature T_g, so it has a higher tendency to stay glassy during storage, compared to other nonreducing disaccharides such as sucrose. When it is used as a protective matrix in a dry form, the high viscosity of the glassy state slows chemical reactions such as free radical oxidation compared to the crystalline state (Crowe et al. 1998). Finally, trehalose is also able to act as an effective thermoprotectant against protein denaturation (Eleutherio et al. 1993). However, the high price of trehalose inhibits its use in starter culture production on an industrial scale.

Trehalose was reported to preserve *L. acidophilus* during vacuum drying (Conrad et al. 2000). The presence of borate, which could cross-link the trehalose, thereby raising T_g of the dry matrix even further, significantly enhanced the protective ability of trehalose during drying and storage. Trehalose and sucrose were superior to glucose and maltodextrin in maintaining the cell viability of LAB strains *Enterococcus faecium* and *L. plantarum* in both freeze-drying and FB drying (Strasser et al. 2009). Nag and Das (2012a) used trehalose and lactose as cryoprotectants for *L. casei* during freeze-drying. By contrast, an extensive study on the effect of different sugars (sucrose, maltose, lactose, trehalose, glucose, fructose, and sorbitol) on the viability of *L. plantarum* immediately after FB drying found that only sorbitol and maltose, and to a lesser extent, sucrose, were effective (Linders et al. 1997a). However, the moisture contents after drying were high—mostly >10%. In a further study of the air drying of *L. plantarum*, Linders et al. (1997c) confirmed that sorbitol and maltose improved the postdrying activity from 44% for the control sample to 79% and 66%, respectively, whereas the addition of trehalose resulted in an activity of only 30%. The positive effect of maltose and sorbitol was thought not to be due to their ability to lower membrane phase transition temperature (T_m) but through their free radical scavenging activity. Selmer-Olsen et al. (1999a) studied the effect of various protective solutes and found that nonfat milk solids and betaine showed the best protective effect during FB drying and storage of *L. helveticus*, while the protective effect of glycerol was better during dehydration than during storage. Skim milk powder (SMP) or reconstituted skim milk powder (RSMP) are other carrier matrices that have been shown to improve cell survival during dehydration, especially during spray drying. It was suggested that this effective protection may be related to the lactose in SMP, where lactose interacts with cell membrane and helps to maintain the membrane integrity in a manner similar to other nonreducing disaccharides such as trehalose and sucrose (Fu et al. 2012). However, Santivarangkna et al. (2006) found that lactose when used as a solid carrier did not exert any effect in improving the survival of *L. helveticus* during vacuum drying compared to the control where cell pellets were dried without carrier. As the major components of skim milk are lactose, casein, and whey proteins, whether or not the significant protective effect of RSMP is due to the presence of milk proteins rather than the lactose remains unexplored.

Combination of a protectant and a bacterial culture often results in encapsulation of the bacteria, and microencapsulation has been extensively explored as a means of preserving viability during dehydration, storage, and transit of the gastrointestinal tract (e.g., De Vos et al. 2010; Favaro-Trinidade et al. 2011). Nag et al. (2011) developed a microencapsulation technique for *L. casei* 431 using a protein–polysaccharide complex.

Currently, freeze-drying is the most popular industrial method for the preservation of LAB, but spray drying and fluid bed drying, and occasionally vacuum drying, are also being used (Santivarangkna et al. 2007).

11.6.1 Freeze-Drying

Freeze-drying is the preferred method for culture collections worldwide, including the American Type Culture Collection (ATCC) and the National Collection of Type Cultures (NCTC) (Morgan et al. 2006). Freeze-dried cultures have advantages in that they are easy to handle and have low shipping and storage costs. A typical freeze-drying process consists of two steps: cells are first frozen at –196°C and then dried by sublimation under high vacuum. Even though freeze-drying is the most preferred method of preservation, a huge loss of viability during the freezing step (Uzunova-Doneva and Donev 2000) and storage at temperatures above refrigerated temperatures (Conrad et al. 2000) has been reported in the literature. Destabilization of cell membranes and their integral proteins was reported to be the main cause for cell injury during freezing and thawing (Conrad et al. 2000). However, a few protectants were found to be highly effective over others in achieving higher viability during the freezing step and storage compared to controls dried without protectant (Carvalho et al. 2004).

11.6.2 Spray Drying

Because of high costs and energy consumption of freeze-drying as well as volume limitations within a freeze-drying container, spray drying is considered a good alternative long-term preservation method for starter and probiotic cultures (Morgan et al. 2006). The speed of drying and continuous production capability of the spray dryer at relatively low cost is very useful for drying large amounts of LAB. Spray drying of probiotics, however, causes huge declines in viability during drying and storage, and thus has been less practiced commercially. Probiotic cultures are heat sensitive; thus, spray drying must be mild enough to avoid damaging them but sufficiently efficient to yield a powder with moisture content below 4% (wet basis), which is required for storage stability (Peighambardoust et al. 2011).

Spray drying produces a dry powder by atomizing the liquid at high velocity and directing the spray of droplets into a flow of hot air, for example, 150°C–200°C. The atomized droplets have a very large surface area in the form of millions of micrometer-sized droplets (10–200 μm), which results in a very short drying time when exposed to hot air in a drying chamber. The temperature of spray-dried particles is limited to the wet-bulb temperature by the evaporative cooling effect, and thus thermal inactivation is limited. An increase in inlet air temperature has only a slight effect on cell viability (Peighambardoust et al. 2011). After the initial evaporation, the particle surface becomes dry, the drying rate is diffusion limited, and the temperature of the particle can rise. During this stage, the extent of thermal inactivation depends on the drying parameters such as outlet temperature, residence time, and feed rate. Many researchers have obtained higher viability of microbial cells at lower outlet temperatures, but an outlet air temperature that is too low may result in high residual moisture contents that exceed the level required for prolonged powder storage life and stability (Santivarangkna et al. 2011). Desmond et al. (2002) reported higher stability of spray-dried powdered *L. paracasei* during storage when lower outlet air temperatures were used. Following spray drying, 49% of the bacteria survived at an outlet temperature of 85°C–90°C compared to 4% at an outlet temperature of 95°C–100°C.

11.6.3 FB Drying

An FB is a bed of solid particles with a stream of air or gas blowing upward through the particles at a rate high enough to set them in fluidlike motion. Particles are freely suspended in the air stream and simultaneously dehydrated by rapid exchange of heat and mass between gas and particles, which minimizes overheating. FB dryers offer similar advantages compared to spray dryers that include large-scale, low-cost continuous production (Bayrock and Ingledew 1997). Drying time in an FB could be longer than that in spray drying, but heat inactivation can be minimized and easily controlled by using relatively low air temperatures. FB drying of microorganisms has been studied for yeast, and the process is successfully employed in the production of commercial dry yeast (Santivarangkna et al. 2011).

Only a few studies have been made on lactobacilli. The drawback is that only granular materials can be dried, and therefore cells must be entrapped or encapsulated in support materials such as skim milk, potato starch, alginate, or casein. The low water activity of the support material may pose an osmotic shock leading to a reduction in viability after mixing. Mille et al. (2004) studied the viability of *L. plantarum* after mixing and FB drying with casein powders of different water activity (a_w). When *L. plantarum* cells were mixed with extremely desiccated casein powder ($a_w \leq 0.1$), viability was reduced to 2.5%, but when the water activity of the casein powder was 0.75, 100% viability was achieved.

Larena et al. (2003) compared three methods of drying: freeze-drying, spray drying, and FB drying on the viability of *Penicillium oxalicum*. Results showed 100% viability after freeze-drying and FB drying but only 20% viability after spray drying. Stability trials showed a drop of at least 50% viability in freeze-dried bacteria after 30 days at room temperature irrespective of protective agents used, but FB-dried *P. oxalicum* remained 100% viable after 30 days at room temperature. Viability dropped to 40% after 60 days and then remained constant up to 180 days.

Spray drying on an FB was studied with *L. casei*. In this process, the concentrated cells were sprayed by a nozzle over the carrier materials (lactose and NaCl particles), which were fluidized in the dryer. The granules with the culture on the surface were dried by counter flow air in an FB chamber. In this study, the product obtained was more hygroscopic than the pure carriers, and the viability was only in the range of 20%. It was assumed that the low viability was due to the cell concentration process (microfiltration) or the osmotic effects between the carrier materials and the cells (Zimmermann and Bauer 1990).

11.6.4 Vacuum Drying

Vacuum drying under conventional conditions (temperature range between 30°C and 80°C) generally leads to huge cell loss due to heat damage. But heat stress can be minimized by further reducing the chamber pressure to values just above the triple point of water, which leads to low product temperatures close to 0°C. This process is referred to as controlled low-temperature vacuum (CLTV) dehydration (Bauer et al. 2012). CLTV dehydration is a well-suited dehydration method for heat- and oxygen-sensitive LAB as the temperature can be maintained as low as possible without freezing, and heat and vacuum are applied to promote the evaporation of water. Its other advantages include lower drying time and operational cost compared to freeze-drying. The basic vacuum dryer consists of a chamber containing heated shelves. Trays containing the wet materials are placed on the shelves, and water is removed in a vacuum pump and condensed at a condenser. Studies on the storage stability of vacuum-dried LAB are rare (Foerst et al. 2012). Vacuum-dried *L. paracasei* F19 was found to be stable at refrigeration temperatures for a storage period of three months, but a huge decline in its viability was observed when stored at 37°C (Foerst et al. 2012).

At the low drying temperatures of FB and vacuum drying, heat inactivation is negligible, but dehydration inactivation may impose serious problems. It has been reported that between 8°C and 25°C, microorganisms exhibit a higher sensitivity to the osmotic pressure caused by dehydration compared to temperatures >25°C and <8°C (Laroche and Gervais 2003; Mille et al. 2005). The reason for this trend was related to the membrane phase transition in the reported temperature range, which could lead to an increased sensitivity of the lipid bilayer to the water flow caused by diffusion.

11.7 STORAGE AND PACKAGING

The storage stability of probiotic culture over its shelf life is critical as cell death occurs not only during processing but also during storage. The method of storage and packaging is significant as the dried culture must be protected from heat, oxygen, light, and moisture. Factors including temperature (Strasser et al. 2009; Teixeira et al. 1995), moisture level (Linders et al. 1997a), protective carrier (Selmer-Olsen et al. 1999a; Strasser et al. 2009), preadaption of cells to stresses (Prasad et al. 2003), and oxidative stress (Teixeira et al. 1996) have been investigated for their effects on the maintenance of cell activity during the storage period. The mechanism of cell death during storage is different from that in a thermal drying process. Since there is usually no heat stress during storage, the main stresses that cells suffer are the low water potential and the oxidative stress. The loss of cell viability at elevated temperatures (above refrigeration temperature) is mainly related to the formation of reactive oxygen species (ROS), such as O_2^- (superoxide), and OH^- (hydroxyl radical), in the presence of oxygen, and concomitant fatty acid oxidation and DNA damage (Selmer-Olsen et al. 1999b). During storage, natural degradation of macromolecules that are essential to life can occur, considering that *in vitro* lipids and proteins undergo oxidation and denaturation, respectively, during prolonged storage. Teixeira et al. (1996) reported lipid oxidation of spray-dried *L. bulgaricus* after prolonged storage. Lipid oxidation can cause an increase in membrane permeability and also affect the enzymatic activities associated with the membrane (Castro et al. 1995).

Cell viability during storage is generally considered to be inversely related to the storage temperature with much higher cell death at ambient and higher storage temperatures than at refrigerated temperatures (Corcoran et al. 2004; Silva et al. 2002; Strasser et al. 2009; Teixeira et al. 1995). The moisture level of the dried cells also plays a key role in the maintenance of cell activity, and a high moisture level during storage is disadvantageous (Selmer-Olsen et al. 1999b; Strasser et al. 2009). Another possible cause for cell death during storage, particularly if the residual moisture level is too high, may be the continuation of metabolic activities in the cell, even at a very slow rate. The cells then eventually suffer a natural death (Fu and Chen 2011). Under low moisture conditions, microorganisms can attain an anabiotic state in which metabolic processes are temporarily halted, but cells remain viable and can be stored for long periods of time. The type of protective carrier used is also important as protective carriers can stabilize the cellular structures in a glassy state with high viscosity, which restricts molecular movement and inhibits the transport of harmful free radicals (Selmer-Olsen et al. 1999a). Preadaption of cells could lead to the accumulation of compatible solutes and the production of chaperone proteins, which alleviate the osmotic stress and stabilize macromolecular structures (such as proteins, DNA, and RNA). Thus, the factors that could protect cells during storage are similar to those that exhibit protective effects during drying. Utilization of these techniques has been successful in maintaining the viability of *L. casei* 431 during prolonged ambient storage (Nag and Das 2012b; Das et al. 2012).

Regarding the role of oxygen during storage, vacuum packaging can retain better storage stability of the cells than direct exposure to air (Chavez and Ledeboer 2007; Hernandez et al. 2007). This could be due to the fact that accumulation of ROS within a cell could cause an irreversible damage to the cell components. Teixeira et al. (1995) found the evidence of damage to the cell wall, cell membrane, and DNA during storage of *L. delbrueckii* ssp. *bulgaricus*. The use of ascorbic acid and monosodium glutamate as antioxidants on storage stability has also been reported by Teixeira et al. (1995). These antioxidants had a protective effect at 4°C; however, an increase in cell death at 20°C was observed. It was supposed that in addition to its antioxidant property, ascorbic acid could also have pro-oxidant property and could possibly generate hydroxyl radicals that can attack and oxidize biological molecules. By contrast, the use of oxygen-scavenging agents during storage did not show any significant effect in protecting cell viability (Chavez and Ledeboer 2007; Wang et al. 2004).

11.8 CONCLUDING REMARKS

Modern consumers are increasingly interested in their personal health and expect the food that they eat to be healthy or even capable of preventing illness. Gut health in general has shown to be the key sector for functional foods in Europe. The probiotic yogurt market is now well established with its beneficial effects on human heath well accepted. New product categories containing probiotic bacteria outside the dairy sector will certainly be a key research and development area for future functional food markets. However, the viability and stability of probiotics has always remained a marketing and technological challenge for industrial producers, particularly under ambient conditions. Probiotic foods should contain specific probiotic strains and maintain a suitable level of viable cells during the product's shelf life. Before probiotic strains can be delivered to consumers, they must first be able to be manufactured under industrial conditions, and then survive and retain their functionality during storage as frozen or dried cultures, and also in the food products into which they are finally formulated.

Lactobacilli are some of the most widely used probiotics, and *L. casei*, which was isolated from the human gut and is a generally-regarded-as-safe (GRAS) organism, is one of the most popular. The probiotic culture must be grown to get the maximum concentration of viable cells, generally close to 10^{10} cfu ml^{-1}, which entails finding the optimum medium and conditions for enhancing its growth. In general, the MRS medium has been found the best, but a variety of cheaper substrates have also been successful.

The resistance of the cultured cells to the subsequent and potentially debilitating processes of isolation, freezing or heating, dehydration, and ambient storage can be strengthened in a number of ways. Exposing the cells to osmotic, acid, thermal, or starvation conditions alters their metabolism to enhance their resistance. It can also induce the synthesis of compatible solutes, which can ameliorate osmotic shock, or stress proteins, which are chaperone vital cell constituents and prevent miscoiling and protein denaturation. However, this must be achieved without causing an excessive mortality of the cells. The addition of protectants, generally saccharides, can stabilize membrane lipids during dehydration and also form glasses around the dry cells that shield them from oxidative and other destructive processes.

Harvesting the bacteria during the stationary phase in an uncontrolled fermentation, when the concentration of lactic acid is high and nutrient levels are exhausted, coincides with two stress conditions, namely, high acid and nutrient depletion, and has been found to yield cells with good survival during drying and storage (Taneja 2013).

Fluidized bed (FB) drying offers a low-energy and cost-efficient alternative to freeze-drying that is currently the most common method of preservation for LAB cultures. Moreover, FB drying can be carried out at low temperatures compared to the potentially lethal temperatures of spray drying. FB drying is extensively used in the yeast industry, but very few studies have described its use to preserve LAB.

The maximization of cell survival during a thermal drying process needs to take into account the combined effect of heat and dehydration damages, which requires a balanced consideration of multiple factors including the intensity of each stress (e.g., temperature and water potential), the time of cells exposed to each stress, and the changing rate of these stresses (e.g., the drying rate and the rate of temperature variation). During drying, microorganisms undergo an increasing osmotic stress condition as the water activity decreases.

A better understanding of all these factors has helped in achieving the necessary robustness of bacteria required for various industrial processes during large-scale production of shelf-stable commercial cultures and their application in various food products. Currently, industrial demand for technologies ensuring probiotic stability in foods stored at ambient temperature remains strong. The development of optimized culture media for probiotics and processing techniques that ensure stability during dehydration and storage is now a real possibility and will allow commercial-scale production of probiotic cultures for use in an increasing variety of products.

REFERENCES

Aguirre-Ezkauriatza, E.J., J.M. Aguilar-Yanez, A. Ramirez-Medrano, and M.M. Alvarez. 2010. Production of probiotic biomass (*Lactobacillus casei*) in goat milk whey: Comparison of batch, continuous and fed-batch cultures. *Bioresource Technology* 101 (8):2837–2844.

Alvarez, M.M., E.J. Aguirre-Ezkauriatza, A. Ramirez-Medrano, and A. Rodriguez-Sanchez. 2010. Kinetic analysis and mathematical modeling of growth and lactic acid production of *Lactobacillus casei* var. rhamnosus in milk whey. *Journal of Dairy Science* 93 (12):5552–5560.

Avonts, L., E. Van Uytven, and L. De Vuyst. 2004. Cell growth and bacteriocin production of probiotic *Lactobacillus* strains in different media. *International Dairy Journal* 14 (11):947–955.

Axelsson, L. 1998. Lactic acid bacteria: Classification and physiology. In *Lactic Acid Bacteria: Microbiology and Functional Aspects*, eds. S. Salminen and A. von Wright. New York: Marcel Dekker, pp. 1–72.

Bai, D.M., Q. Wei, Z.H. Yan, X.M. Zhao, X.G. Li, and S.M. Xu. 2003. Fed-batch fermentation of *Lactobacillus lactis* for hyper-production of L-lactic acid. *Biotechnology Letters* 25 (21):1833–1835.

Bauer, S.A.W., S. Schneider, J. Behr, U. Kulozik, and P. Foerst. 2012. Combined influence of fermentation and drying conditions on survival and metabolic activity of starter and probiotic cultures after low-temperature vacuum drying. *Journal of Biotechnology* 159 (4):351–357.

Bayrock, D. and W.M. Ingledew. 1997. Mechanism of viability loss during fluidized bed drying of baker's yeast. *Food Research International* 30 (6):417–425.

Beney, L. and P. Gervais. 2001. Influence of the fluidity of the membrane on the response of microorganisms to environmental stresses. *Applied Microbiology and Biotechnology* 57 (1/2):34–42.

Bernardez, P.F., I.R. Amado, L.P. Castro, and N.P. Guerra. 2008. Production of a potentially probiotic culture of *Lactobacillus casei* subsp *casei* CECT 4043 in whey. *International Dairy Journal* 18 (10/11):1057–1065.

Bevilacqua, A., M.R. Corbo, M. Mastromatteo, and M. Sinigaglia. 2008. Combined effects of pH, yeast extract, carbohydrates and di-ammonium hydrogen citrate on the biomass production and acidifying ability of a probiotic *Lactobacillus plantarum* strain, isolated from table olives, in a batch system. *World Journal of Microbiology & Biotechnology* 24 (9):1721–1729.

Brinques, G.B., M.D. Peralba, and M.A.Z. Ayub. 2010. Optimization of probiotic and lactic acid production by *Lactobacillus plantarum* in submerged bioreactor systems. *Journal of Industrial Microbiology & Biotechnology* 37 (2):205–212.

Carvalho, A.S., J. Silva, P. Ho, P. Teixeira, F.X. Malcata, and P. Gibbs. 2004. Relevant factors for the preparation of freeze-dried lactic acid bacteria. *International Dairy Journal* 14 (10):835–847.

Castro, H.P., P.M. Teixeira, and R. Kirby. 1995. Storage of lyophilized cultures of *Lactobacillus bulgaricus* under different relative humidities and atmospheres. *Applied Microbiology and Biotechnology* 44 (1/2):172–176.

Chavez, B.E. and A.M. Ledeboer. 2007. Drying of probiotics: Optimization of formulation and process to enhance storage survival. *Drying Technology* 25 (7/8):1193–1201.

Codex Alimentarius. 2003. Standard for fermented milks. Codex standard 243-2003.

Conrad, P.B., D.P. Miller, P.R. Cielenski, and J.J. de Pablo. 2000. Stabilization and preservation of *Lactobacillus acidophilus* in saccharide matrices. *Cryobiology* 41 (1):17–24.

Corcoran, B.A., R.P. Ross, G.F. Fitzgerald, P. Dockery, and C. Stanton. 2006. Enhanced survival of GroESL-overproducing *Lactobacillus paracasei* NFBC 338 under stressful conditions induced by drying. *Applied and Environmental Microbiology* 72 (7):5104–5107.

Corcoran, B.M., R.P. Ross, G.F. Fitzgerald, and C. Stanton. 2004. Comparative survival of probiotic lactobacilli spray-dried in the presence of prebiotic substances. *Journal of Applied Microbiology* 96 (5):1024–1039.

Costa, E., J. Usall, N. Teixido, N. Garcia, and I. Vinas. 2000. Effect of protective agents, rehydration media and initial cell concentration on viability of *Pantoea agglomerans* strain CPA-2 subjected to freeze-drying. *Journal of Applied Microbiology* 89 (5):793–800.

Crowe, J.H., J.F. Carpenter, and L.M. Crowe. 1998. The role of vitrification in anhydrobiosis. *Annual Review of Physiology* 60:73–103.

Crowe, J.H., F.A. Hoekstra, and L.M. Crowe. 1992. Anhydrobiosis. *Annual Review of Physiology* 54:579–599.

Csonka, L.N. and A.D. Hanson. 1991. Prokaryotic osmoregulation—Genetics and physiology. *Annual Review of Microbiology* 45:569–606.

De Man, J.D., M. Rogosa, and M.E. Sharpe. 1960. A medium for the cultivation of lactobacilli. *Journal of Applied Bacteriology* 23:130–135.

Desmond, C., C. Stanton, G.F. Fitzgerald, K. Collins, and R.P. Ross. 2002. Environmental adaptation of probiotic *Lactobacilli* towards improvement of performance during spray drying. *International Dairy Journal* 12 (2/3):183–190.

De Vos, P., M. Faas, M. Spasojevic, and J. Sikkemma. 2010. Encapsulation for the preservation of functional and targeted delivery bioactive food components. *International Dairy Journal* 20(4): 292–302.

Ding, S.F. and T.W. Tan. 2006. L-lactic acid production by *Lactobacillus casei* fermentation using different fed-batch feeding strategies. *Process Biochemistry* 41 (6):1451–1454.

Eleutherio, E.C.A., P.S. Araujo, and A.D. Panek. 1993. Protective role of trehalose during heat-stress in *Saccharomyces cerevisiae*. *Cryobiology* 30 (6):591–596.

FAO/WHO. 2001. Health and nutritional properties of probiotics in food including powder milk with live lactic acid bacteria. *Report from FAO/WHO Expert Consultation*. October 1–4. Córdoba, Argentina.

Favaro-Trinidade, C.S., R.J.B. Heineman, and D.L. Pedroso. 2011. Developments in probiotic encapsulation. *CAB Review: Perspectives in Agriculture, Veterinary Science, Nutrition and Natural Resources* 6:1–8.

Foerst, P., U. Kulozik, M. Schmitt, S. Bauer, and C. Santivarangkna. 2012. Storage stability of vacuum-dried probiotic bacterium *Lactobacillus paracasei* F19. *Food and Bioproducts Processing* 90 (C2):295–300.

Friesen, T., G. Hill, T. Pugsley, G. Holloway, and D. Zimmerman. 2005. Experimental determination of viability loss of *Penicillium bilaiae* conidia during convective air-drying. *Applied Microbiology and Biotechnology* 68 (3):397–404.

Fu, N. and X.D. Chen. 2011. Towards a maximal cell survival in convective thermal drying processes. *Food Research International* 44 (5):1127–1149.

Fu, N., M.W. Woo, F.T. Moo, and X.D. Chen. 2012. Microcrystallization of lactose during droplet drying and its effect on the property of the dried particle. *Chemical Engineering Research & Design* 90 (1A):138–149.

Fung, W.Y., Y.P. Woo, and M.T. Liong. 2008. Optimization of growth of *Lactobacillus acidophilus* FTCC 0291 and evaluation of growth characteristics in soy whey medium: A response surface methodology approach. *Journal of Agricultural and Food Chemistry* 56 (17):7910–7918.

Gervais, P. and P.A. Marechal. 1994. Yeast resistance to high levels of osmotic pressure—Influence of kinetics. *Journal of Food Engineering* 22 (1–4):399–407.

Giard, J.C., A. Hartke, S. Flahaut, A. Benachour, P. Boutibonnes, and Y. Auffray. 1996. Starvation-induced multiresistance in *Enterococcus faecalis* JH2-2. *Current Microbiology* 32 (5):264–271.

Ha, M.Y., S.W. Kim, Y.W. Lee, M.J. Kim, and S.J. Kim. 2003. Kinetics analysis of growth and lactic acid production in pH-controlled batch cultures of *Lactobacillus casei* KH-1 using yeast extract/corn steep liquor/glucose medium. *Journal of Bioscience and Bioengineering* 96 (2):134–140.

Hartke, A., S. Bouche, J.C. Giard, A. Benachour, P. Boutibonnes, and Y. Auffray. 1996. The lactic acid stress response of *Lactococcus lactis* subsp LACTIS. *Current Microbiology* 33 (3):194–199.

Hayakawa, K., H. Sansawa, T. Nagamune, and I. Endo. 1990. High density culture of *Lactobacillus casei* by a cross-flow culture method based on kinetic-properties of the microorganism. *Journal of Fermentation and Bioengineering* 70 (6):404–408.

Hernandez, A., F. Weekers, J. Mena, E. Pimentel, J. Zamora, C. Borroto, and P. Thonart. 2007. Culture and spray-drying of *Tsukamurella paurometabola* C-924: Stability of formulated powders. *Biotechnology Letters* 29 (11):1723–1728.

Kets, E.P.W. and J.A.M. de Bont. 1994. Protective effect of betaine on survival of *Lactobacillus plantarum* subjected to drying. *FEMS Microbiology Letters* 116 (3):251–255.

Kets, E.P.W., P.J.M. Teunissen, and J.A.M. de Bont. 1996. Effect of compatible solutes on survival of lactic acid bacteria subjected to drying. *Applied and Environmental Microbiology* 62 (1):259–261.

Kopp-Hoolihan, L. 2001. Prophylactic and therapeutic uses of probiotics: A review. *Journal of the American Dietetic Association* 101 (2):229–241.

Kosin, B. and S.K. Rakshit. 2006. Microbial and processing criteria for production of probiotics: A review. *Food Technology and Biotechnology* 44 (3):371–379.

Larena, I., P. Melgarejo, and A. De Cal. 2003. Drying of conidia of *Penicillium oxalicum*, a biological control agent against *Fusarium* wilt of tomato. *Journal of Phytopathology* 151 (11/12):600–606.

Laroche, C. and P. Gervais. 2003. Achievement of rapid osmotic dehydration at specific temperatures could maintain high *Saccharomyces cerevisiae* viability. *Applied Microbiology and Biotechnology* 60 (6):743–747.

Lievense, L.C., M.A.M. Verbeek, T. Taekema, G. Meerdink, and K. Vantriet. 1992. Modeling the inactivation of *Lactobacillus plantarum* during a drying process. *Chemical Engineering Science* 47 (1):87–97.

Lievense, L.C., M.A.M. Verbeek, K. Vantriet, and A. Noomen. 1994. Mechanism of dehydration inactivation of *Lactobacillus plantarum*. *Applied Microbiology and Biotechnology* 41 (1):90–94.

Liew, S.L., A.B. Ariff, A.R. Raha, and Y.W. Ho. 2005. Optimization of medium composition for the production of a probiotic microorganism, *Lactobacillus rhamnosus*, using response surface methodology. *International Journal of Food Microbiology* 102 (2):137–142.

Linders, L.J.M., G.I.W. deJong, G. Meerdink, and K. Van't Riet. 1997a. Carbohydrates and the dehydration inactivation of *Lactobacillus plantarum*: The role of moisture distribution and water activity. *Journal of Food Engineering* 31 (2):237–250.

Linders, L.J.M., G. Meerdink, and K. Van't Riet. 1996. Influence of temperature and drying rate on the dehydration inactivation of *Lactobacillus plantarum*. *Food and Bioproducts Processing* 74 (C2):110–114.

Linders, L.J.M., G. Meerdink, and K. Van't Riet. 1997b. Effect of growth parameters on the residual activity of *Lactobacillus plantarum* after drying. *Journal of Applied Microbiology* 82 (6):683–688.

Linders, L.J.M., W.F. Wolkers, F.A. Hoekstra, and K. Van't Riet. 1997c. Effect of added carbohydrates on membrane phase behavior and survival of dried *Lactobacillus plantarum*. *Cryobiology* 35 (1):31–40.

Miao, S., S. Mills, C. Stanton, G.F. Fitzgerald, Y. Roos, and R.P. Ross. 2008. Effect of disaccharides on survival during storage of freeze dried probiotics. *Dairy Science & Technology* 88 (1):19–30.

Mille, Y., L. Beney, and P. Gervais. 2005. Compared tolerance to osmotic stress in various microorganisms: Towards a survival prediction test. *Biotechnology and Bioengineering* 92 (4):479–484.

Mille, Y., J.P. Obert, L. Beney, and P. Gervais. 2004. New drying process for lactic bacteria based on their dehydration behavior in liquid medium. *Biotechnology and Bioengineering* 88 (1):71–76.

Morgan, C.A., N. Herman, P.A. White, and G. Vesey. 2006. Preservation of micro-organisms by drying: A review. *Journal of Microbiological Methods* 66 (2):183–193.

Nag, A. and S. Das. 2012a. Effect of trehalose and lactose as cryoprotectants during freeze drying, in vitro gastro-intestinal transit and survival of microencapsulated freeze dried *L. casei 431* cells. *International Journal of Dairy Technology* 66 (2):162–169.

Nag, A. and S. Das. 2012b. Delivery of probiotic bacteria in long life ambient stable foods using a powdered food ingredient. *British Food Journal* 115 (9):1329–1341.

Nag, A. and S. Das. 2012c. Improving ambient temperature stability of probiotics with stress adaptation and fluidized bed drying. *Journal of Functional Foods* 5 (1):170–177.

Nag, A., K.-S. Han, and H. Singh. 2011. Microencapsulation of probiotic bacteria using pH-induced gelation of sodium caseinate and gellan gum. *International Dairy Journal* 21(4) 247–253.

O'Callaghan, J. and S. Condon. 2000. Growth of *Lactococcus lactis* strains at low water activity: Correlation with the ability to accumulate glycine betaine. *International Journal of Food Microbiology* 55 (1–3):127–131.

Ooi, L.G. and M.T. Liong. 2010. Cholesterol-lowering effects of probiotics and prebiotics: A review of in vivo and in vitro findings. *International Journal of Molecular Sciences* 11 (6):2499–2522.

Palmfeldt, J., P. Radstrom, and B. Hahn-Hagerdal. 2003. Optimisation of initial cell concentration enhances freeze-drying tolerance of *Pseudomonas chlororaphis*. *Cryobiology* 47 (1):21–29.

Peighambardoust, S.H., A.G. Tafti, and J. Hesari. 2011. Application of spray drying for preservation of lactic acid starter cultures: A review. *Trends in Food Science & Technology* 22 (5):215–224.

Poddar, D., Nag, A., Das, S., and H. Singh. 2012. Stabilisation of probiotics for industrial application. In *Innovation in Healthy and Functional Food*, eds. D. Ghosh, D. Das, D. Bagchi, and R.B.Smarta. Boca Rato, FL: CRC Press, pp. 269–304.

Polak-Berecka, M., A. Wasko, M. Kordowska-Wiater, M. Podlesny, Z. Targonski, and A. Kubik-Komar. 2010. Optimization of medium composition for enhancing growth of *Lactobacillus rhamnosus* PEN using response surface methodology. *Polish Journal of Microbiology* 59 (2):113–118.

Poolman, B. and E. Glaasker. 1998. Regulation of compatible solute accumulation in bacteria. *Molecular Microbiology* 29 (2):397–407.

Prasad, J., P. McJarrow, and P. Gopal. 2003. Heat and osmotic stress responses of probiotic *Lactobacillus rhamnosus* HN001 (DR20) in relation to viability after drying. *Applied and Environmental Microbiology* 69 (2):917–925.

Rijkers, G.T., W.M. Vos, R. Brummer, L. Morelli, G. Corthier, and P. Marteau. 2011. Health benefits and health claims of probiotics: Bridging science and marketing. *British Journal of Nutrition* 106:1291–1296.

Robert, H., C. Le Marrec, C. Blanco, and M. Jebbar. 2000. Glycine betaine, carnitine, and choline enhance salinity tolerance and prevent the accumulation of sodium to a level inhibiting growth of *Tetragenococcus halophila*. *Applied and Environmental Microbiology* 66 (2):509–517.

Santivarangkna, C., M. Aschenbrenner, U. Kulozik, and P. Foerst. 2011. Role of glassy state on stabilities of freeze-dried probiotics. *Journal of Food Science* 76 (8):R152–R156.

Santivarangkna, C., B. Higl, and P. Foerst. 2008. Protection mechanisms of sugars during different stages of preparation process of dried lactic acid starter cultures. *Food Microbiology* 25 (3):429–441.

Santivarangkna, C., U. Kulozik, and P. Foerst. 2006. Effect of carbohydrates on the survival of *Lactobacillus helveticus* during vacuum drying. *Letters in Applied Microbiology* 42 (3):271–276.

Santivarangkna, C., U. Kulozik, and P. Foerst. 2007. Alternative drying processes for the industrial preservation of lactic acid starter cultures. *Biotechnology Progress* 23 (2):302–315.

Schiraldi, C., V. Adduci, V. Valli, C. Maresca, M. Giuliano, M. Lamberti, M. Carteni, and M. De Rosa. 2003. High cell density cultivation of probiotics and lactic acid production. *Biotechnology and Bioengineering* 82 (2):213–222.

Selmer-Olsen, E., S.E. Birkeland, and T. Sorhaug. 1999a. Effect of protective solutes on leakage from and survival of immobilized *Lactobacillus* subjected to drying, storage and rehydration. *Journal of Applied Microbiology* 87 (3):429–437.

Selmer-Olsen, E., T. Sorhaug, S.E. Birkeland, and R. Pehrson. 1999b. Survival of *Lactobacillus helveticus* entrapped in Ca-alginate in relation to water content, storage and rehydration. *Journal of Industrial Microbiology & Biotechnology* 23 (2):79–85.

Shah, N.P. 2007. Functional cultures and health benefits. *International Dairy Journal* 17: 1262–1277.

Silva, J., A.S. Carvalho, H. Pereira, P. Teixeira, and P.A. Gibbs. 2004. Induction of stress tolerance in *Lactobacillus delbrueckii* ssp. *bulgaricus* by the addition of sucrose to the growth medium. *Journal of Dairy Research* 71 (1):121–125.

Silva, J., A.S. Carvalho, P. Teixeira, and P.A. Gibbs. 2002. Bacteriocin production by spray-dried lactic acid bacteria. *Letters in Applied Microbiology* 34 (2):77–81.

Soccol, C.R., L.P.D. Vandenberghe, M.R. Spier, A.B.P. Medeiros, C.T. Yamaguishi, J.D. Lindner, A. Pandey, and V. Thomaz-Soccol. 2010. The potential of probiotics: A review. *Food Technology and Biotechnology* 48 (4):413–434.

Strasser, S., M. Neureiter, M. Geppl, R. Braun, and H. Danner. 2009. Influence of lyophilization, fluidized bed drying, addition of protectants, and storage on the viability of lactic acid bacteria. *Journal of Applied Microbiology* 107 (1):167–177.

Suokko, A., M. Poutanen, K. Savijoki, N. Kalkkinen, and P. Varmanen. 2008. ClpL is essential for induction of thermotolerance and is potentially part of the HrcA regulon in *Lactobacillus gasseri*. *Proteomics* 8 (5):1029–1041.

Taneja, N. 2013. Maximising viability of *Lactobacillus paracasei* ssp. *paracasei L. casei 431* during processing and ambient storage. MTech thesis, Massey University, Palmerston North, New Zealand.

Teixeira, P., H. Castro, and R. Kirby. 1995. Spray drying as a method for preparing concentrated cultures of *Lactobacillus bulgaricus*. *Journal of Applied Bacteriology* 78 (4):456–462.

Teixeira, P., H. Castro, and R. Kirby. 1996. Evidence of membrane lipid oxidation of spray-dried *Lactobacillus bulgaricus* during storage. *Letters in Applied Microbiology* 22 (1):34–38.

Tymczyszyn, E.E., R. Diaz, A. Pataro, N. Sandonato, A. Gomez-Zavaglia, and E.A. Disalvo. 2008. Critical water activity for the preservation of *Lactobacillus bulgaricus* by vacuum drying. *International Journal of Food Microbiology* 128 (2):342–347.

Tymczyszyn, E.E., A. Gomez-Zavaglia, and E.A. Disalvo. 2007. Effect of sugars and growth media on the dehydration of *Lactobacillus delbrueckii* ssp. *bulgaricus*. *Journal of Applied Microbiology* 102 (3):845–851.

Uzunova-Doneva, T. and T. Donev. 2000. Influence of the freezing rate on the survival of strains *Saccharomyces cerevisiae* after cryogenic preservation. *Journal of Culture Collections* 3: 78–83.

Veld, J. and R. Havenaar. 1991. Probiotics and health in man and animal. *Journal of Chemical Technology and Biotechnology* 51 (4):562–567.

Wang, Y.C., R.C. Yu, and C.C. Chou. 2004. Viability of lactic acid bacteria and bifidobacteria in fermented soymilk after drying, subsequent rehydration and storage. *International Journal of Food Microbiology* 93 (2):209–217.

Wood, J.M., E. Bremer, L.N. Csonka, R. Kraemer, B. Poolman, T. van der Heide, and L.T. Smith. 2001. Osmosensing and osmoregulatory compatible solute accumulation by bacteria. *Comparative Biochemistry and Physiology. Part A, Molecular & Integrative Physiology* 130 (3):437–460.

Yoo, I.K., H.N. Chang, E.G. Lee, Y.K. Chang, and S.H. Moon. 1996. Effect of pH on the production of lactic acid and secondary products in batch cultures of *Lactobacillus casei*. *Journal of Microbiology and Biotechnology* 6 (6):482–486.

Zacharof, M.P., R.W. Lovitt, and K. Ratanapongleka. 2010. The importance of lactobacilli in contemporary food and pharmaceutical industry: A review article. *Proceedings of the 2010 International Conference on Chemical Engineering and Applications,* Singapore, February 26–28, 2010.

Zhang, Y., W. Cong, and S.Y. Shi. 2010. Application of a pH feedback-controlled substrate feeding method in lactic acid production. *Applied Biochemistry and Biotechnology* 162 (8):2149–2156.

Zimmermann, K. and W. Bauer. 1990. Fluidized bed drying of microorganisms on carrier material. In *Engineering and Food, Volume 2: Preservation Processes and Related Techniques*, eds. W.E.L. Spiess and H. Schubert. London: Elsevier Applied Science, pp. 666–678.

12 Drug–Dietary Supplement Interactions

Noriaki Yohkoh

CONTENTS

12.1 INTRODUCTION

Drug–dietary supplement interaction is defined as a phenomenon that the drug action is weakened or strengthened or the side effect such as toxicity is modulated when a dietary supplement is taken together with a pharmaceutical drug. Many incidents involving them have been currently reported[1–4] and attracted social attention. One reason for this is the recent development and distribution of a variety of supplements, and also the trend of self-medication so that the occasion of (self-)administration of dietary supplements is increased. However, as medications progress, new types of drugs are provided, and thus the number and types of drug used in clinic are also increased. Therefore, the chances of taking both together are increased whether it is liked or not. Another cause is the increase in potency of recently developed drugs because the pharmacokinetics of stronger drugs are considered more sensitive to interaction with dietary supplements. In the past, drug potencies were much lower, and a typical dosage per administration was 100 mg to 1 g. Today, a typical dosage is in the range of 100 μg to a few milligrams. Given these factors, the influence of dietary supplement intake on the primary and side effects of medications cannot be ignored any longer.

The number of currently available dietary supplements is enormous, and their quality and composition vary greatly between suppliers due to a lack of strict legal regulation. In fact, hardly any detailed investigations on the safety and efficacy of dietary supplements have been conducted. In addition, the indications for many supplements are not clear and their efficacies are often not scientifically proven. In short, there is a severe lack of scientific evidence supporting the safety and efficacy of dietary supplements. There are also nearly 20,000 approved prescription drugs available in Japan at this moment, and when over-the-counter (OTC) drugs are included, this number rises to over 30,000. The range of possible drug–dietary supplement interactions is therefore astronomical, resulting in a lack of advancement in the relevant scientific and statistical research. Along with the already confirmed combinations of interactions, there is an abundance of disorganized information.

Against this background, it is prudent to excise caution when combining drugs and dietary supplements. Drugs with severe possible side effects, drugs with narrow margins of safety between their effective and excessive dosages, and drugs with clinical reports of interaction with dietary supplements should be researched. Some examples of drugs to be taken with special caution are warfarin, oral diabetes medications, digitalis, β-blockers, and β-agonists.

Although the current opinion of mixed usage is often negative, dietary supplements can provide an effective support to medication, such as ameliorating drug side effects or supplementing nutrients.

12.2 MECHANISMS OF DRUG–DIETARY SUPPLEMENT INTERACTIONS

Drugs and dietary supplements interact through two mechanisms: pharmacokinetic and pharmacological. In pharmacokinetic interaction, dietary supplements influence the absorption, distribution, metabolism, and excretion of drugs, thereby altering their efficacy and toxicity, whereas in pharmacological interaction, dietary supplements directly enhance or antagonize drugs.[5,6]

12.2.1 Pharmacokinetic Interaction

When administered orally, most drugs are absorbed by the intestine, transported throughout the body via the blood, metabolized primarily by the liver, and excreted by the kidney in urine. Intake of dietary supplements may alter this sequence of events, resulting in the increase or decrease of the drug's primary effect or side effects. This mechanism is called pharmacokinetic interaction.

12.2.1.1 Examples of Influence on Drug Absorption

Dietary supplements containing minerals such as calcium, iron, magnesium, and zinc form insoluble chelates with fluoroquinolones and tetracycline antibiotics, inhibiting their intestinal absorption.[7] Furanocoumarin—the bitter substance in grapefruit juice—and its derivatives (e.g., bergamottin and dihydroxybergamottin) strongly inhibit drug metabolism enzymes in the intestine, resulting in the increased absorption of calcium channel blockers such as felodipine and nisoldipine.[6]

12.2.1.2 Examples of Influence on Drug Metabolism

St John's wort, a dietary supplement, upregulates a certain type of drug metabolism enzyme in the liver. As a result, the metabolism of many drugs such as digoxin, warfarin, theophylline, cyclosporine, and disopyramide is enhanced, lowering their blood titer and reducing their efficacy.[1,8]

12.2.2 Pharmacological Interaction

When a drug and a dietary supplement have the same or opposite effects, their combined use results in a pharmacological interaction. A dietary supplement with the same effect as the drug will enhance the drug's effect (synergy), and a dietary supplement with the opposite effect as the drug will diminish the drug's effect (antagonism). Some classes of drug's function by altering the body's level of certain biological materials or nutrients. A dietary supplement may therefore influence the drug's effects or side effects by altering the levels of these biological materials and nutrients as well.

12.2.2.1 Examples of Drug Synergy (Enhanced Effect or Side Effect)

The use of dietary supplement containing vitamin A is contraindicated with etretinate (prescribed for keratosis) or tretinoin (used to treat malignant tumors). Because these drugs are vitamin A derivatives, their use combined with vitamin A supplement will cause a side effect similar to hypervitaminosis A. Extract of *Ginkgo biloba* leaves has an anticoagulant effect, so its combined use with aspirin, ticlopidine, or warfarin promotes hemorrhage through enhancement of the drug's anticoagulant effects.[1,7] Lactotripeptides, dried bonito oligopeptide, and sardine peptide inhibit the activity of angiotensin-converting enzyme (ACE); their combined use with ACE inhibitors such as captopril, enalapril, or imidapril may therefore result in a synergistic effect, causing hypotension.

12.2.2.2 Examples of Drug Antagonism (Reduced Effect or Side Effect)

As warfarin regulates blood coagulation by reducing the blood titer of vitamin K, the use of dietary supplements also containing vitamin K (such as chlorella or Aojiru [suspension of vegetable leaf powder made from such as kale]) will diminish warfarin's effects. The antirheumatic drug methotrexate impairs the function of folic acid, so dietary supplements containing folic acid may diminish not only the effect of methotrexate but also its side effects.[1] Isoniazid is prescribed for tuberculosis, but as a side effect causes peripheral neuropathy by depleting vitamin B_6 in the peripheral nervous system. This side effect, however, can be prevented by the supplementary intake of vitamin B_6.

12.2.3 Drug-Induced Shifts in Nutrients

To maintain proper health, sufficient dietary intake of nutrients such as vitamins and minerals is essential. Some drugs lower the blood titer of certain nutrients by inhibiting their absorption or enhancing their excretion. In such cases, it is often necessary to take vitamin or mineral supplements to avoid deficiencies. As these supplements prevent or reduce the drugs' side effects, the drug action can be categorized as interaction. Conversely, drug-induced nutrient shifts can be toxic, as is the case with an excess of fat-soluble vitamins or minerals such as magnesium and iron. Elevation in the blood titer of these substances may cause symptoms similar to overdose; thus, their use requires caution.

12.2.3.1 Drug-Induced Reduction and Depletion of Nutrients

The long-term use of furosemide (a loop diuretic), probenecid (a uricosuric prescribed for gout), and metformin (a biguanide antihyperglycemic) may reduce and eventually deplete vitamin B_1 blood levels. Anticonvulsants such as phenytoin reduce the blood titer of folic acid. Many drugs such as cholestimide (an anion exchange resin agent), cimetidine and famotidine (histamine H_2 receptor antagonists), and omeprazole (a proton pump inhibitor) reduce the absorption of zinc, causing taste impairment.

12.2.3.2 Drug-Induced Augmentation and Overdose of Nutrients

Some drugs may raise the blood titer of certain nutrients by enhancing their absorption or inhibiting their excretion. For example, simvastatin, atorvastatin, and oral contraceptive pills increase the blood titer of vitamin A. Spironolactone and triamterene (potassium-sparing diuretic) raise an internal magnesium concentration.

12.2.4 Interactions

Currently, little empirical information about drug–dietary supplement interactions is known, and the available information[1–14] is summarized in Tables 12.1 through 12.3. Table 12.1 lists the dietary supplements containing vitamins, Table 12.2 lists the dietary supplements containing minerals, and Table 12.3 lists the supplements that fall in other categories.

TABLE 12.1
Dietary Supplements Containing Vitamins

Dietary Supplement	Drug	Interaction
Vitamin A	Keratosis medication (etretinate), leukemia medication (tretinoin)	Induction of side effects similar to vitamin A overdose due to vitamin A mimetic effects of etretinate and tretinoin
	Anticoagulant (warfarin)	Increased efficacy of warfarin aided by anticoagulant-enhancing effect of vitamin A
	Anticancer drugs (paclitaxel)	Enhanced side effects of myelosuppression by paclitaxel
	Tetracycline antibiotics (minocycline, etc.)	Increased intracranial pressure (induction of headache)
Vitamin B_6	Anti-Parkinson (levodopa)	Reduced intracerebral concentration of levodopa leading to decreased efficacy
	Antiepileptic drug (phenitoin, phenobarbital)	Reduced efficacies of phenytoin and phenobarbital due to their enhanced metabolization and excretion
	Vitamin E	Decreased blood concentration of vitamin E
	Antiarrhythmics (amiodarone)	Increased side effect (light sensitivity) of amiodarone
Vitamin C	Carbonic anhydrase inhibitors, diuretics (acetazolamide)	Increased formation of kidney/urinary tract stones
	Anticoagulants (warfarin)	Reduced efficacy of warfarin
	Tetracycline antibiotics (minocycline, etc.)	Increased absorption and blood concentration of tetracycline antibiotics
	Antianemia (iron tablets)	Increased effects/side effects due to enhanced iron absorption
	Female sex hormone (estrogen)	Increased blood estrogen concentration
	Vasopressors (ubidecarenone)	Enhanced antioxidative effect
Vitamin D	Digitalis (digoxin, digitoxin, etc.)	Enhanced efficacy
	Thiazide diuretics (hydrochlorothiazide)	Induction of hypercalcemia
	Psoriasis treatment (calcipotriene)	Enhanced effects/side effects of calcipotriene
	Antiarrhythmic (verapamil)	Reduced efficacy of verapamil
	Antihypertensive drugs (diltiazem)	Reduced efficacy of diltiazem
	Antacids (preparations containing aluminum, calcium, and magnesium)	Enhancement of absorption, induction of hyperaluminemia, hypercalcemia, and hypermagnesemia
Vitamin E	Anticoagulant (warfarin), antiplatelet drug (ticlopidine, aspirin, etc.)	Enhancement of anticoagulant effect
	Immunosuppressants (cyclosporin)	Increased cyclosporin effect/side effect due to enhanced cyclosporin absorption
	Vasopressors (ubidecarenone)	Enhanced antioxidative effect
Vitamin K	Anticoagulants (warfarin)	Reduced efficacy of warfarin
Niacin	Uricosuric drugs, gout (allopurinol, probenecid)	Reduced efficacies of allopurinol and probenecid
	Statins (simvastatin, atorvastatin, etc.)	Increased risk of rhabdomyolysis
	Oral antidiabetics (glimepiride, pioglitazone, etc.)	Reduced effectivity of oral antidiabetic agent
Folic acid	Anticancer drug (fluorouracil)	Enhanced toxicity of fluorouracil
	Anticancer drug (methotrexate)	Reduced effects/side effects of methotrexate
	Anticancer drug (capecitabine)	Enhanced side effects (diarrhea, vomiting) of capecitabine
	Antiepileptic drugs (phenytoin, primidone, phenobarbital)	Reduced efficacy of antiepileptic drugs

TABLE 12.2
Dietary Supplements Containing Minerals

Dietary Supplement	Drug	Interaction
Calcium	Osteoporosis treatment (bisphosphonate drugs)	Decreased efficacy due to inhibited bisphosphonate absorption
	New quinolone and tetracycline antibiotics	Reduced efficacy of antibiotics due to inhibited absorption
	Prostate cancer treatment (estramustine)	Reduced absorption and efficacy of estramustine
	Thyroid hormone (levothyroxine)	Reduced absorption and efficacy of levothyroxine
	Psoriasis treatment (calcipotriene)	Enhanced effects/side effects of calcipotriene
	Digitalis (digoxin, digitoxin, etc.)	Enhanced efficacy of digitalis
	Calcitriol	Hypercalcemia due to enhanced calcium absorption
	Thiazide diuretics (hydrochlorothiazide, etc.)	Induction of hypercalcemia
Magnesium	Osteoporosis treatment (bisphosphonate drugs)	Decreased efficacy due to inhibited bisphosphonate absorption
	New quinolone antibiotics, tetracycline antibiotics (minocycline, etc.)	Reduced efficacy of antibiotics due to inhibited absorption
	Digitalis (digoxin, digitoxin, etc.)	Decreased efficacy of digitalis
	Statins (lovastatin, etc.)	Reduced blood titer of statins due to inhibited absorption
	Antiepileptic drug (gabapentin)	Reduced gabapentin blood titer due to inhibited absorption
	Hyperkalemia treatment (polystyrene sulfonate)	Reduced potassium-exchange function by polystyrene sulfonate, induction of metabolic alkalosis
	Calcitriol	Induction of hypermagnesemia
	Potassium-sparing diuretics (spironolactone, etc.)	Induction of hypermagnesemia
Iron	Osteoporosis treatment (bisphosphonate drugs)	Decreased efficacy due to inhibited bisphosphonate absorption
	New quinolone antibiotics, tetracycline antibiotics	Reduced efficacy of antibiotics due to inhibited absorption
	Cephalosporin antibiotic (cefdinir)	Reduced efficacy of cefdinir due to inhibited absorption
	Anti-Parkinson (levodopa), antihypertensive (methyldopa), anti-rheumatic/immunosuppressant (penicillamine), thyroid hormone (levothyroxine)	Reduced efficacies of drugs due to the interference of their absorption by iron
Copper	Antirheumatic/immunosuppressant (penicillamine)	Reduced efficacy of penicillamine due to inhibited absorption
	Immunosuppressant (mycophenolate mofetil)	Reduced efficacy of mycophenolate mofetil due to inhibited absorption
	Osteoporosis treatment (bisphosphonate drugs)	Decreased efficacy due to inhibited bisphosphonate absorption

(*Continued*)

TABLE 12.2
(Continued) Dietary Supplements Containing Minerals

Dietary Supplement	Drug	Interaction
Zinc	Osteoporosis treatment (bisphosphonate drugs)	Decreased efficacy due to inhibited bisphosphonate absorption
	New quinolone antibiotics, tetracycline antibiotics	Reduced efficacy of antibiotics due to inhibited absorption
	Antirheumatic/immunosuppressant (penicillamine)	Reduced efficacy due to inhibited penicillamine absorption
	Potassium-sparing diuretics (spironolactone, etc.)	Increased blood titer of zinc
Aluminum	Osteoporosis treatment (bisphosphonate drugs)	Decreased efficacy due to inhibited bisphosphonate absorption
Iodine	Angiotensin II receptor blockers (losartan, telmisartan, etc.)	Increased serum potassium concentration due to inhibited potassium excretion
	ACE inhibitors (captopril, enalapril, etc.)	Increased serum potassium concentration due to inhibited potassium excretion
	Antithyroid drug (thiamazole)	Reduced thyroid function
	Psychiatric medication, mood stabilizer (lithium)	Reduced thyroid function
	Antiarrhythmic agent (amiodarone)	Increased plasma concentration of iodine

ACE, angiotensin-converting enzyme.

TABLE 12.3
Other Major Dietary Supplements

Dietary Supplement	Drug	Interaction
Alfalfa	Anticoagulant drug (warfarin)	Enhanced anticoagulant effect of warfarin
	Estrogenic hormone (estrogen)	Reduced efficacy of estrogen
	Immunosuppressants (cyclosporine, tacrolimus, etc.)	Reduced efficacy of immunosuppressants
Aloe	Anticoagulant (warfarin)	Enhanced anticoagulant effect of warfarin
	Digitalis (digoxin, digitoxin, etc.)	Enhanced side effects of digitalis due to hypokalemia
	Thiazide diuretics (hydrochlorothiazide, etc.), loop diuretic (furosemide)	Hypokalemia
	Corticosteroids (prednisolone, dexamethasone, etc.)	Hypokalemia
	Laxatives	Enhanced efficacy of laxative
Bitter orange (*Aurantii pericarpium*)	Caffeine	Enhanced effects/side effects of caffeine
	Antidepressant (MAO inhibitor)	Enhanced effects/side effects of antidepressant
	Sedative (midazolam)	Enhanced effects/side effects of midazolam
	Antihypertensive agent (felodipine)	Enhanced effects/side effects of felodipine
Black cohosh	Anticancer drug (cisplatin)	Reduced efficacy of cisplatin
	Statin (atorvastatin)	Enhanced risk of liver disease

TABLE 12.3
Other Major Dietary Supplements

Dietary Supplement	Drug	Interaction
Blond psyllium (*Plantago psyllium*)	Antiepileptic drugs (carbamazepine), mania treatment (lithium carbonate), digitalis (digoxin, digitoxin, etc.), iron tablets, anticoagulants (warfarin)	Reduced efficacies of carbamazepine, lithium, digitalis, iron, and warfarin due to inhibited absorption
	Laxative	Enhanced efficacy of laxative due to the laxative effect of blond psyllium
	Oral antidiabetics (chlorpropamide, glimepiride, pioglitazone, etc.)	Enhanced reduction of blood glucose by the oral antidiabetic agent
Blueberry	Insulin, oral antidiabetics (chlorpropamide, glimepiride, pioglitazone, etc.)	Enhanced reduction of blood glucose
L-Carnitine	Anticoagulants (warfarin, acenocoumarol)	Enhanced anticoagulant effects of warfarin and acenocoumarol
	Thyroid hormone (levothyroxine)	Reduced efficacy of thyroid hormone
Casein peptides	Antihypertensives	Enhanced efficacy of antihypertensive agents
Chitosan	Anticoagulant drug (warfarin)	Enhanced anticoagulant effect of warfarin
	Antiepileptic drug (sodium valproate)	Reduced blood titer of valproate
Chlorella	Anticoagulant (warfarin)	Reduced efficacy of warfarin
	Immunosuppressants (cyclosporine, tacrolimus, etc.)	Reduced efficacy of immunosuppressants
Chondroitin	Anticoagulants (warfarin), antiplatelet drugs (ticlopidine, aspirin, etc.)	Enhancement of anticoagulant effect
Coenzyme Q10	Anticoagulants (warfarin)	Weakening of anticoagulant effect
	Antihypertensive agents (captopril, amlodipine, etc.)	Enhanced efficacy of antihypertensive agents
	Oral antidiabetics (chlorpropamide, glimepiride, pioglitazone, etc.)	Enhanced reduction of blood glucose by the oral antidiabetic agent
	Statins (lovastatin, etc.)	Prevention and reduction of statin's side effects, such as muscle ache
Dong Quai	Anticoagulants (warfarin), antiplatelet drug (ticlopidine, aspirin, etc.)	Enhancement of anticoagulant effect
DHA	Antihypertensive agents (captopril, amlodipine, etc.)	Enhanced efficacy of antihypertensive agents
	Anticoagulants (warfarin), antiplatelet drugs (ticlopidine, aspirin, etc.)	Enhancement of anticoagulant effect
	Statins (lovastatin, etc.)	Enhancement of statin efficacy
	Laxatives	Enhanced efficacy of laxative
EPA	Antihypertensive agents (captopril, amlodipine, etc.)	Enhanced efficacy of antihypertensive agents
	Anticoagulants (warfarin), antiplatelet drugs (ticlopidine, aspirin, etc.)	Enhancement of anticoagulant effect
	Statins (lovastatin, etc.)	Enhancement of statin efficacy
	Laxatives	Enhanced efficacy of laxative
Fig	Vitiligo vulgaris medication (methoxsalen)	Increased risk of light sensitivity due to methoxsalen
	Insulin, oral antidiabetics (chlorpropamide, glimepiride, pioglitazone, etc.)	Enhanced reduction of blood glucose

(*Continued*)

TABLE 12.3
(Continued) Other Major Dietary Supplements

Dietary Supplement	Drug	Interaction
Garlic	Anticoagulants (warfarin), antiplatelet drugs (ticlopidine, aspirin, etc.)	Enhancement of anticoagulant effect
	Vitamin E	Enhancement of anticoagulant effect
	Immunosuppressant (cyclosporin)	Reduced immunosuppressive effect due to enhanced cyclosporin degradation
	Ethyl icosapentate	Enhanced efficacy of icosapentate
	Anti-AIDS medication (saquinavir)	Reduced saquinavir efficacy due to enhanced drug metabolism
Ginkgo	NSAID (ibuprofen)	Enhancement of anticoagulant effect by ibuprofen
	Anticoagulants (warfarin), antiplatelet drugs (ticlopidine, aspirin, etc.)	Enhancement of anticoagulant effect
	Vitamin E	Enhancement of anticoagulant effect
	Oral antidiabetics (chlorpropamide, glimepiride, pioglitazone, etc.)	Enhanced reduction of blood glucose
	Migraine treatment (dihydroergotamine)	Increased risk of dual subdural hematoma
	Antihypertensive (nifedipine)	Enhanced side effects of nifedipine
	Thiazide diuretics (hydrochlorothiazide, etc.)	Reduced efficacy of thiazide diuretics
	Antiepileptic drug (phenitoin, carbamazepine)	Reduced anticonvulsive effect
	Theophylline	Increased induction of seizures
Ginseng (*Panax*)	Anticoagulants (warfarin), antiplatelet drugs (ticlopidine, aspirin, etc.)	Enhancement of anticoagulant effect
	Insulin, oral antidiabetics	Enhanced reduction of blood glucose
	Corticosteroids (prednisolone, dexamethasone, etc.)	Increase of blood corticosteroid titer
	Immunosuppressants (cyclosporin, etc.)	Decreased immunosuppressant effect
	Breast cancer treatment (tamoxifen, toremifene)	Reduced efficacy of tamoxifen and toremifene due to the estrogen-mimetic effect of carrots
	Loop diuretics (furosemide)	Decreased diuretic effect
	Immunosuppressants (cyclosporin, azathioprine)	Decreased immunosuppressant effect
Glucosamine	Anticoagulants (warfarin), antiplatelet drugs (ticlopidine, aspirin, etc.)	Enhancement of anticoagulant effect
Isoflavones	Conjugated estrogens	Enhanced efficacy of estrogen; increased pulmonary embolism
	Breast cancer treatment (tamoxifen, toremifene)	Reduction of tamoxifen and toremifene efficacy due to the estrogen-mimetic effect of isoflavones
Kava	Anxiolytics (alprazolam)	Excessive sleepiness, fatigue, and disorientation
	Anti-Parkinson medication (levodopa)	Reduced efficacy of levodopa
	Sedatives (clonazepam, lorazepam, zolpidem, etc.)	Excessive sleepiness

TABLE 12.3
Other Major Dietary Supplements

Dietary Supplement	Drug	Interaction
Licorice	Anticoagulant (warfarin)	Reduced efficacy of warfarin
	Digitalis (digoxin, digitoxin, etc.)	Enhancement of digitalis' side effects due to hypokalemia
	Diuretics (furosemide, hydrochlorothiazide, etc.)	Hypokalemia
	Steroids (prednisolone, dexamethasone, etc.)	Hypokalemia
	Estrogenic hormone (estrogen)	Reduced efficacy of estrogen
	Antihypertensive agents (captopril, amlodipine, etc.)	Reduced efficacy of antihypertensive agent
Parsnip root (*Pastinaca sativa*)	Vitiligo vulgaris medication (methoxsalen)	Increased risk of light sensitivity caused by methoxsalen
Red yeast	Statins (lovastatin, etc.)	Enhanced effects/side effects due to serum cholesterol-reducing effect
Saw palmetto (*Palmier nain*)	Estrogenic hormone (estrogen)	Enhanced effects/side effects of estrogen
	Anticoagulants (warfarin), antiplatelet drugs (ticlopidine, aspirin, etc.)	Enhancement of anticoagulant effect
	Iron drugs (ferrous citrate, ferrous sulfate)	Decreased absorption of iron due to chelate formation
St John's wort (*Hyperici herba*)	Anticoagulants (warfarin), antiepileptic drugs (phenitoin, carbamazepine), antiarrhythmic agents (disopyramide, amiodarone, etc.), bronchodilators (theophylline, aminophylline, etc.), antidepressants (amitriptyline, etc.), female sex hormones (estradiol), statins (simvastatin, atorvastatin, etc.), antifungal drugs (itraconazole, ketoconazole, etc.), histamine antagonists (fexofenadine, etc.), immunosuppressants (cyclosporin), other drugs that are metabolized by drug-metabolizing enzymes	Reduced efficacy of drugs as a result of enhanced drug metabolism, caused by the upregulation of drug-metabolizing enzymes by St John's wort
	Psychostimulant drug (methyphenidate)	Decreased effect of methylphenidate
	Antidepressants (paroxetine, fluvoxamine, trazodone, etc.)	Increased side effects such as cardiac abnormalities, tremors, and anxiety due to increased serotonin in the brain
	Migraine treatments (sumatriptan, zolmitriptan)	Increased side effects such as cardiac abnormalities, tremors, and anxiety due to increased serotonin in the brain
	Antitussives (dextromethorphan)	Increased side effects such as cardiac abnormalities, tremors, and anxiety due to increased serotonin in the brain
	Nonnarcotic analgesics (pentazocine, tramadol)	Increased side effects such as cardiac abnormalities, tremors, and anxiety due to increased serotonin in the brain
	Tetracycline antibiotics (minocycline, etc.)	Increased risk of light sensitivity
	Antiarrhythmic agent (procainamide)	Increased side effects due to enhanced absorption of procainamide

(*Continued*)

TABLE 12.3
(Continued) Other Major Dietary Supplements

Dietary Supplement	Drug	Interaction
Tryptophan	Antidepressants (amitriptyline, paroxetine, trazodone, etc.), antitussives (dextromethorphan), nonnarcotic analgesics (pentazocine, tramadol)	Increased side effects such as cardiac abnormalities, tremors, and anxiety due to increased serotonin in the brain
	Sedatives (lorazepam, zolpidem, etc.)	Increased side effects such as severe sleepiness
Turmeric (*Curcuma*)	Anticoagulants (warfarin), antiplatelet drugs (ticlopidine, aspirin, etc.)	Enhancement of anticoagulant effect
	Oral antidiabetics (chlorpropamide, glimepiride, pioglitazone, etc.)	Enhanced reduction of blood glucose by the oral antidiabetic agent

AIDS, acquired immunodeficiency syndrome; DHA, docosahexaenoic acid; EPA, eicosapentaenoic acid; MAO, monoamine oxidase inhibitors; NSAID, nonsteroidal anti-inflammatory drug.

REFERENCES

1. Japan Medical Association, Japan Pharmaceutical Association, and Japan Dental Association (Eds.). (2013). *Natural Medicine Comprehensive Database: Consumer Version 2013* (digital version). Tokyo: Dobunshoin Co., Ltd.
2. Shizuoka Pharmaceutical Association (Ed.). (2008). *Dietary Supplements and Specified Functional Food Q&A for Self-Improvement*. Tokyo: Nanzando Co., Ltd.
3. Fujimura, T. (2006). *Drug–Food Interactions That Everyone Should Know*. Osaka, Japan: Nagai Shoten Co., Ltd.
4. Food, Recreational Product and Drug Interaction Research Group. (1998). *Food, Recreational Products and Drug Interaction*. Tokyo: Jiho Co., Ltd.
5. Josai University. (2007). *Comprehensive Food–Drug Interaction: Basics and Application*. Tokyo: Kazan Co., Ltd.
6. Uchida, S. and Yamada, S. (2007). Food, dietary supplement and drug interactions. *Bunseki*, September edition, pp. 454–460.
7. Sawada, Y. and Satoh, H. (2012). Interaction between prescription drugs and OTC drugs, health food and dietary supplements. *Farumashia* (48): 1062–1066.
8. Sawada, Y. and Ohtani, H. (2002). Interaction of medicine and food—Clinical mishaps due to the encounter of food, recreational products and drugs. Drug enhancement of detoxification diminishes the pharmacological effect; Drugs and St. John's wort (1). Commentary. *Iyaku Journal (Medicine & Drug Journal)* (38): 1791–1799.
9. Josai University. (2007). *Handbook of Food–Drug Interactions*. Tokyo: Maruzen Co., Ltd.
10. Hori, M. (2007) Guaranteed usefulness! Encyclopedia of dietary supplements—Recommendation on their use and interactions with drugs. Hokendohjinsha Inc.
11. Tokuyama, S. (2003). Disease-dependent effects of dietary supplement and their interaction with medications (4): Osteoporosis. Commentary. *Clinics & Drug Therapy* (22): 52–56.
12. Sawada, Y., Miki, A., Sata, H., and Ohtani, H. (2002). Interaction of medicine and food—Clinical mishaps due to the encounter of food, recreational products and drugs. Drugs annulling the effects of vitamins; Drugs and coenzyme Q10-containing food, recreational products, health food and dietary supplements. Commentary. *Iyaku Journal (Medicine & Drug Journal)* (40): 2064–2069.
13. Tokuyama, S. (2002). Dietary supplement seminar for clinicians—Disease-dependent effects of dietary supplement and their interaction with medications: hypertension. Commentary. *Clinics & Drug Therapy* (21): 1006–1011.
14. Sawada, Y. and Ohtani, H. (2000). Interaction of medicine and food—Toxic effects of the encounter of food, recreational products and drugs. Food and recreational products turn medicines into poisons; Drugs and grapefruit juice (1). Commentary. *Iyaku Journal (Medicine & Drug Journal)* (36): 500–507.

13 Journey from Pharmaceuticals to Food

Role of Evidence-Based Approach

R.B. Smarta

CONTENTS

13.1 CLEARING THE DECK

Achieving and maintaining a competitive advantage in any industry neither is nor has it been easy. Both pharmaceuticals and foods sectors have independent as well as related challenges as both are at the different stages of a life cycle.[1] Pharmaceutical sector is at a maturity stage, whereas nutraceutical and food sectors in some countries are at growth and infancy stages. Hence, the challenges faced by both sectors need to be looked at separately and together as the population has become more sophisticated, demanding, and expect product innovations; legislation also needs to cope with the environment and become more reflective and complex.

The challenges become even greater for the pharmaceutical industry: experiencing difficult economic conditions in the past few years and coming to a point of falling over a patent cliff. Top management teams across the world had to start concentrating on minimizing costs and identifying alternate revenue models.[2] These efforts have served only to magnify the challenges.

These challenges have been faced by the music and film industries earlier; they are hard nuts to crack in every sector and arguably even harder to crack for businesses in the food and pharmaceutical sectors. Both sectors are dependent on the health of human beings because human life is precious. Companies in these two sectors are governed by extremely rigorous standards of quality, safety, and legislation, especially in the areas of health and hygiene.

Today, pharmaceutical industry requires total reforms right from developing alternate revenue models to derive new focus on investments and behavior. To curb all the existing unethical practices, there is a need for better governance and transparency between the regulators and the manufacturers.

13.2 CONSUMER HEALTH

Legislation and regulations are the major concerns for most companies in the food and pharmaceutical sectors. The need to comply with the Food and Drug Administration (FDA), the Medicines and Healthcare Products Regulatory Agency (MHRA) rules, the National Food Authority (NFA) guidelines of codex, other food safety standards regulatory bodies, and sanctioning of claims by authorities are essential. Every new product across every part of the process, from intrinsically clean environments to end-of-batch safety and quality, must be evaluated for strict compliance before implementation. This can make the process quite tough.

This is of utmost importance as the decision to buy would be taken by a consumer. Consumers may get influenced and as they are looking for options and in need due to their health concerns may decide to choose without a proper diagnosis! Hence, providing quality, safe products and adequate information to make informed decision is an onus of the industry.

13.3 EXISTENCE AND GROWTH OF PHARMACEUTICAL SECTOR

Changing market scenario, steep increase in research and development (R&D) cost, continuity in patent expiry, increasing health-care cost, and the disease burden of every country have compelled providers to reduce health-care cost. In addition, the educated and health-conscious patients as consumers are moving toward wellness!

Today, common man fears sickness, yet he lacks the much-needed awareness about the healthy habits and health benefits of any product. Thus, he gets influenced to go for over-the-counter products, or based on his experiences, sometimes he goes back to the traditional medicines as they are often cost effective and more easily available.

With the changing disease pattern and the lifestyle diseases, consumers are shifting toward prevention. Utility of both medicines and food products inclusive of nutraceutical products today coexist for the complete wellness of a patient/consumer. Usually for chronic diseases, physicians are making patients aware of lifestyle changes and advising them to make necessary changes in their lifestyle and food.

13.4 CHALLENGES FACED BY PHARMACEUTICAL SECTOR

Due to these changes in the disease pattern, consumer education, ambition of remaining fit, drying up of patented product pipeline, expectations of patients, and both consumer and pharma sectors are facing challenges.

1. *Patent expiries giving rise to generic medicines*: An increasing number of patents have expired or are in the process of expiration in the past and coming years. These patent expiries are likely to fuel the growth of the global generic medicines market by 2017. The inevitable patent expiry of blockbuster drugs will work in favor of generic medicine manufacturers. They will get a further boost by the judicious cost-containment efforts of the government and health-care service providers all over the world.[3]

Even generic medicines will need to comply with new legislations, regulations, and pricing issues/standards meticulously. The stage at which generic medicines will operate would be a newly developed platform of affordable medicines for each population of the world. As a result, it is possible that generic medicines will have to deal with differential pricing for different countries.

Challenges are in the form of drying up of new product pipeline, selecting new revenue models for generic medicines, and declining margins due to nature of generic medicines and different regulatory framework of each country.

2. *Choosing adaptable and flexible ways of corporate operations*: Having an experience of marketing medicines with more margins for a niche group of patients with selective and particular diseases, companies will have to change to a model such that it takes maximum benefit from the market evolution. It has to shift from serving a niche group to a bigger group of patients with a wide variety of disorders or diseases. They will have to operate at lower margins and perhaps at lower costs! It is quite a difficult task, but Teva, Mylan, and big generic pharmaceutical companies have taken up this route and they are making evident for the industry that adaptability with changing environment is the choice. The challenge is to change the style of operations and to orient management mindsets to suit the new environment.
3. *Moving away from product-centric to "patient-centric" approach as a reform*: As the times change, pharmaceutical companies should come up with strategies keeping in mind the patient pull approach. In addition, to gain higher consumption, ideally there should be a shift from the R&D department and product centricity toward patient centricity.

   In developed nations, the emphasis has to be on quality health care for the elderly as these nations have a higher mix of aged population. Therefore, lifestyle or chronic diseases, occurring more often in the population over the age of 60, are a grave challenge for developed nations.[4]
4. *Need for innovations in research and development*: Higher investment and lesser output is a reality of R&D today. Hence, R&D needs to be revamped using different models to become cost effective. A consortium is certainly a wise choice. It involves pairing of many companies wherein they put their resources and focus on a particular patient-centric goal. As a consortium, it standardizes the process. Hence, a consortium may work in favour of companies as a group investment and the outcome can be shared among them. It is a challenge for every player in pharmaceuticals.

   Emerging markets are most likely to experience a rise in the chronic diseases to a larger extent compared to communicable diseases. In the case of developing countries, the health-care systems face a greater burden of communicable diseases. Additionally, if we take a look at the management of health around the world, health management in most countries is done by their respective governments, whereas in India it is not the responsibility of the central government but the state government in spite of having an annual health budget.

   Another observation indicates that in developed countries such as the United States, the United Kingdom, and Japan, there is a rise in the older population that has stunted the growth of pharmaceuticals to as low as 2%–3% unlike in emerging nations. In addition, the cost of medication is very high all over the world, which is not the case in India. The common man suffers due to changing lifestyles as well as the hereditary issues passed down to him. These issues bring to light the need to stay healthy and fit. A large number of people want to look toward the alternative cures moving away from the conventional treatment and the ways to prevent occurrence of diseases and disorders.

13.5 EMERGING ROLE OF NUTRITION ALONG WITH PHARMACEUTICAL SECTOR

Amino acids have added growth momentum in pharmaceutical prescription market and acted as a growth driver for the past few years. Physicians welcomed amino acids—an essential ingredient of nutrition!

After careful deliberation, the industry came to realize that amino acids can open up new business opportunities such as nutraceuticals. Thus, it would not be wrong to say that pharmaceutical industry has adopted a shift toward nutraceuticals to grow and stay in the race to be on the top.

Total parenteral nutrition (TPN) and external nutrition (EN) have been the facts of life in hospitals, but now through amino acids it has become evident that nutrition has an essential role alongside pharmaceuticals for the benefit of the patients.

Taking a step ahead makes it clear that neither nutraceuticals nor pharmaceuticals are separate from the other, but they coexist for the betterment of patients and medical consumers.

13.6 CHALLENGES FACED BY FOOD SECTORS

1. *Chances of self-medication*: The awareness about the nutraceutical products is increasing every day. With this hype, a large number of people are gaining knowledge about the perceived health benefits these products offer for preventing or treating some major health disorders such as cardiovascular diseases or arthritis. This, however, may have a flip side that is often overlooked. There may be a chance that patients turn to these nutraceutical products in lieu of taking treatment for a certain serious disorder they are experiencing. Therefore, it would certainly be risky if a consumer starts self-medicating for the disorder experienced as he/she may be exposed to the quality issue for that particular nutraceutical product as well instead of turning to much-needed medical assistance.[2]
2. *Evolving regulatory compliances*: As the field of nutraceuticals has many issues, regulations are either evolving or harmonizing all over the world. Globally, the stress is upon having pertinent regulations that are made with guidance for policy makers over high-quality scientific, technical, and regulatory nuances.[5]
3. *Right pricing*: Since these nutrition products are made from the natural plant extracts and other such ingredients, they turn out to be quite expensive as these extraction processes are extremely tedious and time consuming. Hence, the issues of pricing become a major challenge for the nutraceutical and dietary supplement products. To get the right pricing, it is a general observation that such products are overpriced for a common man. In emerging or developing countries, affordability becomes an important criterion for consumption. Therefore, it is an obvious conclusion that consumers are hesitant to buy overpriced products even though they can actually be beneficial. Nutraceuticals need the right pricing to gain momentum in any market.
4. *Appropriate claims*: Various regulatory bodies in different countries around the world have created their own set of claims which nutraceutical manufacturers need to follow when making health claims on labels and market them in any country. The claims for labeling and marketing are becoming increasingly clear due to the evolving regulations.

   Manufacturers have to be careful when marketing their products so that they do not make any claims that can be misleading and false. This is because overclaiming or wrong claiming will lead to their products being vulnerable, and concerned regulatory bodies would respond in a legal way.

   Regulatory bodies now lay emphasis on the importance of having science-based claims that do not mislead the consumer in any way.[6]

   Stringent regulations have ensured that health claims are watertight so manufacturers have to be careful that they do not make claims over and above the existing ones for their product when marketing it. Thus, the result would be a law suit if the product does not offer any health benefits as claimed by the label.

5. *Misleading advertising and promotions*: Although it is related to health claims and their promotion to consumers, irresponsible advertising may end up misleading the consumer.

As a consumer, "taste" campaign may increase consumption, but in such products, emphasis is on the science behind the product that needs more exposure. Consumer's safety, quality, and effectiveness is the uppermost concern. There are advertising guidelines and compliances that have to be governed.[7]

13.7 NURTURING DOUBLE VISION COEXISTENCE MODEL

The existing model of nutraceuticals is pharmaceutical driven. It focuses on cure for diseases or ailments for their customers in the sick-care sector. However, where the needs and trends in the environment are changing, the focus of the nutraceutical model will have to shift from sick care to health care (i.e., preventive and promotion aspects).

In the nutraceuticals domain, there exists peripheral opportunities for chronic lifestyle diseases and ailments prevailing in the sick-care sector as well as preventive opportunities in the wellness sector. The pharmaceutical-driven model does not address the need for both prevention and promotion aspects.

With the emergence of lifestyle changes, addition of chronic lifestyle diseases and ailments such as obesity, tuberculosis, diabetes, arthritis, malaria, and cholera, the diseases can be managed through preventive efforts. Although lifestyle diseases cannot be completely cured, they can be well managed through the nutraceutical medium. In a densely populated country such as India where the lifestyle diseases are catching up in the urban and rural areas, the middle-class population drives the need for conscious preventive and curative disease management rather than just curative aspects.

There has to be a parallel opportunity between the "needs of the sick-care and wellness market" and the "opportunities at the peripheral for nutraceuticals." Transition from sick-care to health-care market is the need of the hour for disease management (Figure 13.1).

Physicians may not really understand nutrition as they are not formally trained; however, a growing number of physicians are realizing the need to supplement their medical treatment with a good, balanced diet and nutraceutical products.[13] In most health-care facilities, stress is being given not just on the right medication but also on offering the right food for quick and effective recovery of a patient. This is more often being done in the case of patients suffering from cardiovascular disorders. In addition, in the case of diabetics, doctors recommend a strict diet along with the prescribed medication. Hospitals and big health-care centers have dieticians and nutritionists as a part of their team that treats patients so as to create the most ideal environment for his/her recovery.

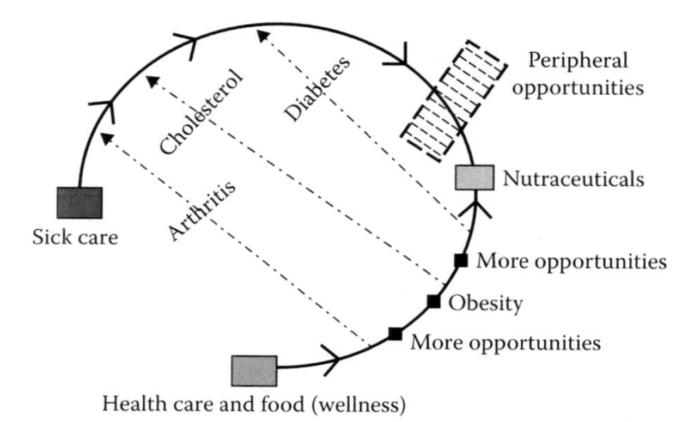

FIGURE 13.1 Conceptual nutraceutical model. (Conceptual model by Smarta, R.B.)

Lolimbrāja in Vaidyajeeva stated clearly that

पथ्ये सति गदार्तस्य किमौषध निषेवणै : ।
पथ्येऽसति गदार्तस्य किमौषध निषेवणै : ।। वैदयजीवन - लोलिंबराज

(In fact for those who follow the right diet where is the need for medicines and for those who do not follow the right diet what is the use of medicines?)

Ancient wisdom from Ayurveda clarifies that right diet is sufficient to take care of any disorders of faulty lifestyle. This means right diet takes care of health and thereby no need of medicines.

13.8 MICROBIOME THEORY—RELATIONSHIP WITH DYSFUNCTION AND DISORDER

Approximately 100 trillion bacteria of several hundred species dwell in the human body. These microbes help the human body carry out their bodily functions with ease. Microbes reside in the human body or the host, and the host gets many health benefits in lieu of the provided food and shelter.

For instance, in case of a person experiencing irritable bowel syndrome (IBS), there is an imbalance in the microbiome count. This can simply be treated by the intake of probiotics. Probiotics fall under the nutraceutical space, indicating the rising need for such products to be included in the daily diet.

The idea is to develop good diagnostic tools and combine them with drugs and nutrition. It can be said that the largest drug in one's repertoire is the food that he/she eats three times a day, every day of their life. Instead of inventing things, one must try to see how they can work with the components—the nutrients—that are in the existing food.

13.9 SHIFT TO INTEGRATED (HOLISTIC) MEDICINES

Integrated medicine is a new paradigm in health care that focuses on the synergy and deployment of the best aspects of diverse systems of medicine including modern medicine, Homeopathy, Siddha, Unani, Yoga, and Naturopathy in the best interest of the patients and the community.

The increasing public demand for traditional medicines has led to considerable interest among policy makers, health administrators, and medical doctors regarding the possibilities of bringing together traditional and modern medicine. Traditional medicine looks at health, disease, and causes of diseases in a different way. The integration of traditional medicine with modern medicine may imply the incorporation of traditional medicine into the general health service system.[8] The purpose of integrated medicine is not simply to yield a better understanding of differing practices, but primarily to promote the best care for patients by intelligently selecting the best route to health and wellness.

Surveys and other sources of evidence indicate that traditional medical practices are frequently utilized in the management of chronic diseases. Traditional medicine presents a low-cost alternative for rural and semi-urban areas where modern medicine is inaccessible.

An approach to harmonizing activities between modern and traditional medicine will promote a clearer understanding of the strengths and weaknesses of each and will encourage the provision of the best therapeutic option for patients.[9]

13.9.1 Advantages of Integrated Medicine

The advantages of integrated medicine are as follows:

- It provides the widest array of options to patients (one in three adults in the United States use at least one complementary or alternative medical [CAM] therapy).
- It provides an opportunity to combine the "best of both conventional medicine and complementary alternative medicine."

- It provides cost-effective treatment options.
- It results in better patient outcomes, measured in terms of symptom relief, functional status, and patient satisfaction.
- It focuses on holistic health and well-being.

Technology is seen as one of the three important drivers of increasing health-care accessibility. Selection and adoption of appropriate technology often make a critical difference in the success of health-care reform and reengineering, and it has the capability to revolutionize the way health care is delivered.

Therefore, it would not be wrong to conclude that using traditional medicines and modern medicines together can be a great therapeutic option for the patients. These two different schools of medicine will work in harmony so as to complement each other's strengths and weaknesses, providing the best of health to the individual undergoing treatment.

Of course, the amount and complexity of legislation is one thing—its rate of change is another. Simply staying abreast can be a major task in itself.

13.10 TREATMENT OF MEDICAL PROBLEMS WITH NUTRITION-BASED PRODUCTS

There are three kinds of diseases: (1) metabolic syndrome (disorders that increase the risk of diabetes and heart problems), (2) gastrointestinal health (all the problems that happen in your intestinal tract), and (3) a more challenging is an evidence to prove. We believe that there is more evidence that good nutrition could help manage cognitive decline. After an extensive research, Fruitflow, an extract from tomatoes, was launched recently. It can be added to yogurts, beverages, and other food products. It is the only functional food ingredient addressing platelet aggregation by keeping them smooth, and by this, it avoids aggregation in the blood vessels.

Another notable example of nutrition for improved health is Ubiquinol. It is a safe, cost-effective ingredient and the key to male infertility. Ubiquinol or Coenzyme Q10 has now been used in a wide range of nutraceuticals either alone or in combination with other substances to create "all-round" fertility products.

13.11 CONVERGENCE OF PHARMACEUTICALS AND NUTRACEUTICALS

We believe that there is a convergence between food and pharmaceuticals. Chronic diseases need more than just the traditional pharmaceutical approach. Our food and nutrition expertise can help create a new industry where nutrition plays a bigger role in helping people who live with difficult chronic medical conditions.

With the aging population, getting people healthy at the later stage of life is difficult and costly. We believe that health care is going to be in real difficulty if we do not start looking at it in a different way.

This is in turn compelling the policy makers, health administrators, and medical doctors to look to these alternatives. The way traditional medicines perceive health, diseases, and their causes is very different. Treating a patient using both modern medicine and traditional medicine goes on to imply that the general health service system can include traditional medicines.

The immense turmoil in the pharmaceutical industry has pushed them to look toward greener pastures to tap into. The nutraceutical and functional foods and beverages market is currently experiencing tremendous growth potential and is a big attraction for pharmaceutical companies.[10]

There was a rise of 44% in the retail sales value of fortified/functional foods between 2005 and 2010 across 32 markets where Euromonitor International (London, UK) conducted its detailed health and wellness research. For functional bottled water, the global value went up by 70% in

2006–2011. However, the sales value for vitamins and dietary supplements rose by almost 50%. It has been estimated that the global nutraceuticals market is likely to reach US$250 billion by 2015 as per a report by Global Industry Analysts (San Jose, CA).

With great advances in biotechnology, there is great interest in using food for medical purposes. Using natural herbs and foods for treating ailments is a very old tradition in many cultures. Interest is now expanding toward food with medicinal properties (enriched foods). Probiotics, prebiotics, functional foods, clinical foods, and nutraceuticals are promoted as "good-for-you foods" either as a part of the daily diet or for a particular disorder treatment such as improving digestion and bone density.

13.12 WAY FORWARD

13.12.1 Case for Evidence-Based Products

13.12.1.1 Pre-License Activities

Pharmaceutical organizations want to understand the causes of diseases and to develop compounds that can be tested in the laboratory for their effectiveness in treating a disease. Once these targets are identified by the research and development team, they are usually passed to a team responsible for turning these targets into medicines that can be submitted for clinical trials. Once the complex clinical trial process is completed, the medicines can be submitted for regulatory approval, and if granted, they can be sold to health-care providers. End to end, this process of discovering and developing a new molecular entity (NME) takes an average of 13.5 years and (capitalized) costs of US$1.8 billion.

The number of targets that hit the market as medicines is a small percentage that is fast diminishing. During target identification, scientists work to understand the cause of disease by looking at pathways, for example, how the disease begins and spreads. They investigate gene expression, protein mechanisms, and many other highly complex aspects of the causes of disease.

Advances in genomics, proteonomics, and protein modeling are transforming the discovery process. At the same time, the sheer volume and structure of the data is challenging the leaders in the field to develop shared ontologies to express the meaning of the data.

It is in this area of intensive scientific research that collaboration and transformative partnerships between the pharmaceutical companies themselves can take place. A common ontology allows a shared understanding of the semantics of the original data. The cloud offers to host large volumes of complex data in collaborative platform and at relatively low cost. This is an area where Logica (Reading, UK) and our partner ecosystem have a definite, key role to play.

At the end of the discovery process, the objective is to turn target compounds into medicines. This is the costly part of the process, and later when a target falls out of the process, greater is the loss incurred by the company. It is therefore vital that those medicines entering the later stage of clinical trials, with live patients, are targeted at a narrow, highly specific cohort.

13.12.1.2 Evidence Creation Activities

An evidence-based approach is also being adopted by nutraceuticals for regulatory compliance that can be ideally seen through the following examples. Fruitflow is a natural, scientifically substantiated ingredient that contributes to healthy blood flow in the body. In 10 human trials, consumption of Fruitflow has been proven to maintain healthy platelet aggregation and improve blood flow.

Ubiquinol/Coenzyme Q10 is an essential electron carrier in the body's mitochondrial respiratory chain. Male infertility is a worldwide health issue. For the first time, clinical usefulness of Ubiquinol for the treatment of idiopathic male infertility has been documented. The benefit of the study has resulted in improved semen quality and it is a safe, cost-effective, and discrete treatment of reduced sperm motility.

Innovation of Life (TNO)'s patented *in vitro* gastrointestinal models (TIM 1 and TIM 2) are unique gastrointestinal models that are flexible and accurate, and produce highly reproductive results. Nutritional and functional properties of foods and ingredients can be assessed under simulated physiological digestion conditions. The systems are very well validated and offer broad versatility in experimental strategies and goals.

For over 15 years, TIM models have been used successfully and the study results have been described in over 30 scientific peer-reviewed publications. The models are FDA approved. TIM experiments are free from ethical constraints and are conducted without harming animals or humans. Thus, the nutraceutical industry is working on a clinical evidence-based approach, which is sure to work to their advantage.

13.12.2 Case for "Care" of Consumers

Cycle of care can be defined as the method that carefully analyzes the current state of the health-care process, and it can be designed as per any particular medical condition. It also enables us to judge how one can bring about cost-effectiveness and enhance clinical outcomes, accessibility, and quality of care for the masses.

There are different phases in the care cycle, which have a close link with the stage of the disorder/disease. The phases of a care cycle are as follows: prevention, screening, diagnosis, treatment, management, and surveillance.[11]

The final phase of a care cycle is surveillance in which a consumer/patient has to see his/her doctor for checkups at foretold intervals of time. This is an extremely vital phase as it would keep the doctor informed about any possibilities of a recurrence of the disease. A perfect example of this would be a cancer patient because it can be observed that there are regular follow-ups post cure.

Therefore, it would be apt to say that a cycle of care ideally takes care of an individual's health condition. In the sick/ill patient's it transits to being fit and fine in the process.

The various phases of cycle of care are a part of health care such that the two concepts are not very different from the other. Health care can be simply defined as preventing, treating, and managing of an illness and at the same time ensuring complete wellness through preservation of mental and physical well-being. This can be done with ease through the cooperation of medical and allied health-care professionals. In a number of emerging and developing economies, the health-care systems are facing major risks due to excessive demands and costs, low quality of care as well as restricted access that is not properly coordinated.

It can be seen that there is a relation between the cycle of care and the health outcomes. This is because the quality of care, the cost of treatment as well as the access are the major attributes when trying to determine whether the actual health outcomes are in sync with the optimum achievable outcome for the same. There are three different levels of health care, namely, primary, secondary, and tertiary.

In 1978, the World Health Organization (WHO) defined primary health care as essential health care based on practical, scientifically sound, and socially acceptable method and technology; made universally accessible to all in the community through their full participation at an affordable cost; and geared toward self-reliance and self-determination.[12]

When medical specialists offer a service and are not the first ones to be contacted by a patient, it is known as secondary patient care. More often a patient is referred to secondary care by his/her primary care provider (usually his/her general practitioner [GP]) and is delivered in clinics or hospitals.

Tertiary care is a highly specialized consultative method of treatment in which a patient is referred from the primary or secondary health-care professional or even inpatients in hospitals. Often, tertiary care is given by personnel and in facilities for the further medical investigations and treatment. For instance, tertiary care services are often needed when treating cancer patients and cardiac surgery patients, and also include advanced neonatology services.

13.12.3 Case for Branding

People fundamentally enjoy having the power to choose between comparable options. In the absence of brands with a strong image, the consumer will naturally prefer to buy the cheapest product. Brands must fulfill the promises made to the market, or the consumers will no longer support them with future purchases. Competitive position—innovation in product marketing—differentiation will provide a competitive edge to a brand.

Developing customer-focused products that address the specific needs of different customer demographics and then focusing on credibility building, lowering prices of products can help brands to be different.[13]

Increasing awareness of a product among physicians and consumers through advertising and education in context of increasing consumer sophistication leads to more opportunities for shelf space in mainstream retailers.

There have been very few nutraceutical or functional food brands. One example that clearly stands out is Yakult. It was suggested that in the case of functional foods and beverages, almost all consumer segments would be expected to make trade-offs between functional benefits, and price and carrier attributes given the complex and multifaceted nature of consumer food choice.

13.13 CONCLUSION

Instead of there being a convergence, there is a need for both pharmaceutical and nutraceutical sectors to coexist. It can be observed that in a cycle of care both the sectors are absolutely important for the betterment of society. Pharmaceuticals and nutraceuticals can work hand in hand to ensure complete wellness in the treatment cycle of health care as they could be used at different stages. Evidence is the heart of the product and care design through R&D clinical trials, *in vitro* studies, and human studies. The cost of R&D can be readjusted when both work together, and unlike pharmaceuticals, nutraceuticals can create big healthy brands for better health.

Nutraceuticals can work at the preemptive stage as well as the healing stage, whereas surgery and specialized work will always be needed for treating an ailment a patient is suffering from.

Thus, healthy living must be encouraged where importance is given to responsible nutrition as there is a swift move from the preventive method of treatment to a curative one. The advancements in technology and increasing innovation in nutraceuticals have given rise to more high-quality nutraceutical products. Now, it is possible to even cure diseases with these nutraceutical products. This is especially useful in the case of chronic diseases. The treatment of a chronic disease through pharmaceutical products can be further hastened if it is supported by nutraceutical products.

It is important to remember the fact that in Eastern knowledge systems a patient was treated for an ailment or disorder using the diet. Diet or food eaten was considered as the remedial medicine. Therefore, it can be observed that medication and diet are interwoven, and both pharmaceuticals and nutraceuticals can be used in a synergistic way for curing the patient.

Therefore, nutraceutical space will experience a tremendous amount of growth, but this would not imply that the nutraceutical space would overshadow the pharmaceutical space in any way. This is because for quick and wholistic treatment, both sectors have to coexist in the future.

For the coexistence between the two sectors, it is vital to improve the level of education about nutraceutical products available in the market. This education has to begin with the doctors as they need to understand how the nutraceutical products would not only aid but even cure diseases and disorders. Thus, the syllabus for the doctors has to be amended. Consumer awareness is certainly different for different demographics and those who have access to the Internet and such mediums are aware of all the health benefits nutraceutical products offer. But for those people who do not have access to an advanced medium of communication, health awareness campaigns are a good option. There has to be awareness on all levels, apart from the ones mentioned such as pharmacists and regulators.

REFERENCES

1. Boogaard, P.J. 2012. Facing cross-industry challenges in the food and pharma industries. Scientificcomputing.com (October). http://www.industriallabautomation.com/Facing.php.
2. Lockwood, G.B. 2007. The hype surrounding nutraceutical supplements: Do consumers get what they deserve? *Nutrition*, 23: 771–772.
3. Gopakumar, K.M. and M.R. Santhosh. 2012. An unhealthy future for the Indian pharmaceutical industry? *Third World Resurgence*, 259: 9–14.
4. Ernst & Young. 2011. Global Pharmaceutical Industry Report 2011, Progressions Building Pharma 3.0. http://www.ey.com/Publication/vwLUAssets/Progressions_-_Building_Pharma-3.0/$FILE/Progressions_2011_Final.pdf.
5. Interlink Knowledge Cell. 2011. Regulatory perspective of nutraceuticals in India. Interlink White Paper. http://interlinkconsultancy.com/white_paper/index.html.
6. Gebhart, F. 2012. Retail pharmacists find opportunity in the growing medical food market. *Drug Topics*, December. http://drugtopics.modernmedicine.com/drug-topics/news/modernmedicine/modern-medicine-feature-articles/retail-pharmacists-find-opportunity.
7. Grant Thronton and FICCI. 2012. Nutraconsensus: Emerging insights on Nutraceuticals—Players and policy makers. A whitepaper prepared by Grant Thornton and FICCI for FICCI-HADSA Nutraceuticals 2012 "Regulation, Categorisation and Commercialisation," Mumbai, India, November 6, 2012. http://gtw3.grantthornton.in/assets/reports/GT_FICCI%20Nutraconsensus_Whitepaper.pdf.
8. Smarta, R.B. 2012. Market: Focused innovation in food and nutrition. In: *Innovations in Healthy & Functional Foods*, Boca Raton, FL: CRC Press, pp. 511, 512.
9. The Firshein Center. Integrative medicine and complete wellness. The Firshein Center: For Comprehensive Medicine, New York. www.firsheincenter.com/nutrition-as-medicine.html.
10. PricewaterhouseCoopers. 2012. Food as pharma. http://www.pwc.com/en_GX/gx/retail-consumer/pdf/rc-worlds-newsletter-foods-final.pdf.
11. Verbeek, X.A.A.M. and W.P. Lord. 2007. The care cycle: An overview. http://incenter.medical.philips.com/doclib/enc/fetch/2000/4504/577242/577256/588821/5050628/5313460/5318705/8_Care_Cycle.pdf%3fnodeid%3d5318718%26vernum%3d1.
12. Rao, M. and D. Mant. 2012. Strengthening primary healthcare in India: White paper on opportunities for partnership. *British Medical Journal*. http://www.bmj.com/content/344/bmj.e3151.
13. Harman, C. 2003. *The Global Report on Nutraceuticals*. London: ABOUT Publishing Group.

14 Heme Oxygenase-1 Induction Inhibits Intestinal Inflammation
Role of Food Factors

Yuji Naito, Tomohisa Takagi, Akihito Harusato, Yasuki Higashimura, and Toshikazu Yoshikawa

CONTENTS

14.1 INTRODUCTION

Heme oxygenase (HO) is the rate-limiting enzyme in heme catabolism, a process that leads to the generation of equimolar amounts of biliverdin, free iron, and carbon monoxide (CO) (Maines 1997). HO-1 is highly inducible by a vast array of stimuli, including oxidative stress, heat shock, ultraviolet radiation, ischemia–reperfusion, heavy metals, bacterial lipopolysaccharide (LPS), cytokines, nitric oxide (NO), and its substrate, heme (Shibahara 1988). HO-2 is a constitutive gene expressed in neurons, endothelium, and many other cell types. Although both HO-1 and HO-2 catalyze the identical biochemical reaction, there are some fundamental differences between the two in genetic origin, primary structure, and molecular weight. HO-1, once expressed under various pathological conditions, has an ability to metabolize high amounts of free heme to produce high concentrations of its enzymatic by-products that can influence various biological events and has recently been the focus of considerable medical interest (Abraham and Kappas 2008). HO-1 expression can confer cytoprotection and anti-inflammation in various disease models including gastrointestinal injuries. The cytoprotective effects of HO-1 may be related to the formation of end products because the pharmacological application of CO and biliverdin/bilirubin can mimic the HO-1-dependent cytoprotection and anti-inflammation in many injury models. In this chapter, we provide a new molecular mechanism underlying the function of HO-1 focusing on the differentiation of macrophages and its possible clinical implications, especially in intestinal inflammatory diseases.

14.2 HO-1 EXPRESSION IN BACH1$^{-/-}$ MICE

The transcriptional upregulation of the *ho-1* gene, and subsequent *de novo* synthesis of the corresponding protein, occurs in response to the elevated levels of its natural substrate heme and to a multiplicity of endogenous factors (Ryter and Choi 2009; Takahashi et al. 2009). Many agents that induce HO-1 are associated with oxidative stress in that they (1) directly or indirectly promote the intracellular generation of reactive oxygen species (ROS), (2) fall into a class of electrophilic antioxidant compounds that include plant-derived polyphenolic substances, or (3) form complexes with intracellular reduced glutathione and other thiols. Two enhancer regions located at approximately −4 and −10 kb relative to the *ho-1* transcriptional start site have been identified in the mouse gene. The dominant sequence element of the enhancers is the stress-responsive element (StRE), which is structurally and functionally similar to the Maf-responsive element (MARE) and the antioxidant-responsive element (ARE) (Alam and Cook 2007). Several transcriptional regulators bind these sequences, including nuclear factor erythroid 2-related factor-2 (Nrf2) and Broad complex-Tramtrack-Bric-a-brac (BTB) and cap'n'collar (CNC) homolog 1 (Bach1). Nrf2 contains a transcription activation domain and positively regulates HO-1 transcription, whereas Bach1 competes with Nrf2 and represses transcription (Igarashi and Sun 2006; Morse et al. 2009; Sun Jang et al. 2009).

Bach1 under baseline condition forms a heterodimer with small Maf proteins which represses transcription of the *ho-1* gene by binding to MARE in the 5′-untranslated region of the *ho-1* promoter. Under conditions of excess heme, increased heme binding to Bach1 causes a conformational change and a decrease in DNA-binding activity followed by nuclear export of Bach1, which in turn leads to transcriptional activation of the *ho-1* gene through MARE. Heme also induces nuclear translocation of Nrf2, a partner molecule for the family, and promotes stabilization of Nrf2. Thus, an intracellular heme concentration displaces Bach1 from the MARE sequences by heme binding, which then permits Nrf2 binding to a member of small Maf proteins, ultimately resulting in transcriptional activation of *ho-1* genes.

To confirm the role of Bach1, Professor Igarashi's group (Tohoku University, Tohoku, Japan) produced Bach1 deficiency in mice in 2002 (Sun et al. 2002). Collaborating with them, we also confirmed that *ho-1* is constitutively expressed at higher levels in many tissues of Bach1-deficient mice (Figure 14.1), indicating that Bach1 acts as a negative regulator of transcription of *ho-1* gene. It has been reported that Bach1-deficient mice are resistant to various injuries involving atherosclerosis (Omura et al. 2005), meniscal degeneration (Ochiai et al. 2008), hypertension (Mito et al. 2008), hepatitis (Iida et al. 2009; Inoue et al. 2011), and rheumatoid arthritis (Hama et al. 2012) (Table 14.1).

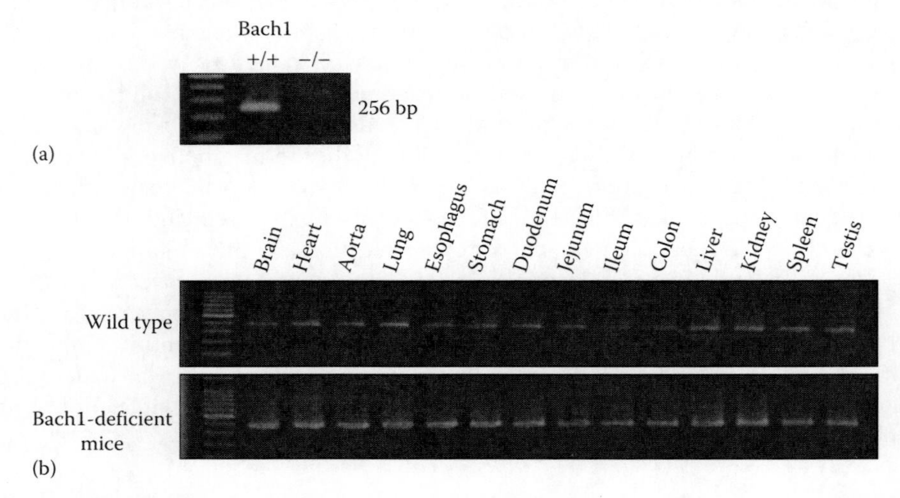

FIGURE 14.1 HO-1 expression in various tissues in Bach1-deficient mice. (a) Expression of Bach1 mRNA in whole blood was examined by RT-PCR using Bach1$^{+/+}$ and Bach1$^{-/-}$ mice. (b) Expression of heme oxygenase-1 (HO-1) mRNA in various tissues was examined by RT-PCR using wild (Bach1$^{+/+}$) and Bach1-deficient mice.

TABLE 14.1
Phenotypes in Bach1-Deficient Mice

Effects	Model	Reference
Inhibit osteoclastogenesis	Rheumatoid arthritis	Hama et al. 2012
Induce M2 macrophages	Crohn's disease	Harusato et al. 2013
Inhibit apoptosis induced by indomethacin	Small intestinal injury	Harusato et al. 2011
Inhibit steatohepatitis	NASH	Inoue et al. 2011
Inhibit intestinal injury induced by indomethacin	Small intestinal injury	Harusato et al. 2009
Inhibit liver injury by GalN and LPS	Hepatitis	Iida et al. 2009
Preserve spinal cord injury	Spinal cord injury	Yamada et al. 2008
Inhibit myocardial hypertrophy and remodeling	Pressure overload	Mito et al. 2008
Inhibit histological degeneration	Meniscal degeneration	Ochiai et al. 2008
Inhibit smooth muscle proliferation	Atherosclerosis	Omura et al. 2005

GalN, D-galactosamine; LPS, lipopolysaccharide; NASH, nonalcoholic steatohepatitis.

14.3 INDOMETHACIN-INDUCED INTESTINAL INFLAMMATION IN BACH1$^{-/-}$ MICE

Development of endoscopic diagnostic techniques for small intestine diseases, such as video capsule endoscopy and balloon enteroscopy, has led to the revelation that nonsteroidal anti-inflammatory drugs (NSAIDs) such as indomethacin and aspirin can induce intestinal mucosal injuries. Although the pathophysiology of NSAID-induced intestinal injuries has been elucidated extensively, practical therapies for such injuries are unavailable. Therefore, it is important to investigate preventive and therapeutic strategies for NSAID-induced intestinal injuries. In 2009, we first reported that Bach1 deficiency ameliorated indomethacin-induced intestinal mucosal injury in mice (Harusato et al. 2009). However, the precise underlying mechanisms remain unclear. Because the administration of NSAIDs is known to induce apoptosis of small intestinal epithelial cells (Omatsu et al. 2010), we have also focused on the relationship between small intestinal injury and NSAID-induced apoptosis, and the role of Bach1.

Terminal deoxynucleotidyl transferase-mediated deoxyuridine triphosphate (dUTP) nick end labeling (TUNEL) assay and cleaved caspase-3 detection were performed using the *in situ* Apoptosis Detection Kit (Takara Biochemicals, Shiga, Japan) and Western blotting, respectively. Few TUNEL-positive cells were detected in the intestinal epithelium of the normal mice. Indomethacin markedly increases the number of TUNEL-positive cells on the top of the villi in the wild-type mice with the destruction of the intestinal structure. However, the number of apoptotic cells and mucosal damage were significantly attenuated in Bach1-deficient mice (Harusato et al. 2011). Western blotting showed the increase of cleaved caspase-3 induced by indomethacin in the wild-type mice, and this increase was suppressed in the Bach1-deficient mice. Furthermore, the administration of tin protoporphyrin (SnPP), an HO-1 inhibitor, to Bach1-dificient mice reversed the decrease in cell number of apoptotic cells as well as the suppression of cleaved caspase-3 levels. From these results, we have concluded that upregulation of HO-1 by Bach1 deficiency is involved in the resistance to indomethacin-induced apoptosis in the Bach1-deficient mice treated with indomethacin. In addition, we have shown that the production of tumor necrosis factor-α (TNF-α), which plays an important role in indomethacin-induced apoptosis in the small intestine (Fukumoto et al. 2011), is markedly suppressed in Bach1-deficient mice after indomethacin administration, indicating the involvement of TNF-α-mediated apoptosis signaling in the pathogenesis of indomethacin-induced small intestinal injury.

Higuchi et al. (2009) have also shown that lansoprazole, a proton pump inhibitor, inhibits indomethacin-induced intestinal injury in rats and that the inhibition is reversed by SnPP.

Because CORM, a CO donor, also ameliorates these injuries, cytoprotective effects of HO-1, in part, exert via CO-dependent manner (Yoda et al. 2010). Many reports have confirmed the anti-inflammatory and cytoprotective effects of HO-1 inducers on small intestinal injuries induced by ischemia–reperfusion, LPS, radiation, and burn shock (Naito et al. 2011). The upregulation of HO-1 expression by the inhibition of Bach1 or directly CO-releasing molecule may be a novel therapeutic strategy for NSAID-induced small intestinal injuries.

14.4 ROLE OF HO-1 IN EXPERIMENTAL COLITIS MODELS

Inflammatory bowel diseases (IBDs), such as Crohn's disease (CD) and ulcerative colitis (UC), are characterized by chronic and relapsing inflammatory disorders of the intestinal tract. The pathogenesis of IBD probably involves a complex interworking of genetic, environmental, and immunoregulatory factors, but the exact etiology of IBD remains unclear. To deal with the complexity of the multifactorial inflammatory process in IBD, a new therapeutic strategy that controls the intestinal immune response through a different approach should be added or combined with the existing treatment options.

Wang et al. (2001) used a rat model of CD induced by 2,4,6-trinitrobenzene sulfonic acid (TNBS) to investigate whether the expression of HO-1 is an endogenous mechanism responsible for host defense against inflammatory injury in colonic tissue. They demonstrated that HO activity and HO-1 gene expression increased markedly after TNBS induction and that administration of tin mesoporphyrin (SnMP), an HO inhibitor, potentiated the colonic damage as well as decreased the HO-1 activity. These results indicate that HO-1 plays a protective role in the colonic damage induced by TNBS enema. Using a dextran sulfate sodium (DSS)-induced colitis model of mice, an animal model of UC, we have demonstrated that HO-1 mRNA is markedly induced in inflamed colonic tissue, whereas HO-2 mRNA is constitutively expressed (Naito et al. 2004). Coadministration with zinc protoporphyrin (ZnPP), an HO inhibitor, also enhanced intestinal inflammation and increased the disease activity index as determined by a calculated score based on changes in body weight, stool consistency, and intestinal bleeding.

Recent investigation has shown that upregulation of HO-1 by several HO-1 inducers significantly reduces the intestinal injuries induced by DSS or TNBS. In these studies, HO-1 inducers increase HO-1 expression in intestinal mucosa and ameliorate mucosal injury as well as inflammatory cell accumulation by decreasing infiltrating neutrophils and lymphocytes via the inhibition of nuclear factor kappaB (NF-κB)-dependent proinflammatory cytokines. To further assess the anti-inflammatory mechanisms, Zhong et al. (2010) have examined whether hemin enhanced the proliferation of regulatory T (T_{reg}) cells and suppressed the production of interleukin (IL)-17 in a DSS colitis model. Flow cytometry analysis has revealed that hemin markedly expands the $CD4^+$ $CD25^+$ $Foxp3^+$ (T_{reg}) population and attenuates IL-17 and T helper 17 (TH17)-related cytokines. It has also been demonstrated that HO-1 exerts immunoregulatory effects by modulating T_{reg} cell function (Brusko et al. 2005), and that HO-1 activity in antigen-presenting cells is important for T_{reg}-mediated suppression, providing an explanation for the apparent defect in immune regulation in HO-1-deficient mice (George et al. 2008).

14.5 HIGH EXPRESSION OF HO-1 IN M2-TYPE MACROPHAGES

To investigate the localization of HO-1, we performed immunofluorescent staining of colonic specimen and we have shown that HO-1-immunopositive cells are found to be mainly F4/80-positive macrophages in the mucosa and submucosa. This finding is consistent with the known localization of macrophages in the gut in Bach1-deficient mice (Harusato et al. 2011). These data suggest that HO-1-positive macrophages may play an important part in the protection of intestinal damage associated with inflammation. Therefore, we performed isolation and analysis of peritoneal macrophages derived from Bach1-deficient and wild-type mice. Functionally, macrophages have at least two distinct subpopulations: the classically (M1) and the alternatively (M2) activated macrophages.

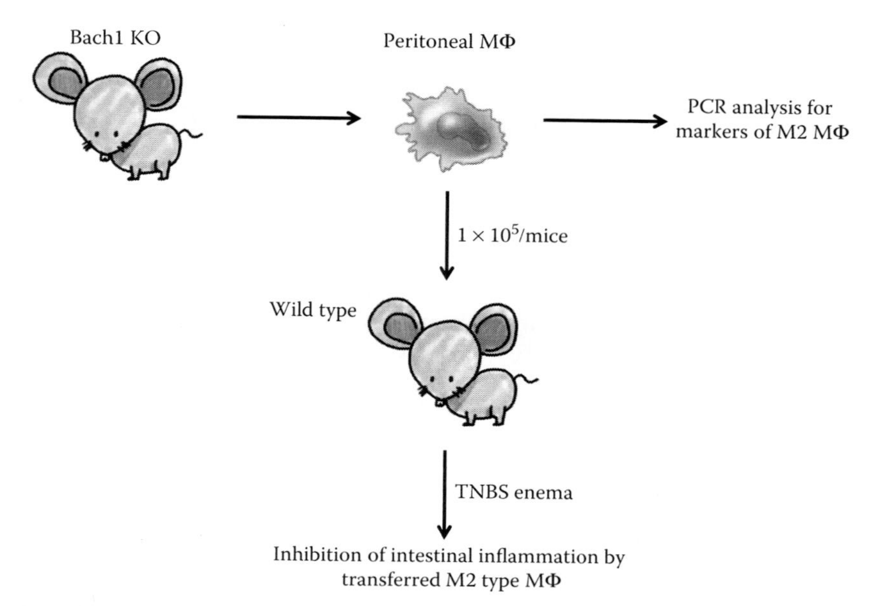

FIGURE 14.2 Transfer of Bach1-deficient macrophages into wild-type mice ameliorates the intestinal inflammation. KO, knock out; Mϕ, macrophage; PCR, polymerase chain reaction; TNBS, 2,4,6-trinitrobenzene sulfonic acid.

M1 macrophages produce NO and TNF-α to clear bacterial, viral, and fungal infections. However, M2 macrophages have an important role in responses to parasite infection, tissue remodeling, angiogenesis, and tumor progression. In the recent study, we first found that peritoneal macrophages derived from Bach1-deficient mice appeared to be M2 macrophages because the expression of M2 macrophage markers including Arginase-1, Fizz1, Ym1, and MRC1 are upregulated (Harusato et al. 2012) (Figure 14.2). In addition, after 6 h incubation with TNF-α, interferon (IFN)-γ production, and inducible NO synthase, expression was inhibited in the macrophages derived from the Bach1-deficient mice compared to those derived from the wild-type mice. After the transfection of Bach1 small interfering RNA (siRNA) into RAW 264 macrophage cells, the expression of M2 macrophage markers such as MRC1 and Fizz1 are upregulated as the expression of HO-1 is upregulated. M2-type macrophages have been recognized as anti-inflammatory macrophages (Gordon and Martinez 2010; Hunter et al. 2010), and upregulation of HO-1 expression, especially in macrophages, might induce a phenotypic shift to M2-type macrophages (Choi et al. 2010; Louvet et al. 2011). Data described as a result of our recent study are consistent with the hypothesis that skewing of macrophages to the M2 phenotype by the mediation of HO-1 plays an important role in the regulation of inflammation.

Furthermore, we also found that adoptive transfer of these HO-1 highly expressed macrophages significantly attenuated the extent of colonic inflammation (Figure 14.2). Recent studies also support that the macrophages with the upregulation of HO-1 correlated with cytoprotection as an antioxidant against various disorders (Gordon and Martinez 2010; Smythies et al. 2005). Therefore, cell therapy transferring HO-1 highly expressed M2 macrophages could possibly be a candidate to regulate intestinal inflammation in human IBD. Finally, administration of HO inhibitor significantly worsened the extent of colitis in Bach1-deficient mice, suggesting that inhibition of HO-1 might have downregulated the phenotypic switch from M1 macrophages to M2 leading to the enhancement of anti-inflammatory property. Our data suggest a novel role of HO-1 and HO-1 highly expressed M2 macrophages in intestinal inflammation. We also demonstrated the crucial role of Bach1 in gastrointestinal tract and also unraveled the mechanism of HO-1-mediated inhibition of experimental colitis in Bach1-deficient mice. It is not well known whether HO-1 induces a phenotypic shift to M2 macrophages in human IBDs. We need to investigate this issue to assess whether induction of

HO-1 in macrophages or adoptive transfer of HO-1 highly expressed M2 macrophages could be a therapeutic option to treat human IBD.

14.6 AGARO-OLIGOSACCHARIDE-INDUCED HO-1 EXPRESSION IN MACROPHAGES

Agar, which is extracted easily from red algae, is widely used as a food and gelling agent in Asian countries. Agarose, the main component of polysaccharides in agar, is hydrolyzed easily to yield oligosaccharides (Figure 14.3). The resultant oligosaccharides termed agaro-oligosaccharides (AGOs) have been investigated widely in terms of structure and bioactivity (Chen and Yan 2005; Chen et al. 2006; Kazlowski et al. 2008). In recent years, previous reports show that AGOs

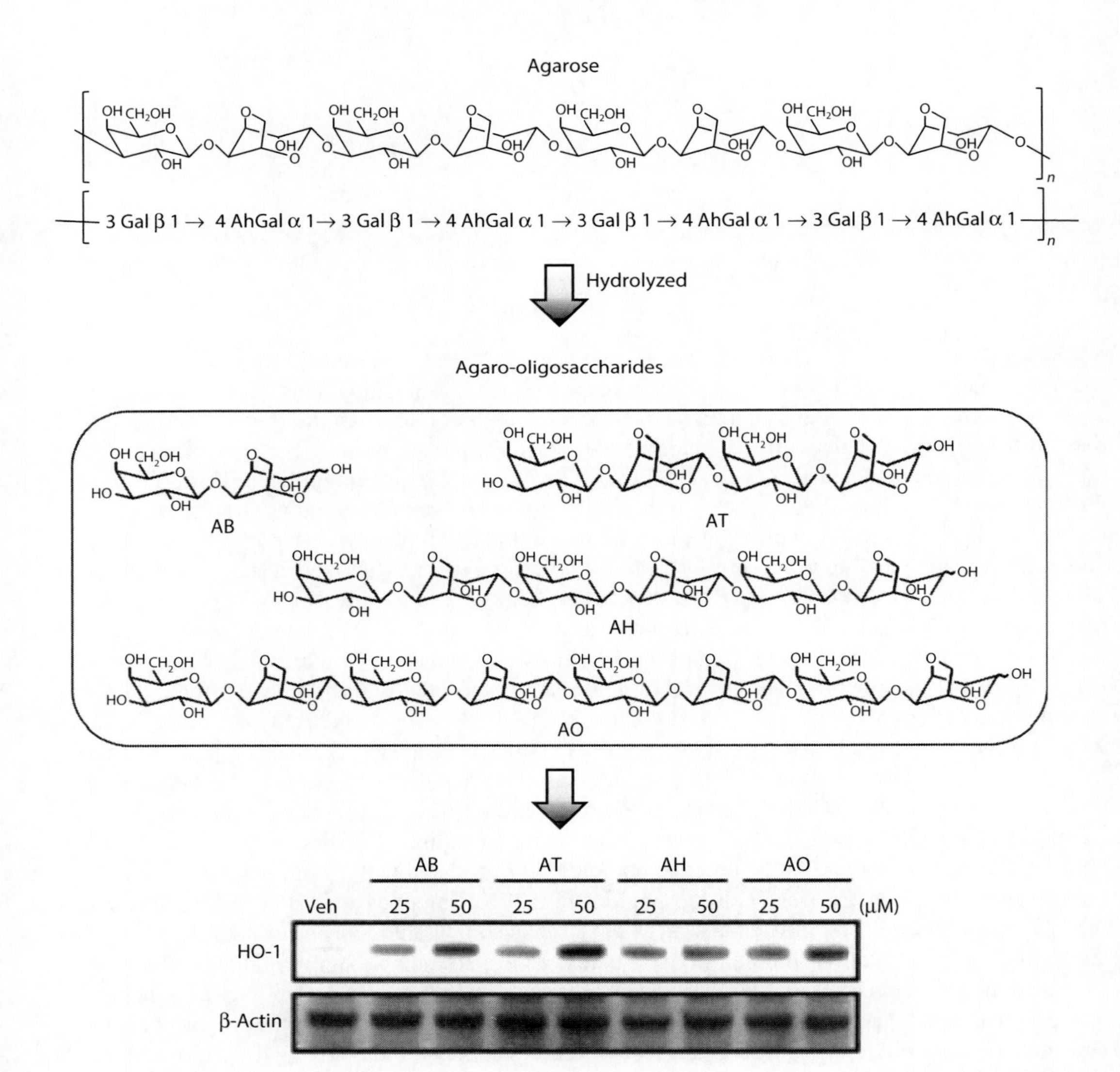

FIGURE 14.3 Agarotetraose (AT) most strongly induces HO-1 among four oligosaccharides produced by hydrolyzation of agar. AB, agarobiose; AH, agarohexaose; AO, agarooctaose. (Adapted and reprinted with permission from Enoki et al. *Biosci. Biotechnol. Biochem.*, 74, 766–770, 2010; Higashimura et al. *J Gastroenterol*, 48, 897–909, 2012.)

induced HO-1 expression in macrophages and that they might engender an anti-inflammatory property (Enoki et al. 2010). Recently, we have investigated the therapeutic effect of AGOs in murine TNBS-induced colitis, which shares both immunological and pathological features with human IBD (Higashimura et al. 2012). Furthermore, we have clarified the molecular mechanism of AGO-mediated induction of HO-1 using mouse macrophage-like cell line.

To study the effect of AGOs on HO-1 expression in the colonic mucosa of mice, the mice were treated with AGOs administered orally for the designated time periods. In response to the administration of AGOs, HO-1 expression was upregulated significantly compared to the vehicle-administered group. Induced HO-1 expression was mainly observed in mononuclear cells in colonic submucosa, and HO-1 immunoreactivity was coexpressed with F4/80, which is the marker protein of macrophage, indicating that AGO administration caused a marked increase in the population of HO-1-postive macrophages. Further, to characterize the molecular mechanism of HO-1 induction by AGOs, we examined HO-1 expression in RAW 264 cells cultured in the presence of AGOs. Depression of Nrf2 using the siRNA approach partially inhibited the upregulation of HO-1 mediated by agarotetraose (AT), a most active ingredient among AGOs (Figure 14.3), in RAW 264 cells. In addition, it has been reported that multiple protein kinase pathways such as extracellular signal-regulated kinase (ERK), p38 kinase, and phosphatidylinositol 3-kinase (PI3K) have been proposed to play a role in HO-1 induction. We have shown that AT promoted the phosphorylation of c-*Jun* amino-terminal kinase (JNK), but not ERK or p38 kinase, and AT-mediated induction of HO-1 was suppressed by SP600125, a specific inhibitor for JNK. Our results indicate that AGO-mediated HO-1 induction requires both functions of Nrf2 and JNK pathways.

Finally, we have investigated the effect of AGO administration on TNBS-induced colitis in mice (Higashimura et al. 2012). Increased colonic damage and myeloperoxidase activity after TNBS treatment were significantly inhibited by AGO administration. The expression of TNF-α mRNA and protein was significantly increased in the colonic mucosa after treatment, and these increases were suppressed by AGOs because SnPP, an HO inhibitor, canceled AGO-mediated amelioration of TNBS-induced colitis, indicating that HO-1 induced by AGOs play a crucial protective role against the TNBS-induced colitis.

14.7 CONCLUSION

The biological significance of HO-1 upregulation in gastrointestinal inflammation remains to be fully elucidated. However, there is no doubt that CO derived from HO-1 exerts significant effects on many pathways of cellular metabolism. In inflamed intestinal cells, CO may inhibit the inflammatory response by consequently influencing the synthesis of cytokines, expression of adhesion molecules, and cell proliferation. Although the mechanisms underlying HO-1 activity on gene expression are not well known, the results obtained in recent years have demonstrated its importance in modulation of the inflammatory reaction. Our recent experimental studies have clearly demonstrated that HO-1 expression is a self-defense mechanism against inflammation via shifting macrophages to the M2 phenotype. These data suggest that HO-1 is a possible therapeutic target in several kinds of gastrointestinal inflammation.

ACKNOWLEDGMENTS

This work was supported by a Grant-in-Aid for Scientific Research (C) to Y. Naito (no. 22590705), and (C) to T. Takagi (no. 22590706) from the Japan Society for the Promotion of Science, and by an Adaptable and Seamless Technology Transfer Program through target-driven R&D to Naito from the Japan Science and Technology Agency. Naito received scholarship funds from Otsuka Pharmaceutical Co., Ltd., Otsuka, Tokyo, Japan, and Takeda Pharmaceutical Co., Ltd., Takeda, Tokyo, Japan.

REFERENCES

Abraham, N.G. and A. Kappas. 2008. Pharmacological and clinical aspects of heme oxygenase. *Pharmacological Reviews*, 60: 79–127.

Alam, J. and J.L. Cook. 2007. How many transcription factors does it take to turn on the heme oxygenase-1 gene? *American Journal of Respiratory Cell and Molecular Biology*, 36: 166–174.

Brusko, T.M., C.H. Wasserfall, A. Agarwal, M.H. Kapturczak, and M.A. Atkinson. 2005. An integral role for heme oxygenase-1 and carbon monoxide in maintaining peripheral tolerance by $CD4^+CD25^+$ regulatory T cells. *Journal of Immunology*, 174: 5181–5186.

Chen, H., X. Yan, P. Zhu, and J. Lin. 2006. Antioxidant activity and hepatoprotective potential of agaro-oligosaccharides in vitro and in vivo. *Nutrition Journal*, 5: 31.

Chen, H.M. and X.J. Yan. 2005. Antioxidant activities of agaro-oligosaccharides with different degrees of polymerization in cell-based system. *Biochimica et Biophysica Acta*, 1722: 103–111.

Choi, K.M., P.C. Kashyap, N. Dutta, G.J. Stoltz, T. Ordog, T. S. Donohue, A.J. Bauer et al. 2010. CD206-positive M2 macrophages that express heme oxygenase-1 protect against diabetic gastroparesis in mice. *Gastroenterology*, 138: 2399–2409.

Enoki, T., S. Okuda, Y. Kudo, F. Takashima, H. Sagawa, and I. Kato. 2010. Oligosaccharides from agar inhibit pro-inflammatory mediator release by inducing heme oxygenase 1. *Bioscience, Biotechnology, and Biochemistry*, 74: 766–770.

Fukumoto, K., Y. Naito, T. Takagi, S. Yamada, R. Horie, K. Inoue, A. Harusato et al. 2011. Role of tumor necrosis factor-alpha in the pathogenesis of indomethacin-induced small intestinal injury in mice. *International Journal of Molecular Medicine*, 27: 353–359.

George, J.F., A. Braun, T.M. Brusko, R. Joseph, S. Bolisetty, C.H. Wasserfall, M.A. Atkinson et al. 2008. Suppression by $CD4^+CD25^+$ regulatory T cells is dependent on expression of heme oxygenase-1 in antigen-presenting cells. *American Journal of Pathology*, 173: 154–160.

Gordon, S. and F.O. Martinez. 2010. Alternative activation of macrophages: Mechanism and functions. *Immunity*, 32: 593–604.

Hama, M., Y. Kirino, M. Takeno, K. Takase, T. Miyazaki, R. Yoshimi, A. Ueda et al. 2012. Bach1 regulates osteoclastogenesis in a mouse model via both heme oxygenase 1-dependent and heme oxygenase 1-independent pathways. *Arthritis & Rheumatology*, 64: 1518–1528.

Harusato, A., Y. Naito, T. Takagi, K. Uchiyama, K. Mizushima, Y. Hirai, Y. Higashihara et al. 2013. BTB and CNC homolog 1 (Bach1) deficiency ameliorates TNBS colitis in mice: Role of M2 macrophages and heme oxygenase-1. *Inflammatory Bowel Diseases*, 19: 740–753.

Harusato, A., Y. Naito, T. Takagi, K. Uchiyama, K. Mizushima, Y. Hirai, S. Yamada et al. 2011. Suppression of indomethacin-induced apoptosis in the small intestine due to Bach1 deficiency. *Free Radical Research*, 45: 717–727.

Harusato, A., Y. Naito, T. Takagi, S. Yamada, K. Mizushima, Y. Hirai, R. Horie et al. 2009. Inhibition of Bach1 ameliorates indomethacin-induced intestinal injury in mice. *Journal of Physiology and Pharmacology*, 60: 149–154.

Higashimura, Y., Y. Naito, T. Takagi, K. Mizushima, Y. Hirai, A. Harusato, H. Ohnogi et al. 2012. Oligosaccharides from agar inhibit murine intestinal inflammation through the induction of heme oxygenase-1 expression. *Journal of Gastroenterology*, 48: 897–909.

Higuchi, K., Y. Yoda, K. Amagase, S. Kato, S. Tokioka, M. Murano, K. Takeuchi, and E. Umegaki. 2009. Prevention of NSAID-induced small intestinal mucosal injury: Prophylactic potential of lansoprazole. *Journal of Clinical Biochemistry and Nutrition*, 45: 125–130.

Hunter, M.M., A. Wang, K.S. Parhar, M.J. Johnston, N. Van Rooijen, P.L. Beck, and D.M. McKay. 2010. In vitro-derived alternatively activated macrophages reduce colonic inflammation in mice. *Gastroenterology*, 138: 1395–1405.

Igarashi, K. and J. Sun. 2006. The heme-Bach1 pathway in the regulation of oxidative stress response and erythroid differentiation. *Antioxidants & Redox Signaling*, 8: 107–118.

Iida, A., K. Inagaki, A. Miyazaki, F. Yonemori, E. Ito, and K. Igarashi. 2009. Bach1 deficiency ameliorates hepatic injury in a mouse model. *The Tohoku Journal of Experimental Medicine*, 217: 223–229.

Inoue, M., S. Tazuma, K. Kanno, H. Hyogo, K. Igarashi, and K. Chayama. 2011. Bach1 gene ablation reduces steatohepatitis in mouse MCD diet model. *Journal of Clinical Biochemistry and Nutrition*, 48: 161–166.

Jang, J.S., S. Piao, Y.N. Cha, and C. Kim. 2009. Taurine chloramine activates Nrf2, increases HO-1 expression and protects cells from death caused by hydrogen peroxide. *Journal of Clinical Biochemistry and Nutrition*, 45: 37–43.

Kazlowski, B., C.L. Pan, and Y.T. Ko. 2008. Separation and quantification of neoagaro- and agaro-oligosaccharide products generated from agarose digestion by beta-agarase and HCL in liquid chromatography systems. *Carbohydrate Research*, 343: 2443–2450.

Louvet, A., F. Teixeira-Clerc, M.N. Chobert, V. Deveaux, C. Pavoine, A. Zimmer, F. Pecker et al. 2011. Cannabinoid CB2 receptors protect against alcoholic liver disease by regulating Kupffer cell polarization in mice. *Hepatology*, 54: 1217–1226.

Maines, M.D. 1997. The heme oxygenase system: A regulator of second messenger gases. *Annual Review of Pharmacology and Toxicology*, 37: 517–554.

Mito, S., R. Ozono, T. Oshima, Y. Yano, Y. Watari, Y. Yamamoto, A. Brydun et al. 2008. Myocardial protection against pressure overload in mice lacking Bach1, a transcriptional repressor of heme oxygenase-1. *Hypertension*, 51: 1570–1577.

Morse, D., L. Lin, A.M. Choi, and S.W. Ryter. 2009. Heme oxygenase-1, a critical arbitrator of cell death pathways in lung injury and disease. *Free Radical Biology & Medicine*, 47: 1–12.

Naito, Y., T. Takagi, K. Uchiyama, and T. Yoshikawa. 2011. Heme oxygenase-1: A novel therapeutic target for gastrointestinal diseases. *Journal of Clinical Biochemistry and Nutrition*, 48: 126–133.

Naito, Y., T. Takagi, and T. Yoshikawa. 2004. Heme oxygenase-1: A new therapeutic target for inflammatory bowel disease. *Alimentary Pharmacology & Therapeutics*, 20: 177–184.

Ochiai, S., T. Mizuno, M. Deie, K. Igarashi, Y. Hamada, and M. Ochi. 2008. Oxidative stress reaction in the meniscus of Bach 1 deficient mice: Potential prevention of meniscal degeneration. *Journal of Orthopaedic Research*, 26: 894–898.

Omatsu, T., Y. Naito, O. Handa, K. Mizushima, N. Hayashi, Y. Qin, A. Harusato et al. 2010. Reactive oxygen species-quenching and anti-apoptotic effect of polaprezinc on indomethacin-induced small intestinal epithelial cell injury. *Journal of Gastroenterology*, 45: 692–702.

Omura, S., H. Suzuki, M. Toyofuku, R. Ozono, N. Kohno, and K. Igarashi. 2005. Effects of genetic ablation of Bach1 upon smooth muscle cell proliferation and atherosclerosis after cuff injury. *Genes to Cells*, 10: 277–285.

Ryter, S.W. and A.M. Choi. 2009. Heme oxygenase-1/carbon monoxide: From metabolism to molecular therapy. *American Journal of Respiratory Cell and Molecular Biology*, 41: 251–260.

Shibahara, S. 1988. Regulation of heme oxygenase gene expression. *Seminars in Hematology*, 25: 370–376.

Smythies, L.E., M. Sellers, R.H. Clements, M. Mosteller-Barnum, G. Meng, W.H. Benjamin, J.M. Orenstein, and P.D. Smith. 2005. Human intestinal macrophages display profound inflammatory energy despite avid phagocytic and bacteriocidal activity. *The Journal of Clinical Investigation*, 115: 66–75.

Sun, J., H. Hoshino, K. Takaku, O. Nakajima, A. Muto, H. Suzuki, S. Tashiro et al. 2002. Hemoprotein Bach1 regulates enhancer availability of heme oxygenase-1 gene. *EMBO Journal*, 21: 5216–5224.

Takahashi, T., H. Shimizu, H. Morimatsu, K. Maeshima, K. Inoue, R. Akagi, M. Matsumi et al. 2009. Heme oxygenase-1 is an essential cytoprotective component in oxidative tissue injury induced by hemorrhagic shock. *Journal of Clinical Biochemistry and Nutrition*, 44: 28–40.

Wang, W.P., X. Guo, M.W. Koo, B.C. Wong, S.K. Lam, Y.N. Ye, and C.H. Cho. 2001. Protective role of heme oxygenase-1 on trinitrobenzene sulfonic acid-induced colitis in rats. *American Journal of Physiology—Gastrointestinal Liver Physiology*, 281: G586–G594.

Yamada, K., N. Tanaka, K. Nakanishi, N. Kamei, M. Ishikawa, T. Mizuno, K. Igarashi et al. 2008. Modulation of the secondary injury process after spinal cord injury in Bach1-deficient mice by heme oxygenase-1. *Journal of Nuerosurgery: Spine*, 9: 611–620.

Yoda, Y., K. Amagase, S. Kato, S. Tokioka, M. Murano, K. Kakimoto, H. Nishio et al. 2010. Prevention by lansoprazole, a proton pump inhibitor, of indomethacin-induced small intestinal ulceration in rats through induction of heme oxygenase-1. *Journal of Physiology and Pharmacology*, 61: 287–294.

Zhong, W., Z. Xia, D. Hinrichs, J.T. Rosenbaum, K.W. Wegmann, J. Meyrowitz, and Z. Zhang. 2010. Hemin exerts multiple protective mechanisms and attenuates dextran sulfate sodium-induced colitis. *Journal of Pediatric Gastroenterology and Nutrition*, 50: 132–139.

Section IV

Innovation versus Regulation

15 Regulatory Framework of Functional Foods in Southeast Asia

Tee E. Siong

CONTENTS

15.1 INTRODUCTION

The increase in consumer interest on functional foods is largely due to the dramatic increase in non-communicable diseases (NCDs) in the Southeast Asian region in the past three decades. Although nutrient deficiencies still exist among the population, the prevalence of obesity, coronary heart disease, diabetes, hypertension, and certain cancers has been increasing over the years. Inappropriate dietary patterns and sedentary lifestyle are the two main causative factors in many NCDs. Health authorities are giving greater emphasis to promoting healthy eating to reduce risk to these diseases. Greater efforts are being made to promote dietary guidelines and to encourage consumption of a wide variety of foods to meet nutritional needs. Consumers are also seeking to find "healthier foods" with greater earnest.

Resulting from intensive research over the decades, foods are now known to perform a tertiary function, besides providing nutrients to nourish the body and prevent deficiencies as well as sensory functions. This tertiary function pertains to regulating the physiological processes of the body and even health promotion. It has been recognized that foods do not merely provide nutrients; there is a new dimension in the relationship between food and health!

A great deal of attention has therefore been given to the potential health significance of components other than nutrients that are found in foods. Much interest has been given to these food components as they are believed to be able to serve physiological roles beyond provisions of basic nutrient requirements, for example, the ability to promote general well-being or even reduce the risk of chronic diseases. Foods containing such components have been termed "functional foods." Such components may be termed "functional components" or "bioactive components."

There has been increasing interest in the potential health significance of other components in foods and functional foods, especially in view of the rising problem of chronic diseases in the region. Various health claims have been made in relation to the potential of functional components to manage body weight, blood lipids, and glucose; improve gut health; enhance immunity; and many other parameters that relate to these diseases.

In response to this situation, regulatory agencies in the region have increased their vigilance over such health claims. As it has been happening elsewhere in the world, rapid developments have taken place in the region.

This chapter first discusses the general aspects of functional foods and health claims, especially in the context of the Southeast Asian region. The definition of health claims used in this chapter and in the regulatory agencies in the region are also discussed. The main part of the chapter reviews the regulatory framework established in countries in the Southeast Asian region in relation to health claims for functional foods. The status of other function claims and disease risk reduction claims in the region is first summarized in several tables. This is followed by more detailed information on the health claims permitted in three countries, which have rather well-established regulatory frameworks for these claims, namely, Republic of Indonesia, Malaysia, and Singapore. Documentation for these countries is also more readily available in the public domain.

15.2 FUNCTIONAL FOODS AND HEALTH CLAIMS IN SOUTHEAST ASIA

To date, there is no unanimously accepted global definition of functional foods. The term may have gained prominence only in recent years in other parts of the world, but in Asia, foods with functional properties have been regarded as an integral part of some cultures for centuries. The International Life Sciences Institute Southeast Asia Region (ILSI SEAR) has been the main driving force in scientific activities on functional foods in the Asian region. In a monograph published by the ILSI (Tee 2004), there was a generally accepted understanding of the essential attributes or the characteristics of functional foods. In this so-called Asian position on functional foods, it was generally agreed that functional foods are foods that, by virtue of physiologically active food components, provide health benefits beyond basic nutrition. These components may be naturally present or be added to the food, and the functional benefits must be scientifically proven. Functional foods must be in conventional food forms and possess sensory characteristics.

The term functional foods is currently not used in most of the relevant regulations or legal systems for food. Similarly, it is also not used in almost all of the regulatory agencies in countries in the Southeast Asian region. Nevertheless, various forms of health claims are permitted in some countries in the region. The regulatory approach by countries in the region toward functional foods is therefore focused on the health claims permitted and their scientific substantiation.

There have been major regulatory developments in health claims in the Southeast Asian region. These claims focus on the role of functional components in improving physiological functions or promoting health. There has been increasing interest among food industries to make health claims on labels. Regulatory agencies continuously monitoring global situation and amending regulations to ensure claims are truthful and not misleading. In keeping with food innovations and renovations, these regulatory agencies have systems in place to review applications for health claims from the food industry.

This chapter reviews the health claim status in eight Southeast Asian countries, namely, Brunei Darussalam, Republic of Indonesia, Lao PDR, Malaysia, the Philippines, Singapore, Thailand,

and Vietnam. Most of the information for this write-up was obtained from the official websites of regulatory authorities of the following countries: Brunei Darussalam (MOH Brunei Darussalam 2010), Republic of Indonesia (NA-DFC 2011), Malaysia (MOH Malaysia 2011), Singapore (AVA Singapore 2011), and Thailand (FDA Thailand 2011). For the other countries, information for this review has been obtained through the most recent of the series of seminars and workshops organized by the ILSI SEAR (2013), held in November 2013 in Jakarta, Indonesia.

15.3 DEFINITION OF HEALTH CLAIMS

For the purpose of this chapter, health claims are as defined in the guidelines established by Codex Alimentarius (2009). All the regulatory agencies in the countries in the Southeast Asian region make reference to the Codex standards and text, although the prevailing national legislations are not exactly the same as this international guide.

Codex guidelines on nutrition claims were first established in 1997. The guide was revised in 2004 with the addition of health claims into the document and became known as the Guidelines for Use of Nutrition and Health Claims. In these guidelines, health claims means any representation that states, suggests, or implies that a relationship exists between a food or a constituent of that food and health. Health claims include the following:

1. Nutrient function claims—These claims describe the physiological role of the nutrient in growth, development, and normal functions of the body.
2. Other function claims—These claims concern specific beneficial effects of the consumption of foods or their constituents in the context of the total diet on normal functions or biological activities of the body. Such claims relate to a positive contribution to health or to the improvement of a function or to modify or preserve health.
3. Reduction of disease risk claims—These claims relate the consumption of a food or a food constituent, in the context of the total diet, to the reduced risk of developing a disease or health-related condition. Risk reduction means significantly altering a major risk factor(s) for a disease or a health-related condition. Diseases have multiple risk factors, and altering one of these risk factors may or may not have a beneficial effect. The presentation of risk reduction claims must ensure, for example, by the use of appropriate language and reference to other risk factors, which consumers do not interpret them as prevention claims.

In the context of functional foods and components, nutrient function claims are not discussed in this chapter. The relevant health claims would be "other function claim" and "reduction of disease risk claim." The "other food constituent or components" mentioned in Subsection 15.1 would include the bioactive components or functional ingredients in food. These are the components of interest and the focus of the health claims.

As a follow-up to the provision of health claims in the Codex guidelines, recommendations on scientific substantiation of health claims were adopted into the Codex system in 2009. These recommendations are now incorporated as an annex in the Guidelines on Nutrition and Health Claims (Codex Alimentarius 2009). Scientific substantiation of health claims are an essential part of the regulatory framework for functional foods.

15.4 STATUS OF HEALTH CLAIMS REGULATIONS IN SOUTHEAST ASIAN COUNTRIES

This subsection provides a summary of the permitted health claims in the Southeast Asian countries. The regulatory framework existing in these countries to review the applications for new health claims are also summarized.

TABLE 15.1
Status of Other Function Claims in Selected Southeast Asian Countries

Country	Other Function Claims
Brunei Darussalam	Not permitted
Republic of Indonesia	Yes, in new health claims regulations 2011 for two components, namely, dietary fiber and plant sterols/stanols
Lao PDR	Not permitted
Malaysia	Twenty-nine claims for a variety of food components (e.g., dietary fibers, nondigestible oligosaccharides, functional carbohydrates, plant sterols/stanols, soy proteins, DHA/ARA, lutein, bifidobacterium), all resulting from petitions from the food industry
The Philippines	Yes, according to Codex
Singapore	Seven specific function claims for infants and children, which include choline, DHA/ARA, nucleotides, and taurine. There are also 10 other function claims permitted, related to four components, namely, collagen, probiotics, prebiotics, and plant sterols/stanols.
Thailand	Not permitted
Vietnam	Yes, according to Codex

ARA, arachidonic acid; DHA, docosahexaenoxic acid.

A review of the status in the eight Southeast Asian countries showed that there are no harmonized health claims regulations in the region. Indeed, there are significant differences in the permitted health claims and the regulatory framework in place in the region.

Other function claims are permitted in five of the Southeast Asian region countries reviewed, namely, Republic of Indonesia, Malaysia, the Philippines, Singapore, and Vietnam (Table 15.1). These claims relate to several bioactive components, for example, several dietary fibers and nondigestible oligosaccharides, plant sterols, and polyunsaturated fatty acids (PUFAs).

Disease risk reduction claims are considered higher level health claims and are permitted only in three of the Southeast Asian region countries, namely, Republic of Indonesia, the Philippines, and Singapore (Table 15.2). These claims are only for a few nutrients or bioactive compounds or food.

Out of the five countries that permitted other function claims and three countries that allow disease risk reduction claims, three of these countries (Republic of Indonesia, Malaysia, and Singapore) publish the permitted claims in the public domain. These are available on their websites and also as printed guides, as indicated in Table 15.3.

These lists are updated from time to time, with new claims added by the regulatory agencies or arising from applications from the food industry. Each application for a health claim has to be

TABLE 15.2
Status of Disease Risk Reduction Claims in Selected Southeast Asian Countries

Country	Other Function Claims
Brunei Darussalam	Not permitted
Republic of Indonesia	Seven types of functional components, positive list approach
Lao PDR	Not permitted
Malaysia	Not permitted
The Philippines	Yes, according to Codex; no positive list published
Singapore	Five nutrient/food-specific claims, positive list approach
Thailand	Not permitted
Vietnam	Not permitted

TABLE 15.3
Positive List of Permitted Health Claims

Country	Permitted Health Claims
Republic of Indonesia	Regulation number HK.03.1.23.11.11.09909 year 2011 of the NA-DFC for the Control of Claims in Labeling and Advertising of Processed Food, accessible from: http://www.pom.go.id.
Malaysia	Updated list in Guide to Nutrition Labelling and Claims (as at December 2010); http://fsq.moh.gov.my.
Singapore	Accessible from the AVA website: http://www.ava.gov.sg/FoodSector/FoodLabelingAdvertisement/.

AVA, Agri-Food & Veterinary Authority of Singapore; NA-DFC, National Agency of Drug and Food Control.

accompanied by scientific substantiation, which will be reviewed by a panel of experts appointed by the regulatory agency. These experts are generally sourced from various institutions and organizations and of different expert background.

Three countries with well-established framework to review applications from the industry are Republic of Indonesia, Malaysia, and Singapore. The expert committees in these countries are as follows:

- Tim Mitra Bestari, National Agency of Drug and Food Control (NA-DFC), Republic of Indonesia
- Expert Group on Nutrition, Health Claims and Advertisement, Food Safety and Quality Division (FSQD), Ministry of Health (MOH) Malaysia
- Advisory Committee on Evaluation of Health Claim, Agri-Food and Veterinary Authority, Singapore

Several other approaches are being undertaken by the Philippines Food and Drug Administration, Thailand Food and Drug Administration, and Vietnam National Institute for Food Control in reviewing applications from the industry. Although these may not be so well structured, these agencies do consult experts from various fields when vetting applications.

All the countries that permit health claims have emphasized that approval for health claims requires evidence from well-designed human intervention trials, which are peer reviewed and published. Other data from observational, *in vitro*, and animal studies may be used for supporting evidence. In those countries with established expert committees, clear instructions for submitting applications using prescribed forms have been provided.

Prior to making health claims, the food component or functional ingredient has to be accepted to be added to food. Regulatory agencies in the three countries mentioned in the above list have established a procedure for the industry to apply for the use of new functional components. The framework for application is as outlined above for applying new health claims. Applications are to be submitted in prescribed forms and must include various required information, including scientific data to prove safety of use and benefits of adding the component to the food.

15.5 DETAILS OF HEALTH CLAIMS IN SELECTED SOUTHEAST ASIAN COUNTRIES

15.5.1 Health Claims in the Republic of Indonesia

The NA-DFC is the competent authority for food and drug control in the Republic of Indonesia (http://www.pom.go.id).

The regulation number HK.03.1.23.11.11.09909 year 2011 is the pertinent regulation on the Control of Claims in Labeling and Advertising of Processed Food. Nutrition and health claims are included in this regulation. The definitions of nutrient function claims, other function claims, and disease risk reduction claims are similar to those in the Codex Alimentarius guidelines.

Processed food that includes health claims as referred to in this Regulation is classified as functional food. This Regulation also defined functional food as a processed food, which contains one or more food components that have been demonstrated through scientific studies to possess specific physiological functions beyond its basic functions, which have been proven to be harmless yet beneficial to health.

This Regulation is the only one in Southeast Asia that uses the term functional foods and provides specific requirements for use of the term:

1. They contain the specified food component in an amount corresponding to the requirements and conditions set out in Appendices IV and V of the said Regulation (these appendices are given in Annexes 1A and 1B).
2. They possess sensory characteristics such as appearance, color, texture or consistency, and flavor that can be acceptable to consumers.
3. They are served and consumed appropriately as food and beverages.

Republic of Indonesia is also adopting a positive list approach. A list of 15 nutrient function claims is provided, which include protein (two function claims), vitamins (ten function claims), and minerals (three claims). These will not be further elaborated in this chapter.

The Regulation has provided a list of other function claims for two components, namely, dietary fiber (three claims) and phytosterols and phytostanols (one claim) (Annex 1A). Disease risk reduction claims for seven types of nutrients/components are also permitted, namely, folic acid, calcium, sugar alcohols/polyols, dietary fiber, phytosterol/phytostanol, soy peptides, and specific protein and soy isoflavones (Annex 1B).

Although not specifically related to this chapter, it is also interesting to note that Republic of Indonesia is the only country in Southeast Asia that permits a glycemic index claim, with high, medium, and low glycemic categories defined.

This Indonesian Regulation has also provided several prohibitions that need to be noted as follows:

1. Processed foods intended for infants are not permitted to make claims, unless otherwise provided for.
2. Foods intended for children aged one to three years are not permitted to make other function claims and disease risk reduction claims, unless otherwise provided for.

The following claims are also prohibited:

1. Statement indicating that consumption of a particular food is able to meet the needs for all essential nutrients
2. Exploiting the fear of consumers
3. Resulting in inappropriate consumption of the particular processed food
4. Implying or indicating that a nutrient or component is able to prevent, treat, or cure a disease

15.5.1.1 Application for Use of New Components and New Health Claims

The NA-DFC has established a system for reviewing applications for the use of new food components and new function claims. Prescribed forms must be used for the applications. Assessment of applications is carried out by the appointed appraisers and/or Tim Mitra Bestari. This is a peer review team comprising experts assigned by the Head of the Agency to conduct assessments, and provide recommendations on applications for the use of new components as well as nutrition and health claims. A number of parameters are examined including history of safe use of the component, physical and chemical properties, allergenic potential, metabolic fate, toxicological studies, and human tolerance studies. For health claims, findings from human clinical trials or observational studies (if intervention studies cannot be conducted) must be submitted to substantiate the intended claims. Data from *in vitro* and animal studies may also be submitted to support the application. The Regulation also provided several pointers in relation to conducting human clinical trials.

15.5.2 Health Claims in Malaysia

The Food Regulations 1985 of Malaysia is under the purview of the FSQD of the MOH Malaysia. The pertinent regulations related to nutrition labeling and claims are Regulations 18A, 18B, 18C, 18D, and 18E. Regulation 18E defines "nutrient function claim" as a nutrition claim that describes the physiological role of the nutrient in the growth, development, and normal functions of the body. Other function claims are also included in this regulation, in line with amendment to Regulation 26 to expand the definition of "added nutrients" to include other food components. These regulations are accessible from the website of the FSQD (http://fsq.moh.gov.my/) (MOH Malaysia 2003).

Regulation 18(6) is also relevant to this discussion as it deals with prohibited claims. Two subclauses are of particular relevance: claims that cannot be substantiated and claims as to the suitability of a food for use in the prevention, alleviation, and treatment or cure of a disease, disorder, or particular physiological condition, except as otherwise permitted.

Health claims in the context of Malaysia are in the form of other function claims, as defined by Codex Alimentarius. Malaysia adopts a positive list approach, with all the health claims ultimately to be listed in Regulation 18E. While waiting for the process of gazettement, these claims are listed in the Guide to Nutrition Labelling and Claims published by the FSQD (MOH Malaysia 2003). These lists are updated from time to time, when new function claims are approved. Indeed, thus far, all the other function claims permitted by Malaysia have arisen from applications by the food industry.

There are to date 29 other function claims (Annex 2A). It can be noted that several of these claims relate to undigestible carbohydrates, functional carbohydrates, and oligosaccharides either singly or in different combinations. Other components include plant sterols, lutein, docosahexaenoic acid (DHA)/arachidonic acid (ARA), sialic acid, and soy protein.

Specific conditions are required to be met to be eligible to make other function claims. Regulation 18E stipulates that the food must contain a minimum amount of the relevant "food component." For specific food components, additional labeling requirements may be required by the authorities to be included, for example, supplementary dietary advice (Annex 2B).

15.5.2.1 Use of New Food Components

Other function claims relate to the beneficial effect of a food component (not nutrients) on the physiological functions or health of a person. Recognizing the potential contributions of other food components to health, Regulation 26 of the Malaysian Food Regulations has been amended to expand the definition of added nutrients to include other food components. Associated with this regulation is Table I of the Twelfth Schedule, which lists the permitted added nutrients.

Other food components, which have been proven to be able to provide health benefits beyond basic nutrition, may be added to food. Food industry would have to apply to the FSQD of the MOH for the component to be recognized as a "nutrient." A procedure has been established for this purpose and this is outlined in Subsection 15.5.2.2.

15.5.2.2 Applications for New Food Components and Function Claims

The FSQD has established a framework to enable the food industry to apply for addition of new food components in foods and the use of new function claims. Applications will be reviewed by an Expert Committee on Nutrition, Health Claims and Advertisement, MOH. Members comprise experts from multigovernment agencies and the academia, and include nutritionists, food scientists, dietitians, and medical doctors.

All applications must be submitted using prescribed forms in which a variety of required information such as detailed information on the food component, its safety of use, and scientific data to substantiate the proposed claim is included. Scientific data should preferably be based on human intervention trials, while epidemiological and experimental studies and review papers may be included as supportive evidences

15.5.3 Health Claims in Singapore

The Agri-Food & Veterinary Authority of Singapore (AVA) administers a total of nine statutes, one of which is the Sale of Food Act. Within this Act is the Singapore Food Regulations. Full details of the Food Regulations are obtainable from the official website of AVA (http://www.ava.gov.sg).

In relation to health claims, Singapore permits 25 nutrient function claims (4 macronutrients, 10 vitamins, 11 minerals) for the general public. These will not be further elaborated in this chapter.

Relevant to this discussion, Singapore permits a number of health claims that may be categorized into four groups and published on the AVA website (http://www.ava.gov.sg). These are listed in Annex 3 and are discussed extensively in this subsection.

In group 1, seven specific function claims are permitted for components used in foods for infants and children, which include choline, DHA/ARA, nucleotides, and taurine (Annex 3A). The second group comprises nine other function claims permitted for general foods. These claims relate to three components, namely, collagen, probiotics, and prebiotics (Annex 3B). The next group comprises a claim relating plant sterols/stanols to lowering blood cholesterol (Annex 3C).

In addition, local food manufacturers and importers may submit applications to the AVA or the Health Promotion Board (HPB) for the use of five nutrient/food-specific diet-related health claims. This is the last group of claims, and they link specific nutrients or foods to lowering the risk of relevant diseases (Annex 3D). These claims relate calcium/vitamin D and osteoporosis; sodium and stroke and heart disease; saturated fat and *trans* fat and heart disease; whole grains, fruits, and vegetables and heart disease; and whole grains, fruits, and vegetables and some types of cancers.

These approved health claims and criteria in Annex 3D have been developed based on Singapore's existing national nutrient claims guidelines formulated by the HPB, with reference taken from currently available guidelines established by major developed countries. Only food products that have been first approved by the HPB to carry the Healthier Choice Symbol may be considered for application of use of these health claims. The HPB will conduct premarket evaluation of applications that are concurrently submitted with applications for the Healthier Choice Symbol. The AVA will conduct premarket evaluation of separate applications.

15.5.3.1 Application for Use of New Health Claims

The AVA has established a system to review applications from the food industry for new function claims. All applications are reviewed by an Advisory Committee on Evaluation of Health Claim based on Codex's recommendations on the scientific basis for health claim. The Committee comprises reputable scientific experts with relevant professional training and experience from various government bodies, tertiary institutions, and industry associations. Applications are submitted using prescribed forms.

The use of new functional ingredients also requires prior approval from the AVA before being allowed to be added to foods. Detailed safety data have to be provided for evaluation, besides data on chemical and physical properties of the component.

15.6 WAY AHEAD

Interest in functional foods in the Southeast Asian countries will further increase, as a result of greater consumer interest in health and wellness. Consumers will continue to seek benefits in foods beyond the basic nutrients to reduce risk to chronic diseases. Product research and development will intensify in keeping with consumer interest. Current focus on scientific substantiation of functional foods will need to be kept up, so as to support regulatory systems and to gain consumer confidence in functional foods and functional ingredients. There should be more human intervention trials on local population groups conducted by research groups in the region.

Regulatory frameworks in the region need to be further strengthened to ensure that health claims are adequately substantiated. At the same time, any regulatory system in place should not hamper innovations in functional foods. The food industry would appreciate transparent evaluation systems, which are conducted systematically and in a timely manner. Several countries in the region already have a well-established regulatory framework for health claims. Other countries will undoubtedly be progressing toward this goal. It is hoped that there would be some degree of harmonization in the regulatory frameworks in the Southeast Asian countries.

An area that has been lacking in this subject of functional foods and health claims is consumer understanding and use of health claims on food labels. This is especially so in the Southeast Asian region. There is a need to have more data of consumer perception toward functional foods and how they utilize health claims. This is to ensure appropriate understanding and usage of claim messages. It is important to emphasize that functional foods are to be taken as part of the daily diet and do not profess to prevent against nutritional diseases.

CONFLICT OF INTEREST

The author declares that he has no conflict of interest.

ANNEX 1A

Other Function Claims Permitted in the Republic of Indonesia

Item Number	Component	Permitted Claim Statements
1	Dietary fiber	1. "Soluble dietary fibre (Psyllium, β glucan from oats, inulin from chicory and pectin from fruits) can help lower blood cholesterol level if accompanied by a diet of low saturated fat and low cholesterol." 2. "Soluble dietary fibre (Psyllium, β glucan from oats, inulin from chicory and pectin from fruits) can help maintain/preserve the function of the digestive tract." 3. "Insoluble dietary fibre can help facilitate bowel movements (laxative), if accompanied by drinking enough water." Requirements: • *Psyllium*, oats, inulin, and pectin shall include the constituent and source components. • The food must contain at least 3 g of dietary fiber per serving. For claim number 1, in addition to meeting the above requirements, the following requirements must also be met: • Maximum total fat of 3 g per serving; if the serving is less than 50 g, then the maximum total fat content shall be 3 g per 50 g. • Maximum saturated fat of 1 g per serving and maximum calories from saturated fat as much as 15%; if the amount per serving is less than 100 g, then the maximum saturated fat content shall be 1 g per 100 g and maximum calories derived from saturated fat at 10%. • Maximum cholesterol of 20 mg per serving; if the serving is less than 50 g, then the maximum cholesterol content shall be 20 mg per 50 g.

(*Continued*)

ANNEX 1A
(Continued) Other Function Claims Permitted in the Republic of Indonesia

Item Number	Component	Permitted Claim Statements
		For claim number 3, in addition to the above-mentioned requirements, the following requirements must be met: • The food must contain at least 3 g or more per day of soluble fiber (oat β-glucan). • The food must contain at least 7 g per day of soluble fiber from *Psyllium* seed husk. Warning: Claims must be accompanied by the following statements: • "Consumption of this food must be accompanied by the consumption of low fat, low saturated fat and/or low cholesterol foods." • "Consumption of this food must be accompanied by a healthy lifestyle."

2	Phytosterols and phytostanols	"Phytosterols and phytostanols help reduce cholesterol absorption from food in the gut when accompanied by a low fat, low saturated fat and low cholesterol diet."
		Requirements:
		• Phytosterols are to be prepared by esterification of a mixture of phytosterols from edible oils with food-grade fatty acids. Phytosterol mixtures must contain at least 80% β-sitosterol, campesterol, and stigmasterol (combined weight).
		• Phytostanols are to be prepared by esterification of a mixture of phytostanols derived from edible oils or by-products from the manufacturing process of craft paper pulp with food-grade fatty acids. Their mixtures must contain at least 80% sitostanol and campestanol (combined weight).
		• The food must contain at least 0.65 g phytosterol per serving for spread and salad dressing or
		• The food must contain at least 1.7 g phytostanol per serving for spread, salad dressing, snack bars, and acidified milk.
		• The claim is applicable for foods that do not require high heat in their preparation.
		• The food must meet the following requirements:
		  • Maximum total fat of 3 g per serving; if the serving is less than 50 g, then the maximum total fat content shall be 3 g per 50 g.
		  • Maximum saturated fat of 1 g per serving and maximum calories from saturated fat as much as 15%; if the amount per serving is less than 100 g, then the maximum saturated fat content shall be 1 g per 100 g and maximum calories derived from saturated fat at 10%.
		  • Maximum cholesterol of 20 mg per serving; if the serving is less than 50 g, then the maximum cholesterol content shall be 20 mg per 50 g.
		• For food products that contain vegetable oils, replacing the terms "phytosterol" and "phytostanol" with "vegetable oil sterol esters and stanol esters" is permitted as long as the said vegetable oil constitutes the only source of sterol/stanol esters in the said food product.
		• Specifically for spread and salad dressing, the fat content may exceed 3 g per 50 g by adding the statement:
		"see nutrition information for fat content" but must still contain 0.65 g phytosterol or 1.7 g phytostanol per 50 g of the food.
		• The food shall meet minimum nutritional requirements, except for salad dressing.

Source: Appendix IV of Regulation number HK.03.1.23.11.11.09909, 2011.

ANNEX 1B

Disease Risk Reduction Claims Permitted in the Republic of Indonesia

Item Number	Component	Permitted Claim Statements
1	Folic acid	• "A nutritionally balanced diet with adequate folic acid intake may reduce the risk of failure of neural tube formation (neural tube defect) in the fetus" • "Adequate intake of folic acid, when accompanied by a nutritionally balanced diet can reduce the risk of failure of neural tube formation (neural tube defect) in the fetus." Requirements: • Consumption of the said food daily is able to meet 100% of the recommended nutrient intake (α-ketoglutarate [RNI]) for folic acid. • The food shall not contain vitamin A in the form of retinol or provitamin A and vitamin D greater than 100% of the daily RNI. • The label of the product must include recommendation for preparation of the product, namely, "The product should preferably be dissolved in boiled water with a maximum temperature of 40°C, because folic acid will be destroyed at high temperature. Warning: Claims shall be accompanied by the following statements: • "The cause of the failure of neural tube formation (neural tube defect) is multifactorial." • "Benefits can only be obtained if consumed from the time of preparing for pregnancy or during pregnancy."
2	Calcium	"Adequate consumption of calcium from an early stage can help delay the occurrence of osteoporosis in the future if accompanied by regular physical exercise and consumption of balanced nutrition." Requirements: • The food shall contain at least 75% of the RNI for calcium per day appropriate for each age group. • The phosphorus content in the said food shall not exceed the calcium content. • Calcium shall not be associated with height increase (bone length). Warning: • Claims shall be accompanied with the statement: "The maximum intake of calcium per day shall be 2500 mg." • In a product containing more than 400 mg calcium per serving, it must be accompanied by a statement that "consumption of more than 2000 mg of calcium per day will not increase the benefit in maintaining bone health."
3	Sugar alcohols/ polyols	"Sugar alcohols/polyols (mannitol, xylitol, maltitol, lactitol) in [name of the food] can help reduce the risk of dental caries, if accompanied by good health habits, one of which is maintaining good dental care." Requirements: • The food does not contain mono- and disaccharides. • A packaging of food with a label surface area of less than or equal to 30 cm² may be permitted to use an abbreviated claim as follows: "This food can help reduce the risk of dental caries." • The claim shall not mention the degree of reduction in the risk of dental caries arising from the consumption of the food containing polyols. • The claim shall not mention that the consumption of food containing polyol is the only way to reduce the risk of dental caries. Warning: Claims shall be accompanied by the statement: "The consumption of more than 20 g of sugar alcohols/polyols per day may cause a laxative effect."

(Continued)

ANNEX 1B
(Continued) Disease Risk Reduction Claims Permitted in the Republic of Indonesia

Item Number	Component	Permitted Claim Statements

4	Dietary fiber	• "A food that is low in fat, in saturated fat and cholesterol that contains soluble fibre (Psyllium, β glucan from oats, inulin from chicory and pectin from fruits), can help reduce the risk of coronary heart disease, which is a disease associated with multiple factors."
		• "Dietary fibre (Psyllium, β glucan from oats, inulin from chicory and pectin from fruits), can help manage blood sugar levels in people with type II diabetes mellitus."
		Requirements:
		• The food shall meet the following requirements:
		  • Maximum total fat of 3 g per serving; if the serving is less than 50 g, then the maximum total fat content shall be 3 g per 50 g.
		  • Maximum saturated fat of 1 g per serving and maximum calories from saturated fat as much as 15%; if the amount per serving is less than 100 g, then the maximum saturated fat content shall be 1 g per 100 g and maximum calories derived from saturated fat at 10%.
		  • Maximum cholesterol of 20 mg per serving; if the serving is less than 50 g, then the maximum cholesterol content shall be 20 mg per 50 g.
		• The food contains at least 0.6 g soluble dietary fiber per serving.
		• It is prohibited to include statements related to cancer of the large intestine (colon).
		• The food contains at least 3 g or more soluble fiber (β-glucan) from oats per day.
		• The food contains at least 7 g soluble fiber from *Psyllium* seed husk per day.

5	Phytosterol and phytostanol	• "Foods containing at least 0.65 grams of phytosterols per serving, when consumed twice a day so that the total daily intake of at least 1.3 grams as part of a diet low in saturated fat and cholesterol, may help reduce the risk of coronary heart disease. One serving of [name of food] provides … grams of phytosterols."
		• "A diet low in saturated fat and cholesterol which includes two servings of the food so as to provide a total daily intake of at least 3.4 grams phytostanol in two meals may help reduce the risk of coronary heart disease. One serving of [name of food] provides ... grams of phytostanol."
		Requirements:
		• The food contains at least 0.65 g of phytosterols per serving for spread and salad dressing.
		• The food contains at least 1.7 g phytostanol per serving for spread, salad dressing, snack bars, and acidified milk.
		• The claim is applicable to foods that do not require high heat in their preparation.
		• The food shall meet the following requirements:
		  • Maximum total fat of 3 g per serving; if the serving is less than 50 g, then the maximum total fat content shall be 3 g per 50 g.
		  • Maximum saturated fat of 1 g per serving and maximum calories from saturated fat as much as 15%; if the amount per serving is less than 100 g, then the maximum saturated fat content shall be 1 g per 100 g and maximum calories derived from saturated fat at 10%.
		  • Maximum cholesterol of 20 mg per serving; if the serving is less than 50 g, then the maximum cholesterol content shall be 20 mg per 50 g.
		• For food products that contain vegetable oils, replacing the terms "phytosterol" and "phytostanol" with "vegetable oil sterol esters and stanol esters" is permitted as long as the said vegetable oil constitutes the only source of sterol/stanol esters in the said food product.
		• Specifically for spread and salad dressing, the fat content may exceed 3 g per 50 g by adding the statement:
		  "see nutrition information for fat content" but must still contain 0.65 g phytosterol or 1.7 g phytostanol per 50 g of the food.
		• The food shall meet minimum nutritional requirements, except for salad dressing.

ANNEX 1B
Disease Risk Reduction Claims Permitted in the Republic of Indonesia

Item Number	Component	Permitted Claim Statements

6	Peptides and specific protein (soy)	"A diet that is low in saturated fat and cholesterol and contains 25 grams of soy protein per day can help reduce the risk of coronary heart disease. One serving of [name of food] provides ... grams of soy protein."
		Requirements:
		• The food shall contain at least 6.25 g soy protein per serving.
		• The food shall meet the following requirements:
		  • Sodium intake should not be more than 120 mg.
		  • Maximum total fat of 3 g per serving; if the serving is less than 50 g, then the maximum total fat content shall be 3 g per 50 g.
		  • Maximum saturated fat of 1 g per serving and maximum calories from saturated fat as much as 15%; if the amount per serving is less than 100 g, then the maximum saturated fat content shall be 1 g per 100 g and maximum calories derived from saturated fat at 10%.
		  • Maximum cholesterol of 20 mg per serving; if the serving is less than 50 g, then the maximum cholesterol content shall be 20 mg per 50 g.
		• The claim shall state the amount of soy protein per serving.

7	Soy isoflavones	"Soy isoflavones (daidzein, daidzin, genistein, genistin) can help lower the blood cholesterol levels, so that it can help reduce the risk of atherosclerosis and coronary heart disease."
		Requirements:
		• The food must contain soy protein or peptide (as a non-isoflavone component) and is not a pure isoflavone.
		• The food must contain at least 5 mg of isoflavones per serving.
		• The food shall meet the following requirements:
		  • Maximum total fat of 3 g per serving; if the serving is less than 50 g, then the maximum total fat content shall be 3 g per 50 g.
		  • Maximum saturated fat of 1 g per serving and maximum calories from saturated fat as much as 15%; if the amount per serving is less than 100 g, then the maximum saturated fat content shall be 1 g per 100 g and maximum calories derived from saturated fat at 10%.
		  • Maximum cholesterol of 20 mg per serving; if the serving is less than 50 g, then the maximum cholesterol content shall be 20 mg per 50 g.
		Warning:
		Claims shall be accompanied by the statement: "Accompanied by consuming foods low in food, low in saturated fat and/or low cholesterol."

Source: Appendix V of Regulation number HK.03.1.23.11.11.09909, 2011; NA-DFC website, http://www.pom.go.id.
Note: Translated from original document that is in Indonesian language by the author and not an official translation.

ANNEX 2A
Other Function Claims Permitted in Malaysia

Functional Component	Health Claim Permitted
Bifidobacterium lactis	• *Bifidobacterium lactis* helps to improve a beneficial intestinal microflora. • *Bifidobacterium lactis* may help to reduce the incidence of diarrhea.
DHA and ARA	• DHA and ARA may contribute to the visual development of infant.
HAMRS	• HAMRS helps to improve/promote colonic/bowel/intestinal function/environment.

(*Continued*)

ANNEX 2A
(Continued) Other Function Claims Permitted in Malaysia

Functional Component	Health Claim Permitted
Isomaltulose	• Isomaltulose is slowly hydrolyzed to glucose and fructose and, therefore, it provides longer-lasting energy compared to sucrose • Isomaltulose is a slowly released source of energy compared to sucrose. • Isomaltulose provides longer-lasting energy compared to sucrose. • Isomaltulose is slowly hydrolyzed to glucose and fructose, compared with sucrose.
Oat-soluble fiber (β-glucan)	• Oat-soluble fiber (β-glucan) helps to lower the rise of blood glucose provided it is not consumed together with other food.
Oligosaccharide mixture containing 90% (wt/wt) GOS and 10% (wt/wt) lcFOS	• Oligosaccharide mixture containing 90% (wt/wt) GOS and 10% (wt/wt) lcFOS is prebiotic. • Oligosaccharide mixture containing 90% (wt/wt) GOS and 10% (wt/wt) lcFOS is bifidogenic. • Oligosaccharide mixture containing 90% (wt/wt) GOS and 10% (wt/wt) lcFOS helps to increase intestinal bifidobacteria and helps to maintain a good intestinal environment. • Oligosaccharide mixture containing 90% (wt/wt) GOS and 10% (wt/wt) lcFOS helps to improve the gut/intestinal immune systems of babies/infants.
Oligofructose-inulin mixture containing 36%–42% oligofructose (DP 2–10) and 50%–56% inulin (DP > 10)	• Oligofructose-inulin mixture containing 36%–42% oligofructose (DP 2–10) and 50%–56% inulin (DP > 10) helps to increase calcium absorption and increase bone mineral density when taken with calcium-rich food.
Patented cooking oil blend	• Patented cooking oil blend helps to increase HDL cholesterol and improve HDL/LDL cholesterol ratio.
Plant sterol/plant stanol/plant sterol ester	• Plant sterol/plant stanol/plant sterol ester helps to lower or reduce cholesterol.
Polydextrose	• Polydextrose is bifidogenic. • Polydextrose helps to increase intestinal bifidobacteria and helps to maintain a good intestinal microflora.
Resistant dextrin/resistant maltodextrin	• Resistant dextrin/resistant maltodextrin is a soluble dietary fiber that helps to regulate/promote regular bowel movement especially of people with a tendency to constipate.
Sialic acid	• Sialic acid is an important component of the brain tissue.
Soy protein	• Soy protein helps to reduce cholesterol.

Source: Guide to Nutrition Labelling and Claims (as at December 2010), Food Safety & Quality Division, Ministry of Health Malaysia.

Note: Other function claims related to D-ribose and calcium 3-hydroxy 3-methyl butyrate monohydrate are in the process of being formally approved.

For all the above claims, words/sentences of similar meaning can also be used:

- Specific conditions required for other function claims, for example, minimum amount of the relevant "food component" that must be present (see Annex 2B).
- Additional labeling requirements, if relevant for specific products.
- Restriction to selected foods, if relevant.

ARA, arachidonic acid; DHA, docosahexaenoic acid; DP, degree of polymerization; HAMRS, high-amylose maize-resistant starch; HDL, high-density lipoprotein; lcFOS, long-chain FOS; LDL, low-density lipoprotein.

ANNEX 2B
Conditions for Making Other Function Claims in Malaysia

Component	Minimum Amount Required	Other Conditions
β-glucan	0.75 g per serving	1. Source of β-glucan shall be from oat and barley. 2. The food to be added with β-glucan shall also contain the total dietary fiber for not less than the amount required to claim as "source": 3 g per 100 g (solids) 1.5 g per 100 ml (liquids) 3. The following statement shall be written on the label: "Amount recommended for cholesterol lowering effect is 3 g per day."
Bifidobacterium lactis	1×10^6 minimum viable cells per gram	Claim only permitted in infant formula, follow-up formula, formulated milk powder for children, and cereal-based food for infant and children
DHA and ARA	A combination of 17 mg/100 kcal DHA and 34 mg/100 kcal of ARA	Claim only permitted in infant formula product
High-amylose maize-resistant starch	2.5 g per serving	
Inulin	1.25 g per serving	This minimum level is for food other than infant formula.
	0.4 g per 100 ml on a ready-to-drink basis	1. This minimum level is specified for infant formula only. 2. The component (inulin and oligofructose/FOS) shall not exceed 0.6 g per 100 ml.
Isomaltulose	15 g per serving	Addition and claim for isomaltulose are not permitted in infant formula.
Lutein	2.5 μg per 100 ml (3.7 μg per 100 kcal)	This minimum level is specified for infant formula only.
	20 μg per 100 ml (30 μg per 100 kcal)	This minimum level is specified for follow-up formula only.
	20 μg per 100 ml (20 μg per 100 kcal)	This minimum level is specified for formulated milk powder for children only.
Oat-soluble fiber (β-glucan)[a]	6.5 g per 100 g	1. Claim for oat-soluble fiber (β-glucan) only permitted in cereal and cereal-based product. 2. Claim only permitted for cereal and cereal-based product where the macronutrient profile (carbohydrate, protein, and fat) complies with RNI for Malaysia. 3. The following statement shall be written on the label of cereal and cereal-based product for making such claim: "For advice regarding consuming this product, consult your health professional."
Oligofructose/FOS	1.25 g per serving	This minimum level is for food other than infant formula.
	0.4 g per 100 ml on a ready-to-drink basis	1. This minimum level is specified for infant formula only. 2. The component (inulin and oligofructose/FOS) shall not exceed 0.6 g per 100 ml.

(*Continued*)

ANNEX 2B
(Continued) Conditions for Making Other Function Claims in Malaysia

Component	Minimum Amount Required	Other Conditions
Oligofructose–inulin mixture containing 36%–42% oligofructose (DP 2–10) and 50%–56% inulin (DP >10)	2 g per serving	Total fructant content in the mixture shall be more than 90% on a dry weight basis.
Oligosaccharide mixture containing 90% (wt/wt) GOS and 10% (wt/wt) lcFOS[b]	0.8 g per 100 ml	1. Claim only permitted in infant and follow-up formulas. 2. The component (oligosaccharide mixture) shall not exceed 0.8 g per 100 ml.
Oligosaccharide mixture containing 90% (wt/wt) galacto-oligosaccharide (GOS) and 10% (wt/wt) lcFOS[c]	0.4 g per 100 ml	1. Claim only permitted in infant formula, follow-up formula, and formulated milk powder for children. 2. The component (oligosaccharide mixture) shall not exceed 0.8 g per 100 ml.
Patented cooking oil blend	Ratio of fatty acid profiles for saturated fatty acid:monounsaturated fatty acid:polyunsaturated fatty acid must be 1:1:1.	The patented cooking oil blend refers to the US patent numbers 5578334 and 5843497.
Plant sterol, plant stanol, or plant sterol ester	0.4 g per serving in a "free basis" form.	1. Types of plant sterol or plant stanol permitted: plant sterol or plant stanol, phytosterol or phytostanol, sitosterol, campesterol, stigmasterol, or other related plant stanol 2. Types of plant sterol esters permitted: campesterol ester, stigmasterol ester, and β- sitosterol ester 3. Amount of plant sterol, plant stanol, or plant sterol ester in a "free basis" form to be added in food shall not exceed 3 g plant sterol or plant stanol per day. 4. Declaration of the total amount of plant sterol, plant stanol, or plant sterol ester contained in the products shall be expressed in metric units per 100 g, per 100 ml, or per package if the package contains only a single portion and per serving as quantified on the label. 5. Only the terms "plant sterol," "plant stanol," or "plant sterol ester" shall be used in declaring the presence of such components. 6. The following statements shall be written on the label of food making such claim: a. "Not recommended for pregnant and lactating women, and children under the age of five years." b. "Persons on cholesterol-lowering medication shall seek medical advice before consuming this product." c. "That the product is consumed as part of a balanced and varied diet and shall include regular consumption of fruits and vegetables to help maintain the carotenoid level." d. "With added plant sterols or plant stanol or plant sterol ester in not less than 10 point lettering."

ANNEX 2B
Conditions for Making Other Function Claims in Malaysia

Component	Minimum Amount Required	Other Conditions
Polydextrose	1.25 g per serving	
Resistant dextrin or resistant maltodextrin	2.5 g per serving	Addition and claim for resistant dextrin or resistant maltodextrin are not permitted in infant formula.
Sialic acid	36 mg per 100 kcal or 24 g per 100 ml	1. The component (sialic acid) shall not exceed 67 mg per 100 kcal (45 mg per 100 ml). 2. Addition and claim only permitted in infant and follow-up formulas. 3. Only natural sialic acid from milk shall be added.
Soy protein	5 g per serving	The following statement shall be written on the label of food for making such claim: "Amount recommended to give the lowering effect on the blood cholesterol is 25 g per day."

Source: Guide to Nutrition Labelling and Claims (as at December 2010), Food Safety & Quality Division, Ministry of Health Malaysia.

[a] In relation to blood glucose claim.

[b] In relation to intestinal immune claim.

[c] In relation to prebiotic, bifidogenic, and intestinal bifidobacteria claims.

ARA, arachidonic acid; DHA, docosahexaenoic acid; FOS, fructo-oligosaccharide; lcFOS, long-chain FOS; RNI, Recommended Nutrient Intake.

ANNEX 3
Health Claims Related to Various Food Components Permitted in Singapore

A. Function Claims Specific to Infant Formula, Infants' Food, and Foods for Young Children (up to Six Years of Age)

1. *Choline* helps support the overall mental functioning.
2. *Docosahexaenoic acid* (DHA) and *arachidonic acid* (ARA) are important building blocks for development of the brain and eyes in infant (only for food for children up to three years of age).
3. *Nucleotides* are essential to normal cell function and replication, which are important for the overall growth and development of infant.
4. *Taurine* helps to support the overall mental and physical development.
5. *Zinc* helps in physical development.
6. Prebiotic blend (GOS and lcFOS)[a] with *zinc* and *iron* supports child's natural defenses.
7. *Nucleotides*[b] support body's natural defenses (only for infant formula targeting infants up to one year of age).

B. Other Function Claims for All Foods

Other function claims may only be used in the exact approved form and not be presented as a product-specific health claim. Truncated or reworded versions that deviate from the intended meaning of the originally approved claim are not acceptable.

Collagen

1. It is a protein in connective tissues found in the skin, bones, and muscles.

Probiotics[c]

1. It helps to maintain a healthy digestive system.
2. It helps in digestion.
3. It helps to maintain a desirable balance of beneficial bacteria in the digestive system.
4. It helps to suppress/fight against harmful bacteria in the digestive system, thereby helping to maintain a healthy digestive system.

(*Continued*)

ANNEX 3
(Continued) Health Claims Related to Various Food Components Permitted in Singapore

Prebiotics[c]

1. Prebiotic promotes the growth of good Bifidus bacteria to help maintain a healthy digestive system.
2. Inulin helps support growth of beneficial bacteria/good intestinal flora in the gut.
3. Oligofructose stimulates the bifidobacteria, resulting in a significant increase of the beneficial bifidobacteria in the intestinal tract. At the same time, the presence of less desirable bacteria is significantly reduced.
4. Inulin helps increase intestinal bifidobacteria and helps maintain a good intestinal environment.

C. Other Function Claims for Specific Foods

Special purpose foods containing phytosterols, phytosterol esters, phytostanols, or phytostanol esters

Plant sterols/stanols have been shown to lower/reduce blood cholesterol. High blood cholesterol is a risk factor in the development of coronary heart disease.

This category of special purpose foods is specified under a new regulation 250A of the Food Regulations.

These products should carry the following mandatory information on their labels:

1. The product is a special purpose food intended exclusively for people who want to lower their blood cholesterol level.
2. Patients on cholesterol-lowering medication should only consume the product under medical supervision.
3. The product may not be nutritionally appropriate for pregnant and breast-feeding women and children under the age of five years.
4. The product should be used as part of a balanced and varied diet, including regular consumption of fruits and vegetables to help maintain carotenoid levels.
5. Consumption of more than 3 g per day of added phytosterols or phytostanols should be avoided.
6. A statement suggests the amount of the food (in grams or milliliters) to be consumed each time (referred to as a serving) and the number of servings suggested to be consumed per day, with a statement of the amount of phytosterols or phytostanols that each serving contains.

D. Nutrient-Specific Diet-Related Health Claims

The following claims, listed in Regulation 9 of the Food Regulations, may be made on prepacked foods that meet the corresponding criteria set out in the Fourteenth Schedule:

1. A healthy diet with adequate calcium and vitamin D, with regular exercise, helps to achieve strong bones and may reduce the risk of osteoporosis. [here state the name of the food] is a good source of/high in/enriched in/fortified with calcium. The criteria for this claim are as follows:
   a. At least 50% of calcium RDA, which is taken as 800 mg and
   b. Low in fat (not more than 3 g fat per 100 g or not more than 1.5 g fat per 100 ml), or fat free (not more than 0.15 g fat per 100 g or 100 ml)
2. A healthy diet low in sodium may reduce the risk of high blood pressure, a risk factor for stroke and heart disease. [here state the name of the food] is sodium free/very low in/low in/reduced in sodium. The criteria for this claim are as follows:
   a. No added salt or
   b. Salt/sodium free (not more than 5 mg sodium per 100 g) or
   c. Very low in salt/sodium (not more than 40 mg per 100 g) or
   d. Low in sodium (not more than 120 mg per 100 g) or
   e. Reduced sodium (if sodium content per reference quantity is not more than 15% of sodium RDA, which is taken as 2000 mg)
3. A healthy diet low in saturated fat and *trans* fat may reduce the risk of heart disease. [here state the name of the food] is free of/low in saturated fats and *trans* fats. The criteria for this claim are as follows:
   a. Low in saturated fat (not more than 1.5 g saturated fat per 100 g and not more than 10% of kilocalories from saturated fat) or
   b. Free of saturated fat (not more than 0.5 g saturated fat per 100 g and not more than 1% of the total fat is *trans* fat) and free of *trans* fat (less than 0.5 g *trans* fat per 100 g) and

ANNEX 3
Health Claims Related to Various Food Components Permitted in Singapore

c. Low in sugar (not more than 5 g per 100 g or not more than 2.5 g per 100 ml), sugar free (not more than 0.5 g per 100 g), or unsweetened or no added sugar and
d. Cholesterol at not more than 100 mg per 100 g and
e. Reference quantity of the food product should not contain sodium in an amount exceeding 25% of sodium RDA, which is taken as 2000 mg.

D. Nutrient-Specific Diet-Related Health Claims

4. A healthy diet rich in whole grains, fruits, and vegetables that contain dietary fiber may reduce the risk of heart disease. [here state the name of the food] is low in/free of fat and high in dietary fiber. The criteria for this claim are as follows:
   a. A product from these food groups—whole grains, fruits, vegetables, or fiber-fortified foods and
   b. Low in fat (not more than 3 g fat per 100 g or not more than 1.5 g fat per 100 ml) or fat free (not more than 0.15 g fat per 100 g or 100 ml) and
   c. High in dietary fiber (not less than 3 g per 100 kcal or not less than 6 g per 100 g or 100 ml) and
   d. With at least 25% of the dietary fiber comprising soluble fiber
5. A healthy diet rich in fiber-containing foods such as whole grains, fruits, and vegetables may reduce the risk of some types of cancers. [here state the name of the food] is free/low in fat and high in dietary fiber. The criteria for this claim are as follows:
   a. A product from these food groups—whole grains, fruits, vegetables, or fiber-fortified foods and
   b. Low in fat (not more than 3 g fat per 100 g or not more than 1.5 g fat per 100 ml), or fat free (not more than 0.15 g fat per 100 g or 100 ml) and
   c. High in dietary fiber (not less than 3 g per 100 kcal or not less than 6 g per 100 g) and
   d. Reference quantity of the food product should not contain sodium in an amount exceeding 25% of sodium RDA, which is taken as 2000 mg.

Source: AVA, http://www.ava.gov.sg.

[a] The combination of GOS and lcFOS present in the product must be in the ratio of 9:1.

[b] The total nucleotide content must be within the range of 72–115 mg/l.

[c] Need to specify the name(s) of the probiotic or prebiotic whenever a claim is made in relation to that probiotic or prebiotic.

RDA, recommended daily allowance.

REFERENCES

AVA Singapore. 2011. Nutrition labelling and claims regulations. Agri-Food and Veterinary Authority Singapore. http://www.ava.gov.sg/FoodSector/FoodLabelingAdvertisement/.

Codex Alimentarius. 2009. Guidelines for use of nutrition and health claims (CAC/GL 23-1997). Food and Agriculture Organization/World Health Organization, Rome, Italy. http://www.codexalimentarius.org/standards/en/.

FDA Thailand. 2001. Notification of the Ministry of Public Health, No. 182 B.E. 2541 (1998) and No. 219 B.E. 2554 (2001). Nutrition labelling. http://iodinethailand.fda.moph.go.th/fda/new/images/cms/top_upload/1147229682_182-41(update).pdf.

ILSI SEAR. 2013. Unpublished working papers for the ILSI SEA region 8th seminar and workshop on nutrition labeling, claims and communication strategies, November 27–28, Jakarta. International Life Sciences Institute South-East Asia Region, Singapore.

MOH Brunei Darussalam. 2010. Supplement to government gazette. Part II. No. 48. October 19, 2010. http://www.moh.gov.bn/download/download/public_health_food_reg_182.pdf.

MOH Malaysia. 2003. Regulation 18E. Nutrient function claim. Food Safety and Quality Division, Ministry of Health Malaysia, Putrajaya, Malaysia. http://fsq.moh.gov.my/v4/images/filepicker_users/5ec35272cb-78/Perundangan/Akta%20dan%20Peraturan/Food_Regs_1985/FR1985_p4.pdf.

MOH Malaysia. 2011. Guide to nutrition labelling and claims (as at December 2010). Food Safety and Quality Division, Ministry of Health Malaysia, Putrajaya, Malaysia. http://fsq.moh.gov.my/v4/images/filepicker_users/5ec35272cb-78/Perundangan/Garispanduan/Pelabelan/GuideNutritionLabel.pdf.

NA-DFC. 2011. Regulation number HK.03.1.23.11.11.09909 year 2011 for the control of claims on labels and advertisements for processed foods. The National Agency of Drug and Food Control, Republic of Indonesia. http://jdih.pom.go.id/produk/PERATURAN%20KEPALA%20BPOM/12_1795_00_x.pdf.

Tee, E.S. 2004. *Functional Foods in Asia: Current Status and Issues*. Singapore: International Life Sciences Institute Southeast Asia Region.

16 Health Claims Regulation and Scientific Substantiation of Functional Foods and the International Comparison

Toshio Shimizu

CONTENTS

16.1 INTRODUCTION

The proportion of senior citizens in relation to the general population is rapidly increasing in developed countries, especially in Japan. With the rapid increase in the number of elderly people and the development of a senior society, chronic diseases such as diabetes, cardiovascular diseases, hypertension, osteoporosis, and cancer are expected to increase. Systematic research on the health benefits of foods started in Japan in 1984. The Japanese Ministry of Education started a project concerning food functionality. The project first defined the concept of functional food; that is, foods have three functions: The primary function is a nutritional function, which is essential to human survival. The secondary function is a sensory function involving both flavor and texture to satisfy the sensory needs. The tertiary function is a physiological function such as regulation of biorhythms, control of aging, the immune system, and body defense beyond nutrient functions. The project defined a functional food as a food having a tertiary function. The project has identified many food components having a tertiary function, and scientific evidence is accumulating regarding the study of the physiological function of food. Bolstered by this scientific evidence regarding the health function of foods, a regulatory system for foods with health claims has been discussed in Japan.

The results of research and development of functional foods prompted the Ministry of Health, Labour and Welfare (MHLW) to establish a regulatory system regarding evaluating functionalities of food. The health claim made on foods is important for promoting the public health as well as communicating between the industry and the consumer. The system of the claims should be regulated by the national regulation based on the scientific substantiation. The Japanese Ministry established a regulatory system of the health claims "Foods for Specified Health Use (FOSHU)" in 1991. The health claims of FOSHU are evaluated product by product by scientific experts committee in the government. The substantiation of health claims in the FOSHU approval system should be based on human intervention studies in addition to *in vivo* animal and *in vitro* studies with statistical analysis. Another health claim in Japan is established by the standardized system "Food with Nutrient Function Claims (FNFC)" in 2001.

The Codex Alimentarius has defined three types of health claims: (1) nutrient function claims, (2) other function claims, and (3) reduction of disease risk claims. The Nutrition Labeling and Education Act (NLEA), which was established in the United States in 1990, defined a claim that describes a relationship between a food or its substance and a disease or health-related condition. The structure/function claim was enacted by the Dietary Supplement Health and Education Act (DSHEA) in the United States in 1994. The European Union established the Regulation on nutrition and health claims made on foods in 2007. There is some similarity in the general concept among the structure/function claims of the DSHEA, the function claims of the Regulation in the European Union, and the FOSHU claims.

16.2 HEALTH CLAIMS REGULATION

16.2.1 Codex Alimentarius

The Codex Alimentarius Commission, which develops harmonized international food standards to protect the health of the consumers and ensure fair trade practices in the food trade, published the guidelines for use of nutrition and health claims.[1] The guideline defined nutrient function claim, other function claims, and reduction of disease risk claims as follows:

> Nutrient function claim—These claims that describe the physiological role of the nutrient in growth, development and normal functions of the body.
>
> Other function claims—These claims concern specific beneficial effects of the consumption of foods or their constituents, in the context of the total diet on normal functions or biological activities of the body. Such claims relate to a positive contribution to health or to the improvement of a function or to modifying or preserving health.
>
> Reduction of disease risk claims—Claims relating the consumption of a food or food constituent, in the context of the total diet, to the reduced risk of developing a disease or health-related condition.

The guideline recommended that health claims must be based on the current relevant scientific substantiation; the level of proof must be sufficient to substantiate the type of claimed effect and the relationship to health as recognized by generally accepted scientific review of the data; and the scientific substantiation should be reviewed as new knowledge becomes available.[2]

16.2.2 Japanese Regulatory System

16.2.2.1 Foods for Specified Health Use

16.2.2.1.1 Background

FOSHU was established as the regulatory system to approve the statements contained on a label regarding the effects of foods on the human body under the Nutrition Improvement Law. FOSHU can have positive effects on human physiological functions, and the related foods are intended to be consumed for the maintenance or promotion of health or special health uses by people who wish to control their health conditions. The health claims of FOSHU must not include medical claims such as claims to "prevent," "cure," "treat," and "diagnose" human diseases. The health claims of FOSHU are evaluated product by product by the experts committee in the government based on the scientific substantiation.

16.2.2.1.2 Scope of FOSHU

The MHLW defined the scope of health claims as follows:

1. Maintain or improve a marker determined by self-diagnosis or health checkup. The example of permitted claim is "This product helps to maintain normal blood pressure, blood sugar, or cholesterol."
2. Maintain (or improve) physiological function and organ function of the human body. The example of permitted claim is "This product enhances the absorption of calcium or this product helps to improve the movement of the bowel."
3. Cause short-term changes in body condition, but not long-term changes. The example of permitted claim is "This product is good for or helps people who feel fatigue."

16.2.2.1.3 Types of FOSHU

The regulatory range of FOSHU was expanded to accept the forms of capsules and tablets in addition to those of conventional foods in 2001. The government added the three new types of FOSHU categories to the current FOSHU in 2005 to answer the request by the manufactures for expanding the FOSHU as well as to do so in harmony with the guideline of Codex Alimentarius concerning to the health claims related to disease risk reduction.[1]

1. *Qualified FOSHU*: The products, which have less sufficient scientific evidence than the current FOSHU, can be allowed to label the health claim with a qualified or conditional phrase, such as "the evidence is not established, but this product may be suitable for a person with slightly elevated blood neutral fat."
2. *Standardized FOSHU*: The product, whose health claim is already established, is granted on the basis of compliance with the separately prescribed standards. Now health claims regarding the gut intestine and the blood glucose are allowed for dietary fiber and oligosaccharide.
3. *Reduction of disease risk FOSHU*: The product with the effective substance evaluated by the MHLW can label the health claim related to risk reduction of disease. At present, foods containing certain amount of calcium are permitted to label the risk reduction of osteoporosis.

16.2.2.1.4 Scientific Substantiation for FOSHU Approval

The essential requirements for FOSHU approval could be summarized as (1) the effectiveness based on scientific evidence; (2) the safety of the product with additional safety studies as well as the information on eating history; and (3) the analytical determination of the effective components for the characterization of the product and the effective component.

The substantiation of effectiveness should be based on the human intervention studies supported by both *in vitro* and animal studies for elucidating the biochemical effectiveness and metabolism. These data should demonstrate statistically significant differences. Basically, a human study should be conducted with the form of the product food in question not only with the purified effective component and the period should be a reasonable term, such as three months as a general rule for FOSHU application. The study should also be well designed, for example, using an appropriate functional marker, an appropriate sample size, and a sufficient number of subjects to prove statistically significant differences. All available literature regarding the related functional components, the related foods, and the related functions should be reviewed. Any new scientific evidence used to support health-related claims must be published by a suitably qualified journal with expert referees who can review the evidence.

Regarding safety, both *in vivo* and *in vitro* studies should be carried out to obtain preliminary data confirming the food's safe intake by subjects before the human study. The safety data for human should be required for at least three times the minimum effective dosage by a reasonable number of subjects during one month because some consumers tend to take more dosage than that of the labeled instruction if the product has a claim concerning health benefits. The literature regarding the related functional components should be reviewed for confirming the safety.

Regarding the analysis for characterization of the product, documentation of the methods for analysis of the related functional components in applied product should submit the suitable and reliable methods of quantitative and qualitative analytical determination. These analytical determinations must be carried out and confirmed before the clinical studies, animal studies, *in vitro* studies, and stability tests.

As an additional documentation, the stability of the related functional components should be confirmed. The effective components and other components in the product should be confirmed to demonstrate the specified amount during the effective life through the use of suitable analytical methods. Additionally, in the case of a product in the form of tablets or capsules, the experiments should be conducted regarding its characteristics of disintegration or dissolution.[3,4]

16.2.2.1.5 Health Claims on FOSHU

Claims for FOSHU are health-related functions that can have beneficial effects on human physiological functions, and the related foods are intended to be consumed for the maintenance or promotion of health or special health uses by individuals who wish to control their health conditions. Existing FOSHU health claims can be classified into the following nine groups according to the approved health claims, the markers in the clinical studies, and the related substances[5]:

1. *Gastrointestinal (GI) condition*: The markers of clinical study are defecation frequency, fecal quantity, bacterial flora in feces, and short-chain fatty acid production. The examples of the claims are "This helps increase intestinal bifidobacteria and thus helps maintain a good GI condition" and "This product helps to improve bowel movements." The effective components are several oligosaccharides and dietary fibers.
2. *Blood pressure*: The markers of clinical study are systolic blood pressure and diastolic blood pressure. The example of the claim is "This is suitable for persons with slightly elevated blood pressure." The effective components are peptides, glycoside, and soy protein.
3. *Serum cholesterol*: The markers of clinical study are low-density lipoprotein (LDL) cholesterol, high-density lipoprotein (HDL) cholesterol, and total cholesterol. The example of the claim is "This helps people decrease serum cholesterol levels." The effective components are soy protein, chitosan, and phytosterol.
4. *Blood glucose*: The markers of clinical study are fasting blood glucose, postprandial blood, and hemoglobin A1c. The example of the claim is "This is helpful for those who are concerned about their blood glucose levels." The effective components are indigestive dextrin, L-arabinose, and polyphenol.
5. *Mineral absorption*: The markers of clinical study are stable isotope or pyridinoline and deoxypyridinoline for calcium absorption, and hemoglobin and serum ferritin for iron absorption. The example of the claim is "This product has high bio-availability for humans and is suitable for supplementing calcium or iron." The effective components are oligosaccharides, phosphopeptides, and heme iron.

6. *Blood lipid*: The markers of clinical study are serum triglyceride, chylomicron, cholesterol, and body fat rate. The example of the claim is "This helps reduce postprandial serum triglyceride levels." The effective components are polyphenol and digesting globin.
7. *Tooth health*: The marker of clinical study is acidity with micro tip set in tooth. The example of the claim is "This product is a low- or non-cariogenic product." The effective components are some sugar alcohols and polyphenols.
8. *Bone health*: The markers of clinical study are urinary bone absorption markers (pyridinoline, deoxypyridinoline) and serum carboxylated osteocalcin. The example of the claim is "This could promote bone calcification." The effective components are vitamin K_2, isoflavones, and milk proteins.
9. *Body fat*: The markers of clinical study are serum triglyceride, chylomicron, cholesterol, and body fat rate. The example of the claim is "This product makes it inhibit the increment of the levels of body fat." The effective components are middle-chain fatty acid and catechin.

16.2.2.2 Food with Nutrient Function Claims

The MHLW enacted a regulatory system for standardized nutrient function claims "Foods with Nutrient Function Claims" and established "Foods with Health Claims," which consists of FOSHU and FNFC in 2001 (Table 16.1).[6]

The Nutrient Function Claims of 12 vitamins and 5 minerals were standardized as follows:

Vitamin A (or beta-carotene) is a nutrient that helps to maintain vision in the dark and to maintain skin and mucosa healthy.

Vitamin D is a nutrient that promotes to absorb calcium in the gut intestine and aids in the development of bone.

Vitamin E is a nutrient that helps to protect fat in the body from being oxidized and to maintain the cell healthy.

Vitamin B_1 is a nutrient that helps to produce energy from carbohydrate and to maintain skin and mucosa healthy.

Vitamin B_2 is a nutrient that helps to maintain skin and mucosa healthy.

Vitamin B_6 is a nutrient that helps to produce energy from protein and to maintain skin and mucosa healthy.

Niacin is a nutrient that helps to maintain skin and mucosa healthy.

Biotin is a nutrient that helps to maintain skin and mucosa healthy.

Pantothenic acid is a nutrient that helps to maintain skin and mucosa healthy.

Folic acid is a nutrient that aids in the red blood cell formation and contributes to the normal growth of the fetus.

Vitamin B_{12} is a nutrient that aids in the red blood cell formation.

Vitamin C is a nutrient that helps to maintain skin and mucosa healthy and has an antioxidizing effect.

TABLE 16.1
Classification and Class Name of Japanese Regulation

	Foods with Health Claims		
Drug (Including Quasidrug)	**FOSHU (Individual Approval System)**	**FNFC (Standard Regulation System for Nutrients)**	**The Other Foods (Including a Part of So-Called Health Foods)**
Medicinal claims	Nutrient content claims Health claims	Nutrient content claims Nutrient function claims	Nutrient content claims

Calcium is a nutrient that is necessary in the development of bone and teeth.
Iron is a nutrient that is necessary in the red blood cell formation.
Zinc is a nutrient that is necessary to maintain normal gustation, helps to maintain skin and mucosa health, participates in the metabolism of protein and nucleic acid, and helps to maintain good health.
Copper is a nutrient that aids in the red blood cell formation and helps the normal activity of many kinds of enzymes and bone development.
Magnesium is a nutrient that is necessary for the development of bone and teeth, helps the normal activity of many kinds of enzymes and energy production, and is necessary to maintain normal blood circulation.

16.2.3 The United States

The NLEA was established in 1990. The Act allowed a claim that describes a relationship between a food or its substance and a disease or health-related condition. This type of claim is similar in concept to the reduction of risk of disease of the Codex Alimentarius, authorized by the US Food and Drug Administration (FDA) that reviewed the scientific literature, for example, "diets high in calcium may reduce the risk of osteoporosis."[7]

The structure/function claim, which was defined by the DSHEA in 1994, is a statement that describes the role of a nutrient or a dietary ingredient intended to affect the bodily structure or function in human or that characterizes the mechanism of a nutrient or a dietary ingredient such as vitamins, minerals, herbs, and amino acids in a dosage form of medicine, such as capsules and tablets. It acts to maintain such a structure/function relationship, for example, "Calcium builds strong bones."[8]

Under the DSHEA, dietary supplement manufacturers can sell their product with their structure/function claims by the notification of the claim they make to the FDA. The manufacturers are responsible for ensuring the accuracy and truthfulness of these claims. The FDA announced by the Guidance for Industry that the "gold" standard is double-blind, randomized, controlled trial (DB-RCT) design.[9]

16.2.4 The European Union

The European Union enforced the Regulation[10] on nutrition and health claims made on food in 2007, which consists of function claims, disease risk reduction claims, and claims referring children's development and health. The function claims defined under Article 13 as "the role of a nutrient or other substance in growth, development and the functions of the body; or psychological and behavioral functions; or slimming and weight control or reduction in the sense of hunger, or an increase in the sense of satiety or the reduction of the available energy from the diet." There are two subcategories of claims in Article 13: (1) The health claims in Article 13.1 shall be "based on generally accepted scientific evidence," which is called to be "general function claim," and (2) the additional claims in Article 13.5 shall be "based on the newly developed scientific evidence," which is called "new function claim." The disease risk reduction claim and claims referring children's development and health are regulated in Article 14.

The European Food Safety Authority (EFSA) makes a scientific judgment on the extent to which a cause-and-effect relationship is established between the consumption of the food or the constituent and the claimed effect (for the target group under the proposed conditions of use) by considering the strength, consistency, specificity, dose–response, and biological plausibility of the relationship. While animal or *in vitro* studies may provide supportive evidence (e.g., in support of a mechanism), human data are central for the substantiation of the claim.[11]

The European Commission published a list of 222 permitted generic function claims that were based on the scientific evaluation by the EFSA in December 2012.[12] The claims for the nutrients such as vitamin, mineral, amino acid, and fatty acid head the list and amount to 80% in the list. Other claims are for dietary fiber (7%) and substitutes or replacement for excess intake food component such as sugar and sodium (5%).

16.2.5 People's Republic of China

The Ministry of Health established the regulation for the Control of Health Food in 1996, in which health food is defined as a food that has the function of regulating the human body but is not used for therapeutic purposes. The effects of health food must be demonstrated by necessary animal and/or human functional tests that demonstrate that the product has a definite and stable health function. Food products that claim health functions shall be reviewed and approved by the Ministry of Health. Human intervention studies are important basic data in this evaluation system. Human intervention studies for already approved health claims are performed by the government research organization. The examples of products and the health claims on approved health foods are as follows: regulating the immune function (royal jelly, soybean oil, shark powder, spirulina, and pollen); delaying the aging process (wolfberry and honey); controlling obesity (soybean oil, orange peel, and shark fin); and improving memory (Chinese yam, poria, pig brain extract, and phospholipids).[13]

16.2.6 Australia and New Zealand

The Food Standard Australia New Zealand (FSANZ) was established to protect the health and safety of the people in Australia and New Zealand by maintaining a safe food supply. A new standard to regulate the health claims on food labels and in advertisements became law on January 18, 2013.[13] Two types of health claims are as follows:

1. *General-level health claims* refer to a nutrient or substance in a food and its effect on a health function. They must not refer to a serious disease or a biomarker of a serious disease. An example of this type of claim is "Calcium is good for bones and teeth."
2. *High-level health claims* refer to a nutrient or substance in a food and its relationship to a serious disease or a biomarker of a serious disease. For example, diets high in calcium may reduce the risk of osteoporosis in people 65 years and over. An example of a biomarker health claim is "Phytosterols may reduce blood cholesterol."

Food businesses wanting to make a general-level health claim will be able to base their claims on one of the more than 200 preapproved food–health relationships in the Standard or the food business may self-substantiate a food–health relationship in accordance with detailed requirements set out in the Standard. Self-substantiation of food–health relationships for general-level health claims must be notified to the FSANZ prior to making the claim on food labels or in advertisements for food.

High-level health claims must be based on a food–health relationship preapproved by FSANZ. There are currently 13 preapproved food–health relationships for high-level health claims listed in the Standard. All health claims are required to be supported by scientific evidence to the same degree of certainty, whether they are preapproved by FSANZ or self-substantiated by food businesses.

16.3 INTERNATIONAL COMPARISON OF SCIENTIFIC SUBSTANTIATION

Initially, scientific substantiation of the effectiveness and safety in human studies is essential. The guidelines for use of nutrition and health claims in the Codex Alimentarius recommended that "health claims must be based on current relevant scientific substantiation and the level of proof must be sufficient to substantiate the type of claimed effect as recognized by generally accepted scientific review of the data, and the scientific substantiation should be reviewed as new knowledge becomes available."[1]

The guidelines for use of nutrition and health claims in the Codex Alimentarius proposed that "Health claims should primarily be based on evidence provided by well-designed human intervention studies."[15] The EFSA published the guidance which announced that human data are central for the substantiation of the claim.[11] The FDA announced that the "gold" standard is DB-RCT design. In the FOSHU approval system, an applicant is required to substantiate the health claim by the statistical analysis of the human intervention studies. These are based on the same concept that human intervention study with high quality is the prime evidence needed to scientifically substantiate claims, supported by animal and *in vitro* studies.

16.4 INTERNATIONAL COMPARISON OF REGULATORY SYSTEMS

16.4.1 General Concept

Initially, scientific substantiation of the effectiveness including human studies is essential. In 2008, the draft guidelines in Codex recommended the following criteria[15]:

- Health claims should primarily be based on the evidence provided by well-designed human intervention studies. Human observational studies are not generally sufficient per se to substantiate a health claim, but where relevant, they may contribute to the total evidence. The data from animal, *ex vivo*, or *in vitro* studies may be provided in support of the relationship between a food or a food constituent and its health claim, but they cannot be considered alone as sufficient per se to substantiate any type of health claim.
- The total evidence, including unpublished data where appropriate, should be identified and reviewed, including the evidence to support the claimed effect, evidence that contradicts the claimed effect, and evidence that is ambiguous or unclear.
- Evidence based on human studies should demonstrate a consistent association between the food or the food constituent and the health claim, with little or no evidence to the contrary.

These statements seem to be in general agreement with the global concepts concerning the approval of health claims. In the FOSHU approval system, an applicant is required to substantiate the health claim by the statistical analysis of the clinical studies, animal studies, and *in vitro* metabolism and biochemical studies. These requirements are similar concept to the Codex guideline.

The secondary importance is the totality of evidence. FOSHU documentation requires that not only the experimental data but also all the available scientific evidence including published scientific literature should be reviewed systematically. As for the way of confirming the benefit and safety of a product in a human study, any existence of undesirable or adverse effects should be additionally examined. Consumers can be confident in the truthfulness of the information they find in claims when a third party has assessed the scientific substantiation behind the claim before the product enters the market.

The White Paper of the European Union declared in 2000[16] that "Consumers have the right to expect information on food quality and constituents that is helpful and clearly presented, so that informed choices can be made." Confirmation of the scientific substantiation of a product's efficacy, any adverse effects, and a view of the totality of evidence can allow consumers to make an "informed choice." Informed choice is also the general concept for FOSHU. The notification of the MHLW declares that consumers should have enough information on the health benefits of products to permit them to choose the products by themselves. Claims and labeling are the most important ways to inform consumers. The results of the clinical studies as well as the animal studies should be published in a journal that is reviewed by expert referees. Health-related claims cannot help consumers make "informed choices" until they have been substantiated scientifically, reflecting the totality of the evidence.

16.4.2 Regulation of Health Claims

16.4.2.1 Classification of Health Claims

Health claims can be classified into three major categories as follows:

1. *Nutrient function claims*: The scientific concepts of Codex defines claims that describe the physiological role of the nutrient in growth, development, and normal body function.[1] The nutrients include three major nutrients, vitamins, and minerals. In the EU regulation, most well-established claims (Article 13.1) are those of nutrients. The Japanese FNFC belongs to this type of claim.
2. *Other function claims or structure/function claims*: Most structure/function claims in the DSHEA are involved in other function claims defined by the Codex.[1] The innovative function claims (Article 13.5) in the EU regulation and the Japanese FOSHU are similar to the category of the other function claims.
3. *Disease risk reduction claims*: The Codex proposed disease risk reduction claims in 2002. There is the same claim in the EU regulation (Article 14). The MHLW added the disease risk reduction claim in FOSHU in 2005. The relationship between disease and food or its substances in the NLEA in the United States is a similar concept to the disease risk reduction claim.

16.4.2.2 Regulatory System

At present, the systems for regulating the health claims of foods could be classified into the following three types:

1. *The standard regulation system*: In the case of the general health claims, which could be standardized by expert authorities, no specific substantiation is required to an individual product before entering into the market because a general claim is based on well-established and generally accepted knowledge based on evidence.
2. *The notification system*: Under the DSHEA in the United States, dietary supplement manufacturers can sell their product with any structure/function claims just by notifying their claims to the FDA, which are not approved but also evaluated by the FDA.
3. *The individual approval system*: Innovative health claims other than general health claims should be evaluated product by product by independent experts. Claims of FOSHU belong to innovative health claims rather than general health claims. Thus, the Japanese FOSHU approval is based on this type of evaluation system.

Health claims can be classified in comparison with international discussion (Table 16.2). Most nutrient function claims could be well-established claims. Most function claims other than nutrient function claims could be innovative claims for the first step of the product development. These innovative claims should be evaluated and approved by the authorities such as the FOSHU system or the EU system. Well-established claims such as nutrient function claims based on well-established and generally accepted knowledge could be standardized by expert authorities without the need for specific substantiation regarding an individual product.

TABLE 16.2
International Comparison of Regulation

	Standard Regulation System	Individual Approval System	Notification System
Nutrient function claim	EU(13.1) FNFC	EU(13.1)	
Other function claim (structure/function)	EU(13.5) S-FOSHU	EU(13.5) FOSHU	US(DSHEA)
Disease risk reduction claim	US(NLEA) EU(14)	DRR-FOSHU	

EU(13.1), Article 13.1 health claims of Regulation (EC) No. 1924/2006; EU(13.5), Article 13.5 health claims of Regulation (EC) No. 1924/2006; DRR-FOSHU, disease risk reduction FOSHU; EU(15), Article 14 health claims of Regulation (EC) No. 1924/2006; FNFC, Food with Nutrition Functional Claim; S-FOSHU, Standardized FOSHU; US(DSHEA), DSHEA in the United States; US(NLEA), NLEA in the United States.

16.5 CONCLUSION

The Japanese scientific project began to substantiate scientific evidence regarding health-related effects of foods on the human body in the early 1980s and defined foods with such effects as "functional food." In 1991, the MHLW established FOSHU as a regulatory system to approve the statements included on the labels of functional foods. For FOSHU approval, not only the experimental data but also all the available scientific evidence including published scientific literature must be reviewed systematically. This is a concept similar to the scientific substantiation and the totality of available scientific evidence in the general guidance on the evaluation of health claims published by the EFSA in 2011.[11]

Such substantiation should be based on experimental studies in human subjects as well as in animals and *in vitro*. The labeling of health-related foods should be based on scientific evidence, which is desired to aiming the international standards. The nutrient function claim was included in the Codex Alimentarius guidelines.[1] Claims of the FNFC in Japan are equivalent to the Codex Alimentarius nutrient function claims. Most statements of the Japanese FOSHU have similarity to the category of the innovative function claims of the European Union or the structure/function claims in the United States. It is desirable that the national governments and the international organizations make efforts to harmonize the international standards regarding a regulatory system of health-related claims for governing the promotion of research, production, and marketing of foods that will support the quality of life for individuals in the coming senior society.

REFERENCES

1. Codex Committee. 2004. Revised. Guidelines for use of nutrition and health claims. http://std.gdciq.gov.cn/gssw/JiShuFaGui/CAC/CXG_023e.pdf.
2. Grossklaus, R. 2009. Codex recommendations on the scientific basis of health claims. *European Journal of Nutrition*, 48(Suppl 1): S15–S22. doi:10.1007/s00394-009-0077-z.
3. Consumer Affairs Agency. 2010. Regulatory systems of health claims in Japan. http://www.caa.go.jp/en/pdf/syokuhin338.pdf.
4. Shimizu, T. 2003. Health claims on functional foods: The Japanese regulations and an international comparison. *Nutrition Research Reviews*, 16: 241–252.
5. Shimizu, T. 2011. Health claims and scientific substantiation of functional foods—Japanese regulatory system and the international comparison. *European Food and Feed Law Review*, 3(6): 144–152.
6. Consumer Affairs Agency. 2010. Nutrition labelling systems in Japan. http://www.caa.go.jp/en/pdf/syokuhin569.pdf.

7. Nutritional Labeling and Education Act (NLEA) Requirements. http://www.fda.gov/iceci/inspections/inspectionguides/ucm074948.htm.
8. US Food and Drug Administration. 2000. Regulation on statements made for dietary supplements concerning the effect of the product on the structure or function of the body. *Federal Register*, 65: 999–1050.
9. US Food and Drug Administration. 2008. Guidance for industry: Substantiation for dietary supplement claims made under Federal FDC Act. http://www.cfsan.fda.gov/~dms/dsclmgu2.html.
10. European Parliament and the Council. 2006. Regulation (EC) No 1924/2006 on nutrition and health claims made on foods. http://eur-lex.europa.eu/LexUriServ/LexUriServ.do?uri=OJ:L:2006:404:0009:0025:EN:PDF.
11. European Food Safety Authority. 2011. General guidance for stakeholders on the evaluation of Article 13.1, 13.5 and 14 health claims. *EFSA Journal*, 9(4): 2135. http://www.efsa.europa.eu/en/efsajournal/doc/2135.pdf.
12. European Commission. 2012. EU Register on nutrition and health claims. http://ec.europa.eu/nuhclaims/resources/docs/euregister.pdf.
13. Wu, J. 2005. Guidance to health food in China. In: *Regulatory System of Control of Chinese Health Foods*. Tokyo, Japan: Nikkei Business Publications, pp. 65–85.
14. Australia New Zealand Food Standards. 2013. Nutrition, health and related claims. http://www.foodstandards.gov.au/consumerinformation/nutritionhealthandrelatedclaims/.
15. Guidelines for Use of Nutrition and Health Claims. 2013. http://www.codexalimentarius.org/standards/list-of-standards/.
16. European Commission. 2000. White Paper on food safety. http://ec.europa.eu/dgs/health_consumer/library/pub/pub06_en.pdf.

17 In Pursuit of Claims
What Works, What Does Not, and Why

Josephine M. Balzac and George A. Burdock

CONTENTS

17.1 INTRODUCTION

The objective of this chapter is to describe the primary challenges associated with identifying, accumulating, and presenting evidence in support of presumptive claim(s). The claims made, however, are sometimes at odds with the views and standards held by the US regulatory agencies and may thus be judged not to be "truthful and not misleading," the standard to which claims are held. The end result is often a challenge to the regulatory agency decision, whether through a meeting with the agency, through an administrative procedure (including, but not limited to a hearing before an administrative law judge), or in a federal court. The issue then becomes one of demonstrating that the claim is adequately supported by sufficient evidence according to the standard cited above—a question that often comes down to reasonableness. What then qualifies as admissible evidence and how should it be presented? Gone are the days of the testimony of flamboyant expert witnesses, whose credentials, personalities, and powerful presentations sway juries, which often resulted in the other side producing equally impressive experts for the traditional in-court "battle of the experts." The court case referred to as "Daubert" [1] put an end to these dramas, mandating that the court use a gatekeeper method to assess the degree of relevance, rigor, and validity of what would be admissible. However, in a classic example of the conflict between theory and practice, the gatekeepers (i.e., the judges) often lack the scientific background required to make the often fine distinctions between "good" and "bad" science. Thus, neither side in our adversarial system has provided much insight into what is required to win a case—good rhetoric or good science? This chapter favors the burden of proof over the burden of persuasion—it is a study of what has worked and what has not worked in terms of admissibility and viability in claims support in and out of the courtroom.

The current industry trend of advertising the beneficial effects of food products is continuing at a high rate as a result of consumers' desire for foods that confer something more than traditional nutritive, satietal, or hedonistic value. Consumers are also becoming more health conscious and are assuming more responsibility for their own well-being through the consumption of higher quality functional foods.* Perhaps intuitively, consumers are following Hippocrates' 2500-year-old dictum "let thy food be thy medicine, thy medicine be thy food." New scientific advances link diet to disease and disease prevention, which creates an opportunity to address public health concerns through food. Some foods are termed "functional" because they provide specific health benefits beyond basic nutrition. By demonstrating physiological benefits, functional foods can often reduce the risk of chronic disease, manage chronic disease, and promote growth and development.

While at the outset of any product development, there are two primary hurdles for new functional ingredients—demonstration of safety and efficacy; these are two distinct concepts and should not be confused. The safety in use of each individual ingredient must either receive premarket approval from the Food and Drug Administration (FDA) or be generally recognized as safe (GRAS), the latter through a history of safe use or by scientific procedures [2]. For efficacy, to make a claim, there must be scientific evidence of the claimed benefit. The required scientific evidence for both hurdles (safety and efficacy) is dictated by regulations, statutes, case law, and evidence rules.

Science and law are very different but need each other when arriving at a rational conclusion for cause and effect to convincingly demonstrate that the results of the study substantiate the claim. The reliability and relevance of scientific evidence is defined in the court decisions referred to as *Daubert*, *Khumo Tires*, and *Joiner*, and is pertinent to enter through the gates of regulatory agencies and the courts. In order to present the data, there must first be an

* Although the US Food and Drug Administration (FDA) has declined to define "functional food" (see also later in this chapter), the Canadian Health Protection Agency recognizes the term and defines functional food as "Ordinary food that has components or ingredients added to give it a specific medical or physiological benefit, other than a purely nutritional effect." http://www.hc-sc.gc.ca/sr-sr/biotech/about-apropos/gloss-eng.php#f (last visited September 29, 2013).

understanding of what type of evidence is required. Additionally, each study or test needs to be well designed and the results must tip the scales in favor of proving the claim. The relevance, rigor, and validity of the evidence to the claim are essential in determining whether the claim is substantiated.

The "gold standard" for demonstrating efficacy is through the application of double-blind, placebo controlled, randomized clinical trials (RCTs). However, for reasons to be discussed later in this chapter, RCTs can be difficult to utilize in the field of nutrition because the necessary parameters and appropriate controls can be difficult to find and even for ethical reasons. The results of the clinical studies are then left to judges or regulatory agencies (the former of whom often lack scientific expertise) to serve as the gatekeepers on deciding whether the evidence is "in" or "out," and in the event the contemplated trial is a challenge to an agency decision, the judge is left to make all the gatekeeper decisions. Consumers' skepticism exists in the accuracy of valuable effects and the food industry is under strict scrutiny by government agencies, courtrooms, and consumers; thus, this chapter was written to level the playing field between the participants involved.

17.2 CLAIMS

17.2.1 Five Types of Claims Allowed

In modern food regulation, there are basically five types of claims (Table 17.1). These claims will be discussed in the following subsections. These claim types are not mutually exclusive and some overlap exists between them because the claim types were conceived independently and over a considerable period of time to meet specific needs mandated by Congress, the courts or required for implementation. For example, structure/function claims (SFCs) were meant to be exclusively for dietary supplements, but were expanded for foods with the FDA *Federal Register* [3] notice that an SFC could only be made for food *if* the "active" provided nutritive value. Likewise, while health claims and qualified health claims (QHCs) were meant for food ingredients, there is a provision in the regulations allowing for a dietary supplement ingredient to carry a health claim on the label, although that ingredient must still meet the regulatory threshold for addition to food (i.e., GRAS or via a food additive petition [FAP]). So then, as described in this subsection, there are nearly as many exceptions as "rules" and, as time passes, there will undoubtedly be more exceptions to address changing circumstances.

It is useful to first review the history of claims to understand the origin of some of the thinking at the FDA (Table 17.2), that is, up until the late 1930s, mislabeling, deception, and outright fraud permeated claims made by the food industry. Much of the deception was based on egregiously false claims generally promoting continued health, strength, and vitality [4–6]. As a result of the poor self-governance within the food manufacturers' community at the time, the FDA adopted a very cynical viewpoint during this period and a skeptical mind-set that persists

TABLE 17.1
Five Types of Claims Allowed for Food and/or Dietary Supplements

1. Nutrient content claims (and FDAMA claims)
2. Dietary guidance claims
3. Health claims (HCs)
4. Qualified health claims (QHCs)
5. Structure/function claims (SFCs)

FDAMA, Food and Drug Administration Modernization Act.

TABLE 17.2
Evolution of Claims

Pre-1940	The age of snake oil, mislabeling, economic fraud, and deception
1941 *et seq.*	The FDA reclassification of vitamins in excess of RDA as drugs
1971	Proxmire amendment limiting the FDA's power to reclassify vitamins to drugs and allowed limited claims for special dietary needs.
1971	MOU between the FTC and the FDA—FTC limits its venue to advertising and leaves labels to the FDA. The FTC standard of "competent and reliable scientific evidence" is adopted by both agencies.
1972	The FTC's decision in the *Pfizer* case established the *reasonable basis* requirements for substantiation of claims.
1984	Kellogg's All-Bran cereal claim to reduce cancer risks
1990	The NLEA permits health claims for food—regulation finalized in 1993
1994	Dietary Supplement Health and Education Act
1997	FDAMA—Congressional attempt to liberalize claims by allowing "authoritative statements"
1999	FDA: (1) SSA status must be agreed to by the FDA and (2) substance–disease relationship imposed rather than substance–claim relationship as originally proposed by Congress.
2003	Court forces the FDA to liberalize claims; the FDA responds with a new category, qualified health claims.
2007	The FDA launches investigation into validity of supporting data on previously agreed upon and supposedly irrefutable SSA for health claims.

FDA, Food and Drug Administration; FDAMA, Food and Drug Administration Modernization Act; FTC, Federal Trade Commission; MOU, Memorandum of Understanding; NLEA, Nutrition Labeling and Education Act; RDA, Recommended Daily Allowance; SSA, significant scientific agreement.

at least somewhat to this day. This skeptical mind-set led to some unnecessary conflict in the late 1930s and up into the first years of the next decade, when there was much press about the miracle of the newly discovered vitamins and recommended daily allowances (RDA) that were set. Seeing these RDAs as a marketing opportunity, manufacturers produced mega-dose vitamins and hyped these products much like we see large doses of vitamin C promoted today. The FDA, now empowered by the 1938 Amendment to the Food Drug and Cosmetic Act (FDCA), took action by classifying any megavitamins (i.e., multiples of the daily recommend value) as drugs to get them off the market.

The struggle about what could be placed on the market was a seesaw between Congress and the FDA until 1971, when Senator Proxmire* of Wisconsin was able to limit the FDA's power to reclassify vitamins as drugs and also allowed limited claims for special dietary needs. This amendment was the beginning of what was termed the "diabetic aisle" in the US grocery stores which has since expanded into a wide selection of specialized foods. In this same year (1971), the Federal Trade Commission (FTC) and the FDA entered into a memorandum of understanding (MOU) by which the FDA would limit its enforcement to label claims and the FTC would police the media. It was also at this time that the FDA agreed to the FTC standard for analysis of a claim—that a claim should stand up to the scrutiny of "competent and reliable scientific evidence."† Another banner event was the 1984 Kellogg's All-Bran cereal claim that the fiber in the cereal could reduce the risk of cancer. It was several years before this claim was accepted by the FDA and had the effect of initiating a series of new laws starting with the Nutrition Labeling and Education Act of 1990, also referred to as the NLEA.

* Edward William Proxmire (1915–2005) served as a US senator (D) from Wisconsin (1957–1989). Senator Proxmire was a frequent critic of the FDA and notably was the father of the monthly "Golden Fleece Award" for flagrant government-related spending.

† More on this standard later.

The NLEA permitted health claims for food. The FDA was not pleased with the NLEA, as the agency had been very comfortable with the position that it could not approve a health claim for a food—that any such claim would make the food a drug. Regardless, Congress passed the law in response to public pressure, but the FDA, not wanting the law in the first place, was very slow to implement it and did not finalize the regulation until three years later. At the same time, the FDA was still antagonistic to the dietary supplement industry, which unfortunately (for the FDA) had powerful friends on the "Hill." The final straw in the tug of war was when the FDA called all supplements "unapproved food additives," much as the agency had called mega-dose vitamins "drugs" more than 50 years before. The upshot was the passage of the Dietary Supplement Health and Education Act (DSHEA), signed into law by President Clinton and becoming effective on October 15, 1994. The DSHEA, the acronym by which it is identified, specifically noted that dietary supplements were foods and not food additives.

In 1997, Congress passed the FDA Modernization Act (FDAMA), part of which was to liberalize the strict interpretation the FDA had made of the NLEA allowing claims. One of the provisions was to allow manufacturers to use "authoritative statements" as a basis for claims rather than petition FDA for use of the claim. We shall see later that the FDA interpreted this provision to allow only for nutrient guidelines and that any use of "authoritative statements" for health claims would only be a secondary consideration by the FDA.

Perhaps at this point, it would be prudent to define some terms in the regulatory lexicon; that is, inasmuch as diligent marketing and sales personnel recognize the need to help laymen better understand regulatory designations, there are others who might want to obfuscate the status of a substance. Perhaps in consideration of the latter, the regulator has only a finite set of terms with which to work, and as *Procrustean* as this may seem, the wisdom of getting everyone on the same page is intuitively obvious (Table 17.3).

It is important to use the right vocabulary when dealing with a regulatory agency. Terms such as nutraceutical or even functional food have no standing in law or regulation and are looked upon

TABLE 17.3
Learning the Lexicon—"Regulatory" versus "Fanciful"[a] Terms

Regulatory Terms	Fanciful Terms
Drug	Nutraceutical
Food ingredient (FA or GRAS[b])	Aquaceutical
Food for special dietary use	Herbal supplement
Medical food	Natural food
Nutrient supplement[c]	Cosmeceutical
Dietary supplement (ingredient)	Functional food
Cosmetic Claims	
Nutrient claim	
Dietary guidance claim	
Health claim (SSA)	
Qualified health claim	
Structure/function claim	

[a] "... an arbitrary or fanciful name which is not false or misleading in any particular" [21 CFR 133.148(e)(2)].
[b] [FFDCA §201(s)].
[c] 21 CFR 170.3(o)(20) [substances] "... that are necessary for the body's nutritional and metabolic processes."
FA, food additive; GRAS, generally recognized as safe; SSA, significant scientific agreement.

as marketing terms. The official lexicon includes the terms drug, food, food additive or GRAS, and dietary supplement* (ingredient). Regarding claims, the five listed in Tables 17.1 and 17.3 are official names, and the first four, for foods or food ingredients, must also be supported by a FAP or GRAS determination. SFCs can be made for either a food ingredient or a dietary supplement ingredient, but as a dietary supplement ingredient, a FAP or GRAS determination is not necessary. Instead, for a dietary ingredient, it would have to be the subject of a New Dietary Ingredient Notification (NDIN) to determine its safety, if it was not already grandfathered.† If a claim is to be made for a dietary supplement, the FDA must be notified of the claim within 30 days of commencement of marketing. Health claims or QHCs must be approved by the FDA before the claim is made and the product marketed.

17.2.2 Nutrient and Dietary Guidance Claims

Before proceeding on to health claims, QHCs, and SFCs, it would be prudent to understand nutrient and dietary guidance claims to forestall incipient confusion. As part of the NLEA of 1990, standard serving sizes of foods and nutrients therein were imposed on manufacturers, giving rise to the familiar "Nutrition Facts" panel on each packaged food we see in stores today [7]. Nutrient content claims on the label are largely informational in nature and are required. Simply stated, regulations require that certain things be listed on the labels in the Nutrition Facts panel including, but not limited to, the total fat (as both saturated and *trans* fat), protein, carbohydrates, and so on, and for other nutritive substances, the percent of nutrient present in a serving addresses that found in an established daily requirement. As manufacturers responded to consumer demand for low fat, fewer calories, and lower sodium, certain modifier phrases were allowed so consumers could respond to a standardized language—such as "free," "no," or "a trivial source of [a nutrient]." There was also a provision to allow for education of consumers that foods such as broccoli were fat free or celery has few calories. Nutrient content claims also allow the manufacturer to point out the benefits of a nutrient, such as "choline functions as a precursor for acetylcholine" and "choline phospholipids act as a methyl donor," and to characterize the level of substance present (in this case, choline) such as "good source of choline" and "rich in choline."

Congressional passage of the FDAMA in 1997 required the FDA to recognize authoritative statements. While this authoritative statement mandate was meant to apply to health claims emanating from the FDA as well as the private bodies analogous to GRAS expert panels, the FDA applied its interpretation to the law by stipulating that the nutrient content claim must be based on an *authoritative statement* from an appropriate scientific body of the US government, the National Academy of Sciences (NAS), or any of the subdivisions of the former that identifies the nutrient to which the claim refers. The claimant must undergo a premarket notification procedure, and if the FDA does not act to prohibit or modify such a claim, the claim may be used 120 days after FDA receipt of the notification. The FDA called these claims "FDAMA claims" (Table 17.4).

The action by the FDA to insist that FDAMA claims must emanate from sources FDA designates as being credible, has a chilling effect on private, academic, or industry nutritional research—that is, the requirement that an authoritative body may only be a government or quasi-government entity engenders considerable bureaucratic inertia, making a new nutrient content claim very difficult.‡ Although there is a regulation [8] allowing for petition for a nutrient content claim, the discovery of a hitherto unknown nutrient is unlikely because of the variety of foods available to all individuals

* Although the FDCA refers to "dietary supplement," the FDA has reinterpreted this term (i.e., dietary supplement) as meaning the final product in commerce; however, the "active" as the "dietary ingredient." That is, the end product is a dietary supplement, which contains one or more dietary ingredients (i.e., actives) and excipients such as the gelatin capsule, binders, colors, tableting agents (e.g., magnesium stearate), and preservatives.

† That is, in use prior to October 15, 1994, as a dietary supplement.

‡ The easily proven nutrient deficiencies have already been identified, such as vitamin C and scurvy, by the Royal Navy in the eighteenth century.

TABLE 17.4
FDAMA Claims (Examples)

- Drinking fluoridated water may reduce the risk of dental caries or tooth decay.
- Diets containing foods that are good sources of potassium and low in sodium may reduce the risk of high blood pressure and stroke.
- Diets low in saturated fat and cholesterol, and as low as possible in *trans* fat, may reduce the risk of heart disease.

Source: All claims are based on authoritative statements. http://www.fda.gov/food/ingredientspackaginglabeling/labelingnutrition/ucm2006874.htm (last visited September 25, 2013).

in the United States and the absence of isolated populations (such as the need for iodine resulting in the development of goiters in land-locked populations with no access to sea food). Further, would it even be possible on humanitarian grounds to conduct an interventional study for a nutrient deficiency? The investigators would likely have to rely on observational studies (including epidemiological studies), which, as we shall see later, the FDA feels are not adequate.

A second type of claim is called "dietary guidance" and, like an FDAMA nutrient claim, must pattern itself after a statement made by an authoritative body. Typical statements include "carrots are good for your health" and "diets rich in fruits and vegetables may reduce the risk* of some types of cancer and other chronic diseases." Typically, dietary guidance statements make reference to a category of foods, rather than a specific substance. Dietary guidance statements tend to focus on general dietary patterns, practices, and recommendations that promote health. The difference between health claims and dietary guidance claims is that health claims describe the relationship between a substance (specific food or food component) and a disease or health-related condition, although dietary guidance statements do not contain both elements [9]. As with FDAMA claims, dietary guidance claims must be made in the context of a determination by an authoritative body (e.g., the National Academy of Sciences for a Recommended Daily Allowance [RDA]) and can be made without FDA review or authorization before use, and, of course, any statements must be truthful and notmisleading.

17.2.3 Health Claims, QHCs, and SFCs

Before getting into the details of the three types of claims—health claims, QHCs, and SFCs—Table 17.5 dissects out the most salient points comparing the claim types to one another. As the reader will see later, there is very little difference between the criteria for a health claim and a QHC. Both require a linkage between a substance and a disease, and both must reduce the incidence (i.e., risk) of a disease. The primary difference between a health claim and a QHC lies in the fact that health claims are awarded a regulation, but QHCs are given only a letter of enforcement discretion. This lesser threshold for QHC is the result of FDA deciding that a qualified claim is based on what FDA calls "emerging evidence," and because it is emerging evidence, there can hardly be significant scientific agreement (SSA). Probably the biggest impediment to the success of qualified claims is the actual wording of the claim required by the FDA. These claims contain

* The phrase "reduce the risk" [of a disease] is a device to mollify critics that might otherwise see the use of a food as a drug claim; that is, in §201(g)(1) of the Food Drug and Cosmetic Act, a drug is defined as follows: (g)(1) The term "drug" means ... (B) articles intended for use in the diagnosis, cure, mitigation, treatment, or prevention of disease in man or other animals; and (C) articles (other than food) intended to affect the structure or any function of the body of man or other animals ...

TABLE 17.5
Overview of Claims

Health Claims	Qualified Health Claims	Structure/Function Claims
Valid linkage of substance and disease	Valid linkage of substance and disease	Maintains structures or functions of a healthy body
Reduces incidence of disease	Reduces incidence of disease	Maintains healthy body
FDA decision and may include SSA	FDA decision alone—emerging evidence	Competent scientists
FDA premarket approval required	FDA premarket approval required	No requirement for premarket approval
Regulation promulgated	Enforcement discretion	N/A
Difficult, time and cost intensive	Difficult, time and cost intensive	Less difficult than health or qualified health claims?
Not widely accepted	Qualifier refers to quality	Wide acceptance

N/A, not available.

so much guarded, qualifying wording, consumers have indicated in the FDA's own surveys that the guarded wording is though to refer to the quality of the product and not the rigor of the evidence supporting the claim; understandably, QHCs have had only marginal success.

SFCs cannot make reference to a disease but can only refer to maintenance or well-being of a body, such as "calcium builds strong bones" as opposed to a reference to osteoporosis. The level of evidence for an SFC (for the FDA) is the "competent scientist," just as it is with the FTC. No premarket approval is required and there is wide acceptance of these claims. Having now a grasp of the highlights, the history and details as to how a claim can be made are given in Subsections 17.2.3.1 and 17.2.3.2.

17.2.3.1 Health Claims and QHCs

As an integral part of the landmark Nutritional Labeling and Education Act (NLEA) of 1990, health claims were permitted for foods, which caused no end of angst within the FDA. It required the FDA to make a 180-degree turn from the FDA mind-set of protecting the public from quackery by prohibiting any claim on a food, to acting as judge in the area of efficacy that was classically deficient in good studies. Heretofore, rejecting a claim for a food embodied only the simple act of declaring the claim to be a drug claim, even though in the definition of a drug [FDCA §201(g)(1)(C)] food was clearly exempted, that is, "articles (other than food) intended to affect the structure or any function of the body of man or other animals." Further, the NLEA erected a significant barrier, forestalling the maneuver by the FDA to impute drug status by adding the following language:

> A food or dietary supplement for which a claim, subject to … is made in accordance with the requirements of … is not a drug solely because the label or the labeling contains such a claim. A food, dietary ingredient, or dietary supplement for which a truthful and not misleading statement is made in accordance with section 343(r)(6) of this title is not a drug under clause (C) solely because the label or the labeling contains such a statement.

Therefore, with this addition, the law was very clear that the presence of a claim on a label did not make the product a drug. Unfortunately, the FDA saw the NLEA as an erosion of its authority and set the stage for a struggle that continues today.

The FDA struggled for several years to adopt the congressional mandate of the NLEA that the FDA authorize a health claim in the labeling of conventional foods only if the agency "… determines, based on the totality of publicly available scientific evidence (including evidence from well-designed studies conducted in a manner which is consistent with generally recognized scientific procedures and principles), that there is significant scientific agreement, among experts qualified by scientific training and experience to evaluate such claims, that the claim is supported by such evidence"[10]. Clearly, this mandate paralleled §201(s) of the FDCA, which allowed experts (government or nongovernment) to determine the safety of food ingredients. However, the FDA was unable to provide a consistent response to petitioners regarding the qualifications for these experts called for in the NLEA. Finally, under a mandate from the court, the FDA issued a guidance document which advised that "Significant scientific agreement meant that 'the validity of the relationship is not likely to be reversed by new and evolving science, although the exact nature of the relationship may need to be refined.'" However, this guidance was withdrawn in 2009 and replaced with another guidance [11] which commented as follows on SSA:

> The SSA standard is intended to be a strong standard that provides a high level of confidence in the validity of the substance/disease relationship. SSA means that the validity of the relationship is not likely to be reversed by new and evolving science, although the exact nature of the relationship may need to be refined. SSA does not require a consensus based on unanimous and incontrovertible scientific opinion. SSA occurs well after the stage of emerging science, where data and information permit an inference, but before the point of unanimous agreement within the relevant scientific community that the inference is valid.

Further:

> In determining whether there is significant scientific agreement, FDA takes into account the viewpoints of qualified experts outside the agency, if evaluations by such experts have been conducted and are publicly available. For example, FDA intends to take into account
>
> - Documentation of the opinion of an "expert panel" that is specifically convened for this purpose by a credible, independent body;
> - The opinion or recommendation of a federal government scientific body such as the National Institutes of Health (NIH) or the Centers for Disease Control and Prevention (CDC); or the National Academy of Sciences (NAS);
> - The opinion of an independent, expert body such as the Committee on Nutrition of the American Academy of Pediatrics (AAP), the American Heart Association (AHA), American Cancer Society (ACS), or task forces or other groups assembled by the National Institutes of Health (NIH);
> - Review publications that critically summarize data and information in the secondary scientific literature.

The FDA accords the greatest weight to the conclusions of federal government scientific bodies, especially when the evidence for the validity of a substance/disease relationship has been judged by such a body to be sufficient to justify dietary recommendations to the public. When the validity of a substance/disease relationship is supported by the conclusions of federal government scientific bodies, the FDA typically finds that the SSA exists.

Therefore, while the intent of Congress was to allow review by any *bona fide* experts in the area, the FDA has reinterpreted this to mean government experts largely of FDA's own choosing.

17.2.3.2 Structure/Function Claims

The reader should be familiar with SFCs [12] to the extent that these claims are limited to the role of the substance to affect the structure or function of a body or provide for general well-being, that the statement must be "truthful and not misleading," and there must be a disclaimer on the label: "This statement has not been evaluated by the Food and Drug Administration. This product is not intended to diagnose, treat, cure, or prevent any disease."

There are three basic types of SFCs, the first of which is a nutrient deficiency claim (Table 17.6). This type of claim addresses the virtue of a substance addressing a deficiency. There are few of

TABLE 17.6
Structure/Function Claims

Statement claims may

1. Describe a benefit related to a classical nutrient deficiency disease, but must also disclose the prevalence of such disease in the United States.
   For example, for the claim "prevents scurvy," the label must disclose incidence of scurvy in the United States.
2. Describe the role of a nutrient or dietary ingredient intended to affect the structure or function in humans.
   For example, "builds strong bones" not "prevents osteoporosis."
3. Describe general well-being from consumption of a nutrient or dietary ingredient.
   For example, "maintains joint health" not "prevents or relieves arthritis."

these claims, primarily because there are few nutrient deficiencies in the United States and this sort of claim requires a disclosure of the prevalence of the deficiency in the United States. For example, a claim extolling the virtue of vitamin C to prevent scurvy would require a disclosure of the incidence of scurvy in the United States; however, while about 14% of men and 10% of women are deficient in vitamin C, clinically verifiable cases of scurvy are exceedingly rare [13], probably on the order of a thousandth of a percent.

The second type of SFC regards the ability of the substance to affect the structure or function of the body, such as calcium, which could be something on the order of the classical "calcium builds strong bones," but could not be "calcium prevents osteoporosis" because using the word "prevents" makes this a drug claim. If the claim were "calcium decreases the incidence of osteoporosis," this would be a health claim.

The third type of claim is a claim describing general well-being. The claim "chondroitin maintains joint health" would be appropriate, but if chondroitin was advertised to "prevent or relieve arthritis," this would be a drug claim; likewise, "reduces incidence of arthritis" would be a health claim.

An SFC does not require premarket approval as does a health claim or a QHC. However, the claim must be submitted to the FDA within 30 days of use. Because the agency that lobbied so hard against allowing claims (in this context, SFC) but has been mandated to administer them, the criteria for a compliant claim has not been clearly elucidated. There remain too many uncertainties in the system. These uncertainties include (1) no prior approval (or acknowledgment of the claim) which sometimes leads to retroactive judgments, (2) multiple agency review, and (3) claim wording that is sometimes Byzantine and plays on the uncertainty of the consumer. To date there have been 487 warning letters issued for dietary supplements, including over 100 whose key words include "false," "misleading," or "misbranded" [14].

17.3 SCIENCE AND LAW

17.3.1 Introduction

In today's society, both courts and regulatory agencies rely heavily on science to resolve important legal issues. In essence, law needs science and they both intersect during the process of determining a compliant claim for a food. It is not possible to have one without the other, and the success of a product can only be achieved when both are fully understood. The law dictates the framework of the scientific evidence that is necessary to prove the benefit embodied in the claim. The science governs whether or not the claim is substantiated. However, science and law differ in both the language they use and the objectives they seek to accomplish [15].

Courts and regulatory agencies resolve the issues of cause and effect or a substance/disease relationship under very different legal regimes [16]. In the judicial context, the plaintiff must satisfy the burden of proof on general causation and the jury makes the ultimate decision. In the regulatory world, the agency must determine, under the authority established by Congress, whether there is sufficient evidence to ensure efficacy of the product such that the labeling (i.e., the claim) is "not false or misleading in any particular" [16]. The ultimate decision maker in a regulatory agency is either a single administrator or a multimember board [16]. However, the FDA often outsources its decision making, especially for health claims, to outside contractors, including but not limited to the Agency for Healthcare Research and Quality (AHRQ). Therefore, while courts often defer initial scientific relevance/validity decisions to agency administrators, these agency judgments have often come under challenge, as will be discussed in Subsections 17.3.2, 17.3.3, and 17.3.5.

The objective of the law is justice and that of science is truth [15,17]. In seeking justice, you can find truth, "yet there are important differences between the quest for truth in the courtroom and the quest for truth in the laboratory. Scientific conclusions are subject to perpetual revision. Law, however, must resolve disputes finally and quickly" [1]. In law, "evidence" is defined by rules that govern what is admissible and what is not [15]. In science, evidence is provided in terms of the probability* of an event, not the certainty the law would most enjoy [15]. Notwithstanding the foregoing, science and law share the same goals and many of the same methods [15]. Both disciplines use evidence to arrive at a rational conclusion [15]. Scientific research and data is the foundation of cause and effect. The design of a study determines the quality of the data and whether or not you can prove your claim. In the end, the results of the tests either prove or disprove the hypothesis. There must be the appropriate type of scientific evidence of sufficient quality, and it must be relevant and reliable.

In both agencies and courts, the "weight of the evidence" approach is frequently used to arrive at a conclusion about the strength of the cause–effect relationship [16]. The "weight of the evidence" has to do with the "quality of the scientific studies, the strength of the cause-effect associations, the overall consistency of the scientific studies, and the biological plausibility of statistical observations" [16]. Therefore, industry needs to affirm that the scientific research it relies on is not just "internally valid, but also relevant to the product being promoted and specific to the benefit being advertised" [18].

17.3.2 The Gatekeepers

Court decisions and regulatory agencies, through their nonbinding guidelines and previous decisions on approvals, provide the core context of most probable requirements for identifying the necessary scientific evidence to enter through the gates. Either the judges or the courts designees, including the administrators at agencies (when the agency is not a party to a case), serve as the gatekeepers and make the ultimate determination on whether or not the evidence meets the requirements to cross the threshold. In the final analysis, it is the evidence admitted by the gatekeeper, which will determine the universe of information allowed for consideration as the sole basis to determine whether a relationship exists between the substance and the claim. While regulatory agencies give primary emphasis to clinical studies (and only to certain types of clinical studies), the agencies give little weight to animal studies, *in vitro* studies, or potential mechanisms (unless these data controvert the clinical data). The gatekeepers, however, may find that nonclinical studies or observational studies (not considered by the FDA) are integral to interpretation of the data, especially in the absence of a rigorous RCT, which is not always possible to produce. Therefore, based on the review of the gatekeeper and the court, a determination is made as to which evidence is "in" or "out."

The Federal Judicial Center and National Research Council of the National Academies created the third edition of the *Reference Manual on Scientific Evidence* (Manual) [19]. The Manual serves

* The standard most often used is that which has a 95% probability of occurring, commonly expressed as $p < .05$ (i.e., 1.0 = certainty). Higher probabilities (i.e., a greater degree of probability that the finding is likely to occur) are sometimes seen with $p < .01$, $p < .001$, and so on.

as a guide to federal judges while exercising their gatekeeping function in assessing the evidence [19]. It is meant "to provide the tools for judges to manage cases involving complex scientific and technical evidence" [19].

Being a gatekeeper places judges and, sometimes, the agencies into increasingly difficult situations as the scientific evidence and testimony becomes more complex. Additionally, a new public awareness of the availability of scientific evidence leaves juries and consumers with a higher expectation of the use of scientific evidence [19]. Trial judges and juries have little or no schooling or training in sciences and are left "as overseers of the complex merger of science and law" [19]. By contrast, regulatory agencies generally have well-credentialed scientists serving as gatekeepers of their own regulations to decide matters within their area of expertise. The FDA has staff, chemists, microbiologists, toxicologists, food technologists, pathologists, molecular biologists, pharmacologists, nutritionists, epidemiologists, mathematicians, sanitarians, physicians, and veterinarians [20]. However, in the adversarial system embodied in our courts and when the agencies are a party to an action, the court and its gatekeepers are left without an agency to rely upon for assistance, as it is the judgment of the agency that is being called into question.

17.3.3 The Federal Trade Commission

The FTC was originally created in 1914 to prevent unfair methods of competition in commerce; however, over the years, Congress gave the agency greater authority by passing additional laws allowing them to police anticompetitive practices [20]. Part of the FTC's work is performed by the Bureau of Consumer Protection (BCP), whose "mandate is to protect consumers against unfair, deceptive or fraudulent practices" [21]. Section 5(a)(1) provides that "unfair methods of competition in or affecting commerce, and unfair or deceptive acts or practices in or affecting commerce, are hereby declared unlawful" [22].

Most deception cases occur when the consumer is injured by being induced to purchase a product they would not have otherwise selected [23]. The FTC requires a showing of three elements in a deception case: (1) There must be a representation, practice, or omission likely to mislead the consumer; (2) the consumer must be interpreting the message reasonably under the circumstances; and (3) the misleading effects must be "material," which is likely to affect consumer's conduct or decision with regard to a product [23]. Importantly, the deception analysis does not require that actual injury be shown, but instead focuses on the risk of consumer harm [23].

The FTC mandate that all advertisement is "truthful and not misleading" necessitates that each claim be substantiated with "competent and reliable scientific evidence." As noted in Subsection 17.2.1, in a 1971 MOU between the FDA and the FTC, the scientific standard of "competent and reliable scientific evidence" for foods was established. In 2008, the FDA issued guidance which publicly announced that the FDA would follow the FTC in requiring the same standard as its substantiation standard for dietary supplements [24].

17.3.4 The Food and Drug Administration

The Department of Health and Human Services [25] (HHS) is the United States' principal agency for protecting the health of all Americans [25]. The FDA is one of the many agencies within the HHS and is "responsible for protecting the public health by assuring the safety, efficacy, and security of human and veterinary drugs, biological products, medical devices, our nation's food supply, cosmetics, and products that emit radiation" [26]. The FDA also has the duty to assist the public in obtaining the "accurate, science-based information they need to use medicines and foods to improve their health" [26]. There are six product-oriented centers, which carry out the mission of the FDA: One of them is the Center for Food Safety and Applied Nutrition (CFSAN), which ensures that the nation's food supply is "safe, sanitary, wholesome, and honestly labeled" [27]. The CFSAN's primary responsibilities, *inter alia*, include regulations and activities dealing with the safety and labeling of foods (which would include claims) and, in addition, industry outreach and consumer education [27].

17.3.5 Regulatory Challenges with Scientific Evidence

The FDA and the FTC share jurisdiction and operate under the MOU cited in Subsections 17.2.1 and 17.3.3. The FTC handles the truth and falsity of all advertising other than labels, and the FDA has primary responsibility over labels (or any product claims within a certain distance from the product, which is considered part of the label). The FTC protects America's consumers [28]. However, the current regulatory regime and strong oversight present unnecessary barriers to the development and marketing of functional foods and novel products.

Recent FTC decisions requested FDA preapproval before making claims, basically deferring its own standards to the FDA. This choice of activity constitutes an unusual discretion on the part of the FTC, especially considering the fact that the FTC Act only requires that claims be truthful and not misleading, not that they meet FDA requirements. However, the FTC–FDA interaction regarding the standard of "competent and reliable scientific evidence" is becoming more prevalent. For example, the recent FTC POM Wonderful complaint* requiring FDA preapproval for future disease efficacy claims was claimed to be necessary due to the complexity of the scientific issues, the expertise of the FDA to assess these type of claims, and the FTC's interest in harmonizing with the FDA.† In the Final Order issued by the Commission, it stated that:

> Although the Commission does not enforce FDA requirements, we note at the outset that our sister federal agency, the Food and Drug Administration, promulgates and enforces regulations regarding investigational new drug approvals, and that those regulations require multiple phases of clinical trials that collectively represent different—and considerably greater—substantiation than the RCTs required here. We note too, that FDA regulations separately require FDA approval of health claims made on behalf of food products, and that approval of such claims requires the submission of well-designed scientific evidence.‡

The difficulties with these decisions is that the FTC has not codified what it constitutes "competent and reliable scientific evidence," but as these recent decisions show, the FTC may have to begin to move in this direction and create a bright-line rule on the level of substantiation needed for health claims. Otherwise, the FTC will continue to apply the substantiation standards in an *ad hoc* and potentially arbitrary fashion, leaving companies questioning the required level of substantiation needed to prove that their claims are true.

On an operational level, both the FTC and the FDA discuss challenges with inadequate resources. However, the FDA often outsources most of the investigatory work on research-supporting claims via HHS to the Agency for Healthcare Research and Quality (AHRQ), which employs subcontractors (often academics) traditionally opposed to industry initiatives [4]. Further, similarly anti-industry (i.e., self-nominated agenda-driven "consumer interest") Non-Profit Organizations and perhaps to some extent even a competing pharmaceutical industry will constantly demand ever higher thresholds for proof of efficacy pricing these (often) nonpatentable products out of the market. These artificially high thresholds for proof of efficacy have the net effect of stifling innovation and infringing on manufacturers' First Amendment freedom of speech [29].§

---

* See further discussion in Subsection 17.5.3.

† In the matter of POM Wonderful LLC and Roll Global LLC, as successor in interest to Roll International Corporation, companies, and Stewart A. Resnick, Lynda Rae Resnick, and Matthew Tupper, individually and as officers of the companies; FTC File No. 082-3122, http://www.ftc.gov/os/adjpro/d9344/index.shtm.

‡ FTC Press Release, FTC commissioners uphold trial judge decision that POM Wonderful LLC; Stewart and Lynda Resnick; others deceptively advertised pomegranate products by making unsupported health claims, http://www.ftc.gov/opa/2013/01/pom.shtm (last accessed January 16, 2013).

§ See further discussion below.

17.4 CLINICAL STUDIES

17.4.1 Introduction

The dramatic increase in food science and technology is creating a silent revolution, but of as much or more importance, as the discovery of interindividual differences in humans. At the pinnacle of the discovery of these differences between humans is the Human Genome Project [30], which has led to the discovery of single nucleotide polymorphisms (SNPs)—small but subtle changes in human DNA that control various functions in the system for which it codes and explains much of what has been found through epidemiology studies and empirical studies in the clinic. The net result is that "[R]ecent scientific advances have further blurred the line between food and medicine, as scientists identify bioactive food components that can reduce the risk of chronic disease, improve quality of life, and promote proper growth and development" [31]. Functional foods, supplements, and other ingredients can provide many benefits that are typically associated with drugs as they have physiologic benefit (although they are not drugs per se) because they are not "intended for use in the diagnosis, cure, mitigation, treatment, or prevention of disease in man or other animals" [32]. However, in the federal government's zeal to protect the consumer against fraudulent or deceptive practices, federal agencies are requiring more data and suggesting the types of clinical studies and end points more associated with drug testing; that is, regulatory agencies are engaged in a steady elevation of the threshold of the scientific evidence required. In discussing evidence that shows the efficacy of a substance in the food context, the focus is on the scientific research and the results of clinical studies.

17.4.2 Specific Study Types

Confusion abounded within the industry for several years following passage of the NLEA and the FDAMA, as little guidance had been produced by the FDA. Finally, in July 2007, the FDA issued a draft "Guidance for Industry: Evidence-Based Review System for the Scientific Evaluation of Health Claims," which was followed by a finalized version in 2009 [11]. The guidance divided the world into two basic types of studies: (1) interventional and (2) observational. The former (interventional) was deemed the preferred type of study because it provided the strongest evidence or a relationship between a substance and a disease.*

When causation is at issue, there are three major types of information: anecdotal evidence, observational studies, and controlled experiments [15]. Anecdotal reports and observational studies are not as reliable and secure in establishing causation. Anecdotal reports are useful as a stimulus to inquire further but not at establishing causation [15]. Additionally, observational studies can establish association but need further work to bridge the gap from association to causation [15].

Controlled experiments are ideal for establishing causation. A double-blind, placebo controlled, randomized clinical trial (RCT) is considered the gold standard for most efficacy experiments and is known as the "true experiment" [15]. The RCT type of study design is often "the best way to ensure that any observed difference between the two groups in outcome is likely to be the result of exposure to the agent" [15].

Randomization minimizes the likelihood that there are differences in relevant characteristics between those exposed to the agent and those not exposed. Researchers conducting clinical trials attempt to use study designs that are placebo controlled (which means that the group not receiving the agent or treatment is given a placebo), and that use double blinding (which means that neither the

* Note that this relationship is at variance with congressional desire for a relationship between a substance and an effect, which would have the net effect of allowing substantially more claims by lowering the bar for cause and effect. That is, if under the congressional mandate, the requirement would be only to show that a substance affected a biomarker (such as blood lipids or inflammatory markers), without showing an underlying relationship to a disease. However, in a recent speech by Dr. Fabricant of the FDA, he noted that a claim of lowering inflammatory markers in the blood was a drug claim even though an inflammatory marker is only a symptom and not a disease per se.

participants nor those conducting the study know which group is receiving the agent or treatment and which group is given the placebo) [15].

Furthermore, randomization increases reliability and decreases bias. It ensures "that the assignment of subjects to treatment and control groups is free from conscious or unconscious manipulation by investigators or subjects." It may not be the only way to ensure protection, but "it is the simplest and best understood way to certify that one has done so" [15,33].

However, the RCT methodology is less utilized in the field of nutrition science because the necessary parameters or appropriate controls may be difficult to find [34]. Similarly, researchers testing probiotics, such as most trials testing foods, have difficulty finding an appropriate control, and therefore, RCTs may not be the gold standard for some studies [34].

Yet, the RCT is the study required by the agencies because they are not acknowledging the fact that RCTs may not be possible or even the most applicable study in the food context—the quest for the "true experiment" is a case of the quest for perfection often defeating the pragmatic and useful—an abandonment of the FDA philosophy of years' past.

Observational studies, however, are of a less desirable type because the FDA holds that these studies cannot provide convincing evidence of cause and effect. According to the guidelines, with interventional studies, variability is minimized, making this type the best; for example, (1) the quantity and quality of the test substance is better controlled, (2) the test subjects are assigned randomly with no difference between intervention and control groups, and (3) potential bias can be reduced by double blinding. Interventional studies may also employ parallel or crossover designs. The FDA recognizes the biggest problem, even with this gold standard study, which is one of "generalizability"; that is, can the effect seen in the study population be extrapolated or "generalizable" to the US population as a whole or even a slightly different demographic?

The FDA's key points mitigating the validity of an observational study for SSA is that (1) observational studies lack the controlled setting of intervention studies because it cannot be determined if the effect is real or coincidental and confounders of disease risk need to be adjusted to minimize bias (e.g., age and heart disease); (2) observational studies may be prospective or retrospective and, of the two types, retrospective studies are vulnerable to error because they depend on subject recall; and (3) observational studies (a) measure *associations* between the substance and the disease and (b) cannot establish cause and effect between an intervention and an outcome. However, observational studies may be used to corroborate the findings of an interventional study.

Of the types of observational studies, the guidance indicates that *cohort*[*] studies are the most reliable of the observational studies and can yield the relative estimates of risk. *Case–control studies*[†] are deemed less reliable than cohort studies, but allow a comparison of subjects with a disease to those without the disease and can yield an estimate of relative risk, the odds ratio.[‡] A *nested*

---

[*] A study design in which one or more samples (called cohorts) are followed prospectively and subsequent status evaluations with respect to a disease or outcome are conducted to determine which initial participants' exposure to characteristics (risk factors) are associated with it. As the study is conducted, the outcome from participants in each cohort is measured and the relationships with specific characteristics determined. http://www.gwumc.edu/library/tutorials/studydesign101/cohorts.html (last accessed October 3, 2013).

[†] A study that compares patients who have a disease or outcome of interest (cases) with those who do not have the disease or outcome (controls), and looks back retrospectively to compare how frequently the exposure to a risk factor is present in each group to determine the relationship between the risk factor and the disease.

Case–control studies are observational because no intervention is attempted and no attempt is made to alter the course of the disease. The goal is to retrospectively determine the exposure to the risk factor of interest from each of the two groups of individuals: cases and controls. These studies are designed to estimate the odds.

Case–control studies are also known as "retrospective studies" and "case-referent studies." http://www.gwumc.edu/library/tutorials/studydesign101/casecontrols.html (last accessed October 3, 2013).

[‡] An odds ratio (OR) is a measure of association between an exposure and an outcome. The OR represents the odds that an outcome will occur given a particular exposure, compared to the odds of the outcome occurring in the absence of that exposure. ORs are most commonly used in case–control studies; however, they can also be used in cross-sectional and cohort study designs as well (with some modifications and/or assumptions). http://www.ncbi.nlm.nih.gov/pmc/articles/PMC2938757/ (last accessed October 3, 2013).

case–control or *case–cohort study* can yield the odds ratio (an estimate of relative risk) but are deemed less reliable than cohort studies but more reliable than case–control studies. *Cross-sectional studies* collect information regarding food (substance) consumption at a single point in time among people with and without the disease and, although it cannot show cause and effect, it can measure survival of those with the disease rather than the risk of developing the disease, but is considered to be less reliable than cohort and case–control studies. *Ecological studies* can compare disease incidence across different populations, but is considered the least reliable type of observational study.

The FDA does not give much credence to review articles or meta-analysis as the original data are not available in the article or meta-analysis. In terms of animal and *in vitro* studies, these might supply corroborative or illustrative information on mechanisms and relationships, but the physiological differences between animals and humans preclude use according to the FDA.

17.4.3 Surrogate End Points

The lack of allowance for no more than a handful of surrogate end points or biomarkers has been a source of disappointment for investigators. First, because the FDAMA mandated that the relationship should be between a substance and a claim and change in a biomarker could be a relatively easy association to make. Second, the FDA allows only few biomarkers, although there are many valid end points that could be used, such as a reduction in plasma hyperhomocysteinemia (high homocysteine level in the blood, indicating a high risk for mortality) [35], but this is not adequate as it is not an accepted biomarker by the National Institutes of Health (NIH) or NAS. Of the few biomarkers used, low-density lipoprotein (LDL) cholesterol, total cholesterol, and blood pressure are approved for showing eventual cardiovascular disease (CVD). In addition, elevated blood sugar and insulin resistance are acceptable markers for type 2 diabetes. These examples are only two of the few permitted because they can be irrefutably connected (at least presently) to specific diseases. However, not all approved biomarkers can be used, even for a very similar claim.

17.4.4 Evaluating Human Studies

17.4.4.1 Evaluating Interventional Studies

In Table 17.7, the comments from the FDA guideline of 2009 have been summarized and a discussion of the points is presented.

Were the subjects healthy or did they have the disease relevant to the health claim? For example, this potential criticism could come into play with a health claim for glucosamine and chondroitin for reduced incidence of arthritis; that is, if some of the study subjects had minor symptoms or symptoms of beginning arthritis and the test substance relieved progression of the disease, this would not be "decreasing the incidence" of the disease, but treating a disease already present and not a basis for a health claim as defined by FDA. In addition, did the subjects have other diseases that might have masked or even enhanced the effect of the test substance? Was there an appropriate control group? This could be difficult when studying a food. For example, exclusion of consumption of a certain type of polysaccharide or mineral might be difficult, if the substance is readily present in

TABLE 17.7
Evaluating Human Interventional Studies

- Were the study subjects healthy?
- Appropriate control group?
- Independence of role of "substance"?
- Relevant baseline data (e.g., on end point)?
- Appropriate statistical analysis?
- What type of biomarker was measured?
- Length of study appropriate?
- Follow-up on dietary advice?
- Demographic of study group?

the environment. By contrast, if the test substance was unique to a readily identifiable food, such as spinach, it would be relatively easy to maintain a good control.

Independence of the role of a substance may be problematic, especially for some minerals, where many minerals may play a similar role, such as the divalent ions, calcium, magnesium, zinc, and manganese. Relevant baseline data are important, for example, was there a proper washout for subjects in a crossover study? The importance of an appropriate statistical analysis should be emphasized. As discussed in Subsection 17.4.3, using an unacceptable biomarker renders a study useless.

The length of the study should be appropriate such that the "reduced incidence of the disease" can be documented statistically, but this could make the study a very long one because the disease has to develop in the control group. The longer the study, the more dropouts occur and the more uncertainty is introduced.

The generalizability of the study is important. If the study was conducted on a specific demographic, such as elderly females and at risk for osteoporosis, can this be generalized to include younger females in the 25–30 year age group? In addition, there is an increasing trend for Japanese Foods for Specified Health Use (FOSHU) foods to be introduced into the United States. Can viability in a Japanese population be extrapolated to the United States, which has a population that is nearly 70% of European origin? In addition, because of study cost considerations, an increasing number of clinical studies are being conducted in China. In either instance, the FDA may well decide that neither studies performed solely with Japanese or Chinese subjects are reflective of the US population.

17.4.4.2 Evaluating Observational Studies

The guidelines have been very critical of human observational studies. They indicate that several questions will be raised in the evaluation of observational studies including, but not limited to, the ability of the study to evaluate the relationship between the disease and the food component; that is, a common weakness in observational studies is that quantification of exposure is difficult (dietary recall is inadequate and inconsistent amounts of the test substance in food). The FDA makes it very clear that observational studies can be proven wrong with an interventional study and cites the finding of β-carotene and the risk of lung cancer in smokers and asbestos workers (where β-carotene had a pro-oxidant, rather than an antioxidant effect).

17.4.4.3 Evaluating the Totality of Scientific Evidence

The guideline makes it very clear that the FDA regards interventional studies as superior to all others because interventional studies provide the strongest evidence of an effect, regardless of observational reports on the same subject—essentially, observational studies have little value in the presence of an interventional study. Further, interventional studies are designed to avoid selection bias and findings due to chance or other confounders of disease. Interventional studies cannot be ruled out by observational studies, but the reverse is not true. Finally, interventional studies trump all other studies.

17.5 COMPETENT AND RELIABLE SCIENTIFIC EVIDENCE

The FTC has been carefully evaluating the food industries' practices by focusing on the messages being transmitted in regard to the health benefit claims in advertisements. The more aggressive the health claim, the more scrutinized the claim will be and the more substantiation required. The FTC is given the authority to file a complaint under Section 5(b), "if it shall appear to the Commission that a proceeding by it in respect thereof would be to the *interest of the public*." [22]. Upon the Bureau of Consumer Protection's review of the advertisements of health-related claims, the FTC can exercise its enforcement discretion on claims that lack the appropriate scientific support. The current substantiation standard under the FTC requires "competent and reliable scientific evidence." However, the definition continues to be quite flexible and the claims are reviewed on a case-by-case basis, each ruling being specific to the facts of each case.

The FTC requires that claims be truthful and not misleading. If a claim specifies the level of substantiation it relies upon, that level of substantiation is required [18,36]. Substantiation of health benefit claims has been under scrutiny by the FTC since its 1972 decision in the *Pfizer* case, which established the *reasonable basis* requirements for substantiation [37]. If the claim does not specify the level of substantiation it relies upon, an advertiser must have a reasonable basis for its claim [18]. A reasonable basis is constituted by (1) the type of claim, (2) the type of product, (3) the consequences if the claim is false, (4) the benefits of a truthful claim, (5) the ease and cost of developing substantiation for the claims, and (6) the level of substantiation experts in the field agree is reasonable [18].

However, for claims relating to health and safety, the FTC defines the *reasonable basis* requirement as "competent and reliable scientific evidence" [38]. The FTC case law defines "competent and reliable scientific evidence" in only general terms such as "tests, analyses, research, studies, or other evidence based on the expertise of professionals in the relevant area, that have been conducted and evaluated in an objective manner by persons qualified to do so, using procedures generally accepted in the profession to yield accurate and reliable results" [38].*

The competent and reliable standard is made to be flexible, and each determination takes into consideration the "particular claims made and the products for which they are made" [18]. The Guide provides examples of questions that advertisers should ask before disseminating a claim:

1. How does the dosage and formulation of the advertised product compared to what was used in the study?
2. Does the advertised product contain additional ingredients that might alter the effect of the ingredient in the study?
3. Is the advertised product administered in the same manner as the ingredient used in the study?
4. Does the study population reflect the characteristics and lifestyle of the population targeted by the advertisement? [18]

17.5.1 Lane Labs

Recently, the flexible definition of "competent and reliable scientific evidence" returned to haunt the FTC in various cases, as it was presented with once again defining the level of substantiation required to adequately support health-related claims. The FTC brought an action against Lane Labs (Vero Beach, FL) alleging a violation of an earlier consent order [39] when it made claims about a calcium supplement product and a supplement intended to improve male fertility without "competent and reliable scientific evidence" [40]. In the original consent decree, the FTC did not define "competent and reliable scientific evidence" and therefore caused considerable debate of whether or not Lane Labs complied with the order. The district court denied the FTC's motion for contempt, finding that Lane Labs "provided credible medical testimony that the products in question are good products and could have the results advertised" [40]. The Third Circuit recently vacated the order of the district court in Lane Labs and remanded for further proceedings to find if the evidence obtained by Lane Labs was in substantial compliance with the consent decree's requirement of "competent and reliable scientific evidence" [41].

17.5.2 Iovate and Nestle

In Iovate Health Sciences Inc. (Oakville, ON) and Nestlé HealthCare Nutrition (Florham Park, NJ), the FTC appears to have modified its substantiation standard with regard to health-related claims and foods. These consent orders defined "competent and reliable scientific evidence," requiring that companies conduct two double-blind, placebo-controlled clinical studies on humans using the advertised

* See also Interstate Bakeries Corporation, Docket C-4042. http://www.ftc.gov/enforcement/cases-proceedings/012-3182/interstate-bakeries-corporation-matter; In the Matter of Ciba-Geigy Corp. and Ciba Self-Medication, Inc., corp, Docket No. 9279.

product or an "essentially equivalent" product to substantiate certain types of claims. The *Iovate* and *Nestlé* orders held that for certain claims, "competent and reliable scientific evidence" is defined as

> At least two adequate and well-controlled human clinical studies of the product, or of an essentially equivalent product, conducted by different researchers, independently of each other, that conform to acceptable designs and protocols and whose results, when considered in light of the entire body of relevant and reliable scientific evidence, are sufficient to substantiate that the representation is true [42,43].

The FTC defined essentially equivalent as

> [A] product that contains the identical ingredients, except for inactive ingredients (e.g., binders, colors, fillers, excipients), in the same form and dosage, and with the same route of administration (e.g., orally, sublingually), as the covered product; provided that the product may contain additional ingredients if reliable scientific evidence generally accepted by experts in the field demonstrates that the amount and combination of additional ingredients is unlikely to impede or inhibit the effectiveness of the ingredients in the Essentially Equivalent Product [42,43].

17.5.3 POM Wonderful

Perhaps, the FTC thought that it had created a *de facto* substantiation standard when it entered into consent decrees with Iovate and Nestlé. On September 27, 2010, in following these orders, the FTC filed an administrative complaint against POM Wonderful LLC (POM) charging POM with deceptive advertising by making claims that POM Wonderful 100% Pomegranate Juice and POMx supplements prevent or treat heart disease, prostate cancer, and erectile dysfunction [44,45]. The FTC asserted that these aggressive claims were not supported by sufficient scientific evidence [45,46].

On January 10, 2013, the FTC issued a final ruling that bars the POM marketers from making any claim that a food, drug, or dietary supplement is "effective in the diagnosis, cure, mitigation, treatment, or prevention of any disease," including heart disease, prostate cancer, and erectile dysfunction, unless the claim is supported by two randomized, well-controlled, human clinical trials [47]. The FTC specifically ordered that

> The RCTs must be randomized and well controlled based on valid end points and conducted by persons qualified by training and experience to conduct such studies, yielding statistically significant results and be double-blinded unless blinding cannot be effectively implemented. Such studies shall also yield statistically significant results, and shall be double-blinded unless Respondents can demonstrate that blinding cannot be effectively implemented given the nature of the intervention.*

The Final Order also prohibits misrepresentations regarding any test, study, or research, and requires "competent and reliable scientific evidence" to support claims about the "health benefits, performance, or efficacy" of any food, drug, or dietary supplement [48]. The POM case is significant because the FTC elevated its standard of substantiation by requiring two RCTs. It stated that it is "axiomatic that only RCTs can establish that a product is proven to treat, prevent or reduce the risk of a specific disease" [49]. The opinion suggests that two RCTs are now required to support any kind of disease prevention or treatment claims and that at least one such RCT is required for more general claims of healthfulness.

With the current state of uncertainty, advertisers making health claims for a food or dietary supplement should consider these FTC requirements and verify that studies are conducted on humans, utilize the same or an essentially equivalent product, and are valid in light of the entire body of relevant scientific evidence.

* POM Decision, January 10, 2013.

17.6 SIGNIFICANT SCIENTIFIC AGREEMENT

The FDA regulatory framework allows different type of health benefit claims, including nutrient content claims, SFCs, health claims, and QHCs. SFCs describe the effect that a substance has on the structure or function of the body and does not make reference to a disease. SFCs are not required to be preapproved or authorized by the FDA. They are required to be truthful and not misleading [50]. The FDA takes the position that it intends to apply the "competent and reliable scientific evidence" standard of substantiation consistent with the FTC [24].

Health claims can be identified into three categories: (1) claims supported by SSA and approved by the FDA, (2) QHCs containing disclaimers describing the state of emerging scientific data and approved by the FDA, and (3) claims approved by an authoritative statement by a US governmental scientific body and notified to the FDA at least 120 days prior to use. The FDA has authority under three sources to determine which health claims may be used on a label for a food or dietary supplement:

1. The 1990 Nutrition Labeling and Education Act (NLEA) provides for the FDA to issue regulations authorizing health claims for foods and dietary supplements after the FDA's careful review of the scientific evidence submitted in health claim petitions.
2. The 1997 Food and Drug Administration Modernization Act (FDAMA) provides for health claims based on an authoritative statement of a scientific body of the US government or the National Academy of Sciences; such claims may be used after submission of a health claim notification to the FDA.
3. The 2003 FDA Consumer Health Information for Better Nutrition Initiative provides for QHCs where the quality and strength of the scientific evidence falls below that required for the FDA to issue an authorizing regulation.

17.6.1 Health Claims

A "health claim" by definition has two essential components: (1) a substance (whether a food, food component, or dietary ingredient) and (2) a disease or health-related condition [51]. The FDA is permitted to uphold the SSA standard and promulgate regulations for health claims "based on the totality of publicly available scientific evidence (including evidence from well-designed studies conducted in a manner which is consistent with generally recognized scientific procedures and principles), that there is significant scientific agreement among experts qualified by scientific training and experience to evaluate such claims, that the claim is supported by such evidence" [52]. The FDA does not believe that the SSA standard is met when the findings are characterized as preliminary results, statements indicating that research is inconclusive, or statements intended to guide future research [53].

The FDA has approved 12 health claims [54]. These claims include calcium and osteoporosis, dietary fat and cancer, sodium and hypertension, dietary saturated fat and cholesterol and risk of coronary heart disease, fiber and cancer, fiber and coronary heart disease, fruits and vegetables and cancer, folate and neural tubes, plant sterol and coronary heart disease, and soy protein and coronary heart disease [55]. Each of these claims has required terms and the FDA has written out several model statements as a guide [55].

17.6.2 Health Claims from a Scientific Body

The FDAMA permits claims based on the current, published authoritative statements from "a scientific body of the United States with official responsibility for public health protection or research directly related to human nutrition ... or the National Academy of Sciences (NAS) or any of its subdivisions" [56]. The FDA considers the National Institute of Health (NIH) and the Centers for Disease Control and Prevention (CDC) and the Surgeon General within U.S Department of Health and Human Services (HHS); and the Food and Nutrition Service, the Food Safety and Inspection Service, and the Agricultural Research Service within the Department of Agriculture as scientific bodies that may be sources of authoritative statements [56].

The FDAMA defines an authoritative statement as follows: (1) "the relationship between a nutrient and a disease or health-related condition" for a health claim or "identifies the nutrient level to which the claim refers" for a nutrient content claim, (2) it is "published by the scientific body" (as identified in Subsections 17.2.2 and 17.6), (3) it is "currently in effect," and (4) it "shall not include a statement of an employee of the scientific body made in the individual capacity of the employee" [56]. In addition, the FDA currently believes that authoritative statements should also (5) reflect a consensus within the identified scientific body if published by a subdivision of one of the federal scientific bodies and (6) be based on a deliberative review by the scientific body of the scientific evidence [56].

17.6.3 QHCs and Free Speech

QHCs, unlike SSA health claims, are not defined or described in the FDCA or FDA regulations. These claims stemmed from the First Amendment cases in the federal courts which are now authority for the FDA [57]. In 2003, the FDA's Consumer Health Information for Better Nutrition Initiative provided for the use of QHCs only when there is emerging evidence for a relationship between a food, a food component, or a dietary supplement and the reduced risk of a disease or health-related condition [58]. The type of evidence presented in these claims is not well enough established to meet the SSA standard required [58]. Each claim contains qualifying language, which reflects the limited amount of evidence supporting the claim (Table 17.8) [58].

In the Pearson cases, dietary supplement makers challenged the FDA's decision to disallow certain health claims as a violation of the First Amendment. The specific claims that the FDA censored are "antioxidants/cancer," "fiber/colorectal cancer," "omega 3/coronary heart disease," and "folic acid/neural tube defects" [18,57]. In analyzing commercial speech, the courts use a four-part test set forth by the US Supreme Court in *Central Hudson Gas and Electric Corp. v. Public Service Commission of New York*. The "Central Hudson" test requires the court to address the following:

1. Whether the speech at issue concerns lawful activity and is not misleading;
2. Whether the asserted government interest is substantial; if so,
3. Whether the regulation directly advances the governmental interest asserted; and
4. Whether it is not more extensive than is necessary to serve that interest [29].

TABLE 17.8
Standardized Qualifying Language for Qualified Health Claims

Scientific Ranking	FDA Category	Examples of Appropriate Qualifying Language
Second level	B	"Although there is scientific evidence supporting the claim, the evidence is not conclusive."
Third level	C	"Some scientific evidence suggests... however, FDA has determined that this evidence is limited and not conclusive."
Fourth level	D	"Very limited and preliminary scientific research suggests... FDA concludes that there is little scientific evidence supporting this claim."

Source: FDA, http://www.fda.gov/Food/GuidanceRegulation/GuidanceDocumentsRegulatoryInformation/LabelingNutrition/ucm053832.htm.

In *Pearson I*, the court ruled that the Central Hudson factors had been satisfied except the fourth. The court determined that the health claims were not misleading, the FTC has a substantial government interest protecting against consumer fraud, and the SSA standard used by the FDA advances that government interest, however, it failed the last part of the test because the SSA standard was more extensive than necessary and therefore restricted commercial speech [57]. Pearson I signifies how the FDA may not ban health claims as misleading simply because scientific literature is inconclusive. The First Amendment requires the FDA to consider whether the use of disclaimers could cure any alleged deception [57]. The court concluded that the FDA could outright ban claims lacking SSA when (1) evidence in support of a claim is "qualitatively weaker" than evidence against a claim and (2) evidence in support of the claims is outweighed by evidence against the claim [57]. The burden is on the government to prove that even a disclaimer would fail to correct for deceptiveness.

Pearson II "distilled the two narrow circumstances suggested by *Pearson I* to mean simply that when 'credible evidence' supports a claim ... the claim may not be absolutely prohibited" [18,59]. In *Whitaker v. Thompson*, the court observed that "any complete ban on a claim would be approved only when ... there was almost no qualitative evidence in support of the claim and where government provided evidence proving that the public would still be deceived even if the claim was qualified by a disclaimer" [18,60]. In *Pearson II*, the court found that a claim for a key ingredient within a multivitamin "cannot be suppressed as inherently misleading when at least one credible study had shown the key ingredient to be effective" [18,59].

The regulations requiring a certain level of scientific evidence is a perfect example of the merging of science and law. However, the agencies must carefully evaluate their decisions on whether the claim is substantiated by the appropriate level of proof. This is true with the "competent and reliable scientific evidence" threshold and the SSA standard. An outright ban of SFCs or health claims could force the agencies to court as violating the First Amendment freedom of speech because it does not meet the standards discussed in this subsection.

17.7 EVIDENCE IN THE COURTS

"For almost seventy years, the admission of expert testimony in the vast majority of federal and state court was governed by the Court of Appeals of the District of Columbia's decision in *Frye v. United States*" [1,61]. In 1923, the Frye Court was faced with the appeal of a trial judge's decision to exclude the testimony of a scientist who performed the systolic blood pressure deception test, "crude precursor" to the polygraph, on behalf of a criminal defendant [62]. The court affirmed exclusion of this testimony and held that the standard to apply in determining the admissibility of expert testimony is whether the opinion is based on scientific technique that is "generally accepted" in the scientific community [62].

The "general acceptance" standard that emerged from the Frye decision was replaced by the majority rule decided by the US Supreme Court decision in *Daubert v. Merrell Dow Pharmaceuticals* [1]. The case began in the district court when two boys had been born with birth defects, and a lawsuit proceeded claiming that the drug Bendectin caused the birth defects [1]. Merrell Dow presented evidence from their expert that no published scientific study demonstrated a link between Bendectin and birth defects [1]. Daubert presented evidence from eight experts stating that Bendectin could cause malformations in human fetuses [1]. Daubert's experts based their opinions on review of unpublished reanalysis of the epidemiological studies, *in vitro* and *in vivo* animal studies, and chemical–structure analyses [1]. These unpublished reanalyses were ruled to be inadmissible because they had not been published or subject to peer review [1]. The Ninth Circuit Court of Appeals affirmed this decision and reasoned that "that reanalysis is generally accepted by the scientific community only when it is subjected to verification and scrutiny by others in the field" [1].

The plaintiffs appealed to the Supreme Court and successfully argued that Frye's "general acceptance" standard had been superseded by the enactment of the Federal Rules of Evidence (FRE) [1]. FRE 702 Rule provides that an expert may only testify if (1) the expert's scientific, technical, or other specialized knowledge will help the trier of fact to understand the evidence or to determine a fact in issue; (2) the testimony is based on sufficient facts or data; (3) the testimony is the product of reliable principles and methods; and (4) the expert has reliably applied the principles and methods to the facts of the case.*

The Supreme Court held that the FRE required the trial judge to "ensure that any and all scientific testimony or evidence admitted is not only relevant, but reliable" [1]. This requires the trial judge to determine at the outset ... whether the expert is proposing to testify to (1) scientific knowledge that (2) will assist the trier of fact to understand or determine a fact in issue. The four reliability factors described by the Daubert Court are (1) whether the methods used can and have been tested; (2) whether the technique has been peer reviewed or published; (3) the potential rate of error of a technique; and (4) whether the technique is generally accepted by the scientific community [1]. These factors are flexible and clearly require at least a basic understanding of the science being presented.

Daubert designated trial judges as gatekeepers who determine the admissibility of evidence. Their role is to inquire whether expert testimony is relevant and reliable. For the testimony to be relevant, the expert's statements must correspond to the facts of the case and "have a valid scientific connection to the pertinent inquiry as a precondition to admissibility" [19]. In verifying reliability, judges decide whether the testimony or evidence is derived from the scientific method [19]. The Supreme Court recognized that science is defined in terms of its method, quoting "Science is not an encyclopedic body of knowledge about the universe. Instead, it represents a *process* for proposing and refining theoretical explanations about the world that are subject to further testing and refinement."†

Daubert shaped the admissibility of scientific evidence. Its impact created two subsequent decisions establishing the *Daubert Trilogy*. The first of these decisions, *GE v. Joiner*, held that "the abuse of discretion standard applied to a trial judge's Daubert determination." This case further affirms the trial judge as the gatekeeper because the appellate court does not review the case *de novo*, or for the first time, it does not substitute its own judgment. In the second decision, *Kumho Tire Co. v. Carmichael*, the Supreme Court clarified that Daubert applies to all forms of expert testimony, stating that FRE 702 "applies its reliability standard to all 'scientific,' technical, and other specified matters within its scope."‡

17.8 THE FUTURE

At the moment, there is a great trend toward approval of functional foods, which are defined as those foods that contain some health-promoting components beyond traditional nutrients. The first example that comes to mind is dietary fiber and, while fiber has a health claim already, it would have been the first candidate of a food that fits the definition of a functional food simply *because* it has little or no traditional nutritive value and, in fact, is indigestible and not even absorbed, but brings great benefit. At present, the FDA recognizes that some substances are bioactive, but have no formalized claims; for example, probiotics or phytosterols (for blocking cholesterol absorption such as Promise® Margarine), Olestra® (which prevents absorption of the cooking oil), and medium-chain diglycerides (that are metabolized in the body as carbohydrates

* FRE 702.

† 509 U.S. 579, 590 (1993) (quoting brief for the American Association for the Advancement of Science and the National Academy of Sciences as Amici Curiae at 7–8), p. 77.

‡ Khumo, 526 U.S. 137 (1998).

TABLE 17.9
Resolution to the Impending Functional Foods Log Jam

Parameter	Health Claims	Functional Food	Dietary Supplement
Safety standard	"Reasonable certainty"	"Reasonable certainty"	"Reasonable expectation"
Addition to food	Yes	Yes	No
Type of claim	Reduction of risk of disease	Reduction in symptom, change in biomarker or clinical finding	Structure/function
Level of evidence supporting efficacy	Gold standard interventional studies	Observational studies, meta-analysis of studies, biomarker change, epidemiological studies, mechanistic evidence, smaller clinical trials	One or more persuasive studies related to maintenance of healthy function or structure
Deciding body	As currently exists (FDA and regulation promulgated)	Expert panel with notification to FDA (no regulation promulgated)	As currently exists
Public disclosure	Safety information, product identification and manufacturer	Safety information, product identification and manufacturer; efficacy information held in a confidential master file for at least five years.	As currently exists

and not fats). These are just a few of the new functional foods currently marketed which are unsuited for the cumbersome and uncertain claims regulatory structure for claims we have at present. The next 10 years will see a tsunami of functional foods with no mechanism in place for dealing with them effectively.

There are four things the FDA can do to streamline this cumbersome system now in place: (1) Create a new category of "Functional Foods": (2) allow the use of expert panel determinations for safety and claims, just as seen with GRAS determinations; (3) initiate an "official" notification system to ensure that consumers get the right information and are not misled by confusing labels; and (4) provide a term of exclusivity as is done for drugs to allow for a return on the money invested in the product development and testing of functional foods (Table 17.9).

In addition, the FDA needs to make some fundamental reassessments for moving SFC forward: (1) Disconnect the mandatory health–disease linkage, (2) allow claims based on changes in biomarkers, and (3) allow claims for specific population subsets.

17.9 CONCLUSION

This chapter has described the types of scientific evidence that is necessary to prove the efficacy of the claimed benefit of a functional food. Consumers are eagerly waiting for the next functional food that will be advertised as having health-related benefits. As has been shown, the required scientific evidence for each of the five claims is dictated by regulations, statutes, case laws, and evidence rules. Each of the claims requires essentially a different standard of evidence: (1) SFC requires "competent and reliable scientific evidence"; (2) health claim requires an SSA among experts; and (3) QHC requires emerging science of a relationship between a food and the reduced risk of a disease or health-related condition. The FDA or FTC ultimately determines whether the evidence provided meets each of the required thresholds. Following these agency decisions, the courts serve as gatekeepers who determine the only information allowed for consideration to determine whether a relationship exists between the substance and the claim. The judges must inquire whether the expert testimony relying on the scientific evidence is relevant and reliable, and whether it is in or out.

There are many obstacles with identifying, accumulating, and presenting the necessary evidence. A company must first understand the type of claim being made, the evidence required for each agency, and whether that evidence would pass muster to enter the gates in a courtroom. Each study or test must be well conducted and designed, utilizing the appropriate controls and parameters. The results must prove the claim.

REFERENCES

1. *William Daubert, et ux, etc., et al. v. Merrell Dow Pharmaceutical.* (1993). 509 U.S. 579, 585, 593, 594, 597.
2. Burdock, G.A. and Carabin, I.G. (2004). Generally recognized as safe (GRAS): History and description. *Toxicology Letters* 150(1):3–18.
3. Federal Register Notice, January 6, 2000 (65 FR 1000). http://www.gpo.gov/fdsys/pkg/FR-2000-01-06/pdf/FR-2000-01-06.pdf.
4. Burdock, G.A. (2013). Commentary: FDA must overcome skepticism toward health claims. *The Tan Sheet* 21:12–13, April 22.
5. U.S. Food and Drug Administration. “The American chamber of horrors,” http://www.fda.gov/AboutFDA/WhatWeDo/History/ProductRegulation/ucm132791.htm (last accessed May 25, 2013)
6. Kallet, A. and Schlink, F.J. (1933). *100,000,000 Guinea Pigs: Dangers in Everyday Foods, Drugs and Cosmetics*. The Vanguard Press, New York, 312 pp.
7. The Library of Congress. http://thomas.loc.gov/cgi-bin/bdquery/z?d101:HR03562:@@@D&summ2=3&|TOM:/bss/d101query.html (last accessed September 25, 2013).
8. 21CFR101.69.
9. 58 FR 2478 at 2487, January 6, 1993.
10. Section 403(r)(3)(B)(i) of the Act [21 U.S.C. 343(r)(3)(B)(i)].
11. Guidance for industry: Evidence-based review system for the scientific evaluation of health claims—Final, http://www.fda.gov/Food/GuidanceRegulation/GuidanceDocumentsRegulatoryInformation/LabelingNutrition/ucm073332.htm (last accessed October 4, 2013).
12. FDCA §403(r)(6).
13. amednews.com. “Scurvy rare, but cases still are popping up.” http://www.amednews.com/article/20080922/health/309229972/7/ (last accessed October 4, 2013).
14. U.S. Food and Drug Administration. http://www.accessdata.fda.gov/scripts/warningletters/wlSearchResult.cfm?qryStr=dietary%20supplement&webSearch=true&displayAll=true (last accessed October 4, 2013).
15. Federal Judicial Center. (2011). *Reference Manual on Scientific Evidence* (3rd edition), The National Academies Press, Washington, DC, pp. 51, 52, 101, 217, 556, 557.
16. McGarity, T.O. and Shapiro, S.O. (2013). Regulatory science in rulemaking and tort: Unifying the weight of the evidence approach, *Wake Forest Journal of Law and Policy*, 3(1): 101, 104, 105.
17. Bromley, D. A. (1998). Science and the Law, in *AAAS Science and Technology Policy Yearbook 1999*. Washington, DC: American Association for the Advancement of Science. (Based on remarks made August 2, 1998 during the 1998 Annual Meeting of the American Bar Association in Ontario, Canada.)
18. Villafranco, J. and Bond, K. (2009). Dietary supplement labeling and advertising claims: Are clinical studies on the full product required? *Food and Drug Law Journal*, 64(1): 43, 46, 57, 58.
19. Dutkiewicz, A. (2011). Book review: Reference manual on scientific evidence, *Thomas M. Cooley Law Review* 28: 343, 346.
20. FDA, Center for Food Safety and Applied Nutrition (CFSAN), http://www.fda.gov/AboutFDA/CentersOffices/OfficeofFoods/CFSAN/default.htm (last accessed September 22, 2013).
21. Federal Trade Commission 90th Anniversary Symposium (September 22–23, 2004), p. 6, http://www.ftc.gov/ftc/history/docs/90thAnniv_Program.pdf.
22. Unfair methods of competition, 15 U.S.C. § 45(a) (1) (2006); *supra* note 95.
23. *In re International Harvester Co., Federal Trade Commission*, 104 F.T.C. 949 (1984).“Actual deceptions of the public need not be shown in FTC proceedings; representations merely having a ‘capacity to deceive’ are unlawful.” Citing *Federal Trade Commission v. Algoma Lumber Co.*, 291 U.S. 67(1934).
24. FDA, Guidance for industry: Substantiation for dietary supplement claims made under Section 403(r)(6) of the Federal Food Drug and Cosmetic Act, December 2008, http://www.fda.gov/Food/GuidanceRegulation/GuidanceDocumentsRegulatoryInformation/DietarySupplements/ucm073200.htm.

25. US Department of Health and Human Services (USDHHS), http://www.hhs.gov/ (last accessed April 27, 2010).
26. About FDA, http://www.fda.gov/AboutFDA/CentersOffices/default.htm (last accessed April 28, 2010).
27. CFSAN—What We Do, http://www.fda.gov/AboutFDA/CentersOffices/CFSAN/WhatWeDo/default.htm (last accessed April 28, 2010).
28. About the Federal Trade Commission, http://www.ftc.gov/ftc/about.shtm (last accessed September 16, 2013).
29. *Central Hudson Gas & Electric Corp. v. Public Service Commission.* (1980). 447 U.S. 557.
30. The Human Genome Project. National Human Genome Research Institute. http://www.genome.gov/10001772 (last accessed October 3, 2013).
31. Institute of Food Technologists Expert Report, Functional foods: Opportunities and challenges, March 24, 2005.
32. Food, Drug, and Cosmetic Act §201(g)(1)(B).
33. Lavori, P.W., Louis, T.A., Bailar, J.C., and Polansky, M. (1983). Designs for experiments—Parallel comparisons of treatment. *The New England Journal of Medicine* 309(21):1291–1299.
34. Farnworth, E.R. (2008). The evidence to support health claims for probiotics. *Journal of Nutrition* 138(6):1212505–1212545. http://jn.nutrition.org/content/138/6/1250S.full.
35. *Circulation*. http://circ.ahajournals.org/content/101/13/1506.full (last accessed October 3, 2013).
36. *re Thompson Med. Co.* (1986). 791 F.2d 189, 194 (D.C. Cir.).
37. *Pfizer Inc*. (1972). 81 F.T.C. 23, 86.
38. *Novartis Corp.* (1999). 127 F.T.C. 580, 725; *supra* note 2.
39. Lane Labs-USA, Inc., No. 00-CV-3174, slip op. (D.N.J. June 29, 2000) (Stipulated Final Order for Permanent Injunction and Settlement of Claims for Monetary Relief), http://www.ftc.gov/os/caselist/9823558/lanelabsordandsettlement.pdf.
40. *FTC v. Lane Labs-USA, Inc.*, No. 00-CV-3174, 2009 WL 2496532, at 8 (D.N.J. August 11, 2009).
41. *FTC v. Lane Labs-USA, Inc.*, 624 F.3d 575, 577, No. 09-3909, 2010 WL 4226509 (3d Cir. October 26, 2010).
42. *FTC v. Iovate Health Sciences USA*, Inc., Case. No. 10-CV-587, slip op. at 7 (W.D.N.Y. July 29, 2010) (Stipulated Final Judgment), http://www.ftc.gov/os/caselist/0723187/100729iovatestip.pdf.
43. Nestlé HealthCare Nutrition, Inc., FTC File No. 092-3087, Agreement Containing Consent Order at 4 (July 14, 2010), http://www.ftc.gov/os/caselist/0923087/100714nestleorder.pdf.
44. FTC Press Release, FTC Complaint Charges Deceptive Advertising by POM Wonderful, http://www.ftc.gov/opa/2010/09/pom.shtm (last accessed September 27, 2010).
45. See Docket No. 9344, In the matter of POM Wonderful LLC and Roll Global LLC (hereinafter POM Wonderful); FTC File No. 082-3122, http://www.ftc.gov/os/adjpro/d9344/120521pomdecision.pdf.
46. FTC Press Release, http://www.ftc.gov/opa/2010/09/pom.shtm; POM Wonderful, http://www.ftc.gov/os/adjpro/d9344/120521pomdecision.pdf.
47. FTC Press Release, http://www.ftc.gov/opa/2013/01/pom.shtm.
48. FTC Press Release, http://www.ftc.gov/opa/2013/01/pom.shtm.
49. Riette van Laack, *FTC v. Pom Wonderful: The Battle Continues*, http://www.fdalawblog.net/fda_law_blog_hyman_phelps/2012/07/ftc-v-pom-wonderful-the-battle-continues.html (last accessed July 16, 2012).
50. 21 U.S.C. 343(r)(6); 21 CFR 101.93.
51. FDA, http://www.fda.gov/Food/GuidanceRegulation/GuidanceDocumentsRegulatoryInformation/LabelingNutrition/ucm064908. htm#health.
52. Section 403(r)(3)(B)(i); 21 C.F.R. § 101.14(c) (1998).
53. FDA, Guidance for Industry: Notification of a Health Claim or Nutrient Content Claim Based on an Authoritative Statement Authoritative Statement of a Scientific Body, http://www.fda.gov/Food/GuidanceRegulation/GuidanceDocumentsRegulatoryInformation/LabelingNutrition/ucm056975.htm (last accessed June 11, 1998).
54. 21 Code of Federal Regulations 101.72 through 101.83.
55. FDA, http://www.fda.gov/Food/GuidanceRegulation/GuidanceDocumentsRegulatoryInformation/LabelingNutrition/ucm064919.htm.
56. FDA, http://www.fda.gov/Food/GuidanceRegulation/GuidanceDocumentsRegulatoryInformation/LabelingNutrition/ucm056975.htm (last accessed October 5, 2013).
57. *Pearson v. Shalala* (*Pearson II*), 164 F.3d 650, 653, 659 (D.C. Circuit 1999).

58. FDA, Guidance for Industry: Interim Procedures for Qualified Health Claims in the Labeling of Conventional Human Food and Human Dietary Supplements, July 2013, http://www.fda.gov/Food/GuidanceRegulation/GuidanceDocumentsRegulatoryInformation/LabelingNutrition/ucm053832.htm (last accessed October 5, 2013).
59. *Pearson v. Shalala* (*Pearson II*), 130 F. Supp. 2d at 114.
60. *Whitaker v. Thompson.* (2002). 248 F. Supp. 2d 1, 10-11 (D.D.C.).
61. Green, E. and Nesson, C. (1983). *Problems, Cases, and Materials on Evidence*, Little Brown, Boston, MA, p. 649.
62. *Frye v. United States*, 293 F. 1013, 1014 (D.C. Cir. 1923).

18 Clinical Research—History of Regulations

A Global Perspective

Kappillil Anilkumar and K.I. Anitha

CONTENTS

18.1 INTRODUCTION

The role of clinical research in health care is significant. The present world, along with technological advancement, is witnessing a progressive trend in the field of therapeutic devices and drug inventions. The quest for more safer, affordable, and easily available pharmaceutical products is driving all countries to adopt new legal frameworks to regulate ethical clinical research. At times, the ever-elastic dynamics and overstretched interpretation of the word ethics make laws more lopsided than for which they were intended to be. The ethical prospects for any law already enacted are reflective of subject protection, confidentiality of data, informed consent, and right compensation for the victim for adverse events. At the same time, the authors feel that an equal amount of concern should be invested with the rights of the sponsors and investigators to keep the balancing act intact. In this chapter, history and evolution of ethical regulations is examined in a global perspective. Invention of a better cure for a disease is more complex, time consuming, and often entwined with continuous audits and overseeing at different levels of authority than invention of a computer software program. While international covenants guided the ethical principles, the respective governments enacted legislations regulating the clinical trials in their countries in compliance with the broader principles so that a particular drug, therapy, or device invented in a country could well be accepted by other countries. Biomedical research on animals and human subjects are carried on with stricter regulatory monitoring in all countries.

18.2 CLINICAL RESEARCH AND CLINICAL TRIAL

Clinical research is scientific expedition directly involving a particular person or a group of people, or using materials from humans, such as their behavior or samples of their tissue, whereas a clinical trial is one type of clinical research that follows a predefined plan or protocol. In clinical trials, participants can not only play a more active role in their own health care but also access to new therapeutic treatments by contributing to the research.[1] A clinical trial is an organized research study designed to investigate new methods of preventing, detecting, diagnosing, or treating an illness or disease and attempt to improve a patient's quality of life.[2] A clinical trial is any research study that prospectively assigns human participants or groups of humans to one or more health-related interventions to evaluate the effects on health outcomes.[3] Research conducted with human subjects (or on materials of human origin such as tissues, specimens, and cognitive phenomena) for which an investigator (or colleague) directly interacts with human subject could also be called clinical research. Clinical trials are sometimes referred to as clinical studies. While clinical trials technically refer only to those clinical studies involving drugs and other therapies aimed at slowing or stopping a disease, the terms are often used interchangeably.[4] The words "research" or "trial" impliedly convey an element of uncertainty in the end results because both are processes of testing the effectiveness of a drug, a device, or a therapy through experimentations. Various factors such as physiological, psychological, tropical, environmental, genetic, and geographical factors can play a very important role in proving a new drug or therapeutic device safe for human beings.

18.2.1 Process of Clinical Trials

Biomedical research can be classified as preclinical and clinical research. Preclinical research focuses on fundamental biological mechanisms in anatomy, biochemistry, cellular biology, immunology, pharmacology, physiology, and so on, and can contribute to the discovery of new medical treatments. However, clinical research involves clinical laboratory testing or investigational studies involving testing of new clinical procedures, new diagnostic tools, and new medicinal products in humans. The process of drug discovery starts from the idea for a new drug and synthesis and testing of new drugs with animal studies. If specific biological activity is found, compounds are made and then more testing is done to evaluate the compound to reach the project status. Thereafter, a drug development period is defined and permission is sought for plan for invention of new drug with the regulatory. After obtaining permission, clinical studies are planned and started. After the clinical trials, new drug applications are submitted to the regulatory or approval for launching the new drug, and on getting approval, the new drug is launched. After launching the new drug in the market, the postmarketing studies commence for finding new clinical usage and also for developing new usage forms and formulae. On an average, a time span of 20 years is required in this process of a new drug discovery and its successful introduction in the market.

18.2.2 Phases of Clinical Trials

Clinical trials are conducted in four phases as follows:

1. *Phase I (human pharmacology).* In this phase, the prime focus is on the safety and tolerability with the initial administration of Investigational New Drug (IND)–maximum tolerated dose (MTD), kinetics, and dynamics. It is primarily aimed at receiving information on acute tolerability and safety, dose–plasma concentration profiles, maximum safe doses and concentrations, routes of metabolism and elimination, and initial estimates of the variability associated with these measurements.
2. *Phase II (therapeutic exploratory trials).* In this phase, the effectiveness for a particular indication in a small group is in focus. The research in phase II emphasizes on establishing clinical efficacy and incidence of side effects in patient population, defines the most appropriate dose schedule, and provides the detailed pharmacological data for optimum use of the drug.
3. *Phase III (therapeutic confirmatory trials).* In this phase, the research is focused on the therapeutic benefit in large number of patients, and the investigation is to check whether the treatment is effective, to compare it with available and established treatment, and to determine the optimum dosage, the frequency of administration, the usefulness of drug in patients, the safety of treatment, and the common adverse reactions of the compound.
4. *Phase IV (postmarketing trials).* Phase IV is designed to reveal adverse reactions related to prolonged usage, drug efficacy in long-term use, new uses, an assessment of misuse or overuse liability, drug interactions, and compatibility with other agents.[5]

18.2.3 Risk Factor

Clinical testing of medicinal products is intended to test its safety and efficacy, and until confirmatory study results prove them safe and effective, the process of clinical research does not end. If the medicinal product is found to be ineffective or has side effects, the trials are terminated at the earliest. The risks are an invariable part of clinical trials of drugs. Apart from the chemical or medicinal properties in a drug, the targeted population of patients can also influence the risks in a drug. The risk aspects could possibly vary from a cancer drug to a drug for common cold. Similarly, the degree of risk can vary depending on the age profile of the patient, and the degree of harm is greater for participants in drug trials who need multidrug treatments.

18.2.4 Stakeholders in Clinical Research

The major stakeholders at the macro level in any clinical research activity is the larger public all over the world as invention of a new drug is beneficial to them. However, at the micro level, the stakeholders are as follows:

1. Sponsors
2. Subjects
3. Investigator team/sites

4. Contract research organizations (CROs)
5. Institutional ethics committees/institutional review boards
6. Regulatory bodies

18.2.5 Methods of Clinical Research

Researchers use various methods in clinical trials such as placebo, single-blind, and double-blind studies. An innocuous pharmacological agent administered to subjects in a controlled clinical trial group is called placebo. At times, an ineffective use of a dug or therapeutic intervention used in human subjects could result in beneficial outcome and this is called placebo effect. In a single-blind study, usually the subject is kept unaware of the interventions throughout the clinical trial, whereas in a double-blind trial, all the participants remain unaware of the intervention assignments throughout the trial.

18.3 HISTORICAL EVOLUTION OF MEDICAL RESEARCH

From the ancient texts of Indian Vedas[6]; the works of Charaka,[7] Susruta,[8] and Chanakya[9]; the ancient texts of Chinese[10] and Egyptian[11] medicines; and the Old Testament of the *Holy Bible*,[12] it could be evident that from time immemorial, man's quest for medicinal treatments for various illnesses grew along with ethical principles in experimentation in new medicines. The history of modern clinical research could be traced from Herophilos[13] (335–280 BCE), Avicenna[14] (980–1037), Edward Jenner[15] (1749–1823), and Semmelweis[16] (1848–1863) to the ongoing research for a permanent cure for human immunodeficiency virus (HIV) in the twenty-first century. While Herophilos introduced scientific inquiry in medicines, Avicenna transferred ethical experimentation into medicine and evolved the idea of a clinical trial,[17] and through the *Canon of Medicine*,[18] he strived to follow that the drug must be free from any extraneous accidental quality and used on a simple, not a composite, disease. Quite ostensibly, he urged that a drug must be tested with two contrary types of diseases, because sometimes a drug cures one disease and the cure of another could be accidental. Ambroise Paré[19] (1510–1590) conducted clinical research using admixture of turpentine, rose oil, and egg yolk to prevent infection of battlefield wounds and invented an alternative to cauterization for treating wounds. William Harvey[20] (1578–1657) defined the circulatory system. Sir Christopher Wren[21] (1632–1723) conducted intravenous injections (in dogs). Richard Lower and Edmund King[22] (1667) administered the first blood transfusion in man. Antonie Philips van Leeuwenhoek[23] (1632–1723) invented a microscope that discovered bacteria, free-living and parasitic microscopic protists, sperm cells, blood cells, microscopic nematodes and rotifers, and much more. John Hunter[24] (1728–1793), a Scottish anatomist and surgeon, described ligation of the femoral artery in the treatment of popliteal aneurysms. The first known controlled clinical trial was conducted by James Lind (1716–1794)[25] in 1747 for a cure for scurvy. John Snow[26] (1813–1858), a British anesthesiologist, in 1854 traced the connection between the quality of water and the cholera outbreak in Soho, England. Though in 1784, King Louis XVI of France appointed Benjamin Franklin to a royal commission[27] to examine the legitimacy of animal magnetism as a medical cure through a single–blind, placebo-controlled trial,[28] the use of placebos was introduced in 1863 by Austin Flint[29] (1812–1886). Edward Jenner's[30] (1749–1823) great gift to the world was his vaccination for smallpox. This disease was greatly feared at the time as it killed one in three of those who caught it and badly disfigured those survived. Ignaz Philipp Semmelweis[31] (1818–1865) researched on puerperal sepsis and his work *Etiology, Concept and Prophylaxis of Childbed Fever*, which was considered as one of the epoch-making books of medical history. Gregor Johann Mendel[32] (1822–1884) known as the father of genetics research conducted plant hybridization leading to a new theory of inheritance. Joseph Lister[33] (1827–1912) used carbolic acid sprayers in surgeries on the basis of his finding that antiseptics prevented postoperation sepsis. Louis Pasteur[34]

(1822–1895), a French scientist who invented vaccines for anthrax and rabies, discovered that *Staphylococcus aureus*[35] was the cause of boils and *Streptococcus pyogenes*[36] was the cause of puerperal sepsis.[37] From 1877 to 1887, Pasteur employed these fundamentals of microbiology in the battle against infectious diseases. He went on to discover three bacteria responsible for human illnesses: staphylococcus, streptococcus, and pneumococcus.[38] The first informed consent in clinical research was introduced by Walter Reed[39] (1851–1902) and the document was executed in Spanish for Antonio Benigno on November 26, 1900. Emil Adolf von Behring[40] (1854–1917) discovered antibiotics (diphtheria antitoxin) and Paul Ralph Ehrlich[41] (1854–1915) described eosinophils and invented arsenic for treatment of syphilis. Sir Austin Bradford Hill[42] (1897–1991) designed the technique of randomization in assigning the study subjects to the treatment groups, and it was at this time streptomycin[43] was discovered as a promising treatment for tuberculosis. The idea of randomization got introduced in clinical trials by 1923.[44] In 1930, Torald Sollmann[45] suggested a placebo control and blinded observation technique as an answer to investigator bias. Sir Alexander Fleming[46] (1881–1955), while working on influenza virus, observed mold on a staph culture plate with a bacteria-free circle around itself and this led to the discovery of penicillin. Robert Koch[47] (1843–1910) introduced Petri dish, use of blood sugar pour platelets to culture bacteria, and discovered that atoxyl is as effective against this disease as quinine is against malaria. Marie Curie's[48] (1867–1934) invention of radium proved that radioactivity is an intrinsic atomic property of matter and her invention pioneered a mobile X-ray unit for the French army in World War I. The concept of randomization and application of statistical methods in clinical research was introduced by Sir Ronald Aylmer Fisher[49] (1890–1962).

18.3.1 Ethics in Medical Research

The word *ethics* is reported to have originated from the middle English word *ethic*, from the old French word *ethique*, and the Greek word *ethika*, and the dictionary meaning[50] of the word is moral principles that govern a person's behavior or the conduction of an activity, and it is a branch of knowledge that deals with moral principles. Among the Western philosophers, the philosophy of ethics broadly belongs to three schools (Aristotelian, Kantian, and Utilitarian). Aristotelians believe that virtues such as justice, charity, and generosity constitute ethical conduct, whereas followers of Immanuel Kant insist that human as rational beings are bound to obey categorical imperative to respect other rational beings. The utilitarians assert that the guiding principle of conduct should be the greatest happiness or the benefit of the greatest number.[51] The Greek physician Hippocrates[52] (460–370 BCE) is considered as the father of medical ethics. He had established science of medicine and distanced it from theology and philosophy. *The Hippocratic Oath*[53] is perhaps the most widely known of Greek medical texts. It requires a new physician to swear upon a number of healing gods that he will uphold a number of professional ethical standards. One of the first cases dealing with medical malpractice law was that of *Slater v. Baker and Stapleton*[54] that was tried in England in 1767 wherein it was established that wherever there is a concept of a professional standard, physicians and surgeons were to be judged by.

18.3.2 The Elixir Sulfanilamide Tragedy

In 1937, Elixir sulfanilamide,[55] an improperly prepared sulfanilamide medicine, caused mass poisoning in the United States which prompted the authorities to enact the Food, Drug, and Cosmetic Act[56] in 1938. Prior to that, there were only the Pure Food and Drug Act of 1906[57] and the Harrison Narcotics Tax Act[58] of 1914 in the United States to ensure the safety of drugs. In a leading case *Webb v. United States*, 249, U.S.96 (1919),[59] the US Supreme Court ruled that the prescription of narcotics for maintenance treatment was not within the discretion of physicians, and thus not privileged under the Harrison Narcotics Act.

18.3.3 The Lubeck Disaster

Prior to World War II, Albert L. Neisser's[60] (remembered mainly as the discoverer of the etiologic agent of gonorrhea) study was held to be unethical medical research for using prostitutes and children with syphilis to test immunity by serum from syphilitic patients without consent in Germany in 1892. In 1930, the testing of the Bacillus Calmette–Guérin (BCG) vaccine for tuberculosis ended in what is popularly known as the Lubeck disaster in Northern Germany in which 73 deaths were reported out of 250 infants who were administered the vaccine[61] and drew the attention of the global community, which called for a stricter regulatory framework for ethical compliance in biomedical research on human beings.

18.3.4 The Nuremberg Code

The aftermath of World War II witnessed the growing concerns on the ethical principles for conduct of clinical research on human beings when Dr. Karl Bandt, a German doctor, and 23 doctors were accused of organizing and participating in war crimes and crimes against humanity in the form of medical experiments and medical procedures inflicted on prisoners and civilians. They were prosecuted in 1946–1947 and the specific crimes charged included more than 12 series of medical experiments concerning the effects of and treatments for high-altitude conditions; freezing; malaria; poison gas; sulfanilamide; bone, muscle, and nerve regeneration; bone transplantation; saltwater consumption; epidemic jaundice; sterilization; typhus; poisons; and incendiary bombs. These experiments were conducted on concentration camp inmates. Karl Brandt and six other defendants were convicted, sentenced to death, and executed; nine defendants were convicted and sentenced to terms in prison; and seven defendants were acquitted.[62] In April 1947, Dr. Leo Alexander had submitted to the Counsel for War Crimes six points defining the legitimate medical research. The trial verdict adopted these points and added an extra four. The 10 points constituted the "Nuremberg Code,"[63] which included such principles as informed consent and absence of coercion, properly formulated scientific experimentation, and beneficence toward experiment participants.

18.3.5 The Manhattan Project and Radiological Maltreatment

During World War II and the early Cold War era, under the Manhattan Project, the US officials conducted radiation experiments on hospital patients, pregnant women, retarded children, and enlisted military personnel; most of them had no knowledge that they were being subjected to radioactive materials. The study was intended to know the impact of health and safety of workers in atomic bomb plants, and the conduct of the study was shrouded in secrecy on the grounds of national security. Following congressional investigations, numerous official reports, scholarly studies, and lawsuits, the government in the 1990s offered apologies and financial compensation to some of the human victims of radiation testing.[64]

18.3.6 The Universal Declaration of Human Rights

The Universal Declaration of Human Rights[65] was adopted by the General Assembly of the United Nations in 1948. The General Assembly adopted in 1966 the International Covenant on Civil and Political Rights.[66] Article 7 of the Covenant states that "No one shall be subjected to torture or to cruel, inhuman or degrading treatment or punishment. In particular, no one shall be subjected without his free consent to medical or scientific experimentation." Based on this declaration, subsequent regulations were evolved for the protection of the rights and welfare of all human subjects of scientific experimentation.

18.3.7 The Thalidomide Infanticide and Kefauver Amendments

Another instance that necessitated a closer scrutiny of ethics in medical research was the thalidomide tragedy. In the late 1950s, thalidomide[67] was an approved sedative in Europe; however, it was *not* approved in the United States by the US Food and Drug Administration (FDA).[68] The drug that was prescribed to control sleep and nausea throughout pregnancy was soon found out to be the cause for 12,000 babies born with severe fetus deformities. Responding to the alarming situation, the US government in 1962 introduced the "Kefauver Amendments"[69] to the Food, Drug and Cosmetic Act to ensure drug efficacy and greater drug safety. Keauver Amendments made it mandatory for the drug manufacturers to prove the safety and efficacy of pharmaceutical products before marketing them for human consumption.

18.3.8 The Helsinki Declaration

In 1964, the World Health Organization (WHO) took initiatives for the protection of individuals in biomedical researches, and the World Medical Association declared mandatory principles regarding the investigator's responsibility and the written informed consent of the human subjects. The declaration known as the 1964 Declaration of Helsinki[70] governs international research ethics and defines rules for "research combined with clinical care" and "nontherapeutic research." It was clearly declared that "In the treatment of a patient, where proven prophylactic, diagnostic and therapeutic methods do not exist or have been ineffective, the physician, with informed consent from the patient, must be free to use unproven or new prophylactic, diagnostic and therapeutic measures, if in the physician's judgment it offers hope of saving life, reestablishing health or alleviating suffering. Where possible, these measures should be made the object of research, designed to evaluate their safety and where appropriate, published. The other relevant guidelines of this Declaration should be followed." Broadly, the Helsinki Declaration evolved the following ethical conduct in clinical research: (1) research on humans must be based on the results from laboratory and animal experimentation; (2) protocols should be reviewed by an independent ethics committee prior to the commencement of research; (3) voluntary informed consent should be obtained from the subject; (4) only medically qualified individuals should conduct research; and (5) the risks should not exceed benefits. The 1964 Declaration of Helsinki underwent several reviews. In 1975, the role of institutional ethics committees/institutional review boards and the protocols for experimentation were included in the Helsinki Declaration. More rationalization of ethical principles was incorporated in the declaration with subsequent periodical reviews in 1983, 1989, 1996, 2000, 2002, 2004, and 2008.

18.3.8.1 The Tuskegee Syphilis Experiment

A major shot in the arm for those who campaigned against unethical clinical trials was the Tuskegee syphilis study (1932–1972),[71] a project by the US Public Health Service in which the participants were 600 low-income African American males. Even though a proven cure (penicillin) was available, the 400 participants who where infected with syphilis were denied treatment for over 40 years.[72] Many of them died of this disease. Timely intervention of other physicians stopped the study in 1973, when it became a political embarrassment[73] for the administration; in 1997 the then US president apologized to the victims. Due to the publicity generated by the tragedy, US authorities enacted the National Research Act (1974),[74] thereby creating the National Commission for the Protection of Human Subjects of Biomedical and Behavioural Research. The commission drafted the Belmont Report,[75] a foundational document for the ethics of human participation in biomedical and behavioral research in the United States.

18.3.8.2 The International Guidelines for Biomedical Research in Human Beings

In 1982, the Council for International Organization of Medical Sciences (CIOMS) and the WHO published the proposed International Guidelines for Biomedical Research in Human Beings indicating the modus of effective application of the ethical principles that govern the conduct of biomedical research in human beings in developing countries with particular reference to their socioeconomic background, laws, and regulations, and the executive infrastructure. The CIOMS, with the cooperation of the WHO, undertook the task and issued two sets of guidelines: (1) in 1991, the International Guidelines for Ethical Review of Epidemiological Studies,[76] and (2) in 1993, the International Ethical Guidelines for Biomedical Research Involving Human Subject, and in 2002, the same was revised,[77] wherein the guidelines for protocols of biomedical research as well as ethical safeguards to be followed were clearly spelled out. What is more significant in the 2002[78] CIMOS proceedings is a clarification on the use of placebo in clinical trials by reaffirming its earlier position that extreme care must be taken in making use of a placebo-controlled trial and that in general this methodology should only be used in the absence of existing proven therapy. Placebo-controlled trial was made ethically acceptable even if proven therapy is available where for compelling scientifically sound reasons the use of placebo is necessary to determine the efficacy or safety of a prophylactic, diagnostic, or therapeutic method; or where a prophylactic, diagnostic, or therapeutic method is being investigated for a minor condition and the patients who receive placebo will not be subject to any additional risk of serious or irreversible harm.

In 1996, to harmonize similar guidelines, norms, and regulations from different countries, the International Conference on Harmonization of Technical Requirements for Registration of Pharmaceuticals for Human Use (ICH) enacted Good Clinical Practice (GCP) regulations that facilitated the mutual acceptance of the clinical data obtained by different countries.

18.4 REGULATIONS IN MAJOR COUNTRIES

18.4.1 The United States

From 1820, with the introduction of The *United States Pharmacopeia*,[79] the first compendium of standard drugs, the regulatory framework for the protection of human beings from spurious and dangerous medicinal preparation gained momentum. The Federal Food, Drug, and Cosmetic Act (1938)[80] as amended from time to time is crucial in controlling and regulating clinical research in the United States. The Public Health Service Act (1944), the Trademark Act of 1946, the Reorganization Plan 1 of 1953, the Poultry Products Inspection Act (1957), the Fair Packaging and Labeling Act (1966), the National Environmental Policy Act of 1969, the Exit Disclaimer (1970), the Controlled Substances Act (1970), the Controlled Substances Import and Export Act (1970), the Egg Products Inspection Act (1970), the Lead-Based Paint Poisoning Prevention Act (1971), the Federal Advisory Committee Act (1972), the Sunshine Act (1976), the Federal Anti-Tampering Act (1983), the Sanitary Food Transportation Act (1990), the Mammography Quality Standards Act (MQSA) (1992), the Bioterrorism Act of 2002, the Project BioShield Act of 2004,[81] and the Food and Drug Administration Safety and Innovation Act (2012) are crucial in controlling and regulating clinical research in the United States.[82] The FDA was established for setting standards governing the composition, operation, and responsibility of clinical research in the United States. It is responsible for prescribing standards for institutional review boards (IRBs) that review clinical investigations involving human subjects and decides on applications for permission to conduct further research or to market regulated products. In other words, the FDA is responsible for ensuring the protection of the rights, safety, and welfare of human subjects who participate in clinical investigations involving articles as well as clinical investigations that support applications for research or marketing permits for products regulated by the FDA, including food and color additives, drugs for human use, medical devices for human use, biological products for human use, and electronic products. The FDA regulations for protection of human subjects can be found under part 50 [21 Code of Federal Regulations (CFR) part 50] and the regulations for the IRBs can be found under part 56 (21 CFR part 56). On January 27, 1981, the FDA adopted regulations on informed consent of human subjects [21 CFR part 50; 46 Federal Register (FR) 8942] and regulations establishing standards for the composition, operation, and responsibilities of the IRBs that review clinical investigations involving human subjects (21 CFR part 56; 46 FR 8958) which were also adopted by the Department of Health and Human Services (HHS; as regulations on the protection of human research subjects [45 CFR part 46; 46 FR 8366]). The FDA and HHS regulations share a common framework. In December 1981, on the basis of the findings in the Presidential Commission's "First Biennial Report on the Adequacy and Uniformity of Federal Rules and Policies, and Their Implementation, for the Protection of Human Subjects in Biomedical and Behavioral Research, Protecting Human Subjects," all Federal departments and agencies were directed to adopt the HHS regulations (45 CFR part 46). In May 1982, the president's science advisor appointed an *ad hoc* committee for the protection of human research subjects and agreed that uniformity of federal regulations on human subject protection is desirable to eliminate unnecessary regulations and to promote increased understanding by institutions that conduct federally supported or regulated research. The committee developed a model policy with the concurrence of all affected Federal departments and agencies, published as a proposal in the *Federal Register* of June 3, 1986 (51 FR 20204) and the FDA concurred with them. In the *Federal Register* of November 10, 1988 (53 FR 45678), the agency proposed to amend its regulations in 21 CFR parts 50 and 56 to conform them to the Federal Policy for the Protection of Human Research Subjects. On December 5, 1996, the FDA amended the then existing informed consent regulations to require that the consent form signed by the subject or the subject's legally authorized representative be dated by the subject or the subject's legally authorized representative at the time consent is given. The FDA also amended its regulation on case histories to clarify what adequate case histories include and that the case histories must document that the informed consent was obtained prior to participation in a study. Through these amendments, the FDA regulated that the consent form carry the date on which it is signed. This was to ensure that the consent document was obtained prior to entry into the study as required by the FDA regulations (21 CFR parts 50, 312, and 812). On November 9, 1998, the FDA introduced certain categories of research that may be reviewed by the IRB through an expedited review procedure. On November 24, 1998 (21 CFR part 54), an amendment was carried out on the requirement of financial disclosure by clinical investigators by redefining the term "clinical investigators."[83] The National Research Act enacted on July 12, 1974 created the National Commission for the Protection of Human Subjects of Biomedical and Behavioral Research (the commission to make recommendations on research involving children, including the purposes of such research, the steps necessary to protect children as subjects, and the requirements for the informed consent of children or their parents or guardians. On March 8, 1983, the HHS published its final rule incorporating the requirements for the protection of children involved as subjects in HHS-conducted or HHS-supported research (48 FR 9814). This rule is codified at 45 CFR part 46 D. These regulations supplemented the basic regulations governing the protection of human subjects involved in research conducted or supported by the HHS. On February 28, 2001, the FDA introduced an interim rule for additional protection of children as subjects in clinical trials (21 CFR parts 50 and 56) to bring the FDA regulations into compliance with provisions of the Children's Health Act of 2000, the 1998 Pediatric Rule, and the pediatric provisions of the Food and Drug Administration Modernization Act of 1997 (the Modernization Act). The 1998 Pediatric Rule required manufacturers to assess the safety and effectiveness of certain drugs and biological products in pediatric patients. The Modernization Act established economic incentives for manufacturers to conduct pediatric studies on drugs for which exclusivity or patent protection is available under the Drug Price Competition and Patent Term Restoration Act or the Orphan Drugs Act. The FDA's draft guidance "Clinical Pharmacogenomics: Premarketing Evaluation in Early Phase Clinical Studies" focuses particularly on use and evaluation of genomic strategies in early drug development and highlights identification of enrichment options for later trials. The FDA's draft guidance for industry and FDA staff on In Vitro Companion Diagnostic Devices define in vitro diagnostic (IVD) companion diagnostic devices that are essential for the safe and effective use of their corresponding therapeutic products. The draft guidance describes the agency's policies for approval and clearance and for labeling companion diagnostics, contemporaneously with approval and labeling of the therapeutic products. The FDA's draft guidance on Adaptive Design Clinical

Trials for Drugs and Biologics considers the case of enrichment approaches introduced only after randomization and based on interim evaluations. Such a retrospective finding would have to be carefully implemented and highly compelling to be accepted without further study. The FDA's guidance on Providing Clinical Evidence of Effectiveness for Human Drug and Biological Products describes the amount and type of evidence needed to demonstrate effectiveness and is applicable to studies using enrichment designs.[84]

18.4.2 The European Union

The central theme of the drug research and drug safety regulations in the European Union was the internationally recognized ethical and scientific requirements contained in the GCP.[85] Compliance with GCP in designing, conducting, recording, and reporting of clinical trials involving human participation was made mandatory not only in trials carried out in the EU countries but also in trials carried out in other countries if they could be qualified to apply for marketing authorization. The "Clinical Trials Directive" 2001/20/EC[86] set the standards for conducting clinical trials. The Commission Directive 2005/28/EC[87] laid down principles and guidelines for GCP to implement the Directive 2001/20/EC. The Directive 2001/83/EC,[88] as amended by 2003/63/EC,[89] regulated marketing authorization of medicinal products for the European market. The Regulation (EC) 726/2004[90] established the European Medicines Agency (EMEA) and centralized procedure. In addition, the guidelines of the ICH[91] are just as important. The main aim of the ICH is the harmonization in the interpretation and the application of technical guidelines and requirements for successful product registration in Europe, Japan, and the United States. For example, ICH E3[92] defined the structure and content of clinical trial study reports, ICH E6(R1)[93] spelt out the GCP, and ICH E10[94] defined the choice of control groups. Both the regulators and the industry follow the legislation and guidelines discussed in this subsection[95] without which their clinical data would not be acceptable to regulatory authorities in Japan, the United States, and other countries.

The 2012 EU guidelines[96] on the requirements for quality documentation concerning biological investigational medicinal products (IMPs) in clinical trials addressed the specific documentation requirements on the biological, chemical, and pharmaceutical quality of IMPs containing biological/biotechnology-derived substances in cases where no "simplified IMP dossier (IMPD)" is submitted; they were made applicable to proteins and polypeptides, their derivatives, and the products of which are components. (These proteins and polypeptides are produced from recombinant or nonrecombinant cell culture expression systems and can be highly purified and characterized using an appropriate set of analytical procedures.)

18.4.3 Australia

The Helsinki Declaration and the EU version of the GCP adopted by the ICH were adopted by Australia to augment regulating the biomedical research ethics. The National Health and Medical Research Council[97] and the Australian Health Ethics Committee are the organizations that regulate clinical research in Australia. The National Health and Medical Research Council Act (1992): The Therapeutic Good Administration division of the Australian Government Department of Health and Ageing is responsible for regulating medicines and medical devices under the Therapeutic Goods Act (TGA) (1989)[98] to meet the quality, safety, and efficacy standards of drugs and medical devices. The therapeutic goods including medical devices that have not been evaluated by the TGA for its safety and efficacy entered into the Australian Register of Therapeutic Goods (ARTG) for general marketing are required to make use of the Clinical Trial Notification (CTN) or Clinical Trial Exemption (CTX) schemes for marketing approval. The *Australian Clinical Trials Handbook* (2006),[99] the health safety regulatory publication of TGA, clearly describes the practice and procedures involved in clinical research with a particular reference to documentation, ethics committee approvals, serious adverse events, clinical trial monitoring and inspections, and clinical data management apart from the roles and responsibilities of the sponsor and investigator. A pioneer program of the National Health and Medical Research Council is Single Ethical Review for Multi-Centre Research. The objective of the National Approach to Single Ethical Review of Multi-Centre Research (National Approach) formally known as the Harmonization of Multi-Centre Ethical Review (HoMER) is to enable the recognition of a single ethical and scientific review of multicenter human research within and/or across Australian jurisdictions.[100]

The Privacy Act No. 119 (1988)[101] incorporating amendments up to Act No. 49 (2004), the Privacy and Personal Information Protection Act (2005),[102] the Information Privacy Act No. 98 (2000),[103] the Genetic Privacy and Non-Discrimination Act (1998),[104] Prohibition of Human Cloning for Reproduction Act (2002),[105] the Research Involving Human Embryos Act (2006),[106] and the Research Involving Human Embryos Regulations (2003)[107] are also key legislations governing clinical research in Australia.

18.4.4 Canada

The Food and Drugs Act (1985) and the regulations in Part C Division 5 govern the field of clinical research on human subjects in Canada. Health Canada is the federal regulatory arm of the Canadian government on all health matters including biomedical research involving human subjects. The Canadian Institute of Health Research (CIHR) Act (2000) was enacted for creating CIHR for biomedical research; clinical research; research respecting health systems, health services, health of populations, and societal and cultural dimensions of health and environmental influences on health; and other research as required. The Assisted Human Reproduction Act (2004) was enacted for the protection of persons who seek to undergo assisted reproduction procedures by which the principle of free and informed consent is made a mandatory provision and fundamental condition of the use of human reproductive technologies. The Canada Consumer Product Safety Act (2010) was enacted to ban products posing *danger to human health or safety*. Canada has initiated engaging its citizens in the decision-making process to enhance the quality, credibility, and accountability of the process for changing their health regulatory system.[108] Health Canada integrated risk management planning including adoption of ICH guideline E2E and pharmacovigilance planning supported by necessary regulations.[109] Its vigilance framework defined by an interrelated set of regulations, policies, guidances, and procedures was integrated and harmonized with the ICH guidelines,[110] whereas long-term plans in the areas of prevention, health promotion, and protection remained Canada's core values.[111] The Biologics and Genetic Therapies Directorate, the Health Products and Food Branch Inspectorate, the Marketed Health Products Directorate, the Natural Health Products Directorate, the Therapeutic Products Directorate, and the Veterinary Drugs Directorate monitored all drug safety information in various clinical research programs.

18.4.5 Japan

Pharmaceutical clinical research in Japan is based on various laws and regulations, consisting mainly of the Pharmaceutical Affairs Law, Pharmacists Law, Law Concerning the Establishment for Pharmaceuticals and Medical Devices Organization, Law Concerning Securing Stable Supply of Blood Products, Poisonous and Deleterious Substances Control Law, Narcotics and Psychotropic's Control Law, Cannabis Control Law, Opium Law, and Stimulants Control Law. In 2004, the Pharmaceutical and Medical Device Agency[112] was created to provide consultations concerning the clinical trials of new drugs and medical devices, and to conduct approval reviews and surveys of the reliability of application data. The National Institute of Biomedical Innovation was established in Japan to make a major contribution to drug research and development by integrating basic research, research on bioresources, and promotion of research and development. The Ministry of Health and Labour Welfare Ordinance No. 24 of February 29, 2008, was enforced on April 1, 2008 (partially enforced on April 1, 2009). This GCP consists of six chapters and 59 articles. It has three main parts: (1) standards for the sponsoring of clinical studies, (2) standards for the management of clinical studies that are related to sponsors, and (3) standards for the conduct of clinical studies that concern the medical institutions performing the clinical studies. Chapter 2 contains procedures to be followed when clinical trials are sponsored or managed in medical institutions by persons who wish to sponsor clinical trials and provisions to be followed when clinical trials are prepared or managed by persons who wish to conduct clinical trials by themselves ("investigator-initiated trials"). Chapter 3 contains provisions to be followed by the sponsor or persons performing clinical trials on their own for the scientific and ethical conduct of clinical trials. Chapter 4 contains provisions to be followed by the medical institutions performing clinical trials scientifically and ethically, constitution of the IRBs, their duties and responsibilities, informed consent, and so on. Chapter 5 contains standards concerning reexamination data in postmarketing studies. Chapter 6 contains standards concerning the sponsoring of clinical trials.[113] The Second Medium Range Plan (2009–2014) is now under way and efforts are being made to shorten the review period, make views more efficient, and promote international harmonization by strengthening ties with Western and Asian countries, and participation in global clinical trials.

18.4.6 India

Clinical trials in India had a promising future until 2013 as the regulations were almost ICH–GCP compliant and in harmony with the US and international standards. The main attraction for India's emergence as an ideal hub for clinical trials and drug development was due to its significant cost reduction advantage and increased speed and productivity of all R&D phases required to bring a safe and effective drug to market. Clinical research in India was supported by large patient pool, well-trained GCP investigators, premier medical institutions, low per-patient trial cost, and a favorable regulatory climate until 2013. In preindependent India, the Opium Act (1878), the Poisons Act (1919), the Dangerous Drugs Act (1930), and the Drugs and Cosmetics Act (1940) were the first of its kind to regulate the safety, manufacture, import, and export of drugs. The Poisons Act and the Dangerous Drugs Act were repealed by the Narcotic and Psychotropic Substances Act. The Drugs and Cosmetics Act (1940) regulated the manufacture, import, export, and use of pharmaceutical products, and covered Allopathic, Ayurvedic, and Unani systems of medicines. It governed the field of pharmaceutical research and the rules came into force in 1945. While the Pharmacy Act of 1948 regulated pharmacy profession and pharmacy education in India, the Drugs and Magic Remedies (Prohibition of Objectionable Advertisements) Rules (1955) controlled drug advertisements. The Drugs (Prices Control) Order (amended in 1995) enabled the government to fix maximum sale price for bulk drugs and medicinal formulation. During this period, the scarcity of essential medicines and growing dependence on import of drugs paved the way for encouraging research and development in the country, the Indian Patent Act (1970) was enacted. The Patent Amendment Act (2005) for manufacturing and marketing of goods was eased and herbal preparations having medicinal values could be patented, produced, and marketed. In 2011, by introducing the Drugs and Cosmetics (First Amendment) Rules, Schedule Y to the Drugs and Cosmetics Act was incorporated to regulate clinical research, making it mandatory the ethical compliance in biomedical research on human beings. Currently, clinical trials in India are regulated by Schedule Y of the Drug and Cosmetics Rules, which is being revised to bring the Indian regulations up to par with internationally accepted definitions and procedures. Schedule Y also defines the requirements and guidelines for import and/or manufacture of new drugs for sale or for clinical trials. The Central Drugs Standard Control Organization (CDSCO) office has issued guidance document on clinical trial inspection effective from November 1, 2011. The Indian Council of Medical Research (ICMR) has adopted international regulatory guidelines issued by the Ethical Guidelines for Biomedical Research on Human Subjects in 2006[114] and the Guidelines for Good Laboratory Practices in 2008.[115] Indian Good Clinical Practices for Clinical Research,[116] Indian Good Manufacturing Practice guidelines[117] (Schedule M of D&C Act), and

Guidance on Clinical Trial Inspection (2010)[118] were released by the CDSCO.[119] The ICMR in collaboration with the National Institute of Medical Statistics developed an online registry, that is, Clinical Trial Registry of India and mandated registration of all clinical trials held in India from 2009. It was also made mandatory that the CROs furnish information as per the WHO International Clinical Trials Registry Platform dataset. Apart from the CDSCO headed by the Drugs Controller General of India, different agencies such as the Department of Biotechnology (DBT),[120] the ICMR,[121] the Central Bureau of Narcotics (CBN), Review Committee on Genetic Manipulation (RCGM), and Genetic Engineering Approval Committee (GEAC) are also involved in regulating clinical research in India. The CDSCO has also issued guidance for conducting clinical trials of biological products[122] and guidance on common technical documentation for the NDA application, which were other initiatives for streamlining the requirements for conducting clinical trial and new drug approval process in India. A pharmacovigilance[123] program was launched by the government to ensure drug safety. In 2007, three clinical trials were approved in India, which grew to 65 in 2008, 391 in 2009, and 500 in 2010. However, in 2012 there were only 262 approved clinical trials and only 6 up to January 23, 2013.[124] The sudden fall in clinical research and drug development activity in India was the result of media reports of ethical violations in some part of the country leading to deaths of persons.[125] The cause was taken up by a nongovernmental organization (NGO) to the Apex Court through a Public Interest Litigation and the same is pending. In the meantime, the CDSCO introduced certain amendments to Drugs and Cosmetics Rules in January for payment of compensation to subjects suffering from adverse events/serious adverse events in clinical trials[126] and from February 2013[127] a new rule was introduced making it mandatory for all ethics committees to be registered; virtually, the ethical reviews of all the ongoing clinical trials were abruptly stopped as the insertion of Rule 122D prohibited unregistered ethics committees reviewing clinical trials and the provison to the rule prescribed 45 days for the existing ethics committees to apply for registration. The definition of clinical trial-related injury, the time frame for reporting, the financial liability of the sponsor, the fixation of quantum and duration of compensation, and so on remained too unrealistic and impractical as they were evolved ignoring altogether the infrastructure required for implementation and without adequate mass consultation and consensus with all stakeholders including legal experts and medical researchers in the country. Except a few NGOs propagating human rights concerns in clinical research, all the stakeholders in the industry criticized the manner in which the amendments were carried out by the CDSCO without larger and wider consultation with the industry, CROs, sites, and the patient groups. As of today, the picture remains blurred as the Drugs and Cosmetics (Amendment) Bill[128] was presented in Rajya Sabha on August 29, 2013, with a separate chapter on clinical trials. The bill proposes to create a Central Drugs Authority, a supreme body constituted with bureaucrats over the Drugs Controller General of India, and it is mandatory that no clinical trial in the country can be initiated or conducted without the prior approval from this authority. Surprisingly, the contradiction of this provision is glaring as no such permission is made necessary to initiate or conduct any bioequivalence or bioavailability studies of approved drugs by the government-owned institutes, hospitals, and autonomous medical or pharmacy institutions for academic or research purposes. It is not known what rationale was behind this kind of legislation in a country where the doctor-to-patient ratio is 1:1700 against 1.6 each in the United States and China, 2.3 in the United Kingdom, and 3.3 in Sweden per 1000 population, and the WHO's minimum stipulation is 1:1000. Moreover, health care in India is plagued by the malaise of missing doctors at the government-run health facilities.[129]

18.4.7 China

China has a rich tradition of making and using traditional medicines. The Drug Administration Law of China enacted in 1985 made it compulsory that marketing authorization is required for all traditional Chinese medicine drugs. In China, the Ministry of Health (MOH) (Mandarin),[130] the Chinese

Association of Science and Technology, the Ministry of Science and Technology, and the State Food and Drug Administration (SFDA)[131] control the food and drug safety administration. From 1986 onward, while products already in the market were allowed to remain in the market if no adverse events had been reported, approval for clinical trials and marketing was made compulsory for new medicines. Compliance with Good Manufacturing Practices and Good Supplies Practices for traditional Chinese medicine, which were promulgated in 1999, became a fundamental requirement.[132] The State Drug Administration (SDA) was established in 1998, directly under the State Council, to improve the regulation of pharmaceutical products and medical devices, including the regulation of clinical trials and the protection of trial subjects in China. In December 2001, the newly revised Drug Administration Law was enacted to ensure the protection of the rights, safety, and welfare of human subjects; ensure compliance with internationally recognized ethical standards and scientific principles for clinical trials; ensure standardization of the clinical trial process and the credibility and the scientific validity of results; and ensure that clinical trials of all drugs in all phases, including in human bioavailability or bioequivalence studies, are performed according to Chinese GCP, including biotechnology products and traditional Chinese medicines.[133] The Department of Drug Registration and the Department of Drug Safety and Inspection are within the structure of the SDA and are jointly responsible for the evaluation and inspection of clinical trials of new drug. There are several specific regulations focused on drug safety surveillance in China.[134] The most important one is the "Regulation for the Administration of Adverse Drug Reaction Reporting and Monitoring," which was issued jointly by the SFDA and the MOH in 1999, and revised and disseminated in March 2004.[135]

18.4.8 New Zealand

In New Zealand, the legislative coverage on health research includes the Health Act (1956) amended up to 2005, the Human Tissue Act (1964) amended up to 1984, the Hazardous Substances and New Organisms Act (1996) amended up to 2005, the Health Research Council Act (1990), Sections 24 and 25 of the New Zealand Bill of Rights Act, Article 10 (1990), the Health and Disability Commissioner Act (1994), the New Zealand Public Health and Disability Act (2000), and Section 16 of the Medicines Act (1981) amended up to 2005.[136] The regulatory bodies responsible for overseeing clinical research in the country are the Health Research Council (HRC) Ethics Committee,[137] the National Ethics Advisory Committee (NEAC),[138] the MOH,[139] the Health and Disability Commissioner,[140] the Advisory Committee on Assisted Reproductive Technology (ACART),[141] the Gene Technology Advisory Committee,[142] the HRC Standing Committee on Therapeutic Trials,[143] and the New Zealand Medicines and Medical Devices Safety Authority (Medsafe)[144] which are the regulatory bodies of various medical research. To achieve specific needs and clarity, clinical researchers are guided by various guidelines, that is, the RMI Guidelines on Clinical Trials: Compensation for Injury Resulting from Participation in an Industry Sponsored Clinical Trial (1997), Guidelines on Ethics in Health Research (2005), NEAC Ethical Guidelines for Observational Studies: Observational Research, Audits and Related Activities (2006), Medsafe's New Zealand Regulatory Guidelines for Medicines, Vol. 3: Interim Good Clinical Research Practice Guidelines (1998), and so on.

18.4.9 Singapore

The geographical and infrastructural advantage of Singapore has been the hallmark of its emergence as a powerful presence in clinical research field. The legal framework[145] within clinical research activities in Singapore is robust with the Medical (Therapy, Education, and Research) Act (1973), the Medicines Act Section 74 (Cap. 176) (1975), the Medical Registration Act (Cap. 174) (1985), the Medicines (Clinical Trials) Regulations (2000), the Human Cloning and Other Prohibited Practices Act (2005), the Country Key Organizations Legislation Regulations Guidelines 63, Health Products Act (2007), and guidelines issued there under for ethical compliance. The MOH[146] and its National

Medical Ethics Committee (NMEC), Bioethics Advisory Committee (BAC),[147] and Singapore Medical Council (SMC)[148] are responsible for the implementation of the regulatory instruments.

18.4.10 United Arab Emirates

The Health Authority of Abu Dhabi (HAAD) was established in 2007 to monitor all operations in health sector including clinical research in the United Arab Emirates.[149] The UAE Medical Liability Law No. 10 (2008) regulates human subject research. As per the law, clinical trials cannot be conducted without license from any one of the authorities, namely, MOH, HAAD, Dubai Health Authority, Dubai Healthcare City, or public universities. The statutory guidelines for conducting clinical research in the United Arab Emirates are Abu Dhabi Health Research Council (Ref. PHP/PHR/R01), Licensing Requirements for Institutional Human Subjects Research (Ref. PHP/PHR/R02), Policy Governing Research Involving Human Subjects (Ref. PHP/PHR/R03), and HAAD Standard Operating Procedures for Research Ethics Committees (REC SOPs) (Ref. PHP/PHR/R04).[150] The Abu Dhabi Health Research Council is the regulatory body to oversee all clinical research activities in the United Arab Emirates which identified 10 top priority areas for clinical research including cardiovascular disease (CVD) prevention and management, road safety, tobacco control, cancer control, mental health, mother and infant health, and musculoskeletal health. The law mandates that all principal investigators must have a valid HAAD health processional license for doing clinical research and successfully undergo HAAD-approved training on research ethics.[151]

Apart from these countries, the emerging clinical trial locations include Brazil, Russia, Argentina, Bulgaria, Chile, Colombia, Croatia, the Czech Republic, Estonia, Hong Kong, Hungary, Latvia, Lithuania, Malaysia, Mexico, Peru, the Philippines, Poland, Romania, South Africa, South Korea, Taiwan, Thailand, Turkey, and Ukraine. According to the 2009 estimates, about 25% of the industry-sponsored clinical trials were located in these countries corresponding to annually 12,500 sites with 50 ethics committee reviews of clinical trials every working day.[152]

18.5 CONCLUSION

A survey of the regulatory literature reveals that a paradigm shift is needed in regulating clinical research focusing on (1) the need for newer drugs and therapeutic devices, (2) inherent risks and probable benefits to the public in each clinical trial, (3) importance of protection to the researchers as well as the subjects in clinical trials, (4) maintenance of confidentiality, and (5) patents. Enacting a number of legislations without an adequately resourced enforcement mechanism is a serious matter that needs consideration. Nutraceutical product research is poised for high growth in the coming decades. The regulatory framework for the nutraceutical product research should be imaginative, dynamic, and emulative of ethical parameters followed in clinical research.

NOTES

1. Clinical Trials & Clinical Research. http://www.nichd.nih.gov/health/clinicalresearch/Pages/index.aspx (last accessed on September 23, 2013).
2. What are clinical trials and why are they important. http://www.imaginis.com/breasthealth/clinical_trials.asp (last accessed on September 23, 2013).
3. Clinical Trials. http://www.who.int/topics/clinical_trials/en/ (last accessed on September 23, 2013).
4. What's the difference between a clinical trial and a clinical study? http://www.alz.org/research/clinical_trials/clinical_trials_alzheimers.asp (last accessed on September 23, 2013).
5. Bhat, P. and Velingkar, V.S. Drug discovery—A general overview. *Indian Journal Pharmaceutical Education*, 37(2): 94–99, 2003.
6. Sharma, P.J. Four vedas—Sanskrit text with Hindi commentary. http://archive.org/details/FourVedas-SanskritTextWithHindiCommentaryByPanditJaydevSharma (last accessed on September 23, 2013).
7. Mehta, P.M. (Ed.): *Charaka Samhita*. Jamnagar, India: Gulab Kunverba Society, 1949.

8. Puthumana, P. Through the mists of time: Sushrutha, an enigma revisited. *Indian Journal of Plastic Surgery*, 42(2): 219–223, 2009. http://www.ncbi.nlm.nih.gov/pmc/articles/PMC2845368/ (last accessed on September 23, 2013).
9. Shamasastry, R. Kautilya's Arthasastra. https://ia601503.us.archive.org/34/items/Arthasastra_English_Translation/Arthashastra_of_Chanakya_-_English.pdf (last accessed on September 23, 2013).
10. Niazi, A.K., Kalra, S., Irfan, A., and Islam, A. Thyroid over the ages. *Indian Journal of Endocrinology and Metabolism*, 15(Suppl 2): S121–S126, 2011. http://www.ncbi.nlm.nih.gov/pmc/articles/PMC3169859/ (last accessed on September 23, 2013).
11. What is ancient Egyptian medicine. http://www.medicalnewstoday.com/info/medicine/ancient-egyptian-medicine.php#.UjgCGD9x2t4 (last accessed on September 23, 2013).
12. Daniel 1:11–16. *The Holy Bible*. Revised standard version. New York: American Bible Society, 1995.
13. Antiqua Medicina: From Homer to Hippocrates. University of Virginia. http://exhibits.hsl.virginia.edu/antiqua/alexandrian/ (last accessed on September 23, 2013).
14. Kheir, N., Al Saad, D., and Al Naimi, S. Pharmaceutical care in the Arabic-speaking Middle East: Literature review and country informant feedback. *Avicenna*, 2, 2013. http://www.qscience.com/doi/pdfplus/10.5339/avi.2013.2 (last accessed on September 23, 2013); http://www.qscience.com/doi/abs/10.5339/avi.2013.2?prevSearch=Avicenna+on+Healthcare&searchHistoryKey=
15. Jenner, E. (1749–1823). http://www.jenner.ac.uk/edwardjenner (last accessed on September 23, 2013).
16. Bellis, M. History of antiseptics. http://inventors.about.com/library/inventors/blantisceptics.htm (last accessed on September 23, 2013).
17. Heath, P. *Allegory and Philosophy in Avicenna (Ibn Sina): With a Translation of the Book of the Prophet Muhammad's Ascent to Heaven*. Philadelphia, PA: University of Pennsylvania Press, 1992. *Project MUSE*. http://muse.jhu.edu/ (last accessed on September 23, 2013).
18. Gonzalez, C.J. Avicenna's canon of medicine. http://archive.org/details/AvicennasCanonOfMedicine (last accessed on September 23, 2013).
19. Pare, A. http://www.comptonhistory.com/compton2/pare.htm (last accessed on September 23, 2013).
20. Harvey, P. http://www.famousscientists.org/william-harvey/ (last accessed on September 23, 2013).
21. Wren, C. Spartacus educational. http://www.spartacus.schoolnet.co.uk/ARwren.htm (last accessed on September 23, 2013).
22. Kaadan, A.N. and Angrini, M. Blood transfusion in history. http://www.ishim.net/Articles/Blood%20Transfusion%20in%20History.pdf (last accessed on September 23, 2013).
23. van Leeuwenhoek, A. (1632–1723). http://www.ucmp.berkeley.edu/history/leeuwenhoek.html (last accessed on September 23, 2013).
24. Hunter, J. (1728–1793). http://www.surgical-tutor.org.uk/default-home.htm?surgeons/hunter.htm~right (last accessed on September 23, 2013).
25. Milne, I. Who was James Lind, and what exactly did he achieve? *JLL Bulletin: Commentaries on the History of Treatment Evaluation*, 2012. www.jameslindlibrary.org (last accessed on September 3, 2013).
26. UCLA. Biography of John Snow. Department of Epidemiology, School of Public Health. http://www.ph.ucla.edu/epi/snow/snowbio.html (last accessed on September 3, 2013).
27. Martin, C. Mesmerized. *Chemical Heritage Magazine*, Fall 2011/Winter 2012, 29(3). http://www.chemheritage.org/discover/media/magazine/articles/29-3-mesmerized.aspx (last accessed on September 23, 2013).
28. Pocock, S.J. *Clinical Trials: A Practical Approach*. Chichester: Wiley, 1983.
29. Biographical dictionary of contemporary American physicians and surgeons: Austin flint biography. http://cdm16694.contentdm.oclc.org/cdm/compoundobject/collection/p16694coll1/id/669/rec/7 (last accessed on September 23, 2013).
30. Jenner, E. History learning site. http://www.historylearningsite.co.uk/edward_jenner.htm (last accessed on September 23, 2013).
31. Semmelweis, I.P. History learning site. http://www.historylearningsite.co.uk/ignaz_semmelweis.htm (last accessed on September 23, 2013).
32. Genetic History, Gregor Mendel's discoveries. http://library.thinkquest.org/C0118084/History/Mendel.htm (last accessed on September 23, 2013).
33. Roediger, W.E. Requirements for medical research-perceptions from Joseph Lister's development of chromic catgut. http://www.ncbi.nlm.nih.gov/pmc/articles/PMC1292956/?page=1 (last accessed on September 23, 2013).
34. Pasteur, L. http://dwb.unl.edu/Teacher/NSF/C11/C11Links/ambafrance.org/HYPERLAB/PEOPLE/_pasteur.html (last accessed on September 23, 2013).
35. Staphylococci. http://www.channing.harvard.edu/4a.htm (last accessed on September 23, 2013).

36. Group A streptococcal infections. http://emedicine.medscape.com/article/228936-overview (last accessed on September 23, 2013).
37. Zakour, N.L.B. et al. Analysis of a *Streptococcus pyogenes* puerperal sepsis cluster by use of whole-genome sequencing. *Journal of Clinical Microbiology*, 50(7): 2224–2228, 2012. http://jcm.asm.org/content/50/7/2224.short (last accessed on September 23, 2013).
38. *Pneumococcal pneumonia*. http://www.medicinenet.com/script/main/art.asp?articlekey=60945 (last accessed on September 23, 2013).
39. Mehra, A. Politics of participation: Walter Reed's yellow-fever experiments. http://virtualmentor.ama-assn.org/2009/04/mhst1-0904.html (last accessed on September 23, 2013).
40. Von Behring, E. Biographical. http://www.nobelprize.org/nobel_prizes/medicine/laureates/1901/behring-bio.html (last accessed on September 23, 2013).
41. Ehrlich, P. Biographical. http://www.nobelprize.org/nobel_prizes/medicine/laureates/1908/ehrlich-bio.html (last accessed on September 23, 2013).
42. Wilkinson, L. Sir Austin Bradford Hill: Medical statistics and the quantitative approach to prevention of disease. *Addiction*, 92(6): 657–666, 1997.
43. Hill, A.B. Memories of the British streptomycin trial in tuberculosis. *Controlled Clinical Trials*, 11: 77–79, 1990.
44. Rosenberger, W.F. and Lachin, J.M. *Randomization in Clinical Trials: Theory and Practice*. New York: Wiley, 2004.
45. Sollmann, S. The evaluation of therapeutic remedies in the hospital. *Journal of the American Medical Association*, 94: 1280, 1930.
46. Sir Alexander Fleming. Biographical. http://www.nobelprize.org/nobel_prizes/medicine/laureates/1945/fleming-bio.html (last accessed on September 23, 2013).
47. Koch, R. Biographical. http://www.nobelprize.org/nobel_prizes/medicine/laureates/1905/koch-bio.html (last accessed on September 23, 2013).
48. Curie, M. Biographical. http://www.nobelprize.org/nobel_prizes/physics/laureates/1903/marie-curie-bio.html (last accessed on September 23, 2013).
49. Sir Ronald Aylmer Fisher. Biographical. http://www-history.mcs.st-andrews.ac.uk/Biographies/Fisher.html (last accessed on September 23, 2013).
50. Collins, C. and Hands, P. *Collins Thesaurus of the English Language—Complete and Unabridged*. 2nd Edition. Glasgow, Scotland: HarperCollins Publishers, 2002.
51. Definition of ethics in English. http://oxforddictionaries.com/definition/english/ethics (last accessed on September 23, 2013).
52. Hippocrates. http://www.britannica.com/EBchecked/topic/266627/Hippocrates (last accessed on September 23, 2013).
53. Greek medicine, The Hippocratic oath. http://www.nlm.nih.gov/hmd/greek/greek_oath.html (last accessed on September 23, 2013).
54. *Slater v. Baker and Stapleton*, 2 Wils. K.B. 359, 360, 95 E.R. 860, 861 (1767).
55. Sulfanilamide disaster, *FDA Consumer Magazine*, June 1981. http://www.fda.gov/AboutFDA/WhatWeDo/History/ProductRegulation/SulfanilamideDisaster/default.htm (last accessed on September 23, 2013).
56. FDA History—Part II. The 1938 Food, Drug, and Cosmetics Act. http://www.fda.gov/AboutFDA/WhatWeDo/History/Origin/ucm054826.htm (last accessed on September 23, 2013).
57. Law, M.T. How do regulators regulate? Enforcement of the Pure Food and Drugs Act, 1907–1938. *Journal of Law, Economics, and Organization*, 22(2): 459–489, 2006.
58. The Harrison Narcotics Act (1914). http://www.erowid.org/psychoactives/law/law_fed_harrison_narcotics_act.shtml (last accessed on September 23, 2013).
59. *Webb v. United States*, 249 U.S. 96 (1919). http://supreme.justia.com/cases/federal/us/249/96/case.html (last accessed on September 23, 2013).
60. Richter, J.H. Albert Neisser; centenary of birth, 1855-January 22-1955. *Archives of Dermatology*, 71(1): 92, 1955.
61. Simona, LUCA and Traian MIHAESCU. History of BCG Vaccine. *Maedica*, 8(1): 53, 2013.
62. *Introduction to NMT Case 1 U.S.A. v. Karl Brandt et al*. http://nuremberg.law.harvard.edu/php/docs_swi.php?DI=1&text=medical (last accessed on September 23, 2013).
63. Grodin, M.A. Historical origins of the Nuremberg code. *Medicine, Ethics and the Third Reich: Historical and Contemporary Issues*, 169–194, 1994.
64. History of ethics. Claremond Graduate University. http://www.cgu.edu/pages/1722.asp (last accessed on September 23, 2013).

65. Universal Declaration of Human Rights (1948). http://www.unic.org.in/items/Other_UniversalDeclarationOfHumanRights.pdf (last accessed on September 23, 2013).
66. International Covenant on Civil and Political Rights. http://ec.europa.eu/justice/policies/privacy/docs/16-12-1996_en.pdf (last accessed on September 23, 2013).
67. What is thalidomide? http://www.thalidomidesociety.co.uk/whatis.htm (last accessed on September 23, 2013).
68. Instituto Nacional de Salud (INS) Peru. Clinical Trials Regulation Backgrounds. http://www.ins.gob.pe/portal/jerarquia/1/947/clinical-trials-regulation-backgrounds/jer.947 (last accessed on September 23, 2013).
69. Jeremy, A., Greene, M.D., and Podolsky S.H. Reform, regulation, and pharmaceuticals—The Kefauver–Harris amendments at 50. *The New England Journal of Medicine*, 367: 1481–1483, 2012. http://www.nejm.org/doi/full/10.1056/NEJMp1210007 (last accessed on September 23, 2013).
70. World Medical Association. Declaration of Helsinki—Ethical principles for medical research involving human subjects. http://www.wma.net/en/30publications/10policies/b3/ (last accessed on September 23, 2013).
71. Brandt, A.M. Racism and research: The case of the Tuskegee Syphilis experiment. In *Tuskegee's Truths: Rethinking the Tuskegee Syphilis Study*, ed. Susan M. Reverby. Chapel Hill, NC: University of North Carolina Press, 2000.
72. Parran, T. Shadow on the land: Syphilis, the white man's burden. In *Tuskegee's Truths*, ed. Reverby, 2000.
73. Heller, J. Syphilis victims in the U.S. went untreated for 40 years. In *Tuskegee's Truths*, ed. Reverby, 2000.
74. The National Research Act 1974. http://history.nih.gov/research/downloads/PL93-348.pdf (last accessed on September 23, 2013).
75. Zimmerman, J.F. The Belmont report: An ethical framework for protecting research subjects. *The Monitor*, Summer 1997. http://www.impactcg.com/docs/BelmontReport.pdf (last accessed on September 23, 2013).
76. International guidelines for ethical review of epidemiological studies (1991). http://www.dib.unal.edu.co/promocion/etica_epidemiologicos_en.pdf (last accessed on September 23, 2013).
77. International ethical guidelines for biomedical research involving human subjects Prepared by the Council for International Organizations of Medical Sciences (CIOMS) in collaboration with the World Health Organization, Geneva, 2002.
78. WMA. Declaration of Helsinki (as amended up to 2012). http://www.fda.gov/ohrms/dockets/dockets/06d0331/06D-0331-EC20-Attach-1.pdf (last accessed on September 23, 2013).
79. *United States Pharmacopeia*. http://www.drugfuture.com/Pharmacopoeia/USP32/pub/data/v32270/usp32nf27s0_preface.html (last accessed on March 23, 2013).
80. United States Code, Federal Food, Drug, and Cosmetic Act (2010). http://www.gpo.gov/fdsys/pkg/USCODE-2010-title21/html/USCODE-2010-title21-chap9-subchapI.htm (last accessed on September 23, 2013).
81. US DHHS. Other laws affecting FDA. http://www.fda.gov/RegulatoryInformation/Legislation/ucm153119.htm (last accessed on September 23, 2013).
82. US DHHS. The Food and Drug Administration Safety and Innovation Act (2012). http://www.gpo.gov/fdsys/pkg/PLAW-112publ144/pdf/PLAW-112publ144.pdf (last accessed on September 23, 2013).
83. US DHSS. Clinical trial guidance documents. http://www.fda.gov/RegulatoryInformation/Guidances/ucm122046.htm (last accessed on September 23, 2013).
84. US DHHS. Food and Drug Administration. Guidance for industry enrichment strategies for clinical trials to support approval of human drugs and biological products. http://www.fda.gov/downloads/Drugs/GuidanceComplianceRegulatoryInformation/Guidances/UCM332181.pdf (last accessed on March 23, 2013).
85. Good Clinical Practice. July 1996 Version. http://ec.europa.eu/health/files/eudralex/vol-10/3cc1aen_en.pdf (last accessed on September 23, 2013).
86. *Official Journal of European Communities*: Directive 2001/20/EC. http://eur-lex.europa.eu/LexUriServ/LexUriServ.do?uri=OJ:L:2001:121:0034:0044:en:PDF (last accessed on September 23, 2013).
87. Good Clinical Practice: Directive 75/318/EEC. http://eur-lex.europa.eu/LexUriServ/LexUriServ.do?uri=OJ:L:2005:091:0013:0019:en:PDF (last accessed on September 23, 2013).
88. European Parliament: Directive No. 2002/98/EC. http://www.edctp.org/fileadmin/documents/ethics/DIRECTIVE_200183EC_OF_THE_EUROPEAN_PARLIAMENT.pdf (last accessed on September 23, 2013).
89. European Parliament Commission: Directive No.2003/63/EC. http://eur-lex.europa.eu/LexUriServ/LexUriServ.do?uri=OJ:L:2003:159:0046:0094:en:PDF (last accessed on September 23, 2013).

90. European Parliament: Regulation EC No. 726/2004. http://eur-lex.europa.eu/LexUriServ/LexUriServ.do?uri=CONSLEG:2004R0726:20120702:EN:PDF (last accessed on September 23, 2013).
91. ICG Guidelines. http://www.ich.org/fileadmin/Public_Web_Site/ICH_Products/Guidelines/Guidelines_Index.pdf (last accessed on September 23, 2013).
92. ICH Harmonised Tripartite Guideline Structure and Content of Clinical Study Reports E3. http://www.ich.org/fileadmin/Public_Web_Site/ICH_Products/Guidelines/Efficacy/E3/E3_Guideline.pdf (last accessed on March 23, 2013).
93. ICH Harmonised Tripartite Guideline for Good Clinical Practice E6(R1). http://www.ich.org/fileadmin/Public_Web_Site/ICH_Products/Guidelines/Efficacy/E6_R1/Step4/E6_R1__Guideline.pdf (last accessed on September 23, 2013).
94. ICH Harmonised Tripartite Guideline Choice of Control Group and Related Issues in Clinical Trials E10. http://www.ich.org/fileadmin/Public_Web_Site/ICH_Products/Guidelines/Efficacy/E10/Step4/E10_ Guideline.pdf (last accessed on September 23, 2013).
95. European Commission: EudraLex—Volume 10 Clinical trials guidelines. http://ec.europa.eu/health/documents/eudralex/vol-10/.
96. European Commission: Guidelines on the requirement of quality documentation concerning biological investigational medicinal products in clinical trials. 2012. http://ec.europa.eu/health/files/eudralex/vol-10/2012-05_quality_for_biological.pdf (last accessed on September 23, 2013).
97. National Health and Medical Research Council Act (1992). http://scaleplus.law.gov.au/html/pasteact/0/379/top.htm.
98. The Therapeutic Goods Act 1989, the Therapeutic Goods Regulations 1990, and the Therapeutic Goods (Medical Devices) Regulations 2002. http://www.comlaw.gov.au (last accessed on September 23, 2013).
99. *Australian Clinical Trial Handbook*. http://www.tga.gov.au/pdf/clinical-trials-handbook.pdf (last accessed on September 23, 2013).
100. NHMRC. Australia: National approach to single ethical review. http://www.nhmrc.gov.au/health-ethics/national-approach-single-ethical-review (last accessed on September 23, 2013).
101. Privacy Act No. 119 (1988). Incorporating Amendments up to Act No. 49 (2004). http://www.comlaw.gov.au/Details/C2006C00121/Download (last accessed on April 20, 2014).
102. Privacy and Personal Information Protection Act (2005). http://www.austlii.edu.au/au/legis/nsw/consol_act/papipa1998464/index.htm.
103. Information Privacy Act No. 98 (2000). http://www.dms.dpc.vic.gov.au/Domino/Web_Notes/LDMS/PubLawToday.nsf?OpenDatabase.
104. Genetic Privacy and Non-Discrimination Act (1998). http://www.aph.gov.au/parlinfo/billsnet/98021.pdf.
105. Prohibition of Human Cloning for Reproduction Act (2002). http://www.austlii.edu.au/au/legis/cth/consol_act/pohcfra2002465/.
106. Research Involving Human Embryos Act (2006). http://www.comlaw.gov.au/ComLaw/Legislation/ActCompilation1.nsf/0/03F95E485D04231DCA2572F80003B1C3/$file/ResearchInvolvingHumanEmbryosAct2002_WD02.pdf.
107. Research Involving Human Embryos Regulations (2003). http://www.comlaw.gov.au/ComLaw/Legislation/LegislativeInstrumentCompilation1.nsf/all/search/5D4B2F1A72744D3DCA257340000A1339AHEC.
108. Blueprint for renewal: Transforming Canada's approach to regulating health products and food. http://www.hc-sc.gc.ca/ahc-asc/branch-dirgen/hpfb-dgpsa/blueprint-plan/index-eng.php.
109. Notice regarding implementation of risk management planning including the adoption of International Conference on Harmonisation (ICH) Guidance Pharmacovigilance Planning—ICH Topic E2E. http://www.hc-sc.gc.ca/dhp-mps/prodpharma/applic-demande/guide-ld/vigilance/notice_avis_rmp_pgr_e2e-eng.php.
110. Product vigilance. http://www.hc-sc.gc.ca/dhp-mps/prodpharma/applic-demande/guide-ld/vigilance/index-eng.php.
111. Health Canada—Core values. http://www.hc-sc.gc.ca/ahc-asc/activit/about-apropos/index-eng.php#val.
112. Pharmaceuticals and Medical Devices Agency. http://www.pmda.go.jp/english/index.Htm.
113. Pharmaceutical administration and regulations in Japan. http://www.jpma.or.jp/english/parj/pdf/2012.pdf (last accessed on September 24, 2013).
114. Indian Council of Medical Research. Ethical guidelines: For biomedical research on human participants. http://icmr.nic.in/ethical_guidelines.pdf (last accessed on September 24, 2013).
115. Indian Council of Medical Research. Guidelines for Good Clinical Practices. http://icmr.nic.in/guidelines/GCLP.pdf (last accessed on September 24, 2013).
116. CDSCO. Good Clinical Practices. http://unpan1.un.org/intradoc/groups/public/documents/apcity/unpan009867.pdf (last accessed on April 20, 2014).

117. CDSCO. Good Manufacturing Practices. http://www.rajswasthya.nic.in/Drug%20Website%2021.01.11/Revised%20Schedule%20%20M%204.pdf (last accessed on April 20, 2014).
118. CDSCO. Guidelines on Clinical Trial Inspections. http://www.cdsco.nic.in/writereaddata/CT%20Inspection%20-11-2-2011.pdf (last accessed on April 20, 2014).
119. Central Drugs Standard Control Organization. Office of Drugs Controller General of India (DCGI). http://cdsco.nic.in.
120. Department of Biotechnology (DBT). http://dbtindia.nic.in.
121. Indian Council of Medical Research (ICMR). Central Ethics Committee on Human Research. http://www.icmr.nic.in/human_ethics.htm.
122. CDSCO, Guidelines for Industry. http://www.cdsco.nic.in/writereaddata/CDSCO-GuidanceForIndustry.pdf (last accessed on April 20, 2014).
123. Indian Pharmacopoeia Commission. Pharmacovigilance. http://ipc.nic.in/writereaddata/linkimages/pvpi-2611733527.pdf (last accessed on April 20, 2014).
124. CDSCO, Number of DCG(I) Approved Clinical Trials (2013). http://www.cdsco.nic.in/writereaddata/DCG(I)%20approved%20clinical%20trial%20registered%20in%20ctri%20website%20(Jan.2013).pdf (last accessed on April 20, 2014).
125. Chandrasekaran, A. Clinical trials continue to decline in India. *LiveMint*, January 6, 2012. http://www.livemint.com/2012/01/06171430/Clinical-trials-continue-to-de.html (last accessed on May 9, 2014).
126. CDSCO, New Amendments to Drugs and Cosmetics Rules (2013). http://cdsco.nic.in/writereaddata/GSR%2063(E)%20dated01%20.02.2013.pdf (last accessed on April 20, 2014).
127. Karwa, M., Arora, S., and Agrawal, S.G. 2013. Recent Regulatory amendment in Schedule Y: Impact on bioequivalence studies conducted in India. http://omicsonline.org/recent-regulatory-amendment-in-schedule-y-impact-on-bioequivalence-studies-conducted-in-india-jbb.1000154.pdf (last accessed on April 20, 2014).
128. Government of India, Gazette Notification of Bill No. LVIII of 2013 introduced in Rajya Sabha. http://www.drugscontrol.org/Bill%20No.%20LVIII%20of%202013.pdf (last accessed on September 24, 2013).
129. Anantha Kumar, U., Doctor who*? The New Sunday Express Magazine*, September 22, 2013 p. 1.
130. China, Ministry of Health (MOH) (Mandarin). http://www.moh.gov.cn/.
131. China, State Food and Drug Administration. http://www.sfda.gov.cn/.
132. CFDA: White Paper: Status Quo of Drug Supervision in China. http://eng.sfda.gov.cn/WS03/CL0757/62153.html (last accessed on April 20, 2014).
133. http://apps.who.int/medicinedocs/pdf/s4923e/s4923e.pdf.
134. http://www.apec-ahc.org/files/tp201002/Session4_LiJinJu.pdf.
135. http://www.sfdachina.com/info/71-1.htm.
136. All New Zealand laws. http://www.legislation.govt.nz/browse_vw.asp?content-set=pal_statutes.
137. Health Research Council (HRC) Ethics Committee. http://www.hrc.govt.nz/.
138. National Ethics Advisory Committee (NEAC). http://www.neac.health.govt.nz/.
139. MOH. http://www.moh.govt.nz/.
140. Health and Disability Commissioner (HDC). http://www.hdc.org.nz/.
141. Advisory Committee on Assisted Reproductive Technology (ACART). http://www.acart.health.govt.nz/.
142. Health Research Council (HRC), Gene Technology Advisory Committee. http://www.hrc.govt.nz/root/pages_regulatory/Gene_Technology_Advisory_Committee.html.
143. HRC, Standing Committee on Therapeutic Trials. http://www.hrc.govt.nz/root/pages_regulatory/Standing_Committee_on_Therapeutic_Trials.html.
144. New Zealand Medicines and Medical Devices Safety Authority (Medsafe). http://www.medsafe.govt.nz.
145. Singapore Laws. http://statutes.agc.gov.sg/.
146. Singapore Ministry of Health. http://www.moh.gov.sg/.
147. Singapore Bioethics Advisory Committee (BAC). http://www.bioethics-singapore.org.
148. Singapore Medical Council (SMC). http://www.smc.gov.s.
149. www.haad.ae.
150. www.haad.ae/research.
151. www.citiprogram.org.
152. http://www.pfizer.com/files/research/research_clinical_trials/ethics_committee_guide.pdf.

Section V

Farm to Clinic Approach/Functional Aspects of Food Factors

19 Role of Seed to Patient Model in Clinically Proven Natural Medicines

Andrea Zangara and Dilip Ghosh

CONTENTS

19.1 INTRODUCTION

Herbal medicines are the most ancient form of health remedies known to mankind. In spite of the great advances achieved in modern medicine, plants still make an important contribution to health care, and several specific herbal extracts have demonstrated to be efficacious for specific conditions. Moreover, at least 120 distinct chemical substances derived from plants are considered as important drugs currently in use, and several other drugs are simple synthetic modifications of natural products (Fabricant and Farnsworth 2001). Plants can be regarded as "living factories" producing a variety of chemical compounds, including primary metabolites important for the growth of the plants (amino acids, proteins, carbohydrates) and secondary metabolites (alkaloids, terpenoids, phenylpropanoids, polyketides, flavonoids, saccharides). All these components may work together to deliver a synergistic effect in the finished product. Certain herbal medicines, because of the complexity of their chemical content and the variety of bioactivities, can provide the polypharmacology which orthodox drugs cannot deliver. Natural products are rarely evaluated in the well-controlled clinical trials that are required to receive the approval by regulatory bodies, and therefore tend to have less "scientific" evidence to support their efficacy. However, all medicinal compounds are chemicals, whether synthesized in plants, animals, or manufacturing laboratories; therefore, all medicinal chemical compounds should be held accountable to similar standards of quality (identity, purity, and stability), clinical effectiveness, and safety, irrespective of their source: "If it is found to be reasonably safe and effective, it will be accepted" (Angell and Kassirier 1998). Reliable and consistent quality is the basis of efficacy and safety of herbal medicinal products (HMPs). Given the nature of products of plant origin, which are highly variable and complex products with numerous biologically active components that are rarely completely identified, therapeutic results and safety issues vary greatly from product to product, even within a single class. Therefore, the evidence of both benefits and risks is specific to the product tested and cannot necessarily be extrapolated to other products as is the case for synthetically derived compounds. For these reasons, and due to the inherent variability of the constituents of herbal products, it is difficult to establish a quality control parameter, and batch-to-batch variation, in the absence of reference standard for identification, can start from the collection of raw materials and increase during storage and further processing. In conclusion, to minimize the variation in final botanical products, standardization of procedures should cover the entire field of study from the cultivation of a medicinal plant to its clinical application.

19.2 HERBAL MEDICINAL PRODUCTS

Plants have been used for medicinal purposes since before recorded history. Over the centuries, diverse cultural groups developed traditional medical systems, such as Ayurveda and traditional Chinese medicine (TCM). In the early nineteenth century, scientists began to use chemical analysis to extract and modify the active ingredients from plants. Later, chemists began making synthetic versions of plant compounds and the use of herbal medicines was substituted by conventional drugs in most industrialized countries. By contrast, many nonindustrialized countries never abandoned medical herbalism and continued to develop their existing traditional medical systems. The World Health Organization (WHO) estimates that 80% of people worldwide rely on herbal medicines for some part of their primary health care (Robinson and Zhang 2011).

The WHO recognizes herbal medicines as valuable and readily available resources, and states that it is necessary to develop a systematic inventory of medicinal plants, introduce regulatory measures, apply good manufacturing practices, and include herbal medicines in the conventional pharmacopeia of each nation (WHO 2000). In the past few decades, public dissatisfaction with the cost of prescription drugs, combined with an interest in returning to natural or organic remedies, has led to a revaluation of the use of herbal preparations in industrialized countries.

HMPs are any medicinal product, exclusively containing as active substances one or more herbal substances, one or more herbal preparations, or one or more such herbal substances in combination with one or more such herbal preparations (WHO 1996), and with this term, we comprehensively refer within this chapter to herbal drugs, herbal preparations, and finished HMPs. HMPs contain complex mixtures of organic chemicals (the phytocomplex) that work together to produce an effect on the body and that are rarely completely identified. The activity of the phytocomplex is usually stronger than the sum of the activity of single active molecules and the presence of substances apparently having no specific activity can have a very important synergistic effect (Williamson 2001). *Ginkgo biloba*, for example, acts in several ways including the ability to decrease oxygen radical discharge and proinflammatory functions of macrophages (antioxidant and anti-inflammatory), reduce corticosteroid production (antianxiety), increase glucose uptake and utilization and adenosine triphosphate production, improve blood flow by increasing red blood cell deformability, decrease red cell aggregation, induce nitric oxide production, and inhibit platelet-activating factor receptors (Chan et al. 2007). Another example is *Panax ginseng*; more than 200 compounds have been identified in ginseng roots, an herbal substance with verified capability to benefit mental and physical capacities, weakness, exhaustion, and immunity (Soldati 2000; Scaglione et al. 2005). The pharmacological activity of ginseng is primarily due to a number of ginsenosides, but polysaccharides are also regarded as pharmacologically active in this plant (Soldati and Tanaka 1984; Li et al. 2009).

HMPs are sold in many forms (fresh, dried, liquid/solid extracts, tablets, capsules, powders, tea bags, topicals) and, as stated by the WHO, may not contain chemically defined substances such as synthetic compounds or chemicals isolated from herbs. They normally do not possess an immediate or strong pharmacological action but may produce more long-lasting benefits in chronic diseases such as dementia (Perry and Howes 2011): Due to the complexity and the synergy of the individual components of the phytocomplex, their combination, often resulting in complex dose- and time-dependent effects, may affect multiple neuronal, metabolic, and hormonal systems that themselves modulate behavioral processes (Zangara and Wesnes 2012).

Although it is generally believed that most herbal preparations are safe for consumption, some herbs such as the most biologically active substances could be toxic with undesirable side effects that are mainly due to active ingredients, contaminants, and/or interactions with other drugs (Walker 2004). Therefore, the use of HMPs should be carefully monitored to ensure patient safety.

19.3 PHYTOTHERAPY

Phytotherapy is commonly defined as the study of the use of extracts of natural origin as medicines or health-promoting agents. It should be perceived as an allopathic discipline, because the effects that are expected from HMPs are directed against the causes and the symptoms of a disease. In Germany, for example, phytotherapy is classified as a regular discipline of natural orthodox science-oriented medicine, and HMPs have to comply with similar scientific requirements as those of the chemically defined substances in terms of quality, safety, and efficacy (Keller 1996). However, modern mass production of natural products as food supplements or herbal medicines often results in remedies that can differ greatly (dosage form, mode of administration, herbal medicinal ingredients, methods of preparation, and medical indications) from the traditions that form the basis for their perceived safety and effectiveness, and from an acceptable quality standard. The potential benefit provided by the use of HMPs as effective medicine is sometimes perceived to be outweighed by the clinical risks associated with the absence of standard levels of biologically active materials from natural plants (Ernst 1998). The WHO has published monographs on the quality, safety, and efficacy of selected medicinal plants, and recommendations on the cultivation of medicinal plants and on the quality control, safety, and efficacy of HMPs (WHO 1993, 2003, 2007a).

19.4 QUALITY CONTROL

Quality control and standardization of HMPs involve several steps that should start with the sourcing of high-quality raw material and the development of criteria for precise identification of the constituents of each product, together with documentation of the role of the constituent combinations. The next step will be to optimize and control the growing conditions and each subsequent step till the finished product to minimize variability and provide pharmaceutical grade HMPs. Finally, it is necessary to establish its efficacy through biological assays and determine its adverse effect profile through literature or from toxicological studies (both short term and long term) followed by controlled clinical trials (Bauer and Tittel 1996). The lack of pharmacological and clinical data on the majority of HMPs represents a major impediment to the acceptance of natural products by conventional medicine.

The quality of plant raw materials can be influenced by accidental botanical substitution (misidentification of plant species) or intentional botanical substitution (deliberate exchange with other, possibly more toxic, plant species) (Bogusz et al. 2002; Van Breemen et al. 2008; Newmaster et al. 2013). The detection of undeclared chemical or synthetic substances or other active ingredients has also been frequently reported (Blumenthal 2002). Several studies have also highlighted different levels of active ingredients in herbal products, for example, an analysis of 25 available ginseng products found a 15- to 200-fold variation in the concentration of the active ingredients ginsenosides and eleutherosides (Harkey et al. 2001). The intentional or accidental presence of toxic heavy metals in herbal medicines in more than the permissible limit set by the national regulatory authorities is another common problem (Obi et al. 2006). Toxic contaminants are reported at all steps beginning from collection of raw materials to manufacturing (Fong 2002). The presence of pesticide residues in herbal materials has also been detected, highlighting the need for harmonization and standardization on the use of pesticides (Liva 2009). The WHO has established a maximum residue limit (MRL) for pesticides in cultivated or wild medicinal plants as well as appropriate methodologies for their detection (WHO 2007b).

HMPs should be analyzed (generally through chromatographic fingerprint) for chemical consistency at various stages of development to ensure the identity and purity of the product and correct quantification of the active ingredients or markers, and the regulators insist on the importance of the qualitative and quantitative methods for characterizing the samples, quantification of the biomarkers and/or chemical markers, and development of fingerprint profiles (EMEA 2008). However, identification of all of a plant's constituents often fails because of the complexity of the plant's chemical structure, and selective analytical methods or reference compounds may not be available commercially (Zhang et al. 2012). To overcome this problem, the German Commission E elaborated plant monographs (Blumenthal et al. 1998), and more recently, the *European Pharmacopoeia* has started a program to identify and prepare the chemical reference substances for the quality control of herbal drugs and their preparations, describing the analytical methods and quality specifications for each product. The *United States Pharmacopeia* is also following a similar approach, highlighting the need to lay down guidelines for a growing number of HMPs (Schwarz et al. 2009).

19.5 VARIABILITY

Plants are highly variable in the content of numerous biologically active components, and depending on their origin, the growth conditions, and the date of harvest, the content of active components in herbal drugs and herbal preparations will be influenced (Bauer 1998). Cultivated medicinal plants have specific advantages over wild-harvested plants: They show smaller variation in their constituents due to greater genetic uniformity, and the main secondary metabolites can be monitored, allowing for the definition of the best period for harvesting and are therefore generally preferred. The risk of misidentification of plants is avoided, and unsustainable harvesting is also limited. Controlled growth systems allow manipulation of the phenotypic variation in the concentration of medicinally important compounds present at harvest to increase potency, reduce toxin levels, and increase uniformity and predictability of extracts (Canter et al. 2005).

However, as the majority of medicinal plants are still harvested from the wild, it is important to use standardized extracts and to insure that strict guidelines on good agricultural and collection practices are followed, as the ones provided by the European Medicines Agency (EMEA 2006), which cover the cultivation of medicinal plants, as well as their harvesting and postharvesting processes.

Herbal drugs are identified by macroscopic and microscopic comparison with authentic material or accurate descriptions of authentic herbs (Ding et al. 2006) and any further tests that may be required (e.g., thin-layer chromatography). It is essential that herbal ingredients are referred to by their binomial Latin names of genus and species. Other important information is the method of extraction and the standardization procedures. These can influence how much of a particular active constituent is present in the herbal product. Some phytochemicals are more soluble in water, whereas others are soluble in alcohol or oil. The method varies from plant to plant, depending on the types of active constituents (Jones and Kinghorn 2012).

Standardized extracts are processed products, in which some specific and known components, recognized to contribute more than others to the therapeutic activity, are adjusted to a given amount within an acceptable tolerance. Standardization is achieved by adjusting the herbal substances/herbal preparations with excipients or by blending batches of herbal substances and/or herbal preparations. All the other components are still present in the extract because the action of the plant may result from the synergistic activity of several constituents. Moreover, often all the constituents of a plant are not yet fully characterized. Different methods of standardization may yield extracts with distinct properties and make the transferability of clinical data from one extract to another almost impossible, unless bioequivalence (or phytoequivalence), and bioavailability trials are used to prove essential similarity (Barnes 2003a, 2003b).

Medicinal products are pharmaceutically equivalent if they contain the same amount of the same active substance(s) in the same dosage forms that meet the same or comparable standards

(Birkett 2003). These parameters of pharmaceutical similarity applied to chemically defined drugs should also be applied to HMPs:

- *Active ingredient-specific parameters (extract):*
  - Herbal substance (quality)
  - Extraction solvent (type/concentration or solvent strength)
  - Production (extraction procedures)
  - Drug-to-extract ratio
- *Finished product-specific parameters:*
  - Weight of native extract per dosage form
  - Weight of constituents that are solely responsible for the therapeutic effect per dosage form
  - Weight of constituents possessing relevant pharmacological properties per dosage form
  - Excipients per dosage form (type)
  - Dosage form (type)
  - Posology (single dose and daily dose)

The concept of phytoequivalence was originally developed in Germany to ensure consistency of HMPs (Tyler 1999). A chromatographic fingerprint for an herbal product should be identified and compared with the profile of a clinically proven reference product to determine most of the phytochemical constituents to ensure the reliability and repeatability of pharmacological and clinical research. Biopharmaceutical studies researching liberation, absorption, distribution, metabolism, and excretion provide important information on the active metabolite, the effective dose, and the bioequivalence of different extracts, and are the link of *in vitro* data to clinical efficacy, as long as the therapeutically relevant compounds are known (Loew and Kaszkin 2002).

Unfortunately, phytoequivalence is very difficult to be obtained as phytochemical profiles are so complex and closely linked to the way the "seed to patient" process has been followed. It is a common experience that batches of medicinal plants with similar specifications, as species and part of plant, may have quite different chemical compositions due to a number of factors (Cañigueral et al. 2008) such as the following:

1. *Inter- or intraspecies variation.* The variation in constituents is mostly genetically controlled.
2. *Environmental factors.* The quality of a herbal ingredient can be affected by environmental factors such as weather, location, and other conditions under which it was cultivated.
3. *Time of harvesting.* For some herbs, the optimum time of harvesting should be specified as the concentrations of constituents in a plant can vary during the growing cycle and even during the course of a day.
4. *Part of plant used.* Different parts of the plant (e.g., roots, leaves) contain a different profile of constituents, and it is not uncommon for an herbal ingredient to be adulterated with parts of the plant not normally utilized.
5. *Postharvesting factors.* Storage conditions and processing treatments can greatly affect the quality of a herbal ingredient. Inappropriate storage after harvesting can result in microbial contamination. The most important stage of postharvest processing is drying, as dried plants can be preserved for prolonged periods of time (WHO 1998).

Substandard HMPs containing the same extract on the label may vary in their content and concentrations of chemical constituents from batch to batch, and even the same manufacturer can market in different periods the products containing different substances although standardized to achieve a high pharmaceutical quality. This variability can result in significant differences in pharmacological activity: involving both pharmacodynamic and pharmacokinetic issues. The use of a combination of herbal ingredients further complicates the issue of variability; for example, in TCM, current criteria use the concentration of several key ingredients or their total content to control the quality and ensure reproducible clinical efficacy (Li et al. 2013). Although the concentration of each bioactive component or the total concentration of all components may meet the requirements of the criteria, the concentration of each of these bioactive components often varies between batches, and in some cases, one or more of the bioactive components may be absent from a preparation. For example, EGb761, a *G. biloba* extract produced by Dr. Willmar Schwabe Pharmaceuticals, Germany, is standardized for 24% "ginkgo flavone" glycosides and 6% terpene lactones. However, analysis of several *G. biloba* commercial products showed remarkable variations in the rutin and quercetin content as well as the terpene lactone contents, although all the products satisfied the regulatory quality control method (Sticher 1993). Another example is St John's wort (*Hypericum perforatum* L.): Analysis of eight St John's wort products available in the United States found that their hyperforin content varied from 0.01% to 1.89%, and only two products contained sufficient hyperforin likely to be required for antidepressant effects. Similarly, the content of the other active component hypericin varied from 0.03% to 0.29%, and for several products, the actual hypericin content did not correlate with that stated on the product label (range 57%–130% of label claim) (Constantine and Karchesy 1998; De los Reyes and Koda 2002).

A representative example of the achieved successful "seed to patient" standardized process is the *Panax ginseng* C.A. Meyer extract G115® (Ginsana SA, Lugano, Switzerland), obtained only from roots aged between 5 to 7 years, and standardized to an invariable 4% ginsenosides content. The availability of this extract, currently the most well-documented *P. ginseng* extract, has made it possible to generate reproducible results in preclinical and clinical studies (Soldati 2000; Scaglione et al. 2005).

19.6 SAFETY AND EFFICACY

The efficacy and true frequency of side effects for most HMPs are not known because the majority have not yet been tested in large clinical trials and because pharmacovigilance systems are much less extensive than those in place for pharmaceutical products. The development process of HMPs should include traditional evidence and observational trials, but must progress to randomized, double-blind, placebo-controlled trials and pharmacovigilance protocols, and become more closely aligned with the development of new chemical entities.

Clinical trials with herbal drugs are feasible, and a significant amount of clinical data is available for some HMPs such as ginseng, ginkgo, and hypericum; however, few well-controlled, double-blind, placebo-controlled trials have been carried out with herbal medicines. In a recent survey, for around 1000 herbal medicines, only 156 of them had clinical trials supporting specific pharmacological activities and therapeutic applications (Cravotto et al. 2010). The major issues with clinical trials with HMPs, as highlighted by a number of meta-analyses, include the lack of standardization and quality control of the products used in clinical trials, the use of different dosages of herbal medicines, inadequate study designs and statistical power, patients not properly selected, difficulties in establishing appropriate placebos, and wide variations in the duration of treatments using herbal medicines (Calixto 2000). Clinical development of HMPs should include dose–response trials, particularly for newly developed products; however, for those products with a well-established tradition of use, they may not be required. However, any claims of efficacy for clinical indications of HMPs must be evidence based and, depending on the level of the claims, in agreement with grades of evidence recognized by the scientific community and health authorities (EMEA 2004).

19.7 INDUSTRY CHALLENGES

The global complementary medicine market has been estimated at US$83 billion annually. Over the past few decades, the complementary medicine sector has evolved from infancy to a major industry subject, which requires complex supply chains, clinical trials, global marketing, and export acumen that are all based on "seed to patient" concept (Ghosh 2013).

While this transformation is unambiguously positive, it has brought with it a set of policy challenges that need to be addressed if the industry is to continue to employ highly skilled workers in the manufacturing sector and maintain recent export momentum. Development of national preventive health agenda, introduction of innovation incentive and intellectual property (IP) protection, and finally substantial investment commitment from the federal government are the major challenges that are considered by the industry to be critical and that require addressing in the short-to-medium-term basis.

19.8 CONCLUSION

One of the biggest hurdles in toxicological assessment of herbal products is the translation of data of isolated phytochemicals to the "real-life" situation in which whole plants or plant extracts are used (Jordan et al. 2010). Herbal product quality has been a constant problem, as species misidentification or substitution occurs, and contamination or adulteration may be present. Since the introduction of current good manufacturing practices (cGMPs), an integral part of regulation, quality problems associated with HMPs should be less frequent. The application of newer technologies, such as omics and predictive toxicology in early drug development, have more often been applied to therapeutic efficacy or for screening toxicity of chemicals. Increasing the quantity and quality of clinical and scientific information on HMPs will reduce uncertainty in assessment and greatly contribute to decision making related to hazard and risk.

To conclude, different concentrations of the proportion of constituents in herbal medicine may lead to great variations in therapeutic results and safety issues, and consequently, the evidence of both benefits and risks is specific to the product tested and cannot necessarily be extrapolated to other products as is the case for synthetically derived compounds or assume that clinical trials with one brand are relevant to any other product. Health practitioners should become more aware of these issues to recommend to their patients only the exact products that have been consistently proven safety and efficacy in clinical trials conducted according to good clinical practice (GCP). The gold standard of supportive evidence is certainly a randomized, placebo-controlled, double-blind clinical trial, but data from other type of studies such as case reports can improve and contribute to the primary evidence coming from more complex and controlled studies.

REFERENCES

Angell, M. and Kassirier, J.P. 1998. Alternative medicine—The risks of untested and unregulated remedies. *The New England Journal of Medicine*, 339: 839–841.

Barnes, J. 2003a. Quality, efficacy and safety of complementary medicines: Fashions, facts and the future. Part I. Regulation and quality. *British Journal of Clinical Pharmacology*, 55: 226–233.

Barnes, J. 2003b. Quality, efficacy and safety of complementary medicines: Fashions, facts and the future. Part II: Efficacy and safety. *British Journal of Clinical Pharmacology*, 55: 331–340.

Bauer, R. 1998. Quality criteria and standardization of phytopharmaceuticals: Can acceptable drug standards be achieved? *Drug Information Journal*, 32: 101–110.

Bauer, R. and Tittel, G. 1996. Quality assessment of herbal preparations as a precondition of pharmacological and clinical studies. *Phytomedicine*, 2: 193–198.

Birkett, D.J. 2003. Generics—Equal or not? *Australian Prescriber*, 26: 85–87.

Blumenthal, M. 2002. Guest editorial: The rise and fall of PC-SPES: New generation of herbal supplement, adulterated product, or new drug? *Integrative Cancer Therapies*, 1: 266–270.

Blumenthal, M., Busse W., Goldberg A. et al. 1998. *The Complete German Commission E Monographs*. Austin, TX: The American Botanical Council.

Bogusz, M.J., Tufail, M., and Hassan H. 2002. How natural are natural herbal remedies? A Saudi perspective. *Adverse Drug Reactions and Toxicological Reviews*, 21: 219–229.

Calixto, J.B. 2000. Efficacy, safety, quality control, marketing and regulatory guidelines for herbal medicines (phytotherapeutic agents). *Brazilian Journal of Medical and Biological Research*, 33: 179–189.

Cañigueral, S., Tschopp, R., Ambrosetti, L., Vignutelli, A., Scaglione, F., and Petrini, O. 2008. The development of herbal medicinal products: Quality, safety and efficacy as key factors. *Pharmaceutical Medicine*, 22: 107–118.

Canter, P.H., Thomas, H., and Ernst, E. 2005. Bringing medicinal plants into cultivation: Opportunities and challenges for biotechnology. *Trends Biotechnology*, 23: 180–185.

Chan, P.C., Xia, Q., and Fu, P.P. 2007. *Ginkgo biloba* leave extract: Biological, medicinal and toxicological effects. *Journal of Environmental Science and Health Part C*, 25: 211–244.

Constantine, G.H. and Karchesy, J. 1998. Variations in hypericin concentrations in *Hypericum perforatum* L. and commercial products. *Pharmaceutical Biology*, 36: 365–367.

Cravotto, G., Boffa, L., Genzini, L., and Garella, D. 2010. Phytotherapeutics: An evaluation of the potential of 1000 plants. *Journal of Clinical Pharmacy and Therapeutics*, 35: 11–48.

De los Reyes, G.C. and Koda, R.T. 2002. Determining hyperforin and hypericin content in eight brands of St. John's wort. *American Journal of Health Systems Pharmacy*, 59: 545–547.

Ding, S., Dudley, E., Plummer, S., Tang, J., Newton, R.P., and Brenton A.G. 2006. Quantitative determination of major active components in *Ginkgo biloba* dietary supplements by liquid chromatography/mass spectrometry. *Rapid Communications in Mass Spectrometry*, 20: 2753–2760.

EMEA. 2006. Guideline on good agricultural and collection practice (GACP) for starting materials of herbal origin. http://www.ema.europa.eu/docs/en_GB/document_library/Scientific_guideline/2009/09/WC500003362.pdf.

EMEA. 2008. Reflection paper on markers used for quantitative and qualitative analysis of herbal medicinal products and traditional herbal medicinal products. http://www.ema.europa.eu/docs/en_GB/document_library/Scientific_guideline/2009/09/WC500003211.pdf.

Ernst, E. 1998. Harmless herbs? A review of the recent literature. *American Journal of Medicine*, 104: 170–178.

European Medicines Agency (EMEA). 2004. Final concept paper on the implementation of different levels of scientific evidence in core data for herbal medicinal drugs. Doc No. EMEA/CPMP/HMPWP/1156/03. London, March 3. http://www.ema.europa.eu/docs/en_GB/document_library/Other/2009/12/WC500018048.pdf.

Fabricant, D.S. and Farnsworth, N.R. 2001. The value of plants used in traditional medicine for drug discovery. *Environmental Health Perspectives*, 109: 69–75.

Fong, H.H. 2002. Integration of herbal medicine into modern medical practices: Issues and prospects. *Integrative Cancer Therapies*, 1: 287–293.

Ghosh, D. 2013. Chinese medicine opportunity. Nutrition Insight, pp. 20–23. http://www.foodingredientsfirst.com/Twofi.html.

Harkey, M.R., Henderson, G.L., Gershwin, M.E., Stern, J.S., and Hackman, R.M. 2001. Variability in commercial ginseng products: An analysis of 25 preparations. *American Journal of Clinical Nutrition*, 73: 1101–1106.

Jones, W.P. and Kinghorn, A.D. 2012. Extraction of plant secondary metabolites. *Methods in Molecular Biology*, 864: 341–366.

Jordan, S.A., David, G., Cunningham, D.G., and Marles, R.J. 2010. Assessment of herbal medicinal products: Challenges, and opportunities to increase the knowledge base for safety assessment. *Toxicology Applied Pharmacology*, 243: 198–216.

Keller, K. 1996. Herbal medicinal products in Germany and Europe: Experiences with national and European assessment. *Drug Information Journal*, 30: 933–948.

Li, X., Chen, H., Jia, W., and Xie, G. 2013. A metabolomics-based strategy for the quality control of traditional Chinese medicine: Shengmai injection as a case study. *Evidence-Based Complementary and Alternative Medicine*. http://www.hindawi.com/journals/ecam/2013/836179/.

Li, X.T., Chen, R., Jin, L.M., and Chen H.Y. 2009. Regulation on energy metabolism and protection on mitochondria of *Panax ginseng* polysaccharide. *American Journal of Chinese Medicine*, 37: 1139–1152.

Liva, R. 2009. Controlled testing. The cornerstone of all quality natural products. *Integrative Medicine*, 8: 40–42.

Loew, D. and Kaszkin, M. 2002. Approaching the problem of bioequivalence of herbal randomized clinical trial. *Phytotherapy Research*, 16: 705–711.

Newmaster, S.G., Grguric, M., Shanmughanandhan, D., Ramalingam, S., and Ragupathy, S. 2013. DNA barcoding detects contamination and substitution in North American herbal products. *BMC Medicine*, 11: 222.

Obi, E., Akunyili, D.N., Ekpo, B., and Orisakwe, O.E. 2006. Heavy metal hazards of Nigerian herbal remedies. *Science of the Total Environment*, 369: 35–41.

Perry, E. and Howes, M.J. 2011. Medicinal plants and dementia therapy: Herbal hopes for brain aging? *CNS Neuroscience and Therapeutics*, 17: 683–698.

Robinson, M.M. and Zhang, X. 2011. Traditional medicines: Global situation, issues and challenges. In *The World Medicines Situation* (3rd Edition), ed. World Health Organization. Geneva, Switzerland: World Health Organization.

Scaglione, F., Pannacci, M., and Petrini, O. 2005. The standardised G115 *Panax ginseng* C.A. Meyer extract: A review of its properties and usage. *Evidence Based Integrative Medicine*, 2: 195–206.
Schwarz, M., Klier, B., and Sievers, H. 2009. Herbal reference standards. *Planta Medica*, 75: 689–703.
Soldati, F. 2000. *Panax ginseng*: Standardization and biological activity. In *Biologically Active Natural Products*, eds. Cutler, S.J., and Cutler, H.G. New York: CRC press, pp. 209–232.
Soldati, F. and Tanaka, O. 1984. *Panax ginseng*: Relation between age of plant and content of ginsenosides. *Planta Medica*, 50: 351–352.
Sticher, O. 1993. Quality of *Ginkgo* preparations. *Planta Medica*, 59: 2–11.
Tyler, V.E. 1999. Phytomedicines: Back to the Future. *Journal of Natural Products*, 62: 1589–1592.
Van Breemen, R.B., Fong, H.H., and Farnsworth, N.R. 2008. Ensuring the safety of botanical dietary supplements. *American Journal of Chinese Medicine*, 87: 509S–513S.
Walker, R. 2004. Criteria for risk assessment of botanical food supplements. *Toxicology Letters*, 49: 187–195.
Williamson, E.M. 2001. Synergy and other interactions in phytomedicines. *Phytomedicine*, 8: 401–409.
WHO. 1996. WHO expert committee on specifications for pharmaceutical preparations. World Health Organization, Geneva, Switzerland. http://apps.who.int/medicinedocs/en/d/Js5516e/.
WHO. 1998. Quality control methods for medicinal plant materials. World Health Organization, Geneva, Switzerland. http://www.gmp-compliance.org/guidemgr/files/WHO_9241545100.PDF.
WHO. 2000. WHO guidelines for methodologies on research and evaluation of traditional medicine. World Health Organization, Geneva, Switzerland. http://apps.who.int/medicinedocs/es/d/Jwhozip42e/.
WHO. 2003. WHO guidelines on good agricultural and collection practices (GACP) for medicinal plants. World Health Organization, Geneva, Switzerland. http://whqlibdoc.who.int/publications/2003/9241546271.pdf.
WHO. 2007a. WHO monographs on selected medicinal plants. World Health Organization, Geneva, Switzerland. http://apps.who.int/medicinedocs/documents/s14213e/s14213e.pdf.
WHO. 2007b. WHO guidelines for assessing quality of herbal medicines with reference to contaminants and residues. Geneva, Switzerland, World Health Organization. http://apps.who.int/medicinedocs/documents/s14878e/s14878e.pdf.
World Health Organization (WHO). 1993. Research guidelines for evaluating the safety and efficacy of herbal medicines. Research Office for the Western Pacific. World Health Organization, Geneva, Switzerland. http://apps.who.int/medicinedocs/es/d/Jh2946e/#Jh2946e.
Zangara, A. and Wesnes, K.A. 2012. Herbal cognitive enhancers: New developments and challenges for therapeutic applications. In *Brain Aging and Therapeutic Interventions*, eds. Thakur, M.K. and Rattan, S.I.S. Dordrecht: Springer, pp. 267–289.
Zhang, J., Wider, B., Shang. H., Li. X., and Ernst E. 2012. Quality of herbal medicines: Challenges and solutions. *Complementary Therapies in Medicine*, 20: 100–106.

20 Innovation in the Food Industry
Industry–Academia Partnership

Amit Taneja, Anwesha Sarkar, and Shantanu Das

CONTENTS

20.1 INTRODUCTION

About 150 years ago, many products and services that we take for granted today such as the internal combustion engine, light bulb, refrigeration, and antibiotics were nonexistent. People were still living more or less like the peasants from the Roman Empire and did not even come close to resembling the way of life today. Therefore, how did such significant change occur in the twentieth century? To a large extent, the main reason behind this social transformation is innovation (Dodgson and Gann 2010). History clearly shows that innovation took place; however, the term *innovation* was not known and used back then. It was not until 1939 that the term "innovation" was created by Schumpeter in the field of economic science. Only in the past 40 years or so, research has been conducted focusing on how the process of innovation takes place and how it can be more successful (Kohn 2009). There is no doubt that inventions and scientific and technological breakthroughs provide inertia to bring change, improvement, and social progress, but these are not the only reasons (Zakaria 2011). Innovation is when these bright ideas are valued, organized, and managed by people and organizations. A new idea/scientific breakthrough being implemented in practice is innovation in progress (Dodgson and Gann 2010; Stabulnieks 2009).

The outcomes of innovation can be in the form of new products and services or can be in the form of organizational processes such as research and development (R&D) and marketing that impact the way products and services are delivered. It can take the shape of knowledge and judgment in an organization or can be a process that supports enhanced organizational learning. The majority of innovations are incremental in nature. They are additions to already existing products and services or modifications to organizational processes. Innovations can sometimes be radical in nature as they can be a new product or service, which is the first of its kind such as nylon or open-source software. However, on rare occasions, innovation can be transformational, which has a revolutionary effect on the way of life, for instance, the light bulb (Dodgson and Gann 2010).

We are living in a world where globalization is creating equality, and it is becoming difficult for organizations to stand out. Technological breakthroughs are being made at a surprisingly quick rate, and therefore, creating wealth from new and bright ideas has made innovation essential (Cavanillas 2009). Furthermore, we live in a world where population is growing at an ever faster pace, and to cope with challenges such as climate change, food security, energy sustainability, health, and education, the opportunities for innovation are immense (Dodgson and Gann 2010).

One feature of innovation is that it can be found in all sectors of business and in every organization. Like every other sector, innovation is an essential strategic instrument for the food industry to achieve competitive advantage and to be successful in the gaining of market share (Gellynck et al. 2007). However, innovation in the food sector is highly challenging and a relatively complex process compared to fast-growing, technology-driven sectors such as the informatics, communications, and automobile or pharmaceutical industries. Experts attribute this complexity at least in part to the overall R&D expenditure in these sectors. Computing and electronics, health care, software, the Internet, and automotive-aerospace sectors together accounted for more than two-thirds of the global bulk R&D expenditure in 2008 in spite of the global financial meltdown. On average, healthcare organizations such as Novartis, Roche, and GlaxoSmithKline spend around 12% of the total sales on R&D compared to 0.8% by food and beverage companies (Jaruzelski and Dehoff 2009).

Besides being low budget and low technology, innovation in the food industry involves a number of stakeholders from different sectors such as agricultural producers, suppliers, food processing industries, machine manufacturing units, food regulators and legislators, customers, and consumers, which often makes alignment between stakeholders difficult (Sarkar and Costa 2008). Another issue with food innovation lies in traditional consumers' eating patterns. Radical or real breakthrough food innovations are not often introduced in the market owing to reluctance to novel food types. For instance, consumer perceptions of risk and uncertainty did not allow penetration of technological breakthroughs such as gene technology and nanotechnology in mainstream food industry (Beckeman and Skjöldebrand 2007). Moreover, food has cultural, traditional, and often religious values, and has evolved over centuries to meet mankind's taste preference. Therefore, making major changes to consumer sensory experience through innovation seems to be unlikely (Boland 2011). Generally, innovation cannot be carried out by a single organization on its own. Association with suppliers and customers is essential. In addition, local government policy, university research support, and legal systems that influence business and local context are crucial to successful innovation (Dodgson and Gann 2010). In all the major elements of food industry, progress depends not only on owning and using knowledge but also on being innovative with the resources available in the form of people, processes, and technology, and how they connect inside and outside the organizational boundary. Given the large number of stakeholders involved in the production of processed food products, cross-boundary innovation and a high degree of communication should be a widespread practice. However, examples of innovation in the food industry are hard to find in the published text. Case studies are mainly available in the form of anecdotal evidence which are reported online (Sarkar and Costa 2008). Thus, the main goal of this chapter is to address this issue and analyze the drivers of innovation with example from all functions keeping a focus on the knowledge created through R&D. We have also discussed how accessing additional knowledge and skills across the boundary of organizations through open innovation (OI) or connected innovation is important with the help of an example from the food industry around the world and especially from New Zealand food industry. Subsection 20.4 endeavors to provide an example of connected innovation from an academic research institute's perspective.

20.2 DRIVERS OF INNOVATION

What drives innovation? From a historical perspective, it becomes clear that the most important driver of innovation is people with energy, imagination, and the keenness to identify and solve problems (Dodgson and Gann 2010; Zakaria 2011). Isaac Merritt Singer, the inventor of the sewing machine, had the energy and imagination to sell his invention to women at a time when women were assumed to be unable to operate any type of machinery (Zakaria 2011). However, the structures and procedures adopted by organizations also have a huge impact on its innovation objectives. Whether it is the structure of teams, the sources to generate ideas or training, and the incentives for employees, organizations have a choice on how they organize themselves to cope with the challenges in being innovative. For instance, DuPont was one of the few companies to invest in basic research in the mid-twentieth century. Their scientists were given vast amount of freedom to work on wider research goals with the objective of establishing or discovering new scientific facts. This approach led DuPont to make many breakthroughs including the development of Kevlar (Kwolek 1974). Kevlar was a lightweight, fire-resistant fiber that was stronger than steel, developed by Stephanie Kwolek, one of DuPont's scientists. Kwolek's wider research objective was to develop a lighter fiber for use in manufacture of fuel-efficient car tires. She discovered Kevlar in 1964 and it was not until 1971 that its commercially successful application of bulletproof armor was introduced. Kevlar has saved the lives of many police officers worldwide and it symbolizes the contribution of DuPont and Kwolek to innovation as it took DuPont 25 years to develop and commercialize this tough fiber. Kwolek's creativity, experience, and knowledge, and the organizational practices of DuPont led her to successful development of many other fibers such as Lycra and Nomex (Dodgson and Gann 2010). Albeit people and organizations are two different entities in the innovation chain, it is hard to overlook their interdependence. In Subsections 20.2.1 and 20.2.2, we discuss them individually to highlight their distinctive importance.

20.2.1 People

Even though not essential, R&D is a significant source of innovation. R&D can be done in fundamental sciences or in other words can begin with basic research driven mainly by people who are curious and aim to increase the knowledge wealth without having an application or end use in mind. Alternatively, if R&D activities are done with a specific practical or industrial use as the aim, they are considered to be applied research. However, there can be a third type of R&D activity that is a combination of the two classes. According to Donald Stokes, this kind of research activity is termed as "use-inspired basic research." James Watson and Francis Crick published an article in *Nature* in 1953 suggesting a new double-helix model of DNA, which at present is universally accepted as the standard model. The scientists noted at the time that their new model had significant importance in biological science. However, no commercial application occurred to them at that time. Interestingly, today the use of the understanding of the DNA model in bioinformatics, forensics, and genetic engineering is all based on Watson and Crick's work.

However, Thomas Edison's (1847–1931) name is taken synonymously with innovation. With over 1000 patents to his name, his notable inventions such as the light bulb and the power distribution system are a few of the many remarkable contributions that he has made to the society. Edison's research was driven by application and end use. He claimed that all his inventions that he created without considering their end use were never perfected. He spent a lot of time thinking and researching on what was the biggest consumer need and then started his R&D to create a prototype. Louis Pasteur, however, always had an application in mind, but his work was always directed toward the quest for fundamental knowledge that led him to create a new field of scientific research now known as medical microbiology. Thus, his research objective fits the third category as described in Subsection 20.2.1 (Dodgson and Gann 2010).

Applied research may have its starting point from the closing stages of a basic research, and the basic research may be carried out to better understand how an existing technology works. The most important attribute of people involved in innovation is creativity and teamwork. In the words of double Nobel laureate Linus Pauling, "the best way to have a good idea is have lots of them."

Bringing together creative people with different ideas and expertise is important for innovation to have a disruptive element. More creative teams tend to have flexibility in objective, whereas less creative teams have objective that must reflect on the bottom line in short-term periods. People in leadership roles also play an important part in the encouragement of creativity and support of new ideas. By creating a support culture, leaders can keep team members motivated by not discouraging them when their research fails. This culture of creativity has been clearly evident in the twentieth century with examples of Kwolek and Edison (Dodgson and Gann 2010). However, is it possible to have a long-term and flexible approach to R&D nowadays?

There is no doubt that the life cycle of innovation has changed drastically in the twenty-first century due to advances in science and technology. It took radio 38 years to get to an audience of 50 million people in the early 1900s. Television took 13 years to accomplish the same, whereas the Internet and Facebook only took four and two years, respectively, to have more than 50 million subscribers. Ten years ago, if you searched for words such as iPhone, Gmail, or YouTube on the Google search engine, it would have shown no results. However, today all these products and services have more than 50 million users. This also means that the new skills, qualifications, and specializations are required to cope with these changes (Cavanillas 2009).

The role of an R&D professional has evolved dramatically in the past 10 years. It seems that a credible scientific background and creativity are no longer enough. This is because the corporate R&D department has now changed from being a closed entity of internally organized group of scientists to a distributed network of innovators even outside the organization. Product life cycle in the food sector has become increasingly small and budgets have become increasingly modest. The ideal R&D professional must have the ability to understand the overall growth objectives of the company and how R&D fits into these objectives to generate innovation-based growth. The new R&D professionals must try to bridge the gaps between science and business. With globalization and emerging markets, the issues of affordability, availability, and localization have become increasingly important and have to be addressed. A high degree of collaboration and speed to market have become the most important factors for successful innovation (Shields et al. 2006).

20.2.2 Organizations

An organization is a social unit with an identifiable boundary. It is a consciously managed and coordinated structure in which members constantly interact with each other and decision making is conducted through a management system to achieve common set of goals. An organization can be a for-profit business, for example, Fonterra Co-operative Group Ltd., New Zealand, or a not-for-profit social business, for example, Bill and Melinda Gates Foundation, Seattle, WA. The responsibility and authority of the organizational members is defined by the structure of every type of organization. This structure also dictates how its members interact with each other. The progress and growth of an organization is dependent on external factors (e.g., market dynamics, government policies, the competitive landscape, etc.) as well as internal factors (team composition, flexibility in work, culture, etc). In many modern organizations involved in R&D, professional bureaucracy and adhocracy have become the preferred organizational structures. Although these contemporary organizational structures are reasonably effective in coping with challenges presented by the fast-paced R&D environments and in facilitating creative thinking and skill development of a vast number of professionals, they have a number of limitations too (Das 2012). The characteristics and limitations of these two modern organizational structures are discussed and some suggestions regarding overcoming these limitations based on published research are provided.

A professional bureaucracy is characterized by the skill of the people (professionals), who work relatively independently without direct supervision by superiors or interference from colleagues, but have to maintain close contact with their clients (Mintzberg 1993). Examples of professional bureaucracies are universities where professors work relatively independently of supervisors and colleagues but remain closely connected with the students. They have highly standardized skill sets that have been developed through university's education and training as well as in professional associations. There are two main disadvantages with such organizations in terms of being innovative. First, the various professional functions/departments tend to pursue their own defined objectives, often placing their individual goals over the organizational interest (Das 2012). Second, in many cases, universities have been very concerned with generating revenue from their research through spin-off companies and intellectual properties (IPs). Even though there have been a few impressive success stories utilizing this model to commercialize innovative research, such activity is only a tiny fraction of the entrepreneurial activity that the universities are capable of delivering (Dodgson and Gann 2010). The purpose of most small- and medium-sized businesses of collaborating with universities involves technical problem solving. Bigger companies, however, are usually looking to engage in a broader dialog with universities to learn more about the upcoming research areas and directions. There is no doubt that universities are the foundation of inventive ideas and knowledge; however, effective ongoing communication of their capabilities is crucial for collaborative university–industrial innovation process.

An adhocratic organizational structure consists of professionals with varied skill sets, with the absence of decision making and proper formalization. An adhocracy is characterized by intensive coordination, high flexibility, and high responsiveness, which make this form of organization better geared to solve new problems (Das 2012). This type of organization structure also tends to be more innovative as members are working toward achieving a team goal rather than an individual objective. An example of an adhocracy is a project team in a multinational corporation. The process of innovation (e.g., to launch a product that does not exist in any of the divisions) in such organizations usually starts with the formation of a cross-functional project team comprising professionals from different departments. Although the team has a project leader, there is no hierarchy within the team. Interestingly, each team in the project is a functional leader and brings its expertise for success of the project. In many cases, the team members are geographically dispersed. The team is geared to solve a new problem and is flexible enough to change direction if changes in the internal and/or external environment occur (Das 2012).

Another example of adhocracy is the Riddet Institute (RI) in New Zealand. Five leading universities and crown research institutes collaborated to set up this institute as a Centre of Research Excellence in food and nutrition. In this model, the most skilful professionals with various areas of expertise throughout New Zealand were pulled together to work on a well-defined research objective. Many of the experts were employed by a university (professional bureaucracies) and pursued their particular interests; however, they collaborated successfully to address the problems that require multidisciplinary expertise under the umbrella of the RI (an adhocracy). Some examples of the RI projects have been provided toward the end of this chapter. As there is no formal reporting relationship, conflict is a common problem that an adhocratic project team faces. Coupling its ambiguities with its interdependences, an adhocracy emerges as a more politicized structure compared with a professional bureaucracy (Mintzberg 1993). However, the main advantage of an adhocracy is its capability of innovation. We mentioned in Subsection 20.2.1 that the creativity and leadership qualities of Edison were quite remarkable, but he is also known for pioneering the way of organizing innovation. Edison was of the firm belief that innovation was not an end result of individual brilliance but of collaboration and due to people working across boundaries creating a supportive environment. His research facility located at Menlo Park was considered to be a playful, nonbureaucratic, and informal environment where socializing was highly promoted. This facility employed 200 people and produced 400 patents in about six years (Dodgson and Gann 2010). The challenge for modern R&D organizations is not only how they choose to be structured, that is, bureaucratic and formal or informal and unrestrictive, but also how they balance between long-term research that may result in new scientific discoveries that may eventfully lead to disruptive innovation and short-term research to solve well-defined technical problems. Short-term research has a consumer-oriented view that is important for growth of organizations; however, neglecting long-term research may relegate valuable innovation opportunities (Dodgson and Gann 2010).

20.3 OPEN INNOVATION/CONNECTED INNOVATION

A relatively new innovation strategy that most firms use to improve their internal R&D capability and to access external collaborators for shorter innovation life cycle along with improved chances of success has been described by Henry Chesbrough as "open innovation" (Dodgson and Gann 2010; Sarkar and Costa 2008). Traditionally, innovation is "closed," carried out within an organization, and limited by the organization's resources. By contrast, using OI, a company uses not only its internal resources but also its external resources, collaborating with research institutions, ingredient and technology suppliers, and sometimes even consumers (Sarkar and Costa 2008). A good example of a company that has used OI extensively in the past decade is the International Business Machines Corporation (IBM). The IBM has always been widely recognized for its innovative approach since its creation of a mass market for the personal computer in the late twentieth century. Even though more than half of the people employed at the IBM are scientists and engineers, the IBM's continual pursuit of innovation is highlighted by its change in innovation strategy from closed innovation to OI. One of their OI strategies launched in 2006, known as the "innovation jam," involved the creation of a web portal inviting staff, their families, business partners, university researchers, and clients to submit innovative ideas on various predefined areas of strategic interest. They received 40,000 entries within three days of the launch that were then refined to 10 smart business ideas. These business ideas were pushed forward with funds equivalent to US$70 million. The company has reported to have generated a market outcome of nearly US$300 million from these 10 businesses (Dodgson and Gann 2010).

Food industry is gradually moving from a closed innovation to an OI paradigm. Major multinational companies (e.g., Procter & Gamble [P&G], Nestlé, General Mills [GM], and Fonterra) have adopted the OI models to create value over the last decade or so. All the OI models used are based on the principle of collaborating with universities, research institutes, key suppliers, and other stake holders to create value for the consumers and shareholders, as well as create wealth for the respective nations. Most noteworthy illustration of OI in practice has been shown by GM that have launched over 40 products leveraging OI in fewer than 10 years after adopting the OI model in 2004. The product launches leveraging OI were also 1.6 times more successful in the marketplace (Starling 2009). OI is not only beneficial for large companies. Studies in Europe showed that small and medium enterprises (SMEs) using OI have a much higher success rate (55.4%) compared with traditional closed innovators (35.2%). In Subsections 20.3.1 through 20.3.3, we provide examples of OI successfully applied within the food industry and some of the challenges for the implementation of OI (Spitzley et al. 2007).

20.3.1 P&G's OI Model

In the past, many leading companies in the world relied on their larger central laboratories to undertake research. They would employ a large number of scientists with massive research money to make discoveries and file patent applications often to discourage competitors. Bell Laboratories in Murray Hill, New Jersey, most notably known for their communication satellites and motion picture technology developments, had 25,000 staff, 30,000 patents filed for the company, and received six noble prizes for their scientific discoveries. One of the key problems of such a way of organization of R&D in today's business environment is that they are generally long term in orientation. Even though radical innovations are often made, the chances of faster value creation are less likely due to the short life cycle of products (Dodgson and Gann 2010).

P&G, in 2000, was in a similar situation where the growth objectives of the company were not being met. P&G's newly appointed CEO, A.G. Lafley, quickly realized that spending more money on internal R&D was not going to translate into a larger market outcome. Lafley was committed to reinventing the company's innovation model. Till then, most of P&G's top innovations were based on internal cross-functional idea generation and R&D. Over the years, before Lafley's appointment, company records indicated that a small number of products/services the company had acquired beyond the company's boundaries were highly profitable innovations compared to their own internally generated innovations. Lafley outlined the company's new innovation strategy betting on OI as the key to future growth. His aim was to source at least half of the innovation from outside the company's boundaries. His strategy was to attempt, in a complementary way, to better leverage the capabilities of the 7500 researchers and support staff that the company employed by enhancing external connections. "Not-invented-here" syndrome was a common problem with R&D departments of most big corporations worldwide in the early-to-mid-twentieth century. However, in the late twentieth century, this slowly started to shift toward "proudly found elsewhere" attitude (Huston and Sakkab 2006).

By connecting outside the boundary of the company, P&G estimated that the company had a network of 1.5 million talented people whose skills could potentially be used by the company even though the internal staff only comprised 7500 researchers and support staff. This meant that each P&G researcher, on average, was connected to 200 equally talented scientists or engineers elsewhere in the world. This newly created OI model was called "connect and develop." This model meant that with a company's clear consumer-oriented research, the identification and development of promising ideas into real products became cheaper and faster. In addition, tapping into local market across the world was then more efficient with increased local knowledge through company's new connections. In 2000, P&G only had 15% of products in their portfolio which were developed with significant collaboration from outside the company. Today, more than 35% of the new products are introduced to markets worldwide through their connect and develop OI model. Their OI strategy has not only helped launch more successful products to the markets but also made significant improvements in other aspects of innovation such as marketing and operations. The company has reported an increase in R&D productivity by ~60%. This strategy had a clear implication for the bottom line of innovation costs as P&G's R&D investment as a percentage of sales has gone down by 1.2% since 2000. The innovation success rate doubled and almost 100 new products were launched, which had a key aspect acquired from outside the company during development (Huston and Sakkab 2006). In 2008, Lafley said that "more than 40% of our innovation comes from outside the United States. People in India, China, Latin America, and some African countries have become part of our social system. Their presence has made us more open, and this helps compensate for our natural tendency to become more insular" (Lafley 2008).

20.3.2 GM's OI Model

GM is one of the world's leading food companies and has operations in more than 100 countries with an annual turnover of US$16 billion. Cheerios, Haagen-Dazs, Nature Valley, Yoplait, Betty Crocker, Pillsbury, Green Giant, Progresso, and Old El Paso are some of the consumer brands owned by GM. GM officially launched its OI program "General Mills Worldwide Innovation Network (G-WIN)" in 2007 to enhance and accelerate its innovation efforts. G-WIN was based around efficiently matching GM business needs with the talented people from outside that had the potential to meet the needs and solve problems. G-WIN focused on food technology, products, packaging, ingredients, and processing, and, since its launch, has connected with thousands of world-class innovators around the globe (Starling 2009).

In 2009, GM improved its G-WIN innovation model by opening up an innovation portal making it easier for external partners to connect with GM. Scientists, researchers, engineers, inventors, and entrepreneurs around the world have access to G-WIN portal where GM lists its top 50 technology challenges for viewers to respond with their innovation solutions. There is also an option of submitting personal novel ideas on the portal. One of the examples of collaboration through G-WIN was when a $200,000 research grant was awarded to researchers at the University of California–Davis in 2011. Their proposal was aimed at mapping the vanilla genome to develop hybrid varieties of vanilla which were more resistant to disease with the ultimate goal of crop improvements and supporting farmers. This was the winning entry in the sustainability challenge set up by GM where universities worldwide were asked to submit their proposal with their best ideas for encouraging sustainability and responsible usage of resources (Markets 2011).

20.3.3 Nestlé's OI Model

Nestlé is the largest food company in the world rolling out innovative products such as Nescafé cappuccino and BabyNes infant nutrition system. In 2009, Nestlé had an R&D expenditure of CHF 2.021 billion (1.91%), which was significantly higher compared to that of its nearest competitor Kraft (US$ 449 million, 1.2%) based on the overall revenues, despite the global economic downturn (Traitler et al. 2011). Although it is difficult to draw a direct correlation between R&D spending and financial returns of a company, it is clear that consumer-focused innovation had been the main driver augmenting the long-term growth of Nestlé.

Like most other multinational organizations, Nestlé depended entirely on its internal R&D capabilities. However, the paradigm shift in view of introducing OI for accelerating codevelopment of sustainable innovation was recognized by Helmut Traitler in 2006 by creation of Nestlé's Innovation Partnerships (INP) (Traitler and Saguy 2009). The "sharing-is-winning" model proposed by Traitler and Saguy (2009) has been embedded into the innovation strategy of Nestlé. As shown in Figure 20.1, it was a model of cocreation and codevelopment with global partners of complementary strengths such as universities, academic research institutes, individual inventors, small start-ups, biotechnology companies, and key strategic industrial suppliers through alliances, cooperation as well as joint ventures. Working with universities was crucial as fundamental research was generally the starting point of potential cutting-edge innovations. However, anecdotally Nestlé's collaboration with suppliers had resulted in faster product launches in the past.

The main objective was value creation along the entire value chain for all the stakeholders, building of goodwill, and establishment of trust and winning respect (Traitler et al. 2011). However, the objectives, risks, and benefits were shared openly and clearly with all the innovation partners from the beginning to create tangible value and future coinnovation opportunities for all participants without additional internal R&D resources. With strong internal R&D capabilities, Nestlé created a win–win situation connecting internal resources with external partners in collaborative projects. This model had enabled Nestlé to build numerous key partnerships with companies such as Barry Callebaut, BASF, Cargill, Cognis, DSM, DuPont, Firmenich, Fonterra, Givaudan, IFF, Kerry, Mane, Symrise, and Tetra Pak, which led to an output of US$200 million in new businesses within three years of implementation (Traitler et al. 2011).

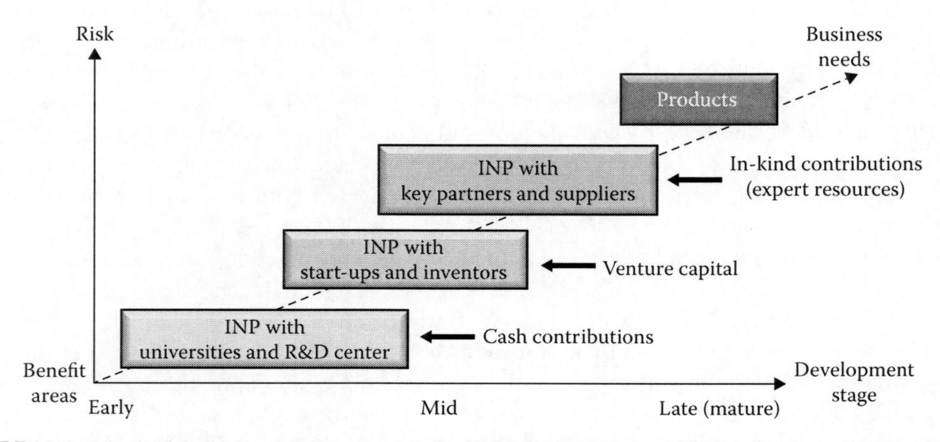

FIGURE 20.1 Typical development stages and risk in "sharing-is-winning" model. (Traitler, H., Watzke, H.J., and Saguy, I.S.: Reinventing R&D in an open innovation ecosystem. *Journal of Food Science*. 2011. 76. R62–R68. Copyright Wiley-VCH Verlag GmbH & Co. KGaA. Reproduced with permission.)

Although the sharing-is-winning model offers coinnovation opportunities for mutual benefits, its implementation in reality was not trivial for Nestlé. Sharing of knowledge without losing confidentiality, ownership of IPs, and lack of alignment were few of the obstacles faced by this giant food player. Initially, Nestlé implemented confidentiality agreements. Later, formalized agreements called master joint development agreements (MJDAs) were introduced which included the scope of the expected collaboration in the first phase (Traitler et al. 2011). The MJDA also comprised a more detailed part describing the project aims and expectations, resources and cost, time schedules, milestones, ownership rights of IPs, and so on. This enabled the innovation partners to smoothly collaborate with Nestlé with better trust and minimized exclusivity conflicts. Thus, INP at Nestlé showed a pioneering approach of implementation of OI for market successes.

However, one of the known risks in the sharing-is-winning OI model was that the definition of value creation was sometimes different for different innovation partners. For example, while value in industries, such as Nestlé, largely included consumer acceptance, growth, market share, profitability, and financial returns to the shareholders, in academia, value creation was mainly in terms of basic knowledge generation, student education, and publications. In addition, IP conflicts were not uncommon where academic partners seek full IP rights on the invention.

Hence, to significantly achieve meaningful outcomes of OI, Saguy (2011a, 2011b) recommended implementing four paradigm shifts for academia–industry collaborations. The paradigm shifts were (1) breaking of the conventional barrier between academia and industry, that is, academia should not only stand out in basic science but also transform the breakthrough invention into a viable concept for the business to support; (2) revision of the IP model to avoid exclusivity conflicts from the beginning; (3) support from management at each stage of innovation to reduce constraints; and (4) finally increased social responsibility by evaluating research's overall societal contribution rather than quantifying with publications, impact factor of journals, and so on.

20.4 OI AND NEW ZEALAND FOOD INDUSTRY

The importance of biological sector, in particular the food industry in New Zealand, is enormous, as it contributes to the majority of gross domestic product (GDP) as well as exports. Globally, New Zealand is regarded as a source of safe and high-quality food. Future economic development of New Zealand will greatly rely on the growth and competitiveness of the food sector, especially export earnings. As the cost of production in New Zealand is relatively higher, the food industry will require significant value addition to reap the maximum benefit from food exports. As SMEs account for 40% of the economy's total output on a value-added basis and 31% of all employment, the participation of SMEs in value creation and export is very much a necessity (Ministry of Primary Industries 2013).

OI programs, in general, require a significant amount of investment. This high level of investment or commitment of resources may be feasible for large corporations but is often a limitation for small- and medium-scale companies. In general terms, a company needs to have the capability to transform a scientific output into a tangible technology or product to be able to harvest economic benefits. It is very important that a company has its own R&D unit and innovation plan or strategy to analyze the trends in the market, identify the gaps and the relevant science with potential to deliver solutions, and finally engage in development and commercialization. However, most small- and medium-scale industries do not have resources or capabilities to carry out these steps and therefore are unable to take full advantage of the scientific/research outputs available to them (Cohen and Levinthal 1990).

New Zealand public sector research in the biological sciences is very strong, and the portion of government R&D funding is higher in New Zealand compared to other countries in the Organisation for Economic Co-operation and Development (OECD) group (MSI 2013). Publically funded scientific research in New Zealand has been known to generate very high-quality output, especially in the biological science area. The total number of scientific publications in the area of biological science from New Zealand was 1716 in 2010 (SCImago 2013). Application of these scientific outputs may solve a range of issues the industry faces and may be able to add value to the products or waste stream. However, in practice, the scientific research outputs are seldom adopted by the industry and even less likely to be adopted by SMEs. The economic benefit from the scientific research remains underachieved.

The productivity and innovation capability of organizations having partnerships with universities is higher than those who do not, and the organizations involved in research in collaboration with universities in particular earn substantial benefits in increased productivity, profitability, and innovation (Coopers and Lybrand 1995). This fact has been highlighted in the innovation policies of three cases presented in Subsections 20.3.1 through 20.3.3, respectively. In recent years, the notion that "ideas and knowledge" are produced by universities and transmitted to industry has changed. As shown by GM, P&G, and Nestlé, the belief now is to cocreate and exchange (Dodgson and Gann 2010). Recent research carried out at the RI has found out that about 70% of New Zealand food companies considered high cost and lack of technical skill as barriers to innovation in functional foods, but less than 20% companies considered New Zealand public research system (universities and Crown Research Institutes) as a source of ideas and knowledge (Khan et al. 2012). It is evident that a more effective communication and technology transfer systems between the research organizations and the industry in New Zealand are required.

We have introduced the RI in Subsection 20.2.2 as a virtual research institute, which combines some of the best talents in food science, technology, and related area. Majority of the initiatives at the RI were in fundamental research. A new product development and innovation group (PDIG) has been established relatively recently to drive the OI agenda and promote collaboration with domestic and international food industries. The last part of this chapter discusses the key strategy that PDIG employed with examples of some noteworthy initiatives. This section is based on the authors' personal experiences as Shantanu Das initiated and led the group for three years and Amit Taneja has been a key member since the inception.

The PDIG at the RI took three different innovation routes but with a common objective, which was to make science and technology available to the consumer through industrial collaboration with a financially sustainable model. The first route was through opportunity identification. A considerable amount of resource was set aside to understand the current trends in health and nutrition, map out the current commercial offerings, and use internal insights to understand the research pipeline. Efforts were directed toward the identification of the whitespace opportunity. By thorough internal capability evaluation and a series of discussions and interviews within the RI, opportunity areas were identified. Following this, objectives were set and resources were put in place to achieve these goals as quickly as possible. One notable aspect of this endeavor was its relatively lower depth and larger width. The core of the work was to prove a scientific principle, that is, besides basic food science, new technology/product development involves a lot of other aspects including scalability, regulatory status, potential applications, costing analysis at different points of the supply chain, and potential collaborations for commercialization. Once the principle was proved, more detailed scientific research was initiated to add the depth of scientific understanding.

An example of this route is the case of the development of a probiotic delivery technology. After the opportunity was identified, the team worked on developing multiple hypothesis based on the information from the industry and academic research as well as their collective experience. These hypotheses were tested rapidly and a working hypothesis was identified. The next piece of work was directed toward proving the principle, specifying the product and processes, and developing applications. By following this route the objective of developing a novel probiotic ingredient was achieved. A process patent was also filed detailing the manufacture of this novel technology (Das et al. 2011a). Interestingly, while working on this part of the research objective, that is, development of a successful technology, PDIG did not divulge in understanding the detailed science behind the technology. However, eventually a number of academic research projects were commissioned to develop that scientific understanding, which successfully answered and explained some of the underlying scientific questions. Potential industrial partners were identified with industry for commercialization. Another example of this route was the development of a patented iron delivery technology, which followed a near-identical development process as described for probiotic delivery technology (Mittal et al. 2012).

The second route of product innovation is through opportunity analysis of existing research outputs. Scientists at the RI had carried out a number of in-depth research projects in certain areas in the past few years, some of which were thought to have high commercial potential. A thorough analysis of the potential application(s) of these research outputs was carried out; all possible applications were screened through different filters, such as product–concept fit, scalability, regulatory issues, existing scientific backup, and further requirement for research, social and environmental issues. Following this process of opportunity analysis, one or two opportunities are identified, which are most practicable for commercialization. An example of this route of innovation involved ovine immunoglobulin research. A PhD project within the RI in this area showed various functionalities, such as upregulation of immune system, growth promotion, balancing intestinal microflora toward positive direction, and upregulation of mucine gene. Through opportunity analysis, four potential applications, such as companion animal's immunity boosting foods, companion animal's dental health products, food for human, and nutraceutical for human through the filters discussed in this subsection, were identified. Companion animal's immunity boosting food was then identified as the one with the highest potential, and application research is now under way to validate the research in product matrix (Balan et al. 2012).

Another example of this innovation route was the commercialization of a technology for producing a high-value ingredient from the meat industry by-products (Das et al. 2011b). Significant amount of research had been carried in this area before the formation of PDIG. Successful opportunity analysis and application development work by PDIG stimulated significant interests in the industry for this technology. Finally, this technology was patented and successfully licensed to a leading New Zealand meat company to further develop and produce commercial high-value ingredients.

The third route of product innovation is through the extension of already commercialized technology existing in the RI's technology pipeline. This was carried out by means of the standard stage-gate process used by food companies. The conception of this process starts with consumer insights mapped out by the marketing unit, which are then translated into a concept and refined with continuous customer feedback. One example of the use of this route for product innovation was the development of omega-3 stick-shot for Speirs Nutritional Ltd., Marton, NZ. The RI's patented omega-3 microencapsulation technology had been launched in 2008 in the form of an emulsion in partnership with Speirs Nutritional (Singh et al. 2006). PDIG developed a range of products based on the concepts developed by Speirs Nutritional and Croda International Plc, East Yorkshire (international marketing partner) marketing teams. The most significant product in the pipeline was a shot, which delivered 500 mg of eicosapentaenoic acid (EPA) and docosahexaenoic acid (DHA) in 5 g of product serve. Ultimately, the spin-off company, Speirs Nutritional Ltd., along with the omega-3 microencapsulation technology was acquired by Croda in 2011. Croda, aware of the significant market growth of omega-3 soft gel capsules, recognized that a differentiated offering in the same product segment could meet the consumer needs of taste and convenience. The microencapsulation emulsion technology (Omelife) complemented Croda's Incromega lipid range that already offered highly concentrated omega-3 ingredients with proven health benefits.

20.5 FUTURE OUTLOOK

The future of food innovation holds huge opportunities and has many challenges. In fact, innovation is the only possible future. Innovation will create new opportunities, without which we are sure to see saturated markets with supply greater than demand, slow economic growth, and social crisis (Cavanillas 2009). There will be challenges as well, where current rules would not apply, textbooks will surely be rewritten, and processes will be changed beyond recognition (Davis 2009). Our world population is headed toward nine billion by 2050. We would need twice the food that is produced today to meet the future demand. Currently, farming practices cannot cope with this. Adding climate change into the mix, you have "crisis on a mass scale" written all over it. History suggests that when the human race was challenged with such a problem, an agricultural revolution brought about the necessary transformation with increased efficiency of crop production. An additional challenge in this area is that a larger portion of food is converted to nonfood material, that is, biofuel (Boland 2011). Out of the major food components, food proteins present a major challenge. Some companies have taken a radical approach of addressing the issue of protein shortage. Food companies such as Beyond Meat, CA, USA are one of the examples of how innovation in the food industry is trying to tackle this problem (Gates 2012). It has been estimated that the production of 1 kg of animal protein (beef, lamb, etc.) requires 100 times more water compared to equivalent quantities of plant-based protein. In terms of energy efficiency, beef contains 206 g of protein per kg but uses 47 MJ of energy to produce a kg of protein. This, when compared with maize, is 10 times less energy inefficient (González et al. 2011). By using pressure and heat treatments, scientists and technologists at Beyond Meat have come up with an innovative way to create a food product that tastes and feels just like chicken in the mouth but is derived completely from plant sources. By substituting for animals as a protein source, much more land can be used to cultivate plant-based sources of proteins and potentially feed more people (Gates 2012). However, the challenge in food innovation does not end with an efficient use of resources to produce enough for everyone. It also needs to focus on the large population of people who suffer from malnutrition. According to the Food and Agriculture Organization of the United Nations (FAO), one-third of the world's population is deficient of one or more micronutrients. Malnutrition can also lead to diseases that continue to burden the local health system. This calls for foods to be supplemented with the right nutrients and also for better education regarding balanced nutrition (Boland 2011). Technologies developed by scientists at the RI for delivering omega-3, probiotics, and iron showcase the innovative approach used to deliver these key components in a variety of commonly consumed foods. Along with undernutrition, another significant issue is the excess consumption of certain food components. Excess intake of sodium is an example of such type of malnutrition, where its consumption is far more than what the human body can utilize. Controlled clinical trials, epidemiological studies, and animal trials have all suggested a strong association between the increased sodium consumption and the incidence of hypertension, a risk factor for heart disease (He and Kelly 2013). One example of food innovation has been the development of salt substitutes that taste just like salt but have no or very low quantity of salt. These have been shown to improve the risk factor for heart disease in clinical trials (Gates 2012). In sum, the future of food innovation is bright because we as humans will always be forced to rethink giving birth to new ideas. The food industry has this unique opportunity not only to use innovation for profit but also for the good of humanity.

REFERENCES

Balan, P., Rutherfurd, S.M., and Moughan, P.J. 2012. Oral healthcare product for companion animals. New Zealand patent application no. 602316.

Beckeman, M. and Skjöldebrand, C. 2007. Clusters/networks promote food innovations. *Journal of Food Engineering*, 79(4): 1418–1425.

Boland, M. 2011. Food of the future. In *Floreat Scientia*, ed. Riddet Institute. Auckland, New Zealand: Wairau Press, p. 256.

Cavanillas, E.S. 2009. The future of innovation is the only possible future. In *The Future of Innovation*, eds. B. von Stamm and A. Trifilova. Surrey: Gower Publishing, p. 502.

Cohen, W.M. and Levinthal, D.A. 1990. Absorptive capacity: A new perspective on learning and innovation. *Administrative Science Quarterly*, 35: 128–152.

Coopers and Lybrand. 1995. *Audit Committee Guide*. New York: Cooper and Lybrand.

Das, S. 2012. *Selected Topics in Management with Examples from the Food Industry.* Saarbrucken, Germany: Lambert Academic Publishing.

Das, S., Nag, A., and Singh, H. 2011a. Process of producing shelf stable probiotic food. New Zealand patent application no. WO2012/026832 A1.

Das, S., Singh, H., Moughan, P.J., Henare, S.J., Cui, J., Wilkinson, B., and Chong, R. 2011b. Meat protein hydrolysate technology. New Zealand patent application nos. 591556, 591557, 591558.

Davis, T. 2009. The future of innovation is for us to decide. In *The Future of Innovation*, eds. B. von Stamm and A. Trifilova. Surrey: Gower Publishing, p. 502.

Dodgson, M. and Gann, D. 2010. *Innovation: A Very Short Introduction*. Oxford: Oxford University Press.

Gates, B. 2012. The future of food. The Gates Notes. http://www.thegatesnotes.com/Features/Future-of-Food#.

Gellynck, X., Vermeire, B., and Viaene, J. 2007. Innovation in food firms: Contribution of regional networks within the international business context. *Entrepreneurship and Regional Development*, 19: 209–226.

González, A.D., Frostell, B., and Carlsson-Kanyama, A. 2011. Protein efficiency per unit energy and per unit greenhouse gas emissions: Potential contribution of diet choices to climate change mitigation. *Food Policy*, 36: 562–570.

He, J. and Kelly, T. 2013. Commentary: Sodium and blood pressure: Never too late to reduce dietary intake. *Epidemiology*, 24: 419– 420.

Huston, L. and Sakkab, N. 2006. P&G's new innovation model. Working Knowledge for Business Leaders, Harvard Business School. http://hbswk.hbs.edu/archive/5258.html.

Jaruzelski, B. and Dehoff, K. 2009. Profits down, spending steady: The global innovation 1000. *Strategy+Business Magazine*. http://www.booz.com/media/uploads/Innovation_1000-2009.pdf.

Khan, R.S., Grigor, J., Winger, R., and Win, A. 2013. Functional food product development—Opportunities and challenges for food manufacturers. *Trends in Food Science and Technology*, 30: 27–37.

Kohn, S. 2009. The future of innovation is in our hands. In *The Future of Innovation*, eds. B. von Stamm and A. Trifilova. Surrey: Gower Publishing, p. 502.

Kwolek, S.L. 1974. Wholly aromatic carbocyclic polycarbonate fibre having orientation angle of less than about 45 degrees. US Patent No. 3819587.

Lafley, A.G. 2008. P&G's innovation culture. *Strategy+Business Magazine*. http://www.strategy-business.com/article/08304?pg=all.

Markets. 2011. General mills awards grant to improve sustainability of vanilla plant. Food Navigator-USA.com. http://www.foodnavigator-usa.com/Markets/General-Mills-awards-grant-for-sustainable-vanilla-research.

Ministry of Primary Industries. 2013. Sector investment plan: Biological industries research fund. http://www.msi.govt.nz/assets/Get-Funded-Documents/2013-science-investment-round/SIP/SIP-Biological-Industries.pdf.

Mintzberg, H. 1993. *Structure in Fives: Designing Effective Organizations*. Englewood cliffs, NJ: Prentice-Hall.

Mittal, V.A., Ellis, A., Das, S., and Singh, H. 2012. Micronutrient fortification process and its usage (Iron fortification for dairy products). New Zealand patent application no. 600756.

Saguy, I.S. 2011a. Academia–industry innovation interaction: Paradigm shifts and avenues for the future. *Procedia Food Science*, 1(1): 1875–1882.

Saguy, I.S. 2011b. Paradigm shifts in academia and the food industry required to meet innovation challenges. *Trends in Food Science and Technology*, 22(9): 467–475.

Sarkar, S. and Costa, A.I.A. 2008. Dynamics of open innovation in the food industry. *Trends in Food Science and Technology*, 19: 574–580.

SCImago Journal & Country Rank. 2013. New Zealand. http://www.scimagojr.com/countrysearch.php?country=NZ

Shields, R.G., Speed, E., Walsh, P., and Wheatley, M.V. 2006. The evolving role of the R&D leader in the consumer packaged goods industry. Spencer Stuart.com. https://www.content.spencerstuart.com/sswebsite/pdf/lib/consumerRD_final.pdf.

Singh, H., Zhu, X.Q., and Ye, A. 2006. Lipid encapsulation. Patent no. WO 2006/115420A1.

Spitzley, A., Rogowski, T., and Garibaldo, F. 2007. *Open Innovation for Small and Medium Sized Enterprises*. Stuttgart, Germany: Stuttgart Fraunhofer-Institute for Industrial Engineering.

Stabulnieks, J. 2009. The future of innovation is a common understanding of the global economic process. In *The Future of Innovation*, eds. B. von Stamm and A. Trifilova. Surrey: Gower Publishing, p. 502.

Starling, S. 2009. General mills expands open source innovation network. FoodNavigator-USA.com. http://www.nutraingredients-usa.com/Suppliers2/General-Mills-expands-open-source-innovation-network.

Traitler, H. and Saguy, I.S. 2009. Creating successful innovation partnerships. *Food Technology*, 63(3): 22–35.

Traitler, H., Watzke, H.J., and Saguy, I.S. 2011. Reinventing R&D in an open innovation ecosystem. *Journal of Food Science*, 76(2): R62–R68.

Zakaria, F. 2011. The future of innovation: Can America keep pace? Time.com. http://www.time.com/time/nation/article/0,8599,2075226,00.html.

21 Curcumin for Prevention and Treatment of Chronic Diseases

An Overview of the Clinical Studies and Evidence-Based Support

Myriam Hinojosa and Bharat B. Aggarwal

CONTENTS

21.1 INTRODUCTION

Curcumin is a compound that was first isolated by Vogel in 1842 from the Indian spice turmeric (*Curcuma longa*) (Shishodia et al. 2005). Its structure was determined by Milobedzka in 1910 and first synthesized by Lampe in 1913, and it was found by Srinivasan to be a mixture of diferuloylmethane, demethoxycurcumin, and bisdemethoxycurcumin (Gupta et al. 2012). This compound is derived from the rhizome of the plant and is 2%–5% by weight of turmeric powder (Figure 21.1) (Shishodia et al. 2005). It turns blackish brown under acidic conditions, red under alkaline conditions, and white when reduced normally in the body (to tetrahydrocurcumin). The PubMed database lists the first published paper on curcumin (published in 1949 in the journal *Nature*) as "Antibacterial action of curcumin and related compounds" (Schraufstatter and Bernt 1949). Only 50 papers were published on curcumin until 1990, but 5755 papers have been published as of the writing of this chapter (almost 1000 in 2012 alone and 400 in 2013 through June), suggesting an extensive investigation in recent years. Of these publications, 2165 are connected to cancer alone, 227 to diabetes, 63 to obesity, 131 to infections, 314 to cardiovascular diseases (CVDs), 153 to Alzheimer's disease (AD), 354 to pulmonary diseases, 283 to colonic diseases, 96 to arthritis, 27 to psoriasis, 62 to pain, and 46 to depression. This indicates an extensive interest in curcumin's role in a wide variety of diseases. Because there is a concern about the bioavailability of curcumin (Anand et al. 2007), 446 citations have appeared to date on various formulations of this compound. Due to a lack of patentability of natural curcumin and perceived problems with its bioavailability, there have been 3617 publications on synthetic curcumin and its analogs with superior activities, such as "Super Curcumin." According to Anand et al. (2008), the search for Super Curcumin is categorized into two types: synthetic analogs or derivatives

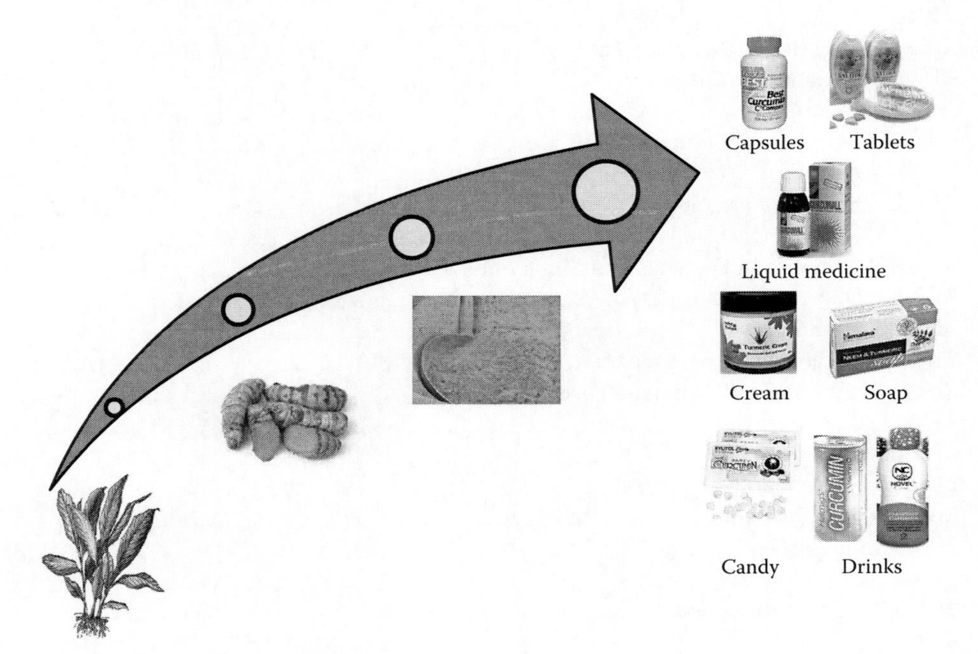

FIGURE 21.1 Translation of curcumin from farm to pharmacy.

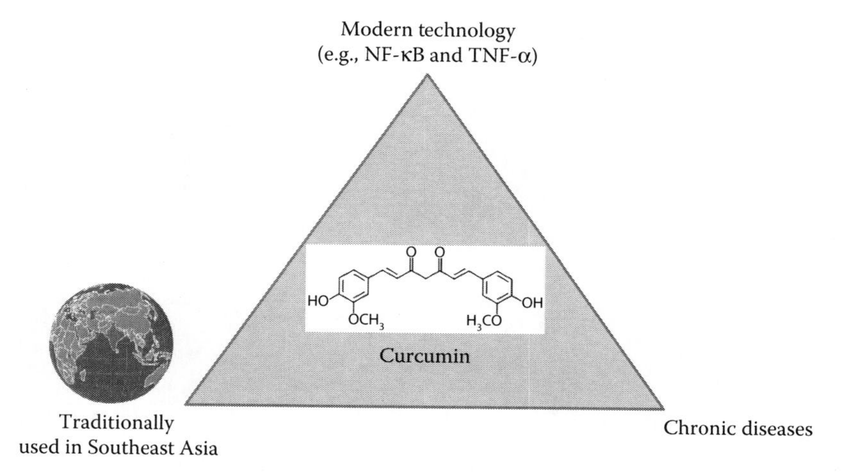

FIGURE 21.2 Translation of curcumin from ancient medicine to modern medicine.

and formulations. It is the ongoing research to find an improved use of curcumin with a better aqueous solubility and higher bioavailability (Anand et al. 2008).

In addition, over 500 companies worldwide currently deal with curcumin directly or indirectly (Goel et al. 2008), indicating the extent of interest (Figure 21.2). The various activities assigned to this compound are indicated by dozens of names assigned to this spice in the traditional Ayurveda medical system of India, each reflecting a different property, such as *jawarantika* (cures fever), *mehagni* (anti-inflammatory), and *rabhargavasa* (dissolves fat). As 612 reviews have been published on curcumin, this chapter primarily focuses on preclinical and clinical reports of this important compound, which has potential for both prevention and treatment of various chronic diseases.

21.2 MECHANISM OF ACTION OF CURCUMIN

Although curcumin has been traditionally used as a regulator of inflammation, the first report to establish the anti-inflammatory action of curcumin was in 1970 when it was compared with aspirin and found to be highly potent (Srimal and Dhawan 1973). Our group was the first to document the anti-inflammatory action of curcumin through the suppression of nuclear factor kappa B (NF-κB) and NF-κB-regulated gene products such as cyclooxygenase (COX)-2, tumor necrosis factor (TNF), and 5-lipoxygenase (LOX) (Singh and Aggarwal 1995; Aggarwal et al. 2006). That curcumin also exhibits antioxidant activity was established in 1976 by Sharma (1976).

The nuclear factor NF-κB is linked to one of the major signaling pathways responsible for most chronic illnesses caused by inflammation. It can be activated by most inflammatory cytokines, Gram-negative bacteria, disease-causing viruses, environmental pollutants, stress, high glucose, fatty acids, ultraviolet radiation, and cigarette smoke (Aggarwal et al. 2004; Kumar et al. 2004; Karin and Greten 2005; Ahn and Aggarwal 2005; Tergaonkar 2006). However, one of the major activators of NF-κB is tumor necrosis factor (TNF)-α, a proinflammatory cytokine that is a major mediator of inflammation and is regulated by the activation of NF-κB (Kim et al. 2012). The pathway between NF-κB and TNF-α is one of the most important pathways responsible for inflammation. It was discovered that curcumin is a potent blocker of NF-κB activation, and the discovery of curcumin as a TNF-α mediator has led to an extensive research on curcumin and its effects on many different chronic diseases (Figure 21.3) (Aggarwal 2003).

Curcumin can inhibit not only TNF-α but also many other proinflammatory cytokines and chemokines, including interleukin (IL)-6, IL-8, macrophage inflammatory protein (MIP)-1a, monocyte chemoattractant protein (MCP)-1, IL-1B, NF-κB, and NF-κB-regulated gene products, all of which can induce inflammation (Sung et al. 2012). It downregulates oncogenic microRNAs and upregulates

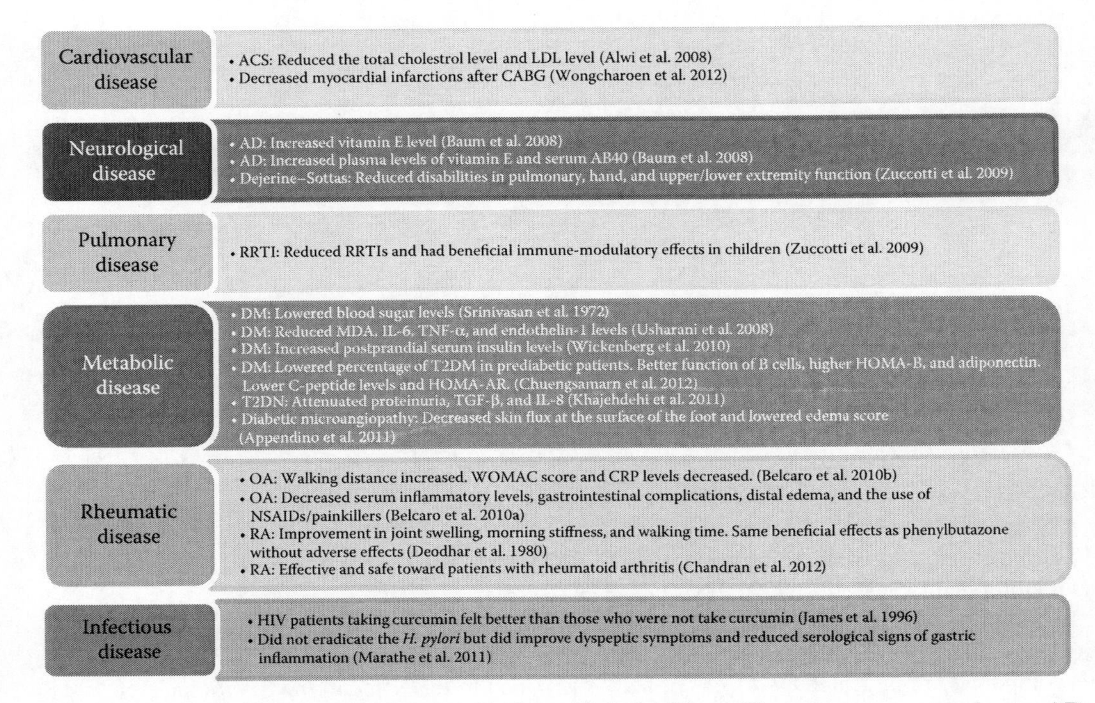

FIGURE 21.3 Lesson learned about curcumin from clinical trials. ACS, acute coronary syndrome; AD, Alzheimer's disease; CABG, coronary artery bypass grafting; CRP, C-reactive protein; DM, diabetes mellitus; *H. pylori, Helicobacter pylori*; HOMA-B, homeostasis model assessment-B; HOMA-AR, HOMA of insulin resistance; IL-6, interleukin-6; LDL, low-density lipoprotein; MDA, malondialdehyde; NSAID, nonsteroidal anti-inflammatory drug; OA, osteoarthritis; RA, rheumatoid arthritis; RRTI, recurrent respiratory tract infection; T2DM, type 2 diabetes mellitus; T2DN, type 2 diabetic nephropathy; TNF-α, tumor necrosis factor-α; TGF-β transforming growth factor-β.

tumor-suppressive microRNAs (Gupta et al. 2013a). The inhibition of NF-κB results in the suppression of cell survival, proliferation, invasion, and angiogenesis, all of which can lead to cancer and other chronic diseases (Kunnumakkara et al. 2008). Due to its ability to inhibit TNF-α and NF-κB, curcumin is a potential therapeutic that may help treat many different inflammatory diseases (Figure 21.4) (Aggarwal et al. 2013).

21.3 ROLE OF CURCUMIN IN CANCER

The mechanistic studies discussed in Subsection 21.2 suggest that curcumin has a potential to prevent and treat cancer (Hasima and Aggarwal 2012). However, the first evidence of its role in cancer was documented in 1985 (Kuttan et al. 1985; Wargovich et al. 1985). In 1987, the first clinical indication of curcumin's anticancer activities in humans was demonstrated by Kuttan, when he conducted a clinical trial to test the effects of topical curcumin on 67 patients with cancer lesions. He found that the curcumin produced symptomatic relief in smell, itching, lesion size, and pain. Since then, many more clinical trials have been performed on the effects of curcumin on cancers such as colorectal cancer (CRC), pancreatic cancer, breast cancer, prostate cancer, multiple myeloma, lung cancer, cancer lesions, and head and neck squamous cell carcinoma (HNSCC) (Gupta et al. 2013b).

21.3.1 Colorectal Cancer

CRC is the second leading cause of cancer deaths in the United States, and there are not many effective treatments against this cancer. However, several clinical trials have demonstrated that curcumin can have positive effects against this cancer.

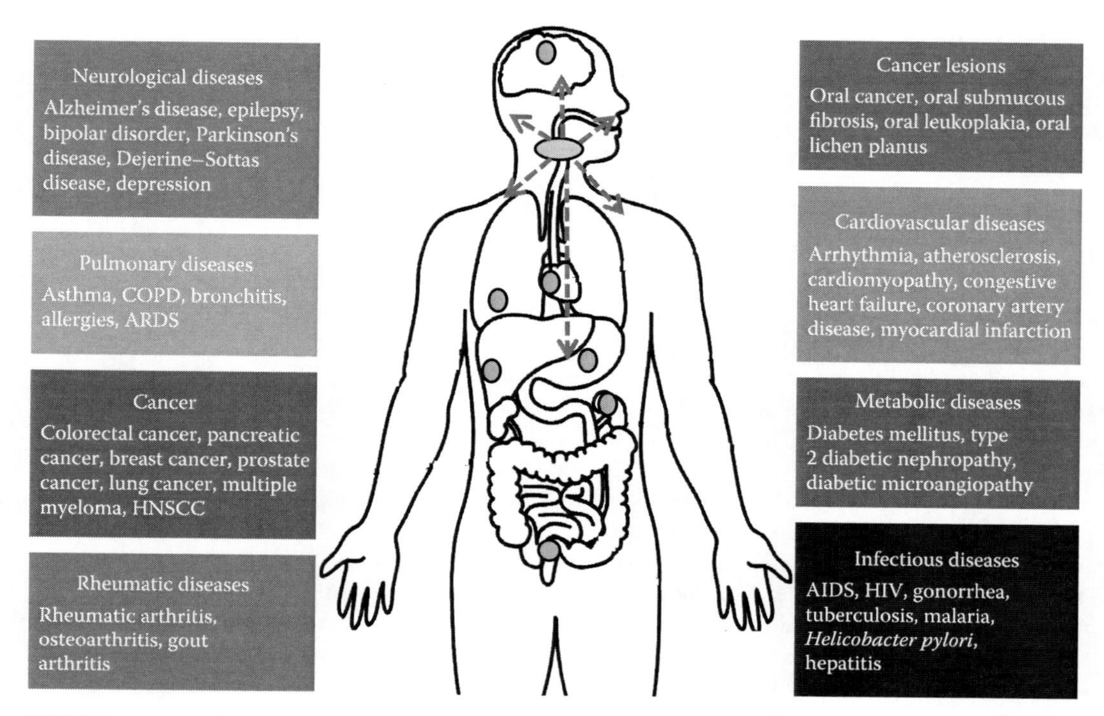

FIGURE 21.4 Role of curcumin in chronic diseases. AIDS, acquired immunodeficiency syndrome; ARDS, acute respiratory distress syndrome; COPD, chronic obstructive pulmonary disease; HIV, human immunodeficiency virus; HNSCC, head and neck squamous cell carcinoma.

In one study, 15 patients with advanced CRC were given *Curcuma* extract orally for four months at doses containing 35–180 mg of curcumin. Patients treated with 36 mg of curcumin for 29 days had a 59% decrease in lymphocytic glutathione (GSH) *S*-transferase activity, which is a marker of DNA adduct formation. DNA adduct formation is the bonding of a piece of DNA to a cancer-causing chemical, which can lead to carcinogenesis. Because curcumin decreased the lymphocytic GSH *S*-transferase activity, this may mean that it can be a potential therapeutic for CRC (Sharma et al. 2001). In another clinical study, 15 patients with advanced CRC refractory to standard chemotherapies were given curcumin doses of 0.45–3.6 g/day up to four months. The 3.6 g dose led to a decrease of 62% in inducible prostaglandin E_2 on day 1 and 57% on day 29 of consumption. After completion of the experiment, a 3.6 g dose of curcumin was recommended for the prevention of cancers outside the gastrointestinal tract (Sharma et al. 2004). In a third study, patients with CRC were given curcumin capsules for one week at three different doses: 3.6, 1.8, or 0.45 g/day. Trace levels of curcumin were identified in the tissue samples, and the levels of oxidative DNA adduct 3-(2-deoxy-β-di-erythro-pentafuranosyl)-pyr[1,2-α]-purin-10(3H)one (M1G), which can become carcinogenic if not repaired, decreased in malignant colorectal tissue. This study also demonstrated that a daily dose of 3.6 g of curcumin can have successful results in CRC patients (Garcea et al. 2005).

Another clinical trial was performed in patients who have familial adenomatous polyposis (FAP), a genetic disorder that may lead to colon cancer. Nonsteroidal and COX-2 inhibitors have been found to be effective in reducing adenomas in this syndrome; however, these drugs produce many adverse effects. To test the effects of curcumin on patients with FAP, five patients were given 480 mg of curcumin and 20 mg of quercetin (a flavonoid found in many plants, fruits, and grains) three times a day. After six months, these patients had fewer polyps and smaller polyps, showing that curcumin can be effective in treating patients with CRC and FAP (Cruz-Correa et al. 2006). In a different clinical trial, patients with CRC who had not undergone surgery were given 360 mg of curcumin for 10–30 days. This led to an increase in body weight, apoptotic cells, and p53 in tumor cells, and a decrease in TNF-α serum levels (He et al. 2011). These clinical trials all demonstrated the potential of curcumin in treating patients with CRC, but more clinical studies are needed to prove that curcumin is effective in preventing and treating this cancer.

21.3.2 Pancreatic Cancer

Pancreatic cancer is the fourth most common cause of cancer death in the world (Hariharan et al. 2008). Oxidative stress is believed to be one of the major causes of pancreatic cancer, so antioxidants are given to patients to try to improve their condition (Kanai et al. 2011). A study in India was performed to test the effects of curcumin and piperine on 20 patients with tropical pancreatitis, a disease that can lead to pancreatic cancer. The patients were given either 500 mg of curcumin with 5 mg of piperine or placebo for six weeks. The study found a significant decrease in the erythrocyte malondialdehyde (MDA) levels, a marker of oxidative stress, and an increase in GSH, an antioxidant, in the patients who received curcumin. Curcumin did not lessen pain, but the study concluded that it may reverse lipid peroxidation in patients with tropical pancreatitis (Durgaprasad et al. 2005). In another clinical study of advanced pancreatic cancer patients, 21 patients were given 8 g of curcumin daily until disease progression, with an evaluation of the cancer every two months. Two of these patients showed clinical biological activity, with one having ongoing stable disease for more than 18 months and the other having a brief tumor regression with increases in serum cytokine levels. Downregulation of NF-κB and COX-2 was observed in some patients after they ingested curcumin. The study concluded that curcumin is well tolerated and has a biological activity in some patients with pancreatic cancer (Dhillon et al. 2008). In a third study, 17 patients with advanced pancreatic cancer were given 8 g of curcumin per day along with 1000 mg/m^2 of gemcitabine, a nucleoside analog used in chemotherapy, three times a week for four weeks. Of the 11 patients who completed the study, one showed a partial response, four had stable disease, and six had tumor progression. The authors of this study concluded that 8 g per day of curcumin with gemcitabine was above the tolerated dose (Epelbaum et al. 2010). However, a more recent study in 2011 by Kanai showed that 8 g/day of curcumin with gemcitabine was safe and well tolerated (Kanai et al. 2011).

21.3.3 Breast Cancer

Breast cancer is the second most common cause of cancer death in women (Sinha et al. 2012). Docetaxel is a microtubule inhibitor commonly used in the early stages of breast cancer along with other chemotherapeutic agents. The tolerability of docetaxel with curcumin was tested in 14 patients with advanced metastatic breast cancer. The maximum tolerable dose of curcumin was 8 g/day, but for curcumin administered with docetaxel, the tolerable dose was 6 g/day for seven consecutive days every three weeks (Bayet-Robert et al. 2010).

21.3.4 Prostate Cancer

Prostate cancer occurs in one of every seven men, making it the most common malignancy in men. The prostate-specific antigen (PSA) test is one method used to diagnose cases of prostate cancer. A high PSA level may indicate that the patient is at risk of developing prostate cancer. A clinical trial was performed in 85 men who had high PSA levels but no prostate cancer to evaluate the effects of curcumin and soy isoflavones on serum PSA levels. These patients were given either a curcumin and an isoflavone supplement or placebo for six months: 43 were given 100 mg of curcumin and 40 mg of isoflavones, and 42 were given placebo. PSA levels decreased in the patients with PSA levels greater than 10 ng/ml, indicating that curcumin and isoflavones could control serum PSA levels. The study concluded that this may be due to the interaction between curcumin and isoflavones in suppressing PSA production (Ide et al. 2010).

21.3.5 Multiple Myeloma (Plasma Cell Myeloma)

Multiple myeloma is the second most common hematologic cancer in the United States. It is an incurable disease for most patients even though there are treatments with bortezomib, thalidomide, and lenalidomide. Monoclonal gammopathy of undetermined significance (MGUS) is a common premalignant plasma cell proliferative disorder with a risk of progression to multiple myeloma. A pilot study was conducted to determine the effects of curcumin on plasma cells and osteoclasts in 26 patients with MGUS: 17 patients were given curcumin at the start of the study and then were given placebo after three months, while nine patients were given placebo initially and then changed to curcumin. Ten patients showed a decrease in the paraprotein load with paraprotein greater than 20 g/L, and five of these patients had a 12%–30% reduction in paraprotein levels while receiving curcumin (Golombick et al. 2009). Another experiment evaluated the tolerability, safety, and clinical efficacy of curcumin on 29 patients with asymptomatic multiple myeloma. Curcumin was given at doses of 2, 4, 6, 8, or 12 g/day either alone or with 10 g of bioperine for 12 weeks. Of the 29 evaluable patients, 12 continued the treatment after 12 weeks and five continued the treatment for one year with stable disease. Curcumin downregulated the activation of NF-κB, signal transducer and activator of transcription (STAT)-3, and COX-2 in most of the patients (Vadhan-Raj et al. 2007). These two studies show the therapeutic potential of curcumin against MGUS and multiple myeloma.

21.3.6 Lung Cancer

Lung cancer accounts for 27% of all cancer deaths in the world. A study was performed in 16 chronic smokers and six nonsmokers to test the effects of turmeric, which is thought to reduce the risk of getting lung cancer. Patients given turmeric at 1.5 g/day for 30 days showed a significant reduction in urinary excretion of mutagens in the smokers. This study suggests that turmeric can act as an antimutagen and can be effective in reducing the risk of lung cancer in smokers (Polasa et al. 1992).

21.3.7 Cancer Lesions

Cancer lesions including oral cancer, oral submucous fibrosis, oral leukoplakia, and oral lichen planus are all associated with tobacco chewing. Patients with cancer lesions demonstrate an increase in the number of micronuclei in their oral mucosal cells and in circulating lymphocytes. For this reason, the number of micronucleated oral mucosal cells can be used as a biomarker for cancer lesions and precancers, as well as for evaluating the effects of therapeutic agents (Gupta et al. 2013b). One study investigated the effects of turmeric oil and turmeric oleoresin in patients with oral submucous fibrosis. The extracts offered protection against the benzo[*a*]pyrene-induced increase in micronuclei in circulating lymphocytes (Hastak et al. 1997). In another study, patients with submucous fibrosis were treated for three months with a daily dose of one of the following: 600 mg of turmeric oil and 3 g of turmeric, 600 mg of turmeric oleoresin and 3 g of turmeric, or 3 g of turmeric only. The three treatments decreased the number of micronucleated cells; however, turmeric oleoresin was the most effective in reducing the number of micronuclei in oral mucosal cells (Hastak et al. 1997). Another study evaluated the effects of curcumin on patients with resected urinary bladder cancer, arsenic-associated Bowen's disease of the skin, uterine cervical intraepithelial neoplasm, oral leukoplakia, and intestinal metaplasia of the stomach. Twenty-five patients received doses of curcumin higher than 8 g/day for three months. The findings demonstrated that curcumin at 8 g/day was safe and it has chemopreventative potential against cancerous lesions (Cheng et al. 2001). A different clinical trial evaluated the effects of curcuminoids in 100 patients with oral lichen planus. Participants received either 2 g/day of curcuminoids or placebo for seven weeks. During the first week, all participants received 60 mg/day of prednisone, which is an immunosuppressant drug. Curcuminoids were well tolerated by the patients, but it was calculated that if curcuminoid had a better outcome than the placebo, it would be by less than 2%, so the experiment was ended earlier than planned

(Chainani-Wu et al. 2007). Since the previous study used the extracts of turmeric, turmeric oil, and turmeric oleoresin, and this study used curcumin, the difference in results may be due to this. In another recent study, patients with precancerous lesions who received 1 g curcumin tablets for one week had an increase in vitamin C and E levels and a decrease in MDA and 8-hydroxy-2′-deoxyguanosine (8-OHdG) in the serum and saliva (Rai et al. 2010).

21.3.8 Head and Neck Squamous Cell Carcinoma

HNSCC is the most common cancer worldwide. It induces oral, laryngeal, and pharyngeal malignancies, with about 40% of cases arising in the oral cavity. Studies have indicated the role of NF-κB, IL-6, IL-8, and vascular endothelial growth factor (VEGF) in the pathogenesis of the disease, so targeting these signaling molecules may be an effective way to target HNSCC (Aggarwal et al. 2004). Thirty-nine patients participated in a clinical trial to determine the effects of curcumin against HNSCC. Saliva was collected from the patients 1 h after they chewed two curcumin tablets for five minutes. Curcumin treatment led to a reduction in IκB kinase-beta (IKKβ) kinase activity, and 8 of 21 patients showed a reduction in IL-8 levels. The authors concluded that IκB kinase-beta (IKKβ) kinase could be used as a biomarker for detecting the effect of curcumin in patients with HNSCC (Kim et al. 2011).

All of the clinical trials discussed in Subsections 21.3.1, 21.3.4, and 21.3.7 were performed to determine the effects of curcumin on different cancers. Curcumin was effective in most of these studies, demonstrating its potential as a therapeutic agent against cancer (Figure 21.5).

21.4 ROLE OF CURCUMIN IN CVDs

CVDs are the number one cause of death in the United States and include arrhythmia, atherosclerosis, cardiomyopathy, congestive heart failure, coronary artery disease, myocardial ischemia, and myocardial infarction, among other diseases. Inflammation and oxidative stress play major roles in most CVD and may lead to atherogenesis. Atherosclerosis, or the hardening of arteries, is caused by oxidative damage, which affects lipoproteins, the walls of blood vessels, and subcellular membranes, and in addition, the oxidation of low-density lipoproteins can lead to the development of atherosclerosis. Proinflammatory cytokines regulated by NF-κB are activated by cardiopulmonary bypass and cardiac global ischemia and reperfusion (I/R) (Yeh et al. 2005). Proinflammatory cytokines can cause cardiomyocytic injury. Chronic transmural inflammation and proteolytic destruction of medial elastin play important roles in the development of abdominal aortic aneurysms (McCormick et al. 2007). NF-κB also regulates C-reactive protein (CRP), which is an inflammatory marker and a known predictor of CVD (Kawanami et al. 2006). Curcumin has been shown to be effective against CVD in several experiments and by numerous mechanisms.

Several studies have shown that curcumin protects the heart from I/R injury. Perhaps one of the earliest activities linked to curcumin was its ability to lower blood cholesterol levels (Rao et al. 1970), and in 1985, Srivastava et al. (1986) showed that curcumin exhibits antithrombotic activity.

21.4.1 Myocardial Ischemia

Myocardial ischemia, the restriction of blood supply causing a lack of oxygen supply to the heart, can lead to myocardial infarctions and other cardiovascular complications. Curcumin was administered to test its effects on myocardial ischemia induced by the ligation of the left descending coronary artery. It protected the test animals from developing low heart rates and blood pressure after ischemia and prevented ischemia-induced elevation in MDA and lactate dehydrogenase (LDH) (Srivastava et al. 1986). A similar study of the effects of curcumin on isoprenaline-induced myocardial ischemia in rat myocardium concluded that curcumin protected rat myocardium against ischemic insult because it reduced the levels of xanthine oxidase, superoxide anion, lipid peroxidase, and myeloperoxidase, and increased the levels of superoxide dismutase (SOD), catalase (CAT),

FIGURE 21.5 Modulation of biomarkers by curcumin in clinical trials for cancer. COX-2, cyclooxygenase-2; GSH, glutathione; IL-8, interleukin-8; MDA, malondialdehyde; NF-κB, nuclear factor kappa B; PSA, prostate-specific antigen; STAT-3, signal transducer and activator of transcription-3; TNF-α, tumor necrosis factor-α. [a] Adapted from Cruz-Correa, M. et al., *Clinical Gastroenterology and Hepatology*, 4, 1035–1038, 2006; [b] Adapted from Garcea, G. et al., *Cancer Epidemiology Biomarkers and Prevention*, 14, 120–125, 2005; [c] Adapted from He, Z.Y. et al., *Cancer Investigation*, 29, 208–213, 2011; [d] Adapted from Durgaprasad, S. et al., *Indian Journal of Medical Research*, 122, 315–318, 2005; [e] Adapted from Vadhan-Raj, S., *Blood*, 110, 357a, 2007; [f] Adapted from Ide, H. et al., *Prostate*, 70, 1127–1133, 2010; [g] Adapted from Golombick, T., *Clinical Cancer Research*, 15, 5917–5922, 2009; [h] Adapted from Hastak, K., *Cancer Letters*, 116, 265–269, 1997; [i] Adapted from Kim, S.G., *Clinical Cancer Research*, 17(18): 5953–5961, 2011; [j] Adapted from Rai, B., *Journal of Oral Science*, 52, 251–256, 2010; [k] Adapted from Sharma, R.A. et al., *Clinical Cancer Research*, 7, 1894–1900, 2001; [l] Adapted from Sharma, R.A. et al., *Clinical Cancer Research*, 10, 6847–6854, 2004.

GSH peroxidase, and GSH *S*-transferase activities (Manikandan et al. 2004). Given these different results, more experiments are needed to determine curcumin's effects on myocardial ischemia.

21.4.2 Acute Coronary Syndrome

Acute coronary syndrome occurs when the blood supply to the myocardium is cut off. ST elevation myocardial infarction, non-ST elevation myocardial infarction, and unstable angina are three clinical conditions involving the coronary arteries. A clinical trial was performed to test the effects of curcumin on total cholesterol, high-density lipoprotein (HDL), and triglyceride levels in 75 patients with acute coronary syndrome. The patients were separated into three groups and received one of the following dosages: 45 mg/day, 90 mg/day, or 180 mg/day. These workers concluded that low-dose curcumin (15 mg/day three times a day) reduced total cholesterol and low-density lipoprotein (LDL) cholesterol levels but increased the HDL cholesterol more than did high-dose curcumin (Alwi et al. 2008). The study suggests that curcumin improved the lipid profile of patients with acute coronary syndrome, but more clinical trials are needed to demonstrate whether curcumin can suppress this syndrome.

Numerous *in vitro* and *in vivo* studies have shown the potential of curcumin against CVD. Yeh et al. (2005) found that NF-κB and the upregulation of proinflammatory genes are involved in the regulation of cardiopulmonary bypass and cardiac global I/R and that curcumin can inhibit the NF-κB activation leading to the suppression of cardiac proinflammatory cytokines. A follow-up study by the same workers demonstrated that curcumin reduced the IL-8, IL-10, TNF-α, and cardiac troponin 1 levels, as well as the appearance of apoptotic cardiomyocytes in rabbits (Yeh et al. 2005). A study performed with rats showed that curcumin can prevent isoproterenol-induced myocardial infarction and that it decreased necrosis (Nirmala and Puvanakrishnan 1996). Curcumin has also been found to have a hypocholesterolemic effect on atherosclerotic rabbits and inhibit lipoperoxidation of subcellular membrane LDL oxidation (Ramirez-Tortosa et al. 1999). It reduces oxidative stress and attenuates fatty streak development in such rabbits owing to lower plasma lipid peroxide levels and higher α-tocopherol and coenzyme Q levels, which reduce the damage in the thoracic and abdominal aorta (Quiles et al. 2002). Olszanecki et al. fed a Western diet, consisting of 21% fat by weight, 0.15% of cholesterol by weight, and no colic acid, to apolipoprotein E (apoE)/LDL receptor double-knockout mice and fed them a low dose of curcumin to test its effects on atherosclerosis. In mice fed with 0.3 g of curcumin daily for four months, atherogenesis was inhibited while cholesterol and triglyceride levels remained unaffected (Olszanecki et al. 2005).

A study of the effects of curcumin on inhibiting platelet-derived growth factor (PDGF) found that curcumin is a potent inhibitor of PDGF-stimulated vascular smooth muscle cell (VSMC) functions and that it plays a critical role in regulating events after vascular injury. Ramaswami et al. (2004) showed that curcumin blocks homocysteine-induced endothelial dysfunction in porcine coronary arteries, and Parodi et al. (2006) found that curcumin suppressed the development of experimental abdominal aortic aneurysms and reduced aortic wall expression of several cytokines, chemokines, and proteinases known to mediate aneurismal degeneration. Although many experiments have considered the effects of curcumin on CVD *in vitro* and *in vivo*, more clinical trials are needed to determine the effects of curcumin in humans. The effect of curcuminoids on the frequency of acute myocardial infarction after coronary artery bypass grafting (CABG) in humans was tested by Wongcharoen. The 121 patients undergoing CABG received either placebo or 4 g/day of curcuminoids given three days before surgery and continued until five days after the surgery. The percentage of myocardial infarctions that occurred in the hospital decreased from 30% to 13.1% in the patients who were receiving curcuminoids. This study concluded that the antioxidant and anti-inflammatory effects of curcuminoids decreased myocardial infarction associated with CABG (Wongcharoen et al. 2012).

21.5 ROLE OF CURCUMIN IN NEUROLOGICAL DISEASES

Neurological diseases such as AD, epilepsy, bipolar disorder, Parkinson's disease, Dejerine–Sottas disease, and depression are all caused by inflammation and TNF-α (Aggarwal and Harikumar 2009). To determine whether curcumin is effective in inhibiting AD and Dejerine–Sottas disease, clinical and *in vivo* trials have been performed.

21.5.1 Alzheimer's Disease

AD is a progressive neurodegenerative disorder and is the most common form of dementia affecting people over the age of 65. Accumulation of iron, zinc, and copper can lead to an abundance of the antioxidants that control and prevent reactive oxygen species (ROS). AD involves inflammation, oxidative damage, and amyloid beta-peptide (Abeta) accumulation (Gupta et al. 2012). The inflammation is characterized by an increased expression of inflammatory cytokines and activated microglia. Current treatments for this disease have numerous adverse effects, and hence there is ongoing research into new and effective treatments.

The safety and tolerability of curcumin was tested in the United States in patients with mild to moderate AD. Thirty-three patients were randomly assigned to a placebo group, low-dose curcumin (2 g/day), or high-dose curcumin (4 g/day). The observations for this study have not yet been published (Ringman et al. 2005). In another study, 34 patients with AD were given either a dose of 1 or 4 g of curcumin or placebo. Although the Mini-Mental State Examination (MMSE) score was not improved in patients treated with curcumin and the level of serum AB40 was not affected, curcumin administration did result in increased vitamin E levels with no adverse effects (Baum et al. 2008). Further clinical trials of the effects of curcumin on patients with AD may lead to a better understanding of curcumin's therapeutic effects on this disease.

Curcumin is known to suppress oxidative damage, inflammation, cognitive deficits, and amyloid accumulation. Curcumin has been shown to mediate its effects against AD through eight mechanisms in the *in vitro* studies.

1. It can protect PC12 rat pheochromocytoma and normal human umbilical vein endothelial cells from Abeta-induced oxidative stress (Kim et al. 2001).
2. It lowered the levels of oxidized proteins and IL-1B elevated in the brains of APPsw mice (transgenic mice bearing the "Swedish mutation" for human amyloid precursor protein) (Lim et al. 2001).
3. It inhibited the formation and extension of fAbeta from Abeta (1-40) and Abeta (1-42) and destabilized preformed fAbeta (Ono et al. 2004).
4. It reduced the levels of amyloid and oxidized proteins, and prevented cognitive deficits in animal models of AD. Since curcumin binds the redox-active metals iron and copper, it may also suppress inflammatory damage by preventing metal induction of NF-κB (Baum and Ng 2004).
5. It inhibits aggregated and disaggregated fAbeta40, and also decreases Abeta formation (Yang et al. 2005).
6. It inhibits peroxidase, which plays a major role in the cytopathologies of AD (Atamna and Boyle 2006).
7. It enhanced Abeta uptake by macrophages of patients with AD. Bone marrow-derived dendritic cells (BMDCs) may correct immune defects in patients with AD and provide a different approach to AD immunotherapy (Fiala et al. 2007).
8. It labels amyloid pathology *in vivo* and leads to a reversal of structural changes in dystrophic dendrites. Curcumin has the ability to disrupt existing plaques and partially restore distorted neuritis in a mouse model (Garcia-Alloza et al. 2007).

After these studies, there was a six-month clinical pilot study in which 34 patients with AD were tested to investigate the effects of curcumin on humans. Subjects received 0, 1, or 4 g of curcumin per day. The serum levels of curcumin were found to reach a maximum of 250 nM at 1.5 h when given with food and 270 nM at 4 h when given with water. The curcumin group showed increased plasma concentrations of vitamin E and increased serum AB40, indicating curcumin's ability to disaggregate amyloid-beta (AB-deposits) in the brain (Baum et al. 2008). More clinical trials are needed to determine how curcumin affects patients with AD.

21.5.2 Dejerine–Sottas Disease

Dejerine–Sottas disease is a severe degenerative, neurological disorder characterized by generalized weakness that can progress to severe disability, loss of sensation, curvature of the spine, and mild hearing loss. It is caused by defects in genes of axons and myelin, and affects myelin such as myelin P0 (MPZ), peripheral myelin protein 22 (PMP22), periaxin gene (PRX), and early growth response gene 2 (EGR2) (Gupta et al. 2013b). A study was performed in a 15-year-old Caucasian girl with Dejerine–Sottas disease. The patient was given 1.5 g of oral curcumin daily for four months and then 2.5 g/day for the next eight months. After one year of curcumin administration, the patient showed a slight increase in knee flexion and foot strength, while the disabilities in pulmonary function, hand function, and upper/lower extremities were either stable or reduced. The neurophysiologic findings of the patient were unchanged (Burns et al. 2009). However, the study suggested the efficacy and safety of curcumin in patients with Dejerine–Sottas disease.

21.6 ROLE OF CURCUMIN IN PULMONARY DISEASE

Asthma, chronic obstructive pulmonary disease, bronchitis, allergies, and recurrent respiratory tract infections (RRTIs) are all pulmonary diseases. The effect of curcumin against these pulmonary diseases has been tested in several *in vivo* and clinical trials.

21.6.1 Asthma

Asthma is a proinflammatory disease that is normally mediated by inflammatory cytokines, and eotaxin, MCP-1, and MCP-3 play major roles in asthma. *In vitro* and *in vivo* experiments have shown that curcumin can help clear constricted airways and increase antioxidant levels (Gupta et al. 2012). Ju et al. (1996) examined the effect of dietary fats and curcumin on immunoglobulin E (IgE)-mediated degranulation of intestinal mast cells in Norway rats. They showed that the influence of curcumin on rat chymase II was not prominent, suggesting that dietary ingredients influence the synthesis of IgE and degranulation of mast cells. Another study found that a high dose of curcumin significantly enhanced IgG levels but did not result in different delayed-type hypersensitivity or natural killer cell activity. Kobayashi et al. (1997) studied the effect of curcumin against allergic diseases by examining the production of IL-2, IL-5, granulocyte macrophage-colony stimulating factor, and IL-4 by lymphocytes from atopic asthmatics in response to house dust mites, and they found that curcumin inhibited *Dermatophagoides farinae* (Df)-induced lymphocyte proliferation and IL-2 production.

21.6.2 Recurrent Respiratory Tract Infections

RRTIs are common diseases in children, and although several anti-inflammatory and antibacterial drugs are used to treat these infections, they may result in adverse effects and the selection of drug-resistant microorganisms. A clinical study of the effects of curcumin on RRTIs was performed on 10 healthy children with RRTIs. The patients received 1 g of lactoferrin and curcumin every 8 h for four weeks. At the end of the experiment, it was found that lactoferrin and curcumin reduced RRTIs and had beneficial immunomodulatory effects in the children (Zuccotti et al. 2009).

21.6.3 Chronic Obstructive Pulmonary Disease

Chronic obstructive pulmonary disease is a long-term disease that narrows the airways over time. The inflammation that occurs during this pulmonary disease activates the NF-κB pathway. Changes in the balance of histone acetylation and deacetylation occur via posttranslational modification of histone deacetylases (HDACs). Curcumin has been shown to restore HDAC activity and corticosteroid function (Marwick et al. 2007; Biswas and Rahman 2008).

21.7 ROLE OF CURCUMIN IN DIABETES AND METABOLIC DISEASES

21.7.1 Diabetes Mellitus

Diabetes mellitus is a prevalent hyperglycemic disorder that can affect the brain, the kidney, the heart, the liver, and other organs. Inflammation and the inflammatory cytokines NK-κB, nuclear factor (erythroid-derived 2)-like-2 (Nrf2), and peroxisome proliferator-activated receptor (PPAR)-γ play major roles in diabetes. TNF and NF-κB have been linked with insulin resistance. Curcumin can suppress blood glucose levels, increase the antioxidant status of pancreatic B cells, and enhance activation of PPAR-γ (Gupta et al. 2013c). That curcumin could lower blood glucose levels in humans was shown as far back as 1971 (Srinivasan 1972). A man who had diabetes for 16 years began a daily regimen of 5 g of turmeric powder, and his blood glucose decreased from 140 to 70 mg/dl. His insulin levels were lowered and the 5 g dosage of turmeric was continued, and his blood glucose level remained low; however, when turmeric ingestion was discontinued, his blood glucose increased to 140 mg/dl again. This study suggested that turmeric may have positive effects on patients with diabetes (Srinivasan 1972). A different clinical study was performed by Usharani et al. on 72 patients with type 2 diabetes. The patients were assigned to receive 300 mg of curcumin twice a day, 10 mg of atorvastatin (a drug used to lower blood cholesterol), or placebo for eight weeks. Curcumin treatment resulted in reductions in MDA and the inflammatory cytokines IL-6, TNF-α, and endothelin-1 (Usharani et al. 2008). Another study tested the effects of curcumin on the postprandial plasma glucose and insulin levels of 14 healthy participants and found that curcumin increased postprandial serum insulin levels but did not affect plasma glucose levels, meaning that it might have an effect on insulin secretion (Wickenberg et al. 2010). A recent study of the efficacy of curcumin in delaying the onset of type 2 diabetes was performed in 240 prediabetic patients. The patients were given either 1.5 g/day of curcumin or placebo capsules. After nine months of treatment, 16.4% of participants in the placebo group had a diagnosis of type 2 diabetes, compared with none of the patients in the curcumin-treated group. The patients who received curcumin had a better overall function of B cells, higher homeostasis model assessment (HOMA)-B and adiponectin, and lower C-peptide levels and HOMA of insulin resistance (HOMA-AR). This study concluded that curcumin may have beneficial effects on prediabetic patients (Chuengsamarn et al. 2012). Interestingly, reformulation of curcumin mixed with yogurt exhibited antidiabetic activity, improving significantly most of the markers that assess experimental diabetes (Gutierres et al. 2012).

21.7.2 Type 2 Diabetic Nephropathy

Type 2 diabetic nephropathy may commonly lead to end-stage renal disease, which is associated with high levels of mortality and morbidity. Proteinuria (an excess of serum proteins in the urine) and transforming growth factor (TGF)-β can contribute to end-stage renal disease. A clinical study of 40 patients with type 2 diabetic nephropathy tested the effects of turmeric on serum and urinary TGF-β, IL-8, and TNF-α and on proteinuria. The patients were divided into the placebo group and the trial group, who received 500 mg of turmeric three times a day. After two months of study, the authors concluded that short-term turmeric supplementation can attenuate proteinuria, TGF-β, and IL-8 in patients with type 2 diabetic nephropathy (Khajehdehi et al. 2011). Long-term trials on such patients are needed to determine whether turmeric's effects are short or long lasting.

21.7.3 Diabetic Microangiopathy

Diabetic microangiopathy is a disease in which the capillary walls become thick and weak, slowing the flow of blood. Poorly controlled diabetic patients with hyperglycemia, which induces biochemical changes in cells, can have retinal, renal, or neural complications, sometimes progressing into advanced periodontal disease.

A clinical study was performed to test whether Meriva, a formulation of curcumin, could improve diabetic microangiopathy in patients who had not been using insulin for at least five years. The patients were treated with the best treatment protocol for their disease, in addition to 1 g/day of Meriva for four weeks. After four weeks, there was indication of improvement in the microangiopathy due to a decrease in skin flux at the surface of the foot and in the edema score. This experiment suggests that Meriva may be beneficial for patients with diabetic microangiopathy (Appendino et al. 2011).

21.8 ROLE OF CURCUMIN IN RHEUMATIC DISEASES

More than 100 rheumatic diseases occur, and the most common forms in the Western world are gout arthritis, osteoarthritis, and rheumatoid arthritis. Rheumatic diseases are those that affect the joints and connective tissue, causing inflammation, pain, and fatigue. Curcumin was first shown to exhibit an activity against rheumatoid arthritis in 1980 (Deodhar et al. 1980). Several clinical trials and *in vivo* experiments have investigated the effects of curcumin against these rheumatic diseases.

21.8.1 Gout Arthritis

Gout arthritis is caused by the presence of crystals of monosodium urate in the joints, bones, and soft tissues. It is treated with nonsteroidal anti-inflammatory agents, oral or intravenous colchicines, and oral, intravenous, or intra-articular glucocorticoids; although these agents are effective, they may have severe side effects. For this reason, natural products, such as curcumin, are being examined to find if they are effective in treating this rheumatic disease (Aggarwal and Harikumar 2009; Aggarwal and Sung 2009).

21.8.2 Osteoarthritis

Osteoarthritis is the second most common arthritis in the world and is caused by articular cartilage failure induced by genetic, metabolic, biochemical, and biomechanical factors. It is also the leading cause of disability in the Western world (Aggarwal and Harikumar 2009). A study of 50 patients with osteoarthritis tested the effects of Meriva at a dose of 200 mg/day. The symptoms of osteoarthritis were measured by using the Western Ontario and McMaster Universities Arthritis Index (WOMAC) score, an indicator of pain level, after the patients walked on a treadmill; inflammatory status was measured by the levels of CRP. After three months of Meriva treatment, the walking distance increased from 72 to 332 m, CRP levels decreased significantly, and the WOMAC score was decreased by 58%. These results showed the efficacy of Meriva in the management of osteoarthritis (Belcaro et al. 2010a). The same scientists then performed a follow-up experiment using a larger group of 100 patients for eight months. After eight months, the WOMAC score had decreased by more than 50% and the walking distance in patients receiving curcumin was three times greater than for control patients. In addition, serum inflammatory levels, gastrointestinal complications, distal edema, the use of nonsteroidal anti-inflammatory drugs (NSAIDs) or painkillers, and the need for hospital admissions and consultations were decreased significantly. This experiment indicates that Meriva treatment is worth considering in the treatment of long-term osteoarthritis (Belcaro et al. 2010b).

21.8.3 Rheumatoid Arthritis

Rheumatoid arthritis is a chronic proinflammatory disease which results from joint stiffness and swelling that leads to decreased apoptosis. It occurs more often in women than in men. It is known that inflammatory cytokines such as TNF, IL-6, and IL-1, and chemokines such as COX-2, 5-LOX, and matrix metallopeptidase (MMP)-9 are the factors for rheumatoid arthritis, as well as smoking and stress (Gupta et al. 2013c). In a clinical trial, curcumin's efficacy was compared with that of phenylbutazone, a prescription drug used to treat arthritis. The patients were separated into two groups and received either curcumin (1.2 g/day) or phenylbutazone (0.3 g/day) for two weeks. Curcumin had the same beneficial effects as phenylbutazone, including improvement in joint swelling, morning stiffness, and walking time, with no adverse effects (Deodhar et al. 1980). In another study, curcumin was compared with diclofenac sodium. Forty-five patients received either 0.5 g of curcumin or the same dosage of curcumin in addition to 0.05 g of diclofenac. Curcumin was found to be effective and safe in these patients with rheumatoid arthritis (Chandran and Goel 2012).

21.9 ROLE OF CURCUMIN IN INFECTIOUS DISEASES

Curcumin has been studied for possible effects on various infectious diseases, including sexually transmitted diseases, tuberculosis, *Helicobacter pylori* infection, malaria, hepatitis, and others spread by bacteria or pathogenic biological agents. That curcumin could affect different kinds of bacteria was first reported in 1949 (Schraufstatter and Bernt 1949).

21.9.1 Gonorrhea

Gonorrhea is the second most prevalent sexually transmitted infection in the United States and is linked with an increased risk of bladder cancer. *Neisseria gonorrhoeae* suppresses T-helper 1 (Th1) and Th2 responses, and enhances Th17 responses via a mechanism involving TGF-β and regulatory T cells. Therefore, blocking TGF-β would alleviate the suppression of antigonococcal responses and would allow for protective immunity. Gonorrhea also activates NF-κB, leading to inflammation. Because vitamin D is known to support the innate immune system, decrease TGF-β and NF-κB activation, and induce LL-37, vitamin D taken together with curcumin might enable the immune system to combat drug-resistant gonorrhea (Youssef et al. 2013). A test in which curcumin and vitamin D are taken together to determine curcumin's effects on gonorrhea is needed to test this hypothesis.

21.9.2 Tuberculosis

Tuberculosis is an infectious disease caused by *Mycobacterium tuberculosis*. Inhibitors of efflux pumps are potential agents to treat tuberculosis; therefore, since curcumin is an inhibitor of efflux pumps, it may serve as a lead compound for generating drugs against tuberculosis (Marathe et al. 2011).

21.9.3 Human Immunodeficiency Virus

Human immunodeficiency virus (HIV) is an infectious viral disease that leads to acquired immunodeficiency syndrome (AIDS). Curcumin has shown to interfere with the replication machinery of the HIV by inhibiting the HDAC1/NF-κB pathway and HIV integrase. Several reports have indicated that curcumin may be effective in treating AIDS. Inhibition of HIV long-terminal repeats (Barthelemy et al. 1998), HIV protease (Vajragupta et al. 2005), and p300/cyclic adenosine monophosphate (cAMP) response element-binding protein (CREB)-binding protein-specific acetyltransferase and repression of the acetylation of histone/nonhistone proteins and histone acetyltransferase-dependent chromatin transcription (Balasubramanyam et al. 2004) are the reasons why curcumin can be effective against AIDS (Aggarwal and Harikumar 2009). A clinical trial examined the effectiveness of curcumin as an antiviral agent against AIDS. Of the 40 AIDS patients examined, 23 were assigned to either a high dose of curcumin (2.5 g/day) or a low dose. The treatment was continued for eight weeks, but no reduction in viral load associated with curcumin was noticed. However, the patients that were taking curcumin reported feeling better than did controls (James 1996).

21.9.4 *Helicobacter pylori*

Helicobacter pylori is one of the most widespread infectious agents and can cause peptic ulcers. The most common treatment for this infection is the use of proton pump inhibitors and antibiotics, but these treatments may cause adverse effects. A clinical trial tested the effects of a seven-day nonantibiotic therapy for eradication of *H. pylori* infection. The therapy included treatment of 25 patients with 30 mg of curcumin, 100 mg of bovine lactoferrin, 600 mg of *N*-acetylcysteine, and 20 mg of pantoprazole twice a day for seven days. The results indicated that 12% of the patients were cured of the *H. pylori* infection, with significant decreases in the overall severity of symptoms. However, the *H. pylori* infection (IgG-Hp) did not decrease after two months, so the authors concluded that the therapy was not effective in eradicating *H. pylori*. Improvements in dyspeptic symptoms and a reduction of serological signs of gastric inflammation were observed after two months, so additional studies are needed to determine the potential of curcumin in the management of *H. pylori* infection (Marathe et al. 2011).

21.9.5 Malaria

Malaria is a mosquito-borne infectious disease caused by the protist *Plasmodium*. One study found that curcumin has antimalarial effects: Curcumin administered orally for five consecutive days reduced the parasite burden by more than 80% (Reddy et al. 2005).

21.9.6 Hepatitis

Hepatitis is an infectious disease that causes inflammation of the liver. A study of mice with paracetamol (acetaminophen) overdose demonstrated that curcumin prevents acetaminophen (APAP)-induced hepatitis by decreasing oxidative stress and liver inflammation and by restoring GSH, which improves liver histopathology (Somanawat et al. 2013).

21.10 CONCLUSION

Although curcumin is just one of the 300 components of turmeric (Li et al. 2011), it has been found within the past two decades to be the most important in its anti-inflammatory activities (Gupta et al. 2013c). More than a decade ago, its anti-inflammatory properties were discovered, leading to its use as a therapeutic agent against several chronic diseases. Today, curcumin is being researched extensively because of its ability to regulate many proinflammatory agents, including transcription factors, cytokines, protein kinases, and enzymes. The effects of curcumin on chronic diseases caused by inflammation such as CVD, neurological, pulmonary, metabolic, and infectious diseases are being investigated *in vivo* and in clinical trials. The first article based on curcumin's effects on human diseases was written by Oppenheimer (1937) and, to date, more than 5000 articles have been written about its therapeutic potential against chronic diseases and its ability to target multiple signaling pathways. Curcumin is inexpensive and can inhibit multiple proinflammatory pathways, and additional research and more clinical trials may help scientists learn more about the therapeutic potential of curcumin against various chronic diseases.

REFERENCES

Aggarwal, B.B. 2003. Signalling pathways of the TNF superfamily: A double-edged sword. *Nature Reviews Immunology*, 3(9): 745–756.

Aggarwal, B.B., Gupta, S.C., and Sung, B. 2013. Curcumin: An orally bioavailable blocker of TNF and other pro-inflammatory biomarkers. *British Journal of Pharmacology*, 169(8): 1672–1692.

Aggarwal, B.B. and Harikumar, K.B. 2009. Potential therapeutic effects of curcumin, the anti-inflammatory agent, against neurodegenerative, cardiovascular, pulmonary, metabolic, autoimmune and neoplastic diseases. *International Journal of Biochemistry & Cell Biology*, 41(1): 40–59.

Aggarwal, B.B., Sethi, G., Ahn, K.S., Sandur, S.K., Pandey, M.K., Kunnumakkara, A.B. et al. 2006. Targeting signal-transducer-and-activator-of-transcription-3 for prevention and therapy of cancer: Modern target but ancient solution. *Annals of the New York Academy Sciences*, 1091: 151–169.

Aggarwal, B.B. and Sung, B. 2009. Pharmacological basis for the role of curcumin in chronic diseases: An age-old spice with modern targets. *Trends in Pharmacological Sciences*, 30(2): 85–94.

Aggarwal, S., Takada, Y., Singh, S., Myers, J.N., and Aggarwal, B.B. 2004. Inhibition of growth and survival of human head and neck squamous cell carcinoma cells by curcumin via modulation of nuclear factor-kappaB signaling. *International Journal of Cancer*, 111(5): 679–692.

Ahn, K.S. and Aggarwal, B.B. 2005. Transcription factor NF-kappaB: A sensor for smoke and stress signals. *Annals of the New York Academy Sciences*, 1056: 218–233.

Alwi, I., Santoso, T., Suyono, S., Sutrisna, B., Suyatna, F.D., Kresno, S.B. et al. 2008. The effect of curcumin on lipid level in patients with acute coronary syndrome. *Acta Medica Indonesiana*, 40(4): 201–210.

Anand, P., Kunnumakkara, A.B., Newman, R.A., and Aggarwal, B.B. 2007. Bioavailability of curcumin: Problems and promises. *Molecular Pharmaceutics*, 4(6): 807–818.

Anand, P., Sundaram, C., Jhurani, S., Kunnumakkara, A.B., and Aggarwal, B.B. 2008. Curcumin and cancer: An "old-age" disease with an "age-old" solution. *Cancer Letters*, 267(1): 133–164.

Appendino, G., Belcaro, G., Cornelli, U., Luzzi, R., Togni, S., Dugall, M. et al. 2011. Potential role of curcumin phytosome (Meriva) in controlling the evolution of diabetic microangiopathy. A pilot study. *Panminerva Medica*, 53(3 Suppl 1): 43–49.

Atamna, H. and Boyle, K. 2006. Amyloid-beta peptide binds with heme to form a peroxidase: Relationship to the cytopathologies of Alzheimer's disease. *Proceedings of the National Academy Sciences of the United States of America*, 103(9): 3381–3386.

Balasubramanyam, K., Varier, R.A., Altaf, M., Swaminathan, V., Siddappa, N.B., Ranga, U., and Kundu, T.K. 2004. Curcumin, a novel p300/CREB-binding protein-specific inhibitor of acetyltransferase, represses the acetylation of histone/nonhistone proteins and histone acetyltransferase-dependent chromatin transcription. *Journal of Biological Chemistry*, 279(49): 51163–51171.

Barthelemy, S., Vergnes, L., Moynier, M., Guyot, D., Labidalle, S., and Bahraoui, E. 1998. Curcumin and curcumin derivatives inhibit Tat-mediated transactivation of type 1 human immunodeficiency virus long terminal repeat. *Research in Virology*, 149(1): 43–52.

Baum, L., Lam, C.W., Cheung, S.K., Kwok, T., Lui, V., Tsoh, J. et al. 2008. Six-month randomized, placebo-controlled, double-blind, pilot clinical trial of curcumin in patients with Alzheimer disease. *Journal of Clinical Psychopharmacology*, 28(1): 110–113.

Baum, L. and Ng, A. 2004. Curcumin interaction with copper and iron suggests one possible mechanism of action in Alzheimer's disease animal models. *Journal of Alzheimer's Diseases*, 6(4): 367–377.

Bayet-Robert, M., Kwiatkowski, F., Leheurteur, M., Gachon, F., Planchat, E., Abrial, C. et al. 2010. Phase I dose escalation trial of docetaxel plus curcumin in patients with advanced and metastatic breast cancer. *Cancer Biology & Therapy*, 9(1): 8–14.

Belcaro, G., Cesarone, M.R., Dugall, M., Pellegrini, L., Ledda, A., Grossi, M.G. et al. 2010a. Efficacy and safety of Meriva®, a curcumin-phosphatidylcholine complex, during extended administration in osteoarthritis patients. *Alternative Medicine Review*, 15(4): 337–344.

Belcaro, G., Cesarone, M.R., Dugall, M., Pellegrini, L., Ledda, A., Grossi, M.G. et al. 2010b. Product-evaluation registry of Meriva®, a curcumin-phosphatidylcholine complex, for the complementary management of osteoarthritis. *Panminerva Medica*, 52(2 Suppl 1): 55–62.

Biswas, S. and Rahman, I. 2008. Modulation of steroid activity in chronic inflammation: A novel anti-inflammatory role for curcumin. *Molecular Nutrition and Food Research*, 52(9): 987–994.

Burns, J., Joseph, P.D., Rose, K.J., Ryan, M.M., and Ouvrier, R.A. 2009. Effect of oral curcumin on Déjérine–Sottas disease. *Pediatric Neurology*, 41(4): 305–308.

Chainani-Wu, N., Silverman Jr, S., Reingold, A., Bostrom, A., Mc Culloch, C., Lozada-Nur., F. et al. 2007. A randomized, placebo-controlled, double-blind clinical trial of curcuminoids in oral lichen planus. *Phytomedicine*, 14(7/8): 437–446.

Chandran, B. and Goel, A. 2012. A randomized, pilot study to assess the efficacy and safety of curcumin in patients with active rheumatoid arthritis. *Phytotherapy Research*, 26(11): 1719–1725.

Cheng, A.L., Hsu, C.H., Lin, J.K., Hsu, M.M., Ho, Y.F., Shen, T.S. et al. 2001. Phase I clinical trial of curcumin, a chemopreventive agent, in patients with high-risk or pre-malignant lesions. *Anticancer Research*, 21(4B): 2895–2900.

Chuengsamarn, S., Rattanamongkolgul, S., Luechapudiporn, R., Phisalaphong, C., and Jirawatnotai, S. 2012. Curcumin extract for prevention of type 2 diabetes. *Diabetes Care*, 35(11): 2121–2127.

Cruz-Correa, M., Shoskes, D.A., Sanchez, P., Zhao, R., Hylind, L.M., Wexner, S.D. et al. 2006. Combination treatment with curcumin and quercetin of adenomas in familial adenomatous polyposis. *Clinical Gastroenterology and Hepatology*, 4(8): 1035–1038.

Deodhar, S.D., Sethi, R., and Srimal, R.C. 1980. Preliminary study on antirheumatic activity of curcumin (diferuloyl methane). *Indian Journal of Medical Research*, 71: 632–634.

Dhillon, N., Aggarwal, B.B., Newman, R.A., Wolff, R.A., Kunnumakkara, A.B., Abbruzzese, J.L. et al. 2008. Phase II trial of curcumin in patients with advanced pancreatic cancer. *Clinical Cancer Research*, 14(14): 4491–4499.

Durgaprasad, S., Pai, C.G., Vasanthkumar, Alvres, J.F., and Namitha, S. 2005. A pilot study of the antioxidant effect of curcumin in tropical pancreatitis. *Indian Journal of Medical Research*, 122(4): 315–318.

Epelbaum, R., Schaffer, M., Vizel, B., Badmaev, V., and Bar-Sela, G. 2010. Curcumin and gemcitabine in patients with advanced pancreatic cancer. *Nutrition Cancer*, 62(8): 1137–1141.

Fiala, M., Liu, P.T., Espinosa-Jeffrey, A., Rosenthal, M.J., Bernard, G., Ringman, J.M. et al. 2007. Innate immunity and transcription of MGAT-III and Toll-like receptors in Alzheimer's disease patients are improved by bisdemethoxycurcumin. *Proceedings of the National Academy Sciences of the United States of America*, 104(31): 12849–12854.

Garcea, G., Berry, D.P., Jones, D.J., Singh, R., Dennison, A.R., Farmer, P.B. et al. 2005. Consumption of the putative chemopreventive agent curcumin by cancer patients: Assessment of curcumin levels in the colorectum and their pharmacodynamic consequences. *Cancer Epidemiology Biomarkers and Prevention*, 14(1): 120–125.

Garcia-Alloza, M., Borrelli, L.A., Rozkalne, A., Hyman, B.T., and Bacskai, B.J. 2007. Curcumin labels amyloid pathology in vivo, disrupts existing plaques, and partially restores distorted neurites in an Alzheimer mouse model. *Journal of Neurochemistry*, 102(4): 1095–1104.

Goel, A., Jhurani, S., and Aggarwal, B.B. 2008. Multi-targeted therapy by curcumin: How spicy is it? *Molecular Nutrition Food Research*, 52(9): 1010–1030.

Golombick, T., Diamond, T.H., Badmaev, V., Manoharan, A., and Ramakrishna, R. 2009. The potential role of curcumin in patients with monoclonal gammopathy of undefined significance—Its effect on paraproteinemia and the urinary N-telopeptide of type I collagen bone turnover marker. *Clinical Cancer Research*, 15(18): 5917–5922.

Gupta, S.C., Kismali, G., and Aggarwal, B.B. 2013a. Curcumin, a component of turmeric: From farm to pharmacy. *Biofactors*, 39(1): 2–13.

Gupta, S.C., Patchva, S., and Aggarwal, B.B. 2013b. Therapeutic roles of curcumin: Lessons learned from clinical trials. *The AAPS Journal*, 15(1): 195–218.

Gupta, S.C., Patchva, S., Koh, W., and Aggarwal, B.B. 2012. Discovery of curcumin, a component of golden spice, and its miraculous biological activities. *Clinical and Experimental Pharmacology and Physiology*, 39(3): 283–299.

Gupta, S.C., Sung, B., Kim, J.H., Prasad, S., Li, S., and Aggarwal, B.B. 2013c. Multitargeting by turmeric, the golden spice: From kitchen to clinic. *Molecular Nutrition Food Research*, 57(9): 1510–1528.

Gutierres, V.O., Pinheiro, C.M., Assis, R.P., Vendramini, R.C., Pepato, M.T., and Brunetti, I.L. 2012. Curcumin-supplemented yoghurt improves physiological and biochemical markers of experimental diabetes. *British Journal of Nutrition*, 108(3): 440–448.

Hariharan, D., Saied, A., and Kocher, H.M. 2008. Analysis of mortality rates for pancreatic cancer across the world. *HPB (Oxford)*, 10(1): 58–62.

Hasima, N. and Aggarwal, B.B. 2012. Cancer-linked targets modulated by curcumin. *International Journal of Biochemistry and Molecular Biology*, 3(4): 328–351.

Hastak, K., Lubri, N., Jakhi, S.D., More, C., John, A., Ghaisas, S.D. et al. 1997. Effect of turmeric oil and turmeric oleoresin on cytogenetic damage in patients suffering from oral submucous fibrosis. *Cancer Letters*, 116(2): 265–269.

He, Z.Y., Shi, C.B., Wen, H., Li, F.L., Wang, B.L., and Wang, J. 2011. Upregulation of p53 expression in patients with colorectal cancer by administration of curcumin. *Cancer Investigation*, 29(3): 208–213.

Ide, H., Tokiwa, S., Sakamaki, K., Nishio, K., Isotani, S., Muto, S. et al. 2010. Combined inhibitory effects of soy isoflavones and curcumin on the production of prostate-specific antigen. *Prostate*, 70(10): 1127–1133.

James, J.S. 1996. Curcumin: Clinical trial finds no antiviral effect. *AIDS Treat News*, (242): 1–2.

Ju, H.R., Wu, H.Y., Nishizono, S., Sakono, M., Ikeda, I., Sugano, M., and Imaizumi, K. 1996. Effects of dietary fats and curcumin on IgE-mediated degranulation of intestinal mast cells in brown Norway rats. *Bioscience Biotechnology and Biochemistry*, 60(11): 1856–1860.

Kanai, M., Yoshimura, K., Asada, M., Imaizumi, A., Suzuki, C., Matsumoto, S. et al. 2011. A phase I/II study of gemcitabine-based chemotherapy plus curcumin for patients with gemcitabine-resistant pancreatic cancer. *Cancer Chemotherapy and Pharmacology*, 68(1): 157–164.

Karin, M. and Greten, F.R. 2005. NF-κB: Linking inflammation and immunity to cancer development and progression. *Nature Reviews Immunology*, 5(10): 749–759.

Kawanami, D., Maemura, K., Takeda, N., Harada, T., Nojiri, T., Saito T. et al. 2006. C-reactive protein induces VCAM-1 gene expression through NF-kappaB activation in vascular endothelial cells. *Atherosclerosis*, 185(1): 39–46.

Khajehdehi, P., Pakfetrat, M., Javidnia, K., Azad, F., Malekmakan, L., Nasab, M.H. et al. 2011. Oral supplementation of turmeric attenuates proteinuria, transforming growth factor-beta and interleukin-8 levels in patients with overt type 2 diabetic nephropathy: A randomized, double-blind and placebo-controlled study. *Scandinavian Journal of Urology and Nephrology*, 45(5): 365–370.

Kim, D.S., Park, S.Y., and Kim, J.K. 2001. Curcuminoids from *Curcuma longa* L. (Zingiberaceae) that protect PC12 rat pheochromocytoma and normal human umbilical vein endothelial cells from betaA(1-42) insult. *Neuroscience Letters*, 303(1): 57–61.

Kim, J.H., Gupta, S.C., Park, B., Yadav, V.R., and Aggarwal, B.B. 2012. Turmeric (*Curcuma longa*) inhibits inflammatory nuclear factor (NF)-κB and NF-κB-regulated gene products and induces death receptors leading to suppressed proliferation, induced chemosensitization, and suppressed osteoclastogenesis. *Molecular Nutrition Food Research*, 56(3): 454–465.

Kim, S.G., Veena, M.S., Basak, S.K., Han, E., Tajima, T., Gjertson, D.W. et al. 2011. Curcumin treatment suppresses IKKβ kinase activity of salivary cells of patients with head and neck cancer: A pilot study. *Clinical Cancer Research*, 17(18): 5953–5961.

Kobayashi, T., Hashimoto, S., and Horie, T. 1997. Curcumin inhibition of Dermatophagoides farinea-induced interleukin-5 (IL-5) and granulocyte macrophage-colony stimulating factor (GM-CSF) production by lymphocytes from bronchial asthmatics. *Biochemical Pharmacology*, 54(7): 819–824.

Kumar, A., Takada, Y., Boriek, A.M., and Aggarwal, B.B. 2004. Nuclear factor-kappaB: Its role in health and disease. *Journal of Molecular Medicine (Berlin)*, 82(7): 434–448.

Kunnumakkara, A.B., Anand, P., and Aggarwal, B.B. 2008. Curcumin inhibits proliferation, invasion, angiogenesis and metastasis of different cancers through interaction with multiple cell signaling proteins. *Cancer Letters*, 269(2): 199–225.

Kuttan, R., Bhanumathy, P., Nirmala, K., and George, M.C. 1985. Potential anticancer activity of turmeric (*Curcuma longa*). *Cancer Letters*, 29(2): 197–202.

Li, S., Yuan, W., Deng, G., Wang, P., Yang, P., and Aggarwal, B.B. 2011. Chemical composition and product quality control of turmeric (*Curcuma longa* L.). *Pharmaceutical Crops*, 2: 28–54.

Lim, G.P., Chu, T., Yang, F., Beech, W., Frautschy, S.A., and Cole, G.M. 2001. The curry spice curcumin reduces oxidative damage and amyloid pathology in an Alzheimer transgenic mouse. *Journal of Neuroscience*, 21(21): 8370–8377.

Manikandan, P., Sumitra, M., Aishwarya, S., Manohar, B.M., Lokanadam, B., and Puvanakrishnan R. 2004. Curcumin modulates free radical quenching in myocardial ischaemia in rats. *International Journal of Biochemistry and Cell Biology*, 36(10): 1967–1980.

Marathe, S.A., Dasgupta, I., Gnanadhas, D.P., and Chakravortty, D. 2011. Multifaceted roles of curcumin: Two sides of a coin! *Expert Opinion on Biological Therapy*, 11(11): 1485–1499.

Marwick, J.A., Ito, K., Adcock, I.M., and Kirkham, P.A. 2007. Oxidative stress and steroid resistance in asthma and COPD: Pharmacological manipulation of HDAC-2 as a therapeutic strategy. *Expert Opinion on Therapeutic Targets*, 11(6): 745–755.

McCormick, M.L., Gavrila, D., and Weintraub, N.L. 2007. Role of oxidative stress in the pathogenesis of abdominal aortic aneurysms. *Arteriosclerosis, Thrombosis, and Vascular Biology*, 27(3): 461–469.

Narayan, V., Ravindra, K.C., Chiaro, C., Cary, D., Aggarwal, B.B., Henderson, A.J., and Prabhu, K.S. 2011. Celastrol inhibits Tat-mediated human immunodeficiency virus (HIV) transcription and replication. *Journal of Molecular Biology*, 410(5): 972–983.

Nirmala, C. and Puvanakrishnan, R. 1996. Effect of curcumin on certain lysosomal hydrolases in isoproterenol-induced myocardial infarction in rats. *Biochemical Pharmacology*, 51(1): 47–51.

Olszanecki, R., Jawień, J., Gajda, M., Mateuszuk, L., Gebska, A., Korabiowska, M. et al. 2005. Effect of curcumin on atherosclerosis in apoE/LDLR-double knockout mice. *Journal of Physiology and Pharmacology*, 56(4): 627–635.

Ono, K., Hasegawa, K., Naiki, H., and Yamada, M. 2004. Curcumin has potent anti-amyloidogenic effects for Alzheimer's beta-amyloid fibrils in vitro. *Journal of Neuroscience Research*, 75(6): 742–750.

Oppenheimer, A. 1937. Turmeric (curcumin) in biliary diseases. *Lancet*, 229: 619–621.

Parodi, F.E., Mao, D., Ennis, T.L., Pagano, M.B., and Thompson, R.W. 2006. Oral administration of diferuloylmethane (curcumin) suppresses proinflammatory cytokines and destructive connective tissue remodeling in experimental abdominal aortic aneurysms. *Annals of Vascular Surgery*, 20(3): 360–368.

Polasa, K., Raghuram, T.C., Krishna, T.P., and Krishnaswamy, K. 1992. Effect of turmeric on urinary mutagens in smokers. *Mutagenesis*, 7(2): 107–109.

Quiles, J.L., Mesa, M.D., Ramírez-Tortosa, C.L., Aguilera, C.M., Battino, M., Gil, A. et al. 2002. *Curcuma longa* extract supplementation reduces oxidative stress and attenuates aortic fatty streak development in rabbits. *Arteriosclerosis, Thrombosis, and Vascular Biology*, 22(7): 1225–1231.

Rai, B., Kaur, J., Jacobs, R., and Singh, J. 2010. Possible action mechanism for curcumin in pre-cancerous lesions based on serum and salivary markers of oxidative stress. *Journal of Oral Science*, 52(2): 251–256.

Ramaswami, G., Chai, H., Yao, Q., Lin, P.H., Lumsden, A.B., and Chen, C. 2004. Curcumin blocks homocysteine-induced endothelial dysfunction in porcine coronary arteries. *Journal of Vascular Surgery*, 40(6): 1216–1222.

Ramírez-Tortosa, M.C., Mesa, M.D., Aguilera, M.C., Quiles, J.L., Baró, L., Ramirez-Tortosa, C.L. et al. 1999. Oral administration of a turmeric extract inhibits LDL oxidation and has hypocholesterolemic effects in rabbits with experimental atherosclerosis. *Atherosclerosis*, 147(2): 371–378.

Rao, D.S., Sekhara, N.C., Satyanarayana, M.N., and Srinivasan, M. 1970. Effect of curcumin on serum and liver cholesterol levels in the rat. *Journal of Nutrition*, 100(11): 1307–1315.

Reddy, R.C., Vatsala, P.G., Keshamouni, V.G., Padmanaban, G., and Rangarajan, P.N. 2005. Curcumin for malaria therapy. *Biochemical and Biophysical Research Communications*, 326(2): 472–474.

Ringman, J.M., Frautschy, S.A., Cole, G.M., Masterman, D.L., and Cummings, J.L. 2005. A potential role of the curry spice curcumin in Alzheimer's disease. *Current Alzheimer Research*, 2(2): 131–136.

Schraufstatter, E. and Bernt H. 1949. Antibacterial action of curcumin and related compounds. *Nature*, 164(4167): 456.

Sharma, O.P. 1976. Antioxidant activity of curcumin and related compounds. *Biochemical Pharmacology*, 25(15): 1811–1812.

Sharma, R.A., Euden, S.A., Platton, S.L., Cooke, D.N., Shafayat, A., Hewitt, H.R. et al. 2004. Phase I clinical trial of oral curcumin: Biomarkers of systemic activity and compliance. *Clinical Cancer Research*, 10(20): 6847–6854.

Sharma, R.A., McLelland, H.R., Hill, K.A., Ireson, C.R., Euden, S.A., Manson, M.M. et al. 2001. Pharmacodynamic and pharmacokinetic study of oral *Curcuma* extract in patients with colorectal cancer. *Clinical Cancer Research*, 7(7): 1894–1900.

Shishodia, S., Sethi, G., and Aggarwal, B.B. 2005. Curcumin: Getting back to the roots. *Annals of the New York Academy Sciences*, 1056: 206–217.

Singh, S. and Aggarwal, B.B. 1995. Activation of transcription factor NF-kappa B is suppressed by curcumin (diferuloylmethane). *Journal of Biological Chemistry*, 270(42): 24995–25000.

Sinha, D., Biswas, J., Sung, B., Aggarwal, B.B., and Bishayee, A. 2012. Chemopreventive and chemotherapeutic potential of curcumin in breast cancer. *Current Drug Targets*, 13(14): 1799–1819.

Somanawat, K., Thong-Ngam, D., and Klaikeaw, N. 2013. Curcumin attenuated paracetamol overdose induced hepatitis. *World Journal of Gastroenterology*, 19(12): 1962–1967.

Srimal, R.C. and Dhawan, B.N. 1973. Pharmacology of diferuloyl methane (curcumin), a non-steroidal anti-inflammatory agent. *Journal of Pharmacy and Pharmacology*, 25(6): 447–452.

Srinivasan, M. 1972. Effect of curcumin on blood sugar as seen in a diabetic subject. *Indian Journal of Medical Sciences*, 26(4): 269–270.

Srivastava, R., Puri, V., Srimal, R.C., and Dhawan, B.N. 1986. Effect of curcumin on platelet aggregation and vascular prostacyclin synthesis. *Arzneimittel-Forschung*, 36: 715–717.

Sung, B., Prasad, S., Yadav, V.R., and Aggarwal, B.B. 2012. Cancer cell signaling pathways targeted by spice-derived nutraceuticals. *Nutrition Cancer*, 64(2): 173–197.

Tergaonkar, V. 2006. NFκB pathway: A good signaling paradigm and therapeutic target. *The International Journal of Biochemistry & Cell biology*, 38(10): 1647–1653.

Usharani, P., Mateen, A.A., Naidu, M.U., Raju, Y.S., and Chandra, N. 2008. Effect of NCB-02, atorvastatin and placebo on endothelial function, oxidative stress and inflammatory markers in patients with type 2 diabetes mellitus: A randomized, parallel-group, placebo-controlled, 8-week study. *Drugs in R&D*, 9(4): 243–250.

Vadhan-Raj, S., Weber, D., Wang, M., Giralt, S., Alexanian, R., Thomas, S. et al. 2007. Curcumin downregulates NF-κB and related genes in patients with multiple myeloma: Results of a phase ½ study. *Blood*, 110(11): 357a.

Vajragupta, O., Boonchoong, P., Morris, G.M., and Olson, A.J. 2005. Active site binding modes of curcumin in HIV-1 protease and integrase. *Bioorganic and Medicinal Chemistry Letters*, 15(14): 3364–3368.

Wargovich, M.J., Eng, V.W., and Newmark, H.L. 1985. Inhibition by plant phenols of benzo[a]pyrene-induced nuclear aberrations in mammalian intestinal cells: A rapid in vivo assessment method. *Food and Chemical Toxicology*, 23(1): 47–49.

Wickenberg, J., Ingemansson, S.L., and Hlebowicz, J. 2010. Effects of *Curcuma longa* (turmeric) on postprandial plasma glucose and insulin in healthy subjects. *Nutrition Journal*, 9: 43.

Wongcharoen, W., Jai-Aue, S., Phrommintikul, A., Nawarawong, W., Woragidpoonpol, S., Tepsuwan, T. et al. 2012. Effects of curcuminoids on frequency of acute myocardial infarction after coronary artery bypass grafting. *American Journal of Cardiology*, 110(1): 40–44.

Yang, F., Lim, G.P., Begum, A.N, Ubeda, O.J., Simmons, M.R., Ambegaokar, S.S. et al. 2005. Curcumin inhibits formation of amyloid beta oligomers and fibrils, binds plaques, and reduces amyloid in vivo. *Journal of Biological Chemistry*, 280(7): 5892–5901.

Yeh, C.H., Chen, T.P., Wu, Y.C., Lin, Y.M., and Jing Lin, P. 2005. Inhibition of NF-kappa B activation with curcumin attenuates plasma inflammatory cytokines surge and cardiomyocytic apoptosis following cardiac ischemia/reperfusion. *Journal of Surgical Research*, 125(1): 109–116.

Youssef, D.A., Peiris, A.N., Kelley, J.L., and Grant, W.B. 2013. The possible roles of vitamin D and curcumin in treating gonorrhea. *Medical Hypotheses*, 81(1): 131–135.

Zuccotti, G.V., Trabattoni, D., Morelli, M., Borgonovo, S., Schneider, L., and Clerici, M. 2009. Immune modulation by lactoferrin and curcumin in children with recurrent respiratory infections. *Journal of Biological Regulators and Homeostatic Agents*, 23(2): 119–123.

22 Human Clinical Trial for Nutraceuticals and Functional Foods

Chin-Kun Wang

CONTENTS

22.1 INTRODUCTION

Food is necessary and important for human beings. The health benefits of functional foods and nutraceuticals have been greatly emphasized recently. Why are functional foods and nutraceuticals so popular? The major reason is the new challenge of noncommunicable diseases (NCDs) to human health. NCDs, also known as chronic diseases, are not transmitted from person to person. They are of long duration and generally slow progression. The four main types of NCDs are cardiovascular diseases (such as heart attacks and stroke), cancers, chronic respiratory diseases (such as chronic obstructed pulmonary disease and asthma), and diabetes.[1]

All age groups and all regions are affected by NCDs. NCDs are often associated with older age groups, but evidence shows that more than nine million of all deaths attributed to NCDs occur before the age of 60. Children, adults, and the elderly are all vulnerable to the risk factors that contribute to NCDs, whether from unhealthy diets, physical inactivity, exposure to tobacco smoke, or the effects of the harmful use of alcohol. To lessen the impact of NCDs on individuals and society, a comprehensive approach is needed which requires all sectors, including health, finance, foreign affairs, education, agriculture, planning, and others, to work together to reduce the risks associated with NCDs, as well as promote the interventions to prevent and control them.[*,2]

For the prevention and therapy of NCDs, conventional medicine is sometimes helpless. Complementary and alternative medicine (CAM) is an emerging field that affords hope. Defining CAM is difficult because the field is very broad and constantly changing. The US National Center of CAM (NCCAM) defines CAM as a group of diverse medical and health-care systems, practices, and products that are not generally considered part of conventional medicine. Conventional medicine (also called Western or allopathic medicine) is medicine as practiced by holders of doctor of medicine and doctor of osteopathic medicine degrees, and by allied health professionals, such as physical therapists, psychologists, and registered nurses. The boundaries between CAM and conventional medicine are not absolute, and specific CAM practices may, over time, become widely accepted.[2]

CAM practices are often grouped into broad categories, such as natural products, mind and body medicine, and manipulative and body-based practices. Although these categories are not formally defined, they are useful for discussing CAM practices. Some CAM practices may fit into more than one category. The natural products of CAM include use of a variety of herbal medicines (also known as botanicals), vitamins, minerals, and other "natural products." Many are sold over the counter (OTC) as dietary supplements. (Some uses of dietary supplements—taking a multivitamin to meet minimum daily nutritional requirements or taking calcium to promote bone health—are not thought of as CAM.) The "natural products" of CAM also include probiotics—live bacteria (and sometimes yeasts) found in foods such as yogurt or in dietary supplements and live microorganisms (usually bacteria) that are similar to microorganisms normally found in the human digestive tract and that may have beneficial effects. Probiotics are available in foods (e.g., yogurts) or as dietary supplements. They are not the same thing as prebiotics—nondigestible food ingredients that selectively stimulate the growth and/or activity of microorganisms already present in the body. CAM includes many kinds of treatments. Functional foods and nutraceuticals are highly accepted and trusted because of their history and habitual use. "Complementary medicine" refers to the use of CAM together with conventional medicine. Functional foods and nutraceuticals may supplement conventional medicine. However, based on the evidence, the related scientific intervention and related information, especially human clinical trials, are critically required.[3–5]

22.2 DISCOVERY

Foods are the best medicine in the human history. Human foods include animals, plants, algae, microorganisms, and so on. To discover the health benefits, preclinical trials including *in vitro*, cell, and animal studies are used. However, the results are discovery and preclinical data. The cell system is fast and easy to know the reaction, and good to understand the detailed mechanism. Animal models are very practical to see the real response. Whether all the results can be completely responsive to the real situation in human being or not? Evidently, the answer is not so affirmative. However, the preclinical data are very worth for the advanced human clinical trial. In addition, the animal study can give a recommended dosage in the next human study.[6] Basically, discovery is a scientific beginning and is also necessary for the new evidence.

For new drugs development, the processed systems are well established, including the preclinical and various phases. In the preclinical discovery, molecular structure, best combination, screening design, cluster analysis, discriminant analysis, and factor analysis can be well obtained first. The dose-related response (regression) and pharmacological activity can also be provided by the animal study.[7] The next

[*] Glossary of clinical trial terms. NIH Clinicaltrials.gov.

step is to understand the genetic toxicity, animal toxicity, reproductive toxicity, carcinogenic toxicity, and the dose–relationship (regression), and further for absorption, distribution, metabolism, and elimination (ADME). Finally, the formulation to reach its site of action for clinical trial was obtained.[8]

For the scientific evidence, functional foods and nutraceuticals almost run the same way as the new drugs.[9] But very complex composition is still found in functional foods or nutraceuticals. The toxicity is clearly required, but the detailed molecular structures are not completely determined. However, the major contributor or so-called indicator for such a health benefit should be well identified. This will be helpful for the quality control and monitoring in the new products.[10]

22.3 SAFETY AND TOXICITY EVALUATION

Functional food is a food type and for common use, including animal, plant, and microorganism. Nutraceuticals are usually powder, tablet, or capsule type (medicine like). Both functional foods and nutraceuticals contain complex composition and are different from that of new drugs. New drug, usually pure compound, needs very careful evaluation on safety. The safety evaluation for functional foods and nutraceuticals is recommended as four catalogs[11]:

1. If the raw materials of the functional foods or nutraceuticals are traditional foods and the processing is normal or traditional cooking, these are generally recognized as safe (GRAS). Good examples are oat meal, yogurt, and tea.
2. If the raw materials of the functional foods or nutraceuticals are traditional ones, but the processing or treatment is not traditional cooking (e.g., solvent extraction), the data for genotoxicity and 28-day toxicity study are required.
3. If the raw materials are not traditional ones, the safety data for the genotoxicity, 90-day toxicity study, and teratogenicity are required.
4. If the raw materials are not traditional ones and with potential risk to health, the safety data for the genotoxicity, teratogenicity, 90-day toxicity study, carcinogenicity, and reproduction are required.

22.4 CLINICAL TRIALS

A clinical study involves research using human volunteers (also called participants). There are two main types of clinical studies: clinical trials and observational studies.[12]

In a clinical trial (also called an interventional study), participants receive specific interventions according to the research plan or protocol created by the investigators. These interventions may be medical products, such as drugs or devices, procedures, or changes to participants' behavior, for example, diet. Clinical trials may compare a new approach to a standard one that is already available or to a placebo that contains no active ingredients or to no intervention. Some clinical trials compare interventions that are already available to each other. The investigators try to determine the safety and efficacy of the intervention by measuring certain outcomes in the participants. For example, investigators may give a treatment to participants who have high blood pressure to see whether their blood pressure decreases.

Every clinical study is led by a principal investigator. Clinical studies also have a research team that may include doctors, nurses, social workers, and other health-care professionals.[13]

22.4.1 SPONSORSHIP

A full series of trials may include sizable costs, and the burden of paying for all the necessary people and services is usually provided by the sponsor, which may be a governmental organization, a pharmaceutical, a biotechnology company, or a food company.[14]

Many research groups are constantly searching for a variety of people to volunteer to be part of human studies. Most of these people will receive a stipend for participation in the study; however, some studies can be time consuming.[15]

22.4.2 Participating in Clinical Studies

A clinical study is conducted according to a research plan known as the protocol. The protocol is designed to answer specific research questions as well as safeguard the health of participants. It contains the following information[16]:

1. The reason for conducting the study
2. The subjects may participate in the study (the eligibility criteria)
3. The number of participants needed
4. The schedule of tests, procedures, and dosages
5. The length of the study

22.4.3 Institutional Review Boards

Each conducted clinical study must be reviewed, approved, and monitored by an institutional review board (IRB). An IRB is made up of physicians, researchers, and members of the community. The primary function of the IRB is to make sure that the study is ethical and the rights and welfare of participants are well protected. This includes making sure that research risks are minimized and are reasonable in relation to any potential benefits, among other things. The secondary function is to review the scientific merit of the protocol since executing a scientifically unworthy protocol is considered unethical. The IRB also reviews the informed consent document. Before initiating a clinical trial, the investigators must obtain the full and informed consent of the participating subjects. The informed consent is a process by which subjects learn the important facts about a clinical trial to help them decide whether to participate. The information should include the purpose of the study, the tests and procedures used in the study, and the possible risks and benefits. In addition to verbal explanations from the investigator, the subjects receive a written consent form detailing the study. Investigators are obligated to address questions regarding short-term benefits, long-term benefits, short-term risks, long-term risks, and treatment options. Subjects are asked to sign the consent form if they agree to participate in the study. However, informed consent is not an immutable contract; the subjects can terminate participation at any time. It is also important to note that the informed consent process continues throughout the study. If new benefits, risks, or side effects are discovered during the study, the investigators must inform the participants. Study subjects are then asked to sign a new consent form if they decide to continue the participation. Common mistakes committed by the investigator include failing to update patients with new safety information in a timely manner and backdating the signature on the consent form, which are serious violations of Good Clinical Practice (GCP).[17]

In addition to being monitored by an IRB, some clinical studies are also monitored by data monitoring committees (also called data safety and monitoring boards).

22.4.4 Clinical Phases

Clinical trials are testing sets in health research and new drug development which generate safety and efficacy data (or more specifically, information about adverse reactions) for health interventions. They are conducted only after satisfactory information has been gathered on the quality of the nonclinical safety, and health authority/ethics committee approval is granted. The evaluation for the health benefits of functional foods and nutraceuticals is very similar to that of new drugs. Clinical trials used in drug development are sometimes described by phases, including phase 0, I, II, III, and IV[16]:

- Phase 0: Pharmacodynamics and pharmacokinetics
- Phase I: Screening for safety
- Phase II: Establishing the testing protocol

- Phase III: Final testing
- Phase IV: Postapproval studies

22.4.4.1 Phase 0

In phase 0, trials are the first-in-human trials. Single subtherapeutic doses of the study drug are given to a small number of subjects (10–15) to gather preliminary data on the agent's pharmacodynamics (what the treatment does to the body) and pharmacokinetics (what the body does to the treatment).[18]

22.4.4.2 Phase I

Researchers test an experimental drug or treatment in a small group of people (20–80) for the first time to evaluate its safety, determine a safe dosage range, and identify side effects. Pharmacokinetics and pharmacodynamics are well obtained in this phase.

Pharmacokinetics: A procedure for the function of the new drug in the human body, including absorption, distribution, and metabolism.

Pharmacodynamics: An induced biochemical and physical function in the human body, including the reaction time, time course, applied dose, and dosing intervals.[19]

Phase I trials are intended to assess the safety, tolerability, and pharmacokinetics of the investigational treatment. Two most notable, early phase I trials are the *single ascending dose (SAD)* and the *multiple ascending dose (MAD)* studies. The former is designed to determine the single, *maximum tolerated dose (MTD)* that can be given before unacceptable toxicities are experienced by study subjects, and the latter is used to project the safe dose for eventual therapeutic use. Much of the literature has discussed how to extrapolate animal data to the starting dose in humans. In estimating the MTD, the investigators usually start with a low dose and escalate the dose until the signs of toxicity are observed. The escalation scheme must be conservative so as not to overreach MTD, but at the same time be efficient in the number of doses and subjects studied. Typically, a small group of healthy volunteers in several cohorts is selected for the first-stage testing in man. These trials often extend over several half-lives of the drug in an inpatient clinic, where the volunteer subjects are monitored around the clock. Special circumstances may justify the use of real patients who suffer a life-threatening or serious disease for which no good alternative treatment is available. This exception to the rule most often occurs in oncologic or antiviral trials. A subset of the phase I trials may be designated to investigate the effects of food on absorption or to counteract the gastrointestinal side effect of the investigational drug.[19,*]

22.4.4.3 Phase II

The experimental treatment is given to a larger group of people (100–300) to see if it is effective and to further evaluate its safety. In addition, other pharmacological and pharmacokinetics are also observed in this phase. Placebo control is used in this phase. The purpose is to explore whether the treatment exhibits credible efficacy and acceptable safety in actual participants.

If MAD study confirms safety and tolerability of the drug, phase II studies ensue with a larger group of real patients. The purpose is to explore whether the drug exhibits preliminary efficacy in real patients and whether tolerance and side effect profile are still within the acceptable limit. While some regard the phase II trial as a whole, others divide it into phases IIA and IIB. Phase IIA is designed as the dose-ranging study to define an "optimal" dose that is to be adopted for phase IIB in a randomized and controlled fashion. To explore efficacy with high degree of clarity, phase II studies are typically conducted in patients with a specific disease or condition but without complications, for example, hypertension without diabetes. Thus, phase II trials are more restrictive in terms of inclusion/exclusion criteria. The number of patients involved will depend on the specifics of the

* Clinicaltrials.gov. http\\www.clinicaltrials.gov.

disease state. For psychopharmacological agents, a fairly large sample size will be needed to discern the preliminary efficacy; for cancer trials, a relatively small number of patients may be sufficient to estimate the response rate. It is inherently clear that the results of the phase II trial will depend on the quality and adequacy of the phase I study. Likewise, the results from the phase II study are the basis for the expanded phase III trials.[20]

22.4.4.4 Phase III

Adequate and well control are designed in this phase. The treatment is given to large groups of people (1000–3000) to confirm its effectiveness, monitor side effects, compare it to commonly used treatments, and collect information that allows it to be used safely. Usually this is a pivotal, placebo-controlled, and active control study. Further research is required on short-term (weeks) and long-term (one to two years) safety and efficacy. Patients usually include the elderly (>65 years), or patients with liver or kidney disease. New drug application, commercialization, and valid evidence can be got in this phase.[21]

Because of the patient number and the trial duration, phase III trials are costly to run and more difficult to manage, especially for chronic medical conditions. Phase III studies are sometimes divided into phases IIIA and IIIB; placebo is used in the former and the market comparator is used in the latter. While phase III may remain blinded throughout the trial period, there are circumstances where open-label extension is carried out to confirm the efficacy. If efficacy becomes apparent at a certain point of the trial, the study may be terminated at an earlier stage through an interim utility analysis, for example, BiDil® (NitroMed, Inc., Charlotte, NC) trial.[22] More often than premature termination for good causes, many new chemical entities are terminated for development due to lack of efficacy through a utility analysis. This is undertaken to avoid the unnecessary suffering borne by the study subjects, besides the obvious economical consideration. Once the phase III trials are satisfactorily concluded, patient data from the multiple centers are pooled and analyzed for submission to appropriate regulatory authorities for review, along with a large body of information from preclinical investigation and manufacturing. Typically, two well-controlled phase III trials, at a minimum, are required for marketing.

22.4.4.5 Phase IV

After the product is commercialized, the adverse event, pharmacological effects on the elderly, child, pregnant woman, morbidity, mortality, and also the new indication of this product are further evaluated (new possible purpose).

Phase IV trials may result in product recall or voluntary withdrawal from the market. Such examples included cerivastatin (Baycol, The Bayer Corporation, Pittsburgh, PA), troglitazone (Rezulin), and rofecoxib (Vioxx) in the recent past.[23] Some phase IV studies are not considered critical for marketing approval at the time of regulatory review, but clarification is important to warrant a postmarketing follow-up, for example, renal study for drugs that are primarily metabolized or drug interaction studies for drugs that are primarily excreted unchanged. Pediatric study, geriatric study, and gender effect study may also fall into this category, depending on the drug in question. There are phase IV studies that are required for labeling change such as the effects of juice, antacid or certain foods on oral medications, or boxed warning against adverse events. The drug company may also undertake phase IV study as a marketing strategy, in which head-to-head comparison with the competitor product is pursued. Prescription to OTC conversion is a commercial strategy undertaken by some drug makers after years of postmarketing surveillance demonstrating a high degree of drug safety. It is interesting to note that many phase IV studies are phase I in technical nature, but phase IV in terms of time sequence of study execution.*

Functional foods and nutraceuticals are not pure substances as the new drugs, and mostly are edible. In addition, the purpose of functional foods and nutraceuticals is not for disease therapy. They usually play the role on the health promotion, disease prevention, or therapy assessment. Phase II is usually recommended for the evaluation of nutraceuticals and functional foods. For further or advanced study, phases III and IV are required.[24]

* Recalls.gov. http//www.recalls.gov/.

22.4.5 Clinical Trial Design

Technical aspects of phase I trial are somewhat different from those in phases II, III, and IV; however, the scientific principles that are applied to all phases remain largely the same. In phase I clinical studies, safety is the primary end point, whereas efficacy is the primary end point in phases II and III. General statistical principles include randomization, blinding, stratification, and hypothesis testing. Randomization assigns patients to treatment groups by chance, eliminating any systematic imbalance in characteristics between patients who receive the test versus the control. Blinding ensures that neither the patient nor the investigator knows to which group specific patients are assigned. The purpose of blinding is to minimize the chance for patients to receive preferential treatment or subject their data to biased interpretation. If blinding is impossible, an independent evaluator for the outcome measure should be used. Stratification is to ensure that the number of patients assigned to the test and the control groups are balanced with respect to certain important attributes, for example, gender or disease stage. After randomized, controlled trials are concluded, statisticians help determine whether any observed difference in the outcomes between the test and the control is real or simply chance occurrence. The process is called hypothesis testing in which a null hypothesis articulates no difference between the observed means of the test and the control populations. Statistical inferences based on the benefit/risk ratio allow the clinicians to accept, with certain level of confidence, the new treatment that is selected as the intervention of choice. Phase III trials are aimed at the anticipated clinical outcomes of a specific disease or condition; therefore, the most important requirement for phase III trial end point is clinical relevance. Clinical trials typically stipulate only one primary end point, although some disease conditions may require measurement of a coprimary end point. A clinical trial can have and does often have multiple secondary end points. When multiple end points yield conflicting results, conclusions can be obscured.

The identity, strength, quality, and purity are the purposes of a good experimental design. For ethics, crossover design is better than parallel design. In addition, the time for crossover design is much longer than that of the parallel design. To prevent the carryover effect, the proper period for washout is required in the crossover design. For phase I, crossover design is usually adopted. For phases II, III, and IV, parallel design is highly recommended. Double bind is usually much more adopted than single blind. Most studies are by randomization and very few by nonrandomization. Multicenter trials are easy to enroll patients than single-center trials, but center effects (treatment-by-center interactions) should be prevented.[23]

22.4.6 Clinical Trial Protocol

A clinical trial protocol is a document used to gain confirmation of the trial design by a panel of experts and adherence by all study investigators, even if conducted in various countries. The protocol describes the scientific rationale, objective(s), design, methodology, statistical considerations, and organization of the planned trial. Details of the trial are also provided in other documents referenced in the protocol.

The protocol contains a precise study plan for executing the clinical trial, not only to assure safety and health of the trial subjects but also to provide an exact template for trial conducted by investigators at multiple locations (in a "multicenter" trial) to perform the study in exactly the same way. This harmonization allows data to be combined collectively as though all investigators (referred to as "sites") were working closely together. The protocol also gives the study administrators, as well as the site team of physicians, nurses, and clinic administrators, a common reference document for site responsibilities during the trial.

The format and content of clinical trial protocols sponsored by pharmaceutical, biotechnology, or medical device companies have been standardized to follow GCP guidance issued by the International Conference on Harmonization of Technical Requirements for Registration of Pharmaceuticals for Human Use (ICH).

22.4.7 Who Can Participate in a Clinical Study?

Clinical studies have standards outlining who can participate, called eligibility criteria, which are listed in the protocol. Some research studies seek participants who have the illnesses or conditions that will be studied. Other studies are looking for healthy participants and some studies are limited to a predetermined group of people who are asked by researchers to enroll. For ethics problem, the subjects enrolled are usually not patients. Health or subhealth subjects are often included. The definition of health or subhealth subjects means that subjects do not have medication for diseases, or their biochemical cofactors reaching borderline higher than normal health subjects. If the patients are enrolled, they could not be asked to quit from normal treatment. Cotreatment is recommended for this study.[25]

22.4.7.1 Eligibility

The factors that allow someone to participate in a clinical study are called inclusion criteria, and the factors that disqualify someone from participating are called exclusion criteria. These are based on things such as age, gender, the type and stage of a disease, previous treatment history, and other medical conditions.[26]

22.4.7.2 Participant Protection

Informed consent is a process in which researchers provide potential and enrolled participants with information about a clinical study. This information helps people decide whether they want to enroll, or continue to participate, in the study. The informed consent process is intended to protect participants and should provide enough information for a person to understand the risks of, the potential benefits of, and the alternatives to the study. In addition to the informed consent document, the process may involve recruitment materials, verbal instructions, question-and-answer sessions, and activities to measure participant understanding. In general, a person must sign an informed consent document before entering a study to show that he or she was given information on risks, potential benefits, and alternatives, and understand it. Signing the document and providing consent is not a contract. Participants may withdraw from a study at any time, even if the study is not over.[27]

22.4.7.3 Safeguard

There are many other safeguards for people taking part in trials. Some of them are as follows:

1. The trial plan (protocol) is inspected and must be approved by the IRB.
2. A data monitoring committee supervises the running of the trial.
3. Insurance for people must have been arranged. Indemnity cover (insurance) that pays compensation to the participants is required, if any harm should come to them because of the trial. The organization funding the trial has to take out insurance. If they do not, the IRB may not approve the trial. In practice, several different bodies may provide cover for all participants.
4. Keep the privacy of participants protected.

22.4.7.4 The Number of Participants Needed

The total number of the enrolled subjects (n value) is determined according to the end point, alpha value (α), power ($1 - \beta$), estimated mean difference, and estimated standard deviation.

1. *End Point*
   End point depends on the goal of study. It is always established from basic knowledge and well-known theory. It is strongly recommended to put the major one (or significant point)

as the end point. The highly possible results (difference) are estimated after your study (from the cited papers).

2. α *Value*

   α (type I error), the difference is usually 0.05 or 0.01. When $p < .05$, $\alpha = 0.05$. $1 - \alpha = 0.95$ (pr 95%); for one tail, $\alpha = 0.05$; for double tail, $\alpha/2 = 0.025$.

3. *Power value*

   β value (type II error) depends on α and *n* values. Power value is equal to $1 - \beta$. If α is a constant (=0.05), high power value needs high *n* value. Power value is usually ≥0.8 (Figure 22.1).

   In the figure, μ_0 is the original mean and μ is the estimated mean (from cited papers).

4. *Estimated mean difference*

   The difference between original mean and estimated mean is $\mu - \mu_0$.

5. *Estimated standard deviation*

   This is usually determined from the cited papers and enrolled subjects and also past experience.

6. n-*Value determination*

$$n = (\sigma_1^2 + \sigma_2^2)(Z_{1-\beta} + Z_{1-\alpha/2})/(X_2 - X_1)$$

where:

σ is the standard deviation

Z is the standard normal variate, $Z = (X - \mu)/\sigma/\sqrt{n}$

It is usually determined by statistical software. When the *n* value is determined, the subjects rejecting percentage need to be well considered before study.

22.4.7.5 Follow-Up after Treatment in a Trial

The follow-up period is the time after all the participants have been treated until the end of the trial. Trials often go on for months or years after all the participants who took part have finished all their treatment. The researchers need to find out what the long-term effects are, for example, whether the

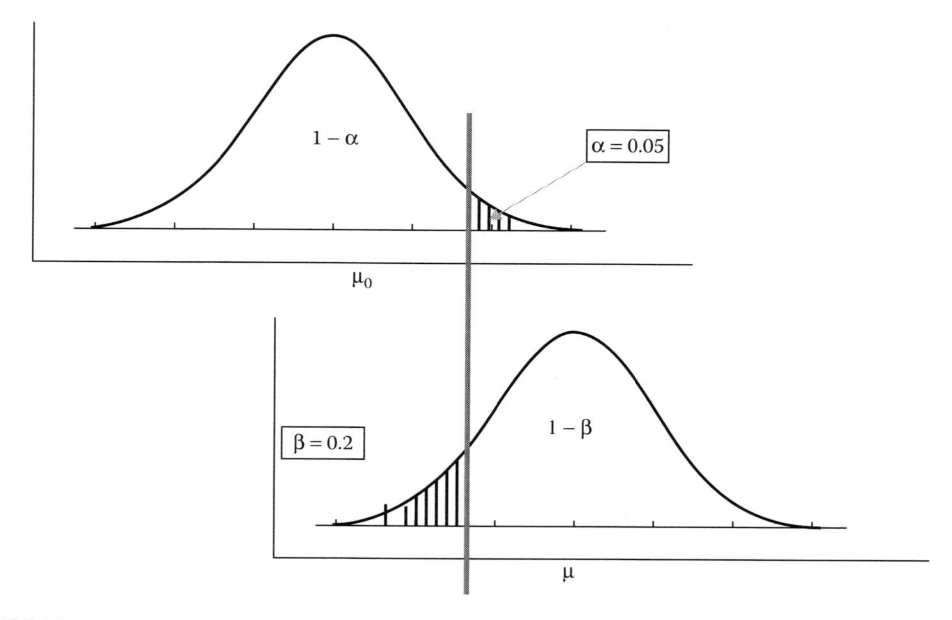

FIGURE 22.1 Statistical power.

new treatment keeps a blood lipid under control for longer than the standard treatment or if there are long-term side effects.[28]

22.5 SUMMARY

Clinical trials represent a win–win situation for all the stakeholders in the drug or nutraceuticals and functional foods development venture. As more and more companies seek to investigate and develop new products, devices, and other technologies, most of which must be tested in a patient population, there is an increasing need for clinical investigators. Academic investigators particularly appreciate clinical studies that can change the way medicine is practiced in the future. For the community practitioners, clinical trials allow them to gain first-hand experience with cutting-edge therapies, which may serve to further strengthen the physician–patient relationship. As professional fee reimbursement continues to shrink, some physicians are looking at alternative ways to generate their income to replace what Medicare and commercial insurers have taken away. One option that is becoming increasingly interesting to them is participation in clinical trials. For the patients, clinical trials can provide attractive options for those who do not have good standard treatments available to them, either because there is not one for their disease or because their treatment ceased being effective. Despite the economical downturn and the continuing merger and acquisition activities, drug makers will continue to fund clinical trials to support their product portfolios. According to the Pharmaceutical Research and Manufacturers Association, clinical trials of drugs account for an average of seven billion dollars over the past several years. Considering only 3%–5% of medical groups who participate in clinical trials, the room for growth is appealing. The largest clinical trial market is indisputably the United States. Owing to pharmaceutical globalization and the ICH, the opportunity for international participation in clinical trials is expanding. India and China have become the formidable forces in the contract research organization (CRO) market, with 11% of the total CRO spending in 2007, and are poised to grow even higher. The international outsourcing trend will continue, driven by economic factors such as cost, efficiency, and regulatory climate. With a high-powered and skillful work force, Taiwan is in a strong position to benefit from the globalized clinical trial business.

REFERENCES

1. Brater, D.C. and Daly, W.J. 2000. Clinical pharmacology in the middle ages: Principles that presage the 21st century. *Clinical Pharmacology and Therapeutics*, 67(5): 447–450.
2. Helene, S. 2010. EU compassionate use programmes (CUPs): Regulatory framework and points to consider before CUP implementation. *Pharmaceutical Medicine*, 24(4): 223–229.
3. ICH Guideline for good clinical practice: Consolidated guidance 1996. International Conference on Harmonization of Technical Requirements for Registration of Pharmaceuticals for Human Use. http://www.ich.org/fileadmin/Public_Web_Site/ICH_Products/Guidelines/Efficacy/E6_R1/Step4/E6_R1__Guideline.pdf.
4. Crossley, M.J., Turner, P., and Thordarson, P. 2007. Clinical trials—What you need to know. *American Cancer Society*, 129(22): 7155.
5. Yamin, K. and Sarah, T. 2010. Seasonality: The clinical trial manager's logistical challenge. Pharm-Olam International (POI). http://www.pharm-olam.com/pdf/POI-Seasonality.pdf (retrieved on April 26, 2010).
6. Yamin, K. and Sarah, T. 2010. Flu, season, diseases affect trials. Applied Clinical Trials Online. http://appliedclinicaltrialsonline.findpharma.com/appliedclinicaltrials/Drug±Development/Flu-Season-Diseases-Affect-Trials/ArticleStandard/Article/detail/652128 (retrieved on February 26, 2010).
7. Bhandari, M. et al. 2004. Association between industry funding and statistically significant pro-industry findings in medical and surgical randomized trials. *Canadian Medical Association Journal*, 170(4): 477–480. http://www.ncbi.nlm.nih.gov/entrez/query.fcgi?cmd=retrieve&db=pubmed&list_uids=14970094&dopt=Abstract (retrieved on May 24, 2007).
8. Hogan & Hartson Pharmaceutical and biotechnology update. 2009. http://www.hoganlovells.com/files/Publication/fac4385e-6c8b-43fb-bf80-a48037906de0/Presentation/PublicationAttachment/09e927e3-95a3-481e-93eb-b79d45891c62/FDA_011309.pdf.

9. Barnett, A. 2003. Revealed: How drug firms "hoodwink" medical journals. *The Observer*, London. http://observer.guardian.co.uk/uk_news/story/0,6903,1101680,00.html (retrieved on May 24, 2007).
10. Pmda.go.jp. Pharmaceuticals and Medical Devices Agency, Japan.
11. Food and Drug Administration Developing products for rare diseases & conditions http://www.fda.gov/forindustry/DevelopingProductsforrareDiseasesConditions/default.htm.
12. Liu, J.J. et al. 2011. Achieving ethnic diversity in trial recruitment. *Pharmaceutical Medicine*, 25(4): 215–222.
13. Life Sciences Strategy Group. 2009. Clinical trial technology utilization, purchasing preferences & growth outlook. Syndicated Publication, May. http://www.lifesciencestrategy.com/newsletter/report-clinical-trial-technology.html.
14. Rang, H.P. et al. 2003. *Pharmacology*. 5th edition. Edinburgh, NY: Churchill Livingstone.
15. Finn, R. 1999. *Cancer Clinical Trials: Experimental Treatments and How They Can Help You*. Sebastopol, CA: O'Reilly & Associates.
16. Curtis, L.M. and Susan, T. 1986. *Clinical Trials: Design, Conduct and Analysis*. New York: Oxford University Press.
17. Shayne, C.G. 2009. *Clinical Trials Handbook*. Hoboken, NJ: Wiley.
18. Chow, S.-C. and Liu, J.P. 2004. *Design and Analysis of Clinical Trials: Concepts and Methodologies*. Hoboken, NJ: Wiley.
19. Pocock, S.J. 2004. *Clinical Trials: A Practical Approach*. New York: Wiley.
20. European Medicine Agency (EMEA). 2007. Guidelines on strategy to identify and mitigate risks for first in human clinical trials with investigational medicinal products. London, July. https://www.tga.gov.au/pdf/euguide/swp2836707en.pdf.
21. Friedman, L.M. et al. 1998. *Fundamentals of Clinical Trials*. 3rd edition. New York: Springer.
22. Temple, R. and Stockbridge, N.L. 2007. BiDil for heart failure in black patients: The US Food and Drug Administration perspective. *Annals of Internal Medicine*, 146: 57–62.
23. De Angelis, C. et al. 2004. Clinical trial registration: A statement from the International Committee of Medical Journal Editors. *JAMA*, 292: 1363–1364.
24. International Committee on Harmonization. 1996. Guidance for industry, E6 Good Clinical Practice: Consolidated guidance. http://www.fda.gov/downloads/Drugs/GuidanceComplianceRegulatoryInformation/Guidances/UCM073122.pdf.
25. Collins. R. et al. 1990. Blood pressure, stroke and coronary heart disease: Part 2. Short-term reductions in blood pressure: Overview of randomized drug trials in their epidemiological context. *Lancet*, 335: 827–838.
26. Van Spall, H.G. et al. 2007. Eligibility criteria of randomized controlled trials published in high-impact general medical journals: A systematic sampling review. *JAMA*, 297(11): 1233–1240.
27. Zhuang, S.R. et al. 2012. Effect of the Chinese medical herbs complex on cellular immunity and adverse effect of breast cancer patients. *The British Journal of Nutrition*, 107: 712–718.
28. Chiu, H.F. et al. 2012. Improvement of liver function in humans using a mixture of *Schisandra* fruit extract and sesamin. *Phytotherapy Research*, 27(3): 368–373.

23 Clinical Outcomes, Safety, and Efficacy of Chinese Herbal Medicines

Daniel Roytas

CONTENTS

23.1 INTRODUCTION

This chapter aims to briefly discuss some pertinent issues related to the safety and quality of Chinese herbal medicine (CHM) from a clinical perspective. It reviews the empirical research regarding the treatment efficacy of CHM, equipping clinicians with understanding and insight into the current evidence supporting the use of CHM in treating a range of health conditions.

An overview of safety issues including adverse drug reactions (ADRs) and herb–drug interactions, synergism, and the databases available to clinicians to review these interactions is discussed. Good manufacturing and agricultural processes and the strategies being employed to minimize the adulteration of herbal medicines will also be explored. Finally, a brief insight into the emerging field of herbomics, which refers to the use of genomic technology to assess plant-based medicines, will be presented.

23.2 CHM: AN OVERVIEW

Traditional Chinese medicine (TCM) encompasses the modalities of acupuncture, CHM, tuina (Chinese therapeutic massage), moxibustion, and qigong (Lu et al. 2012). TCM has been practiced in the People's Republic of China and several neighboring countries for thousands of years, establishing itself as an essential part of the primary health-care system. "Chinese medicine" (CM) is the term that refers to the ongoing development and integration with modern medicine, with a focus on evidence-based medicine practices. One major component of CM practice is the prescription of CHM (Lu et al. 2012). Due to its long history and prominent clinical effect, the integration of CHM into conventional medicine settings as a form of primary health care has been widely accepted by the countries outside of the People's Republic of China such as South Korea and Vietnam (Zhou et al. 2010). Although CHM has gained popularity in recent years, it remains classified as a complementary or alternative medicine practice in most Western countries, including the United States, the United Kingdom, and Australia (Xue et al. 2010).

More than 100,000 unique herbal formulations, 5,000 compounds, 3,500 extracts, and 130 chemical drugs have been derived from over 3,000 Chinese herbs (Qiu 2007). Only a limited number of these formulations have been subjected to randomized controlled trials (RCTs). This absence of evidence has naturally raised questions about the efficacy and public safety of CHM treatment (Wang et al. 2009b). Most of the information regarding the safety and efficacy of CHM treatments is based on historical knowledge and clinical case studies (Wang et al. 2009). Over the past 20 years, CHM has been more rigorously evaluated through RCTs. These studies have begun to investigate more closely the safety, quality, authenticity, and efficacy of CHM treatments (Xue et al. 2010). For example, studies in the 1990s assessing interactions between Danshen and warfarin found the potentiation of warfarin's anticoagulant effect when both substances were used concomitantly (Chan et al. 1995; Lo et al. 2013). Such studies highlighted the potential for significant adverse reactions to occur in patients self-medicating with Chinese herbs, indicating the need for users of CHM to be under the care of a qualified practitioner.

Conducting more safety and efficacy studies of CHM will identify not only the potential adverse effects but also beneficial interactions. By further understanding the potential interactions between CHM and pharmaceutical drugs, CHM practitioners will be able to confidently prescribe herbal treatments to complement Western drug therapy. Research suggests that the joint use of TCM and Western medicines can improve clinical outcomes, expand the scope of treatment, and provide a multifaceted approach for effectively treating complex disease states (Zeng and Jiang 2010). For example, a recently published study has identified that the 20(*S*)-ginsenoside Rh2 (a trace constituent found in ginseng species) can inhibit P-glycoproteins (P-gps), enhancing the bioavailability of P-gp substrates. Such interactions are of great clinical importance as the inhibition of P-gp has the potential not only to enhance the bioavailability of many drugs such as chemotherapeutic agents, but also to reverse the multidrug resistance experienced by individuals undergoing chemotherapy (Zhang et al. 2010).

In 2005, the World Health Organization (WHO) stated that "the safety and efficacy of traditional medicine and complementary and alternative medicines, as well as quality control, have become important concerns for both Health Authorities and the public" (WHO 2005). To date, a significant number of research studies have been conducted to validate traditional CHM practice. Literally, thousands of RCTs have been published on various facets of CHM in an effort to verify whether traditional herbal formulas can be used as safe and effective treatments for a wide range of conditions (Hu et al. 2011). However, due to the vast number of unique herbal formulations, a more comprehensive body of evidence-based research is still required to address the WHO's concerns regarding safety and efficacy. Establishing a strong base of contemporary research literature will play an integral role in the integration of CHM and Western medicine (Hu et al. 2011).

23.3 SAFETY CONCERNS OF CHM

23.3.1 Overview of ADRs

Over the past 20 years, the use of CHM has increased significantly, with more than 1.5 billion users worldwide (Wang et al. 2009b). Approximately 20% of Australians utilize herbal medicine on a regular basis, whereas 7% utilize CHM exclusively (Xue et al. 2007). As the use of CHM by the general public increases, so does the potential for adverse events to occur. In Australia, less than half of the population is aware of the potential safety risks associated with herbal medicines, which is largely due to the classification of herbal medicines as low-risk products (Fan et al. 2012). This misconception may be a key contributing factor in the misuse of herbal medicines, which is an issue of significant concern for regulatory bodies such as the Therapeutic Goods Administration (TGA) (Fan et al. 2012). Although risks are minimized when CHM is used under the supervision of a health-care professional, the potential for adverse effects increases when herbal medicines are not used judiciously or as indicated.

In Australia, although nearly all herbal medicine products are classified as low risk, all therapeutic substances have the potential to cause adverse effects. In the late 1960s, the TGA set up a reporting system so that the records of all adverse drug reactions (ADRs), including those associated with herbal products, could be accurately recorded (Hill 2006). Since the computerization of this database in 1972, the amount of reports quickly increased, providing a key source of information on ADRs for health-care professionals and regulatory bodies.

As of 2010, over 233,300 ADRs had been documented, with 493 associated with the use of an herbal medicine. In 2004–2005 alone, the TGA received a total of 19,729 adverse drug reports, with 1.4% (269) mentioning the involvement of at least one herbal medicine. It should be noted that the involvement of an herbal medicine notified in the ADR report does not necessarily mean that the herbal medicine was responsible for the reactions. Of these 269 reactions, the most commonly reported herbs were *Ginkgo biloba* (37 reports), *Zingiber officinale* (27), *Silybum marianum* (26), *Eleutherococcus senticosus* (23), *Allium sativum* (21), and *Echinacea* spp. (21) (Hill 2006). These figures indicate that while herbal therapies are not completely free of side effects, the likelihood of an adverse event occurring as a direct result of an herbal therapy is unlikely. As herbal medicines are considered "safe" and the majority of ADRs involving herbal medicines are mild in severity, ADRs are less likely to be formally reported. This may indicate the relatively low total number of documented cases compared to pharmaceutical drugs (Shaw et al. 2012).

The majority of all ADRs are reported by health-care professionals such as pharmacists and medical doctors, TCM practitioners are also a potentially useful source of information regarding ADRs associated with herbal medicines (Lin et al. 2009). Most formally trained and qualified practitioners have received a high level of training and are able to identify unexpected side effects and potential toxicity, and can therefore provide accurate and detailed reports of ADRs to agencies such as the TGA through the ADR Reporting System (Shaw et al. 2012). The accurate reporting of ADRs for herbal medicines is important so that appropriate guidelines and warnings can be provided to practitioners and patients. Accurate reporting of ADRs will also allow agencies such as the TGA to develop comprehensive and reliable safety databases, which can be accessed by practitioners to check contraindications and ADRs ensuring patient safety (Shaw et al. 2012).

23.3.2 Herb–Drug Interactions

It is estimated that if all of the adverse events associated with CHM over the past 20 years were multiplied by 100, the number of events would still not be comparable to the number of adverse events associated with pharmaceutical medications in one year (Myers and Cheras 2004). In spite of this fact, medicinal herbs still possess pharmacological activity, and therefore their potential to cause adverse reactions should not be underestimated. Of particular concern are herb–drug interactions, in which an herb may potentiate or inhibit the effect of a drug. Some of these interactions are significant enough to endanger a patient's health (Chan et al. 2010). Herb–drug interactions are largely preventable in which safety data exist; however, many of these interactions have yet to be established (Giveon et al. 2004), a fact highlighted in a growing number of publications (Zhou et al. 2007). Historically, clinical reports of herb–drug interactions have been infrequent with the majority of modern safety data being derived from *in vitro* or animal studies (Hu et al. 2005). However, herb–drug interactions are becoming more commonly reported in clinical trials and by health-care professionals.

Many patients may be apprehensive in disclosing the use of herbal medicines during a general practice consultation, thus increasing the potential for adverse herb–drug interactions to occur (Lin et al. 2009). General practitioners (GPs) identify approximately one adverse event in every 125 consultations as a direct result of an herb–drug interaction. Without discounting the seriousness of adverse events, this is a relatively low frequency, especially when considering the fact that nearly 40% of Australians use prescription drugs and CHM concurrently (Myers and Cheras 2004). CHM practitioners may also encounter similar problems if patients do not disclose the use of prescription drugs or herbal medicines. CHM practitioners can therefore minimize potential adverse reactions by conducting a thorough health history check, using clinical questioning techniques, and informing patients about the risks of herb–drug interactions (Giveon et al. 2004).

23.3.3 Notable Herb–Drug Interactions

This subsection provides a brief overview of known herb–drug interactions. It is recommended that practitioners refer to textbooks or online databases to obtain information on other interactions.

23.3.3.1 Danshen

As mentioned in Subsection 23.2, the use of Danshen (*Salvia miltiorrhiza*) in patients on anticoagulant medications such as warfarin is contraindicated, due to the potentiation of warfarin's anticoagulant effect (Zhou et al. 2012). Danshen is one of the most widely studied herbal medicines and is commonly used in the treatment of cardiovascular and cerebrovascular diseases including hyperlipidemia and acute myocardial infarction (Wu and Yeung 2010). Puerarin is one of the major active constituents of Danshen, which is largely responsible for its activity on the cardiovascular system. However, it is the tanshinone derivatives and salvianolic acids of Danshen that are believed to contribute to the pharmacodynamic effects with warfarin. Tanshinones have been shown to decrease 4-, 6-, and 7-hydroxywarfarin formation and inhibit CYP1A1, CYP2C6, and CYP2C11 enzyme-mediated metabolism of warfarin, increasing plasma warfarin concentrations. Due to the potential for over-anticoagulation and bleeding risk, Danshen and warfarin should only be used under close medical supervision (Zhou et al. 2012).

23.3.3.2 Dong Quai

The herb Dong Quai (*Angelica sinensis*) is commonly prescribed for conditions such as premenstrual syndrome, menopause, fatigue, and hypertension. It has been postulated to interact with a range of blood-thinning agents (Abebe 2002). Dong Quai consists of active constituents that possess antiarrhythmic and antithrombotic activities, such as the coumarin derivatives, osthole, and ferulic acid (Page and Lawrence 1999). These active constituents exert an antithrombotic effect by interrupting platelet polymerization and reducing production of thromboxane A_2. It is this mechanism of action that may potentiate the anticoagulant activity of drugs such as warfarin, aspirin, and nonsteroidal anti-inflammatory drugs (NSAIDs) (Chan et al. 2010). Clinicians should avoid prescribing Dong Quai in patients currently taking blood-thinning medications such as warfarin, as it can increase the international normalized ratio (INR), promote bruising, and increase clotting time (Page and Lawrence 1999).

23.3.3.3 Gan Cao

Gan Cao (*Radix glychyrrizae*), commonly known as licorice, is another herb that clinicians should be cautious in prescribing for patients taking pharmaceutical medications (Asl and Hosseinzadeh 2008). While licorice is one of the most commonly used herbs in TCM formulations, many clinicians may be unaware of the potential herb–drug interactions. Licorice has traditionally been used to treat cystitis, kidney stones, diabetes, and conditions of the respiratory and gastrointestinal systems. It has also exhibited antiviral effects in *in vitro* studies against the Hepatitis virus, Epstein–Barr virus, and human cytomegalovirus (Asl and Hosseinzadeh 2008).

While the active constituents such as triterpenes, flavonoids, isoflavones, and coumarins have been shown to interact with over 100 pharmaceutical drugs, not all interactions are contraindicated (Asl and Hosseinzadeh 2008). For example, the active constituent β-glycyrrhetinic acid is a potent inhibitor of 11β-hydroxysteroid hydroxygenase. This interaction decreases the clearance of prednisolone and increases its bioavailability. Therefore, the concomitant use of licorice with hydrocortisone in the treatment of inflammatory conditions may be clinically beneficial (Schleimer 1991). Conversely, licorice's inhibitory action of 11β-hydroxysteroid hydroxygenase has the potential to cause an aldosterone-like action, resulting in hypokalemia, hypertension, and metabolic alkalosis. Therefore, licorice should be used cautiously in patients exhibiting such symptoms (Omar et al. 2012).

23.3.4 Adulteration of CHMs

Another important safety issue that needs to be highlighted is the adulteration of CHM with pharmaceutical medications. The adulteration of CHM with undeclared synthetic drugs is a problem that can lead to potentially serious adverse effects. An analysis of over 2609 Chinese herbal products prescribed in Taiwanese hospitals found that more than one half contained between two and five synthetic therapeutic substances, while four samples contained at least six (Huang et al. 1997). The most common synthetic substances included NSAIDs, steroids, and analgesics. Other adulterants that have been found in CHMs include phenylbutazone, phenytoin, glibenclamide, corticosteroids, chlopropamide, fenfluramine, and indomethacin (Bogusz et al. 2006). In another report, the analysis of 260 patent CHMs found that almost 10% contained several undeclared synthetic substances (Ernst 2002). Since 1990, adulteration with synthetic substances has been the cause of one documented fatality and six potentially life-threatening events (Ernst 2002).

As recent as 2010, 336 out of 390 warnings for herbal products issued by regulatory authorities in the United Kingdom, the United States, Canada, Singapore, and Australia were due to a direct result of pharmaceutical contamination or adulteration (Shaw 2010). The adulteration of herbal products carrying an AUSTL or AUSTR number has not been identified by the TGA for several years. However, a small number of cases have been recently identified, where nonregistered and nonlisted herbal products adulterated with conventional pharmaceuticals have been unlawfully supplied in Australia. The TGA continues to work with the Australian Customs Service, State and Territory agencies, and product manufacturers and suppliers to eliminate this issue (McEwen and Cumming 2003).

23.4 GOOD MANUFACTURING AND AGRICULTURAL COLLECTION PRACTICE

The majority of CHMs appear to have a wide therapeutic index, which is supported by the broad dosage range of herbs prescribed by clinicians, with few reported toxic effects (Shaw 2010). Although the risk of adverse reactions is minimized while under the care of a qualified practitioner (Chan 2002), toxicity arising from the prescription of poorly grown, manufactured, or adulterated herbal medicines is for the most part, out of the practitioners' control (Shaw 2010). Practitioners rely heavily on manufacturers and suppliers to employ a high standard of good manufacturing practice (GMP) and good agricultural and collection practice (GACP), to be assured that they are receiving a high-quality product (Fan et al. 2012).

While the majority of herbal manufacturers strictly adhere to GMP and GACP, some continue to manufacture fake (or adulterated) herbal medicines, increasing the risk of toxicity and other adverse effects. It is this unscrupulous practice that has significantly contributed to the concerns relating to the safety and efficacy of CHM products (Chan 2003). Many popular herbs are expensive or in short supply. Therefore, it is not uncommon for less expensive, inferior, fake, or adulterated herbs to be incorporated into products and sold to unsuspecting consumers (Huang et al. 1997). For example, the root of *Panax notoginseng* has been substituted with cheaper notoginseng by-products, including the leaf and flower. The leaves and root of *Panax ginseng* are also commonly substituted for the more expensive notoginseng root (Wang et al. 2009a). Products containing substituted herbs do not contain the active constituents of the actual herb. These products are unlikely to elicit the desired or claimed pharmacological activities, thereby negating the expected therapeutic effect (Huang et al. 1997).

Some factors affecting GACP of herbal medicines include the geographical location and the geological origin of the herb; time of planting and harvesting; use of fertilizers, pesticides, and herbicides; weather conditions; collection methods; storage conditions; and processing methods (Shaw 2010). Typical examples of poor GMP and GACP include the misidentification of herbs. For instance, the use of *Caulis aristolochiae manshuriensis* (Guan Mu Tong) has been used instead of Akebiae (Mu Tong). Mu Tong is commonly used with minimal side effects in conditions requiring diuretic and antibacterial activity. However, Guan Mu Tong contains aristolochic acid, which is associated with renal toxicity and carcinogenicity (Zeng and Jiang 2010). The mistaken use of *Radix aristolochiae fangchi for Stephania tetrandra* has led to documented reports of degenerative nephritis and kidney failure, whereas the mistaken use of *Podophyllum* for Chinese gentian has documented evidence of causing degenerative encephalopathy (Zeng and Jiang 2010).

23.4.1 GLOBAL STANDARDIZATION OF GMP AND GACP

To address the quality and safety issues, various health authorities and government agencies, such as the TGA (Australia), the Food and Drug Administration (the United States), and the State Food and Drug Administration (People's Republic of China), have enforced strict regulatory measures upon suppliers and manufacturers of herbal medicines (Fan et al. 2012). In addition to this, the WHO and good practice in TCM (GP-TCM) research in the postgenomic era have been a driving force behind improving the GMP and GACP of CHM. The GP-TCM Work Package 7 (WP7) was established to formulate a framework regarding the GMP and GACP of herbal medicines on a global scale (Fan et al. 2012). The GP-TCM WP7 framework states that manufacturers must produce herbal products that

- Contain the correct ingredients,
- Are of acceptable quality,
- Are free from unacceptable contamination,
- Have evidence to support the claimed shelf life.

The efforts of the GP-TCM WP7 to harmonize the GACP and GMP requirements of key regulatory agencies is of paramount importance, especially if the integration of CHM and Western medicine is to be achieved. At the present time, quality issues appear to be the most challenging aspect of GMP in CHM, far above the concerns of efficacy. For example, some herbal product manufacturers from India and the People's Republic of China continue to use different genera for different batches of the same herbal remedies (Fan et al. 2012). This indicates that these manufacturers are utilizing poor herb identification and analytical techniques, or are not performing them at all (Fan et al. 2012). While this may be due to the fact that determining the efficacy or quality through analysis of individual compounds is contrary to the underlying TCM principles of synergy (Xie et al. 2006), it is more likely due to cost-cutting measures or limited access to specialized analytical equipment.

23.5 ANALYSIS OF HERBAL MEDICINES

23.5.1 Sample Preparation

The most important step in the analysis of herbal medicines is the preparation of a sample, which must be performed before analytical methods such as liquid chromatography can be undertaken. Preparation of herb samples include prewashing, drying or freeze drying, and grinding of the plant material into a homogeneous compound (Ong 2004). From this compound, active constituents are extracted. Depending on the compound being extracted, a variety of methods may be required including sonication, heating under reflux, and Soxhlet extraction. Although these methods are commonly used, they are typically time consuming and often require large amounts of organic solvents. Recent advances in extraction methods utilize elevated temperatures and pressures, significantly reducing the volume of solvents and analyte extraction times. These methods include pressurized liquid extraction (PLE), microwave-assisted extraction (MAE), supercritical fluid extraction (SFE), and accelerated solvent extraction (ASE) (Ong 2004). Once prepared, the analysis of active constituents can be undertaken.

23.5.2 Analytical Methods

One particular analytical method that has effectively identified and therefore reduced the adulteration of CHM is the use of high-pressure liquid chromatography (HPLC)–electrospray ionization–tandem mass spectrometry (LC–ESI–MS–MS) (Bogusz et al. 2006). This analytical method is able to accurately detect foreign chemical substances and alert manufacturers to potential contamination (Bogusz et al. 2006). Although the adulteration of herbal medicines has become less common since the implementation of analytical processes such as LC–ESI–MS–MS, poor-quality products continue to be manufactured and distributed globally. For the most part, the relative scarcity of case reports indicates that adulteration, especially in Australia, is unlikely. Practitioners should be aware of this potential issue and ensure that they procure herbs from reputable suppliers (Ernst 2002).

There are several other analytical methods currently employed by manufacturers to identify herbs and authenticate active constituents for quality purposes. Basic and outdated analytical methods authenticate an herb through the analysis of one or more predetermined compounds that are commonly known as "actives" or "markers." As many CHMs contain the same compounds, such methods are unable to accurately identify a specific plant, let alone provide a quality assessment (Xie et al. 2006). Furthermore, determining the efficacy or quality through analysis of individual compounds is contrary to the TCM principles and such techniques do not adequately meet the complex characteristics of CHM formulations (Xie et al. 2006).

Recent advances in analytical techniques, known as chromatographic fingerprint analysis, allows for the analysis of multiple compounds within a single herb or formulation, which represents a rational approach for the analysis of CHM (Xie et al. 2006). This process utilizes chromatographic techniques such as gas chromatography (GC), HPLC, liquid chromatography–mass spectrometry (LC–MS), and high-performance thin-layer chromatography (HPTLC) to construct specific patterns of recognition for multiple compounds in herbal medicines (Xie et al. 2006). To evaluate DNA polymorphisms, DNA-based molecular techniques such as polymerase chain reactions and sequencing methods are commonly used. These methods can accurately detect the presence or absence of specific markers and the ratios of all compounds within an herb sample. Therefore, chromatographic fingerprint analysis is classified as one of the most powerful approaches for authenticating and evaluating the stability and consistency of CHM products (Xie et al. 2006).

After an official chromatographic fingerprint has been established and an acceptable allowance range has been determined, these parameters can be set as the minimum required specifications for all manufacturers to meet when manufacturing an herbal medicine product (Xie et al. 2006). Once this procedure becomes mandatory and is enforced upon all CHM manufacturers, the overall quality of CHM products can be ensured and clinicians will be assured that the quality and expected effects of all herbal prescriptions will be of a consistent standard.

23.6 IMPACT OF HERBAL INTERACTIONS ON SAFETY AND EFFICACY

Chinese herbs are typically combined with one or more herbs in combinations known as formulas and are the preferred prescription method employed by CHM practitioners (Chen et al. 2006). This is because herbal formulas have distinct advantages over the utilization of a single herb. Although each single herb has its own specific characteristics, when combined or "paired" with another herb, a range of therapeutic actions unique to that combination is achieved (Chen et al. 2006). For example, one herb may potentiate the therapeutic or pharmacological effect of another, reduce its toxicity or side effects, and broaden its application. However, not all combinations are beneficial, as some herbs may reduce the effect of others or increase their toxicity (Shaw 2010). The use of multiple herbs in one formula in CHM is based on the TCM theory known as the "seven relationships between herbs," in which two herbs will interact in one of several ways. These interactions are classified as mutual accentuation (Xiang Xu), mutual enhancement (Xiang She), mutual antagonism (Xiang Wu), mutual counteraction (Xiang Wei), mutual suppression (Xiang Sha), mutual incompatibility (Xiang Fan), and single effect (Dan Xing) (Yang 2010). Single-effect herbs are used by themselves as a treatment, due to the balanced properties and functions they exhibit.

23.6.1 Examples of Seven Relationships between Herbs

23.6.1.1 Mutual Accentuation (Xiang Xu)

A practitioner can utilize one herb to potentiate the therapeutic effect of another. This relationship can be useful for practitioners when they want to achieve a certain response from an herb, without using excessively high doses. For example, the effect of Da Huang (*Radix et Rhizoma Rhei*) to treat constipation is significantly potentiated by Mang Xiao (*Natrii sulfas*) (Yang 2010).

23.6.1.2 Mutual Enhancement (Xiang She)

Mutual enhancement defines the combination of two herbs for their different functions. Practitioners can incorporate one herb into a formula to treat the desired condition such as Zhi Gan Cao (*Glycyrrhiza radix*), and then utilize a second herb such as Da Huang (*Rhei rhizoma*) to enhance or assist the therapeutic effect of the first (Yang 2010).

23.6.1.3 Mutual Antagonism (Xiang Wu)

This combination refers to the ability of an herb to reduce the therapeutic effect of another. There are 18 antagonistic substances identified in CHM. For example, Liu Huang (sulfur) antagonizes Po Xiao (*Glauberis sal*), Wu Tou (*Aconiti radix*) antagonizes Xi Jiao (*Rhinoceri cornu*), and Rou Gui antagonizes Chi Shi Zhi (*Halloysitum rubrum*) (Chan 2003).

23.6.1.4 Mutual Counteraction (Xiang Wei)

Using the mutual counteraction principle, a poisonous or toxic herb can be paired with another herb to reduce or even neutralize its side effects. According to the procedures outlined in the *Chinese Pharmacopoeia*, herbs such as Sheng Jiang (*Zingiberis rhizoma recens*) can be paired with Ban Xia (*Pinelliae rhizoma*) to significantly reduce its toxic effect (Yang 2010).

23.6.1.5 Mutual Suppression (Xiang Sha)

Mutual suppression refers to the ability of an herb to suppress or minimize the toxicity of the second. Herbs such as Zhi Gan Cao (*Glycyrrhiza radix*) can be included in formulations with herbs associated with minor side effects. This is due to Gan Cao's ability to suppress the toxicity of a wide range of herbs, while improving a patient's detoxification processes (Chan 2003).

23.6.1.6 Mutual Incompatibility (Xiang Fan)

Mutual incompatibility refers to 19 different herb pairs that should never be used together, due to the potentially severe adverse effects, which are not caused by either herb when used alone. For example, Gan Cao (*Radix glychyrriza*) should never be combined with Gan Sui (*Euphorbia kansui radix*) (Wang et al. 2012). When used alone, Gan Sui exerts its toxicity through induction of P4502E1 expression. When combined with Gan Cao, this toxic effect is significantly enhanced because Gan Cao also induces P4502E1 expression. Other herbs that are incompatible with Gan Cao include Daji (*Euphorbia pekinensis radix*), Yuan Hua (*Genkwa flos*), and Hai Zao (*Sargassum*) (Table 23.1) (Wang et al. 2012).

In contrast to the "seven relationships theory," a large proportion of the current scientific data investigating the safety and efficacy of CHM is based on the isolated constituents, rather than the

TABLE 23.1
List of Mutual Incompatibilities

Herb	Incompatible Herbs
Gan Cao (*Glycyrrhiza radix*)	Gan Sui (*Euphorbia kansui radix*)
	Da Ji (*Knoxiae radix*)
	Yuan Hua (*Genkwa flos*)
	Hai Zao (*Sargassum*)
Wu Tou (*Aconiti radix*)	Chuan Bei Mu (*Fritillariae cirrhosae bulbus*)
	Gua Lou (*Trichosanthis fructus*)
	Ban Xia (*Pinelliae* rhizome)
	Bai Lian (*Ampelopsitis*)
	Bai Ji (*Bletillae* tuber)
Li Lu (*Veratri nigri radix et rhizoma*)	Ren Shen (*Ginseng radix*)
	Sha Shen–Bei Sha Shen (*Glehniae radix*)/
	Nan Sha Shen (*Adnophorae radix*)
	Ku Shen (*Sophorae flavescentis radix*)
	Xuan Shen (*Scrophulariae radix*)
	Dan Shen (*Salviae miltorrhizae radix*)
	Xi Xin (*Asari herba*)
	Shao Yao–Chi Shao Yao (*Paeoniae radix rubra*)/
	Bai Shao Yao (*Paeoniae radix lactiflora*)

Source: Reprinted from *Chinese Herbal Formulas: Treatment Principles and Composition Strategies*, Yang, Y., Copyright 2010, with permission from Elsevier.

Note: The left hand column represents a single herb that should not be used in a formula with the herbs listed in the right hand column due to potential adverse effects.

whole formula, which are classified as biomedicine and not related to TCM theory (Chen et al. 2006). Instead of assessing the toxic or synergistic effects of a whole formulation or an herb pair, the majority of toxicity studies are purely based on the pharmacological profile of an individual herb, or isolated constituent (Chen et al. 2006). Using this principle, an herbal formulation containing a single toxic herb would be classified as unsafe, when in fact it may pose no safety risk at all when the concepts of synergism or herb pairing are applied. Due to this fact, current safety data of isolated constituents may not accurately reflect the properties of many Chinese herbal formulas (Chen et al. 2006). Modern-day analytical techniques such as chromatographic fingerprint analysis used in conjunction with the "omic" technologies such as herbogenomics may be able to address this issue.

23.6.1.7 Single Effect (Dan Xing)

Traditionally, several single herbs are combined together into a formula, however there are circumstances where an individual (single-effect) herb is used as a stand-alone therapy. A single-effect herb is typically prescribed due to the balanced properties and functions it exhibits. For example, the use of *Ginkgo biloba* has been assessed as a potential treatment in patients with Alzheimer's disease, due to its reported cognition-enhancing effects (Fu and Li 2011).

23.7 EFFICACY OF CHM

In recent years, a dramatic expansion in the number of RCTs assessing CHM has taken place (Manheimer et al. 2009). This is a direct result of the establishment of research institutions such as the National Center for Complementary and Alternative Medicine (NCCAM) within the US National Institutes of Health, which was founded in 1998 (Manheimer et al. 2009). A similar initiative was undertaken by the Australian Government in 2007, with the establishment of the National Institute of Complementary Medicine (NICM). These institutions were designed to support research projects across a wide range of complementary therapies, of which CHM is encompassed (NICM 2007). Over A$7 million in funding was allocated to complementary therapies, with a specific emphasis on the evaluation of herbal medicines (NICM 2007). Furthermore, research institutions such as the Centre for Complementary Medicine Research at the University of Western Sydney (UWS 2013) and the Royal Melbourne Institute of Technology (RMIT) Traditional and Complementary Medicine Research Group (RTCMRG) (RMIT 2013) are dedicated to enhancing the current evidence base of CHMs through an on-going research. Initiatives such as these and others around the world have contributed significantly toward the publication of literally thousands of RCTs in the field of CHM.

A large number of systematic reviews have been undertaken to assess the current evidence of CHMs and to provide direction for future research (Xue et al. 2010). The most well-known and highly regarded producer of up-to-date systematic reviews is the Cochrane Collaboration, which has published over 5000 complete systematic reviews and more than 2000 review protocols on a wide range of subjects (Hu et al. 2011; Manheimer et al. 2009). The Cochrane Collaboration is an international not-for-profit organization that prepares and maintains systematic reviews of an extremely high level to assist health-care practitioners make evidence-based decisions about health care. Compared to other publishers of systematic reviews, Cochrane Reviews have a greater methodological quality, less risk of biased results, and superior outcome measures (Hu et al. 2011; Manheimer et al. 2009).

As Cochrane Reviews provide practitioners with the most accurate, trustworthy, and reliable assessments of the treatment's effect of CHM, this subsection discusses the current state of evidence supporting the use of CHM for a range of health conditions. Although many other systematic reviews have been published on various CHM interventions, only Cochrane Reviews have been included in this subsection as they are peer reviewed at the protocol, complete review stage, and are of comparable or higher quality than the reviews published in other leading journals (Jadad 1998).

23.7.1 Current Evidence of CHM Supported by the Cochrane Collaboration

Seventy-two Cochrane Reviews dedicated specifically to assessing the efficacy of CHMs were identified. Of the 72 reviews, 28 (40%) of the authors' conclusions supported the use of CHM in the treatment of a specific health condition. It should be noted that although these reviews indicated a potentially beneficial effect, the results should be accepted tentatively and applied using professional judgment until more conclusive evidence is established. The majority of the authors highlighted that further high-quality trials need to be undertaken before conclusive recommendations can be made. Table 23.2 shows an overview of current Cochrane Reviews with positive findings of CHM.

23.7.2 Conditions Requiring Further Investigation

Due to insufficient high-quality evidence, CHM was unable to be recommended in 22 (30%) of the 72 reviewed conditions. However, the majority of reviews concluded that many CHM therapies are likely to be validated through further rigorous studies. CHM interventions not currently identified as effective by the Cochrane Collaboration include ectopic pregnancy, polycystic ovarian syndrome, preeclampsia, common cold, chronic kidney disease, stable angina, acute myocardial infarction, adhesive small bowel obstruction, bronchitis, epilepsy, postpartum hemorrhage, cholelithiasis, vascular dementia, intermittent claudication, paraquat poisoning, and macular degeneration.

23.7.3 Unpublished Reviews

Twenty-two reviews (30%) were in the protocol stage, and therefore no abstract or plain language summary was available. The majority of these reviews are likely to be published in the near future and may provide further evidence supporting CHM in the treatment of several conditions. The conditions being assessed are female subfertility, menopausal symptoms, atopic eczema, allergic rhinitis, gastric precancerous lesions, pancreatic cancer, vascular dementia, soft tissue infections, recurrent urinary tract infections, diabetic retinopathy, esophageal cancer, weight loss, chronic fatigue, unexplained miscarriage, colorectal cancer metastasis and relapse, mumps and measles, neonatal jaundice, kidney disease, neonatal jaundice, and heart disease.

23.7.4 Quality of Evidence

The conclusions of all Cochrane Reviews are based on several contributing factors including the risk of bias, heterogeneity of design, setting, dosage, duration, and outcome measures. While the majority of reviews published by the Cochrane Collaboration have found promising evidence to support the use of CHM in a range of conditions, the overwhelming majority of trials included for review are limited by poor methodological quality (Chen et al. 2012; Duan et al. 2008; Gan et al. 2010; Liu et al. 2004, 2013; Zhu et al. 2008). Despite this, many systematic reviews provide preliminary evidence highlighting the potential health benefits associated with CHM, which are likely to be validated through high-quality RCTs. For example, CHM is likely to improve quality of life in chemotherapy patients (Taixiang et al. 2005; Zhang et al. 2007), reduce the incidence of type 2 diabetes (Grant et al. 2009; Liu et al. 2004), and significantly improve the health outcomes of individuals with complications of the cardiovascular system (Chen et al. 2012; Liu et al. 2011; Zheng et al. 2013).

However, many studies concluded that even though a potential for CHM treatment may exist, the current evidence is of such poor methodological quality that no recommendations can be made. For example, a review assessing CHM for the common cold found that while CHM may shorten the symptomatic phase of a cold, no certain recommendations for any kind of CHM could be made due to lack of proper controls and high risk of bias (Wu et al. 2007). In another review, CHM was unable to be recommended for the treatment of diabetic peripheral neuropathy as no well-designed clinical RCTs with objective outcome measures had been conducted (Chen et al. 2011). A third review found that *G. biloba* may be a potentially effective treatment for

TABLE 23.2
Overview of the Literature Currently Available in the Cochrane Database Regarding Chinese Herbal Medicine

Condition	Herb(s)	Dosage	Trial(s) ($n = x$)	Participant(s) ($n = x$)	Results	Comment
Hypercholesterolemia (Liu et al. 2011)	Xuezhikang (XK) Xiaozhiling (XZ)	N/A	22	2130	Compared to inositol nicotinate, XK significantly reduced the total serum cholesterol by an average of 0.9 mmol/L. XZ showed significant differences on HDL-C and LDL-C compared to pravastatin.	XK or XZ may be beneficial in lowering serum cholesterol levels.
Hypertriglyceridemia (Liu et al. 2013)	Zhusuan Huoxue Huoxue Huayu Tongluo Chushi Huayu	N/A	3	170	Compared to benzbromarone, Chushi Huayu reduced serum TG by 1.14 mmol/L. Huoxue Huayu and gemfibrozil reduced serum TG by 2.2 mmol/L compared to gemfibrozil alone. Zhusuan Huoxue reduced serum TG by 0.51 mmol/L compared to fenofibrate.	CHM used alone or in combination with lipid-lowering drugs may have positive effects on reducing serum triglyceride levels.
Primary nephritic syndrome (Feng et al. 2013)	Huangqi formulations	1–10 mL/kg/day for four weeks or 25 g/day for 8–12 weeks or 1.5–2.0 mg/kg/day	9	461	Huangqi significantly increased plasma albumin levels by 6.41 g/dL, reduced urine albumin excretion by 0.57 g/day reduced total cholesterol by 1.70 mmol/L and triglycerides by 0.33 mmol/L compared to prednisone.	Huangqi may be beneficial at improving biochemical markers associated with nephritic syndrome.
Threatened miscarriage (Li et al. 2012)	Shou Tai Pill or individual formulas based on TCM diagnostic patterns	Decoctions consumed three to four times per day; up to three courses (one course = 7–15 days)	44	5100	CHMs combined with Western medicines such as salbutamol, human chorionic gonadotropin, and progesterone were more effective than Western therapies alone to continue pregnancy beyond 28 weeks of gestation.	CHM may be indicated in cases of threatened miscarriage where Western medicines have also been prescribed.
Schizophrenia (Rathbone et al. 2007)	Dang Gui Cheng Qi Tang or Xiao Yao San	100 mL two to three times per day	7	1094	Concomitant use of Dang Gui Cheng Qi or Xiao Yao San and antipsychotics significantly improved mental state and functioning, while reducing side effects.	Schizophrenic patients may benefit more from CHM and antipsychotics than antipsychotics alone.
Primary dysmenorrhea (Zhu et al. 2008)	Individual formulas based on TCM diagnostic patterns	N/A	39	3475	CHM significantly improved pain scores and overall symptoms, and reduced additional medication use compared to NSAIDs and the OCP.	CHM may be beneficial in patients suffering from primary dysmenorrhea.

Primary nephritic syndrome (Chen et al. 2013)	*Tripterygium wilfordii* (TwHF)	1.0–2.0 mg/kg/day for four to eight weeks reducing to <1.0 mg/kg/day for up to 12 months	10	630	Concomitant use of CHM significantly increased complete remission, reduced serum creatinine levels, and side effects than prednisone or cyclophosphamide alone.	TwHF may have an add-on effect on remission in patients with primary nephritic syndrome.
Chronic neck pain (Cui et al. 2010)	Nucis Vomicae Qishe tablets Formula containing Huangqi, Dangshen, Sangqi, Chuanxiong, Lujiao, and Zhimu	Ointment applied three times per day for four weeks 25 tablets two times per day for four weeks 300 mL decoction applied topically two times per day	4	1100	Nucis Vomicae relieved pain better than diclofenac diethylamine emulgel. Qishe tablets relieved pain better than placebo or Jingfukang. The CHM decoction relieved pain better than mobicox or methycobal.	Patients suffering from chronic neck pain due to cervical degenerative disc disease may benefit from oral and topical CHM preparations compared to drug therapy.
Heart failure (Chen et al. 2012)	Shengmai	1 × 250 mL injection per day for 14 days or 10 mL orally two times per day for 50 days	9	600	Shengmai in combination with Western drug treatment improved NYHA scores, ejection fraction, stroke volume, cardiac index, and myocardial contractility than drug therapy alone.	Patients with heart failure may benefit more from CHM and usual drug therapy.
Impaired fasting blood glucose (Grant et al. 2009)	Individual formulas based on TCM diagnostic patterns	N/A	16	1391	CHM in combination with lifestyle interventions is more than twice as likely to maintain normal blood fasting glucose levels (<7.8 mmol/L and 2 h blood glucose <11.1 mmol/L).	Prediabetic patients are less likely to develop diabetes when using CHM and lifestyle modifications.
Pancreatitis (Wang et al. 2005)	Individual formulas based on TCM diagnostic patterns	N/A	15	845	CHM therapy may have a beneficial effect on mortality rates, operative intervention, multiple organ failure, and systemic infection, but not septic complications.	The current evidence is too weak to support the recommendation of any single herb.
Endometriosis (Flower et al. 2012)	Nei Yi formula	Nei Yi pills 10 g twice per day Nei Yi enema 70 mL per day for 12 months	2	158	Compared to danazol or genstrinone, oral Nei Yi pills and Nei Yi enema significantly reduced overall symptom scores, lumbosacral and dysmenorrheal pain, and vaginal nodule tenderness.	Hundred percent of women showed symptomatic improvement, with fewer side effects compared to drug therapy.

(*Continued*)

TABLE 23.2
(Continued) Overview of the Literature Currently Available in the Cochrane Database Regarding Chinese Herbal Medicine

Condition	Herb(s)	Dosage	Trial(s) ($n = x$)	Participant(s) ($n = x$)	Results	Comment
Gastric cancer (Yang et al. 2013)	Individual formulas based on TCM diagnostic patterns	N/A	85	6857	CHM with or without chemotherapy significantly reduced mortality and remission rates, and improved quality of life and leucopenia.	CHM may be useful in patients with gastric cancer; however, further evidence is required to validate its use.
Premenstrual syndrome (PMS) (Jing et al. 2009)	Jingqianping granule Cipher decoction	15 g three times per day for two menstrual cycles 20 g three times per day for 12 days for three menstrual cycles	2	549	Jingqianping significantly eliminated PMS symptoms in 27.7% of participants. The Cipher decoction significantly eliminated PMS symptoms in 30% of participants.	Further studies are required to support the use of CHM for the treatment of PMS.
Colorectal cancer (Taixiang et al. 2005)	Huangqi compound decoction	One to two times per day for six to eight weeks	4	342	Huangqi significantly reduced nausea and vomiting in chemotherapy patients. The rate of leukopenia (WBC <3 × 10^9/L) decreased, while proportions of CD3, CD4, and CD8 T lymphocytes increased.	Huangqi compounds may stimulate immunocompetent cells and decrease side effects in patients treated with chemotherapy.
Angina pectoris (Duan et al. 2008)	Suxiao jiuxin (SJ)	Four to six pills three times per day and 10–15 pills during an angina attack	15	1776	SJ slightly improved ECG measures, reduced symptom frequency and intensity, and reduced diastolic BP compared to nitroglycerin.	SJ may be an effective treatment for angina pectoris with limited side effects.
Atopic eczema (Zhang et al. 2004)	Zemaphyte tea	Two to eight sachets per day for four to eight weeks	4	159	Compared to placebo, Zemaphyte significantly reduced erythema, surface damage, itchiness, and improved sleep quality.	Zemaphyte tea may be an effective treatment for atopic eczema.

Pharyngitis (Huang et al. 2012)	Ertong Qingyan Jiere Koufuye Yanhouling mixture Qinganlan Liyan Hanpian Dandelion soup Yantong capsules	5–15 mL three times per day for five days 20 mL two times per day for five days Two tablets per hour for 14 days. Maximum 20 per day. One serve per day for three days Three capsules three times per day	12	1954	For relieving the symptoms of pharyngitis, Ertong Qingyan Jiere Koufuye was more effective than Fufang Shuanghua Koufuye. Yanhouling mixture was more effective than gentamicin. Qinganlan Liyan Hanpian was more effective than Fufang Caoshanhu Hanpian. Dandelion soup was more effective than penicillin and Yantong capsules were more effective than cephalexin.	CHM may be more efficacious than some drugs, including antibiotics for the treatment of pharyngitis.
Irritable bowel syndrome (IBS) (Liu et al. 2006)	STW 5 STW 5-II Tongxie Yaofang decoction	20 drops three times per day for four weeks 20 drops three times per day for four weeks One dose per day for four weeks	75	7957	Compared to placebo, STW 5, STW 5-II, and Tongxie Yaofang significantly improved IBS symptoms—abdominal pain, diarrhea, and constipation. Twenty-two other CHM formulations improved IBS symptoms over conventional therapy.	CHM may be more effective than convention therapy in IBS treatment with no expected side effects.
Coronary heart disease (CHD) (Zheng et al. 2013)	Xiongshao capsules	Two 250 mg capsules three times per day	4	649	In CHD patients receiving percutaneous coronary intervention (PCI), Xiongshao reduced rates of restenosis, angina pectoris, and adverse effects compared to Western therapy. The combination of Xiongshao and Western therapy was significantly more effective than Western therapy and placebo.	Xiongshao may provide a protective effect by preventing the need for restenosis after a PCI procedure in CHD patients.
Influenza (Jiang et al. 2013)	Ganmao E Shu You injection	3.5 g three times per day for seven days 0.1 g per 250 mL injection 10 mg/kg/day for three to five days	18	2521	Ganmao capsules were found to be more effective than amantadine in decreasing influenza symptoms and reducing recovery time. E Shu You may be as effective as ribavirin in treating influenza.	Ganmao or E Shu You may be a suitable treatment in patients suffering from influenza, with no expected side effects.

(Continued)

TABLE 23.2
(Continued) Overview of the Literature Currently Available in the Cochrane Database Regarding Chinese Herbal Medicine

Condition	Herb(s)	Dosage	Trial(s) (*n* = *x*)	Participant(s) (*n* = *x*)	Results	Comment
Myocarditis (Liu et al. 2012)	*Astragalus membranaceus* Shengmai decoction	4 g two times per day for 14 days 10 mL three times per day for 30 days	20	2177	*Astragalus membranaceus* improved cardiac function, creatine phosphate kinase (CPK) levels, and lactate dehydrogenase (LDH) compared to placebo. Shengmai improved the quality of life, CPK and LDH levels, and reduced abnormal ECG readings compared to supportive therapy.	*Astragalus membranaceus* and Shengmai may improve ventricular premature beat, ECG readings, myocardial enzymes, and cardiac function in myocarditis.
Uterine fibroids (Liu et al. 2013)	*Tripterygium wilfordii* Nona Roguy	40 mg divided dose per day for 12 weeks N/A	21	2222	*Tripterygium wilfordii* reduced average fibroid volume (–23.03 cm^3) and uterine size (–51.25 cm^3) compared to mifepristone. Nona Roguy was as effective as a GnRH agonist at reducing fibroid volume and uterine size. Guizhi Fuling formula and mifepristone were associated with a greater reduction in the fibroid volume and uterine size compared to mifepristone alone.	CHM may be effective at reducing uterine fibroid volume and uterine size; however, further studies are required.
Angina (Wu et al. 2006)	Tongxinluo	Two to four pills three times per day for 2–12 weeks	18	1413	Tongxinluo improved ECG readings indicating ischemia, improved angina symptoms, such as discomfort and chest pain. It appeared to be as effective as isosorbide mononitrate in improving angina symptoms.	Tongxinluo may have possible benefits in unstable angina; however, further research is required.
Breast cancer (Zhang et al. 2007)	Shenmai injection Aifukang formula Shenqi Fuzheng Jiawei Guilu Erxian Dan	340 mL infection for 14 days Four capsules three times per day during chemotherapy 250 mL injection for 10 days One dose per day for two weeks	7	542	Compared to chemotherapy alone, Aifukang combined with chemotherapy reduced nausea and vomiting. Shenqi Fuzheng and Shenmai significantly increased T-lymphocyte subsets (CD3, CD4, CD8). Aifukang and Shenqi Fuzheng significantly improved the quality of life, whereas Jiawei Guilu Erxian Dan reduced liver and kidney toxicity.	Breast cancer patients undergoing chemotherapy may benefit from the concomitant use of CHM.

Hemorrhoids (Gan et al. 2010)	Zhixuekang Zhixue capsule Luohuazizhu Zhixuening	One dose two times per day for 10 days One capsule three times per day for seven days One tablet three times per day for 14 days One dose four times per day for seven days	9	1822	CHM reduced some symptoms associated with hemorrhoids, including hematochezia, congestive hemorrhoidal cushions, and inflammation of the perianal mucosa. In some cases, CHM was more effective than adrenosem, norflaxin, and broad-spectrum antibiotics.	CHM is ineffective at stopping hemorrhoidal bleeding, but may be beneficial at improving various other symptoms.
Type 2 diabetes (Liu et al. 2004)	Holy basil leaves, Xianzhen Pian, Qidan Tongmai, Huoxue Jiangtang Pingzhi, Inolter, Bushen Jiangtang Tang, composite Trichosanthis, Jiangtang Kang, Ketang Ling, Shenqi Jiangtang Yin, Xiaoke Tang, and Yishen Huoxue Tiaogan	Dosages available in reference	66	8302	CHM was found to have a significantly better hypoglycemic response than placebo. It demonstrated a significantly better metabolic control compared to glibenclamide, tolbutamide, and gliclazide. The combination of CHM and drug therapy is more effective than drug therapy alone in diabetic patients. It also showed better hypoglycemic effects compared to diet and behavior change.	CHM may be beneficial in the management of type 2 diabetes; however, further high-quality trials are required.
Hepatitis B (Liu et al. 2001)	Jianpi Wenshen decoction	One dose two times per day for three months	3	307	Compared to interferon, Jianpi Wenshen significantly reduced hepatitis B viral markers and had beneficial effects on serum HBsAg and HBeAg clearance and seroconversion of HBeAg to anti-HBe.	Further research is required until CHM can be recommended as a potential treatment of hepatitis B.

Note: The Condition column represents the condition followed by the authors of the review. The Herb(s) column represents the treatments evaluated in the review which were associated with a positive health outcome. Where applicable, some herbs have the abbreviation included in brackets. The corresponding prescription of the treatments is listed in the Dosage column. The Trial(s) column highlights the number of trials included in the review for the corresponding condition. The Participant(s) column represents the total number of participants included across all of the included trials for that condition. The Results column refers to the treatments identified by the authors as efficacious. The Comment column provides a summary statement relating to clinical outcomes of the herbal therapy.

BP, blood pressure; ECG, electrocardiogram; GnRH, Gonadotropin Releasing Hormone; HBsAg, Hepatitis B Surface Antigen; HBeAg, Hepatitis B "e" Antigen; HDL-C, high-density lipoprotein cholesterol; LDL-C, low-density lipoprotein cholesterol; NYHA, New York Heart Association; OCP, oral contraceptive pill; WBC, white blood cell.

intermittent claudication; however, no recommendations could be made due to the small sample size and inferior quality of the included trials (Nicolaï et al. 2009).

23.8 PRACTITIONER INFORMATION DATABASE

One effective method for practitioners to stay abreast of new evidence regarding herb–drug, herb–herb, and herb–nutrient interactions in clinical practice is through the consultation of practitioner databases. There are various forms of databases readily available for practitioners to utilize, which will guide them to make informed clinical decisions about potential interactions. The World Health Organization Collaborating Centre for International Drug Monitoring (otherwise known as Uppsala Monitoring Centre [UMC]) was formed in 1978 after the realization that no available databases with compiled information on ADRs existed. As of 2011, the UMC database contained over six million drug and herbal reports from over 100 countries. Some of the most commonly reported herbs in this database are *Cannabis sativa* (n = 1057), *G. biloba* (960), *Hypericum perforatum* (713), *Cimicifuga racemosa* (312), *Echinacea purpurea* (302), *Allium sativum* (162), and several ginseng species (125) (Uppsala Monitoring Centre 2013).

Various databases relating specifically to CHM have been developed over the past 10 years. These databases provide detailed compilations of herbal medicines and their phytochemicals (Ehrman et al. 2013; GP-TCM 2011). The Chinese Herbal Constituent Database (CHCD) provides information for over 13,000 compounds found in 300 commonly used herbs. The Bioactive Plant Compounds Database (BPCD) provides information on more than 2500 compounds active against approximately 80 targets, and the TCM Information Database (TCM-ID) has archived information on 1197 formulas, 1098 herbs, and 9852 constituents directly relating to CHM diagnosis and prescription (GP-TCM 2011). These databases represent some of the most comprehensive resources available to clinicians and other TCM health-care professionals not only regarding the constituents of CHMs, but other vital information such as chemical abstract service (CAS) numbers and chemical structures (Ehrman et al. 2013). The CAS number can be used by practitioners in websites such as Toxicology Literature Online (TOXLINE) to search for information on metabolic pathways, as well as pharmacological, physiological, and toxicological effects of herbal constituents. When used appropriately, this information can assist practitioners in making well-informed clinical decisions (Shaw et al. 2012).

There are several other databases that have been developed specifically for practitioners to use in clinical practice. The IMgateway and Natural Standard database contain the most up-to-date information regarding the safety and efficacy of a wide range of herbs and nutrients. These databases are able to identify interactions between nutrients, drugs, and herbs, and provide fully referenced monographs across a range of complementary medicines.

23.9 MODERNIZATION OF CHM WITH HERBOGENOMICS

Herbogenomics is defined as the analysis of the pharmacodynamic effects of a botanical medicine through profiling of the affected genomic or proteomic alterations (Denzler et al. 2010). Pregenomic era analytical methods investigating CHM focused on the isolation of individual active ingredients to understand the potential mechanism of action, efficacy, and toxic effects (Yadav and Dixit 2008). Although concerted efforts were made to modernize CHM using these techniques, they were unable to adequately reflect the complex pharmacological characteristics of CHM. For example, the activity of an herb is the result of several active compounds as well as inert substances. While these inert substances do not appear to have an effect when isolated, they may influence the bioavailability and excretion of an active component (Yadav and Dixit 2008).

Until the recent emergence of omic technologies such as herbogenomics, modern-day analytical methods were unable to provide an answer as to how the complex synergistic interactions and mechanisms of herbal medicines actually worked. Although in its infancy, the application of herbogenomics to CHM is likely to provide novel insights into the intricate workings of herbal medicines, validating the effectiveness of CHM with sound scientific methods (Kang 2008).

23.9.1 Assessing Efficacy and Mechanism of Action

When an herbal medicine is used to treat a particular disease, pharmacological constituents cause alterations at a molecular level, which can be observed through a microarray of mRNAs and protein profiling. These alterations, for example, may cause differential gene expression, which results in the up- or downregulation of one or more genes (Kang 2008). Molecular alterations may also cause posttranslational modifications of proteins, which results in the activation or deactivation of specific regulatory proteins, for example, the CHM formula Han Dan Gan Le has been traditionally used to treat hepatic diseases such as liver fibrosis. Through the study of herbomics, a mechanism of action was able to be elucidated. It was found that Han Dan Gan Le upregulates proteins that are responsible for modulating collagen synthesis, resulting in the fibrolysis of fibrotic hepatic tissue (Kang 2008).

23.9.2 Assessing Toxicity of Herbal Medicines

In addition to providing a mechanistic understanding of the efficacy of CHM, herbogenomics is also able to reveal the potential toxic effects of herbs. Potentially reactive moieties known as "toxicophores" are structural features responsible for the toxic effect of an herb (Ouedraogo et al. 2012). The detection of toxicophores acts as signals, alerting researchers to an herb's potential toxicity. Examples of compounds that exert toxic effects include pyrrolizidine alkaloid esters, which are present in many plants belonging to the Boraginaceae, Asteraceae, and Fabaceae families. These substances are potentially genotoxic and exhibit DNA-binding and cross-linking effects as well as causing chromosomal abnormalities (Ouedraogo et al. 2012).

The use of herbogenomics as a means of validating CHM practices with sound scientific methods is certainly an exciting prospect. This technology will undoubtedly contribute to the discovery of novel active biological compounds and play a major role in proving the existence of synergy. Herbogenomics is likely to be the catalyst for promoting further research in CHM, allowing for the modernization of TCM practice as a whole (Wang et al. 2013).

23.10 CONCLUSIONS

The current evidence supports the use of some Chinese herbal formulations as an effective therapy in the management of a wide range of health conditions. Despite this, the acceptance and integration of such therapies by mainstream medicine is often hindered by concerns of safety and efficacy. These uncertainties have prompted various health and government agencies to develop initiatives to address these concerns. For example, initiatives such as the GP-TCM Work Package 7, spearheaded by the World Health Organization, are likely to generate significant improvements toward the standardization and quality of herbal medicines on a global scale. On a national level, regulatory bodies such as the TGA and FDA aim to ensure that listed herbal medicines are safe, however, due to the complexity of herbal medicines, many pharmacological interactions are still yet to be discovered.

Recent advances in analytical methods and the incorporation of herbogenomic technologies will undoubtedly provide better insights into the efficacy and safety profiles of herbal medicines. However, until further research is undertaken, it is the responsibility of the practitioners to be aware of any known interactions, and to be diligent, not just in the prescription of herbal medicines, but throughout the entire clinical investigation process. This will ensure that any potential or theoretical interactions with other herbs of pharmaceutical medications can be identified, minimizing the likelihood of adverse events to occur.

REFERENCES

Abebe, W. 2002. "Herbal medication: Potential for adverse interactions with analgesic drugs." *Journal of Clinical Pharmacy and Therapeutics*, 27(6): 391–401.

Asl, M.N. and H. Hosseinzadeh. 2008. "Review of pharmacological effects of *Glycyrrhiza* sp. and its bioactive compounds." *Phytotherapy Research*, 22(6): 709–724.

Bogusz, M.J., H. Hassan, E. Al-Enazi, Z. Ibrahim, and M. Al-Tufail. 2006. "Application of LC-ESI-MS-MS for detection of synthetic adulterants in herbal remedies." *Journal of Pharmaceutical and Biomedical Analysis*, 41(2): 554–564.

Chan, E., M. Tan, J. Xin, S. Sudarsanam, and D.E. Johnson. 2010. "Interactions between traditional Chinese medicines and Western therapeutics." *Current Opinion in Drug Discovery & Development*, 13(1): 50–65.

Chan, K. 2002. "Understanding the toxicity of Chinese herbal medicinal products." In *The Way Forward for Chinese Medicines*, Chan, K. and H. Lee, eds. London: CRC Press.

Chan, K. 2003. "Some aspects of toxic contaminants in herbal medicines." *Chemosphere*, 52(9): 1361–1371.

Chan, K., A.C. Lo, J.H. Yeung, and K.S. Woo. 1995. "The effects of Danshen (*Salvia miltiorrhiza*) on warfarin pharmacodynamics and pharmacokinetics of warfarin enantiomers in rats." *The Journal of Pharmacy and Pharmacology*, 47(5): 402–406.

Chen, J., Y. Yao, H. Chen, J.S.W. Kwong, and J. Chen. 2012. "Shengmai (a traditional Chinese herbal medicine) for heart failure." *The Cochrane Database of Systematic Reviews*, 11: CD005052.

Chen, W., Y. Zhang, and J.P. Liu. 2011. "Chinese herbal medicine for diabetic peripheral neuropathy." *The Cochrane Database of Systematic Reviews*, 6: CD007796.

Chen, X., H. Zhou, Y.B. Liu, J.F. Wang, H. Li, C.Y. Ung, L.Y. Han, Z.W. Cao, and Y.Z. Chen. 2006. "Database of traditional Chinese medicine and its application to studies of mechanism and to prescription validation." *British Journal of Pharmacology*, 149(8): 1092–1103.

Cui, X., K. Trinh, and Y.-J. Wang. 2010. "Herbal medicine for cervical degenerative disc disease." *The Cochrane Database of Systematic Reviews*, (1): CD006556.

Denzler, K.L., R. Waters, B.L. Jacobs, Y. Rochon, and J.O. Langland. 2010. "Regulation of inflammatory gene expression in PBMCs by immunostimulatory botanicals." *PLoS One*, 5(9): E12561. doi:10.1371/journal.pone.0012561.

Duan, X., L. Zhou, T. Wu, G. Liu, J. Qiao, J. Wei, J. Ni et al. 2008. "Chinese herbal medicine suxiao jiuxin wan for angina pectoris." *The Cochrane Database of Systematic Reviews*, 1: CD004473.

Ehrman, T.M., D.J. Barlow, and P.J. Hylands. 2013. "Phytochemical databases of Chinese herbal constituents and bioactive plant compounds with known target specificities." *Journal of Chemical Information and Modeling*, 47(2): 254–263.

Ernst, E. 2002. "Adulteration of Chinese herbal medicines with synthetic drugs: A systematic review." *Journal of Internal Medicine*, 252(2): 107–113.

Fan, T.P., G. Deal, H.L. Koo, D. Rees, H. Sun, S. Chen, J.H. Dou et al. 2012. "Future development of global regulations of Chinese herbal products." *Journal of Ethnopharmacology*, 140(3): 568–586.

Feng, M., W. Yuan, R. Zhang et al. 2013. "Chinese herbal medicine Huangqi type formulations for nephrotic syndrome." *The Cochrane Database of Systematic Reviews*, (6): CD006335.

Flower, A., J.P. Liu, G. Lewith, P. Little, and Q. Li. 2012. "Chinese herbs for endometriosis." *The Cochrane Database of Systematic Reviews*, (3): CD006568.

Gan, T., Y.-D. Liu, Y. Wang, and J. Yang. 2010. "Traditional Chinese medicine herbs for stopping bleeding from haemorrhoids." *The Cochrane Database of Systematic Reviews*, 10: CD006791.

Giveon, S.M., N. Liberman, S. Klang, and E. Kahan. 2004. "Are people who use 'natural drugs' aware of their potentially harmful side effects and reporting to family physician?" *Patient Education and Counseling*, 53(1): 5–11.

GP-TCM. 2011. "Good practice in TCM research in the post-genomic era." 223154. http://www.gp-tcm.org/wp-content/uploads/2010/12/D6.5.pdf.

Grant, S.J., A. Bensoussan, D. Chang, H. Kiat, N.L. Klupp, J.P. Liu, and X. Li. 2009. "Chinese herbal medicines for people with impaired glucose tolerance or impaired fasting blood glucose." *The Cochrane Database of Systematic Reviews*, 4: CD006690.

Hill, R.L. 2006. "Pharmacovigilance of herbal medicines in Australia." *Drug Safety*, 29(4): 341–370.

Hu, J., J. Zhang, W. Zhao, Y. Zhang, L. Zhang, and H. Shang. 2011. "Cochrane systematic reviews of Chinese herbal medicines: An overview." *PLoS One*, 6(12): e28696.

Hu, Z., X. Yang, P.C.L. Ho, S.Y. Chan, P.W.S. Heng, E. Chan, W. Duan, H.L. Koh, and S. Zhou. 2005. "Herb–drug interactions: A literature review." *Drugs*, 65(9): 1239–1282.

Huang, W.F., K.C. Wen, and M.L. Hsiao. 1997. "Adulteration by synthetic therapeutic substances of traditional Chinese medicines in Taiwan." *Journal of Clinical Pharmacology*, 37(4): 344–350.

Huang, Y., T. Wu, L. Zeng, and S. Li. 2012. "Chinese medicinal herbs for sore throat." *The Cochrane Database of Systematic Reviews*, (3): CD004877.

Jadad, A.R. 1998. "Methodology and reports of systematic reviews and meta-analyses a comparison of Cochrane reviews with articles published in paper-based journals." *Journal of the American Medical Association*, 280(3): 278.

Jiang, L., L. Deng, and T. Wu. 2013. "Chinese medicinal herbs for influenza." *The Cochrane Database of Systematic Reviews*, (3): CD004559.

Jing, Z., X. Yang, K.M.K. Ismail, X.Y. Chen, and T. Wu. 2009. "Herbal treatment for premenstrual syndrome." *The Cochrane Database of Systematic Reviews*, (1): CD006414.

Kang, Y.J. 2008. "Herbogenomics: From traditional Chinese medicine to novel therapeutics." *Experimental Biology and Medicine*, 233(9): 1059–1065.

Lin, V., P. McCabe, A. Bensoussan, S. Myers, M. Cohen, S. Hill, and G. Howse. 2009. "The practice and regulatory requirements of naturopathy and Western herbal medicine in Australia." *Risk Management and Healthcare Policy*, 2: 21–33.

Liu, J.P., H. McIntosh, and H. Lin. 2001. "Chinese medicinal herbs for asymptomatic carriers of hepatitis B virus infection." *The Cochrane Database of Systematic Reviews*, (2): CD002231.

Liu, J.P., H. Yang, Y. Xia, and F. Cardini. 2013. "Herbal preparations for uterine fibroids." *The Cochrane Database of Systematic Reviews*, 4: CD005292.

Liu, J.P., M. Yang, Y. Liu, M.L. Wei, and S. Grimsgaard. 2006. "Herbal medicines for treatment of irritable bowel syndrome." *The Cochrane Database of Systematic Reviews*, (1): CD004116.

Liu, J.P., M. Zhang, W.Y. Wang, and S. Grimsgaard. 2004. "Chinese herbal medicines for type 2 diabetes mellitus." *The Cochrane Database of Systematic Reviews*, 3: CD003642.

Liu, Z.L., J.P. Liu, A.L. Zhang, Q. Wu, Y. Ruan, G. Lewith, and D. Visconte. 2011. "Chinese herbal medicines for hypercholesterolemia." *The Cochrane Database of Systematic Reviews*, 7: CD008305.

Liu, Z.L., Z.J. Liu, J.P. Liu, and J.S.W. Kwong. 2010. "Herbal medicines for viral myocarditis." *The Cochrane Database of Systematic Reviews*, (7): CD003711.

Lo, A.C., K. Chan, J.H. Yeung, and K.S. Woo. 2013. "The effects of Danshen (*Salvia miltiorrhiza*) on pharmacokinetics and pharmacodynamics of warfarin in rats." *European Journal of Drug Metabolism and Pharmacokinetics*, 17(4): 257–262.

Lu, A., M. Jiang, C. Zhang, and K. Chan. 2012. "An integrative approach of linking traditional Chinese medicine pattern classification and biomedicine diagnosis." *Journal of Ethnopharmacology*, 141(2): 549–556.

Manheimer, E., S. Wieland, E. Kimbrough, K. Cheng, and B.M. Berman. 2009. "Evidence from the Cochrane collaboration for traditional Chinese medicine therapies." *Journal of Alternative and Complementary Medicine*, 15(9): 1001–1014.

McEwen, J. and F. Cumming. 2003. "The quality and safety of traditional Chinese medicines." *Australian Prescriber*, (26): 130–131.

Myers, S.P. and P.A. Cheras. 2004. "The other side of the coin: Safety of complementary and alternative medicine." *The Medical Journal of Australia*, 181(4): 222–225.

NICM. 2007. "National Institute of Complementary Medicine. NICM Background." http://www.nicm.edu.au/about-nicm.

Nicolaï, S.P.A., L.M. Kruidenier, B.L.W. Bendermacher, M.H. Prins, and J.A.W. Teijink. 2009. "*Ginkgo biloba* for intermittent claudication." *The Cochrane Database of Systematic Reviews*, 2: CD006888.

Omar, H.R., I. Komarova, M. El-Ghonemi, A. Fathy, R. Rashad, H.D. Abdelmalak, M.R. Yerramadha, Y. Ali, E. Helal, and E.M. Camporesi. 2012. "Licorice abuse: Time to send a warning message." *Therapeutic Advances in Endocrinology and Metabolism*, 3(4): 125–138.

Ong, E.S. 2004. "Extraction methods and chemical standardization of botanicals and herbal preparations." *Journal of Chromatography B: Analytical Technologies in the Biomedical and Life Sciences*, 812(1/2): 23–33.

Ouedraogo, M., T. Baudoux, C. Stévigny, J. Nortier, J.-M. Colet, T. Efferth, F. Qu et al. 2012. "Review of current and 'omics' methods for assessing the toxicity (genotoxicity, teratogenicity and nephrotoxicity) of herbal medicines and mushrooms." *Journal of Ethnopharmacology*, 140(3): 492–512.

Page, R.L. and J.D. Lawrence. 1999. "Potentiation of warfarin by Dong Quai." *Pharmacotherapy*, 19(7): 870–876.

Qiu, J. 2007. "Traditional medicine: A culture in the balance." *Nature*, 448(7150): 126–128.

Rathbone, J., L. Zhang, M. Zhang, J. Xia, X. Liu, and Y. Yang. 2013. "Chinese herbal medicine for schizophrenia." *The Cochrane Database of Systematic Reviews*, (4): CD003444.

RMIT. 2013. "Traditional and complementary medicine group research." Royal Melbourne Institute of Technology, Melbourne, VIC. http://www.rmit.edu.au/chinese-med/research.

Schleimer, R.P. 1991. "Potential regulation of inflammation in the lung by local metabolism of hydrocortisone." *American Journal of Respiratory Cell and Molecular Biology*, 4(2): 166–173.

Shaw, D. 2010. "Toxicological risks of Chinese herbs." *Planta Medica*, 76(17): 2012–2018.

Shaw, D., L. Graeme, D. Pierre, W. Elizabeth, and C. Kelvin. 2012. "Pharmacovigilance of herbal medicine." *Journal of Ethnopharmacology*, 140(3): 513–518.

Taixiang, W., A.J. Munro, and L. Guanjian. 2005. "Chinese medical herbs for chemotherapy side effects in colorectal cancer patients." *The Cochrane Database of Systematic Reviews*, 1: CD004540.

Uppsala Monitoring Centre. 2013. "Classification and monitoring safety of herbal medicines." World Health Organization, Geneva, Switzerland.

UWS. 2013. "Centre for complementary medicine research." University of Western Sydney, NSW. http://www.uws.edu.au/ssh/school_of_science_and_health/research/research_area_3.

Wang, C.Z., M. Ni, S. Sun, X.-L. Li, H. He, S.R. Mehendale, and C.-S. Yuan. 2009a. "Detection of adulteration of notoginseng root extract with other *Panax* species by quantitative HPLC coupled with PCA." *Journal of Agricultural and Food Chemistry*, 57(6): 2363–2367.

Wang, J., R. van der Heijden, S. Spruit, T. Hankermeier, K. Chan, J. van der Greef, G. Xu, and M. Wang. 2009b. "Quality and safety of Chinese herbal medicines guided by a systems biology perspective." *Journal of Ethnopharmacology*, 126(1): 31–41.

Wang, Q., Z. Guo, P. Zhao, Y. Wang, T. Gan, and J. Yang. 2005. "Chinese medicinal herbs for treating acute inflammation of the pancreas." *The Cochrane Database of Systematic Reviews*, (1): CD003631.

Wang, S., Y. Hu, W. Tan, X. Wu, R. Chen, J. Cao, M. Chen, and Y. Wang. 2012. "Compatibility art of traditional Chinese medicine: From the perspective of herb pairs." *Journal of Ethnopharmacology*, 143(2): 412–423.

Wang, X., A. Zhang, and H. Sun. 2013. "Future perspectives of Chinese medical formulae: Chinmedomics as an effector." *Omics: A Journal of Integrative Biology*, 16(7/8): 414–421.

World Health Organization. 2005. "National policy on traditional medicine and regulation of herbal medicines." WHO report on global safety, Geneva, Switzerland. http://apps.who.int/medicinedocs/en/d/Js7916e/.

Wu, T., R.A. Harrison, X.Y. Chen, J. Ni, L. Zhou et al. 2006. "Tongxinluo (a traditional Chinese medicine) capsules may help patients suffering from unstable angina." *The Cochrane Database of Systematic Reviews*, (4): CD004474.

Wu, T., J. Zhang, Y. Qiu, L. Xie, and G.J. Liu. 2007. "Chinese medicinal herbs for the common cold." *The Cochrane Database of Systematic Reviews*, 1: CD004782.

Wu, W.W.P. and J.H.K. Yeung. 2010. "Inhibition of warfarin hydroxylation by major tanshinones of Danshen (*Salvia miltiorrhiza*) in the rat in vitro and in vivo." *Phytomedicine*: *International Journal of Phytotherapy and Phytopharmacology*, 17(3/4): 219–226.

Xie, P., S. Chen, Y.-Z. Liang, X. Wang, R. Tian, and R. Upton. 2006. "Chromatographic fingerprint analysis—A rational approach for quality assessment of traditional Chinese herbal medicine." *Journal of Chromatography A*, 1112(1/2): 171–180.

Xue, C.C.L., A.L. Zhang, K.M. Greenwood, V. Lin, and D.F. Story. 2010. "Traditional Chinese medicine: An update on clinical evidence." *Journal of Alternative and Complementary Medicine*, 16(3): 301–312.

Xue, C.C.L., A.L. Zhang, V. Lin, C.D. Costa, and D.F. Story. 2007. "Complementary and alternative medicine use in Australia: A national population-based survey." *Journal of Alternative and Complementary Medicine*, 13(6): 643–650.

Yadav, N.P. and V.K. Dixit. 2008. "Recent approaches in herbal drug standardization." *International Journal of Integrative Biology*, 2(3): 195–204.

Yang, Y. 2010. *Chinese Herbal Formulas: Treatment Principles and Composition Strategies*. London: Churchill Livingstone Elsevier.

Zeng, Z.-P. and J.-G. Jiang. 2010. "Analysis of the adverse reactions induced by natural product-derived drugs." *British Journal of Pharmacology*, 159(7): 1374–1391.

Zhang, J., F. Zhou, X. Wu, Y. Gu, H. Ai, Y. Zheng, Y. Li et al. 2010. "20(S)-Ginsenoside Rh2 noncompetitively inhibits P-glycoprotein in vitro and in vivo: A case for herb-drug interactions." *Drug Metabolism and Disposition: The Biological Fate of Chemicals*, 38(12): 2179–2187.

Zhang, M., X. Liu, J. Li, L. He, and D. Tripathy. 2007. "Chinese medicinal herbs to treat the side-effects of chemotherapy in breast cancer patients." *The Cochrane Database of Systematic Reviews*, 2: CD004921.

Zheng, G.H., J.P. Liu, J.F. Chu, L. Mei, and H.Y. Chen. 2013. "Xiongshao for restenosis after percutaneous coronary intervention in patients with coronary heart disease." *The Cochrane Database of Systematic Reviews*, 5: CD009581.

Zhou, L., S. Wang, Z. Zhang, B.S. Lau, K.P. Fung, P.C. Leung, and Z. Zuo. 2012. "Pharmacokinetic and pharmacodynamic interaction of Danshen-Gegen extract with warfarin and aspirin." *Journal of Ethnopharmacology*, 143(2): 648–655.

Zhou, S.F., Z.-W. Zhou, C.-G. Li, X. Chen, X. Yu, C.C. Xue, and A. Herington. 2007. "Identification of drugs that interact with herbs in drug development." *Drug Discovery Today*, 12(15/16): 664–673.

Zhou, X., S. Chen, B. Liu, R. Zhang, Y. Wang, P. Li, Y. Guo, H. Zhang, Z. Gao, and X. Yan. 2010. "Development of traditional Chinese medicine clinical data warehouse for medical knowledge discovery and decision support." *Artificial Intelligence in Medicine*, 48(2/3): 139–152.

Zhu, X., M. Proctor, A. Bensoussan, E. Wu, and C.A. Smith. 2008. Chinese herbal medicine for primary dysmenorrhoea." *The Cochrane Database of Systematic Reviews*, 2: CD005288.

Section VI

Future Trends

24 Traditional Foods and Their Values for Health and Wellness on Evidence-Based Approach

V. Prakash, M.A. Alwar, and M.A. Lakshmithathachar

CONTENTS

24.1 INTRODUCTION

Sustenance of living beings are highly dependent on the food resources that they have access to. There is nothing more precious than food that supports the life of the humans, animals, birds, and the aquatic and the microbial life apart from air and water. There is no medicine that is comparable to food in its vast composition of a variety of biomolecules in one single food let alone a combination of foods! One cannot cure a disease even though the correct medicine is given by ignoring that diet plays a very important role. It is therefore rightly said by physicians that "food is the greatest medicine." Hence, the value of food was known to our ancestors, and there lies the origin of the traditional and ethnic foods that have a great journey through several generations of human beings listed, verified, safety procedures established and which has reached us today. In the current scenario of the modern tools of science, we have a great opportunity to understand the traditional foods and their value for health and wellness through evidence-based approach and make it useful to the society by spreading this knowledge of foods.

It is estimated that corpus of nearly 100,000 medical manuscripts has the legacy in the Indian medical heritage, which perhaps is one of the largest in the world. Of course, these are widely distributed in many heritage institutions in India, and many of them are lost to other countries and are stashed away in museums and other places of private collections and many a times remain undetected and lost after a while. However, a large volume of information that is linked to food is still available with the Indian heritage. Perhaps for lack of a comprehensive catalog and a deep effort to organize subject wise to the vast number, the depth to which these manuscripts have taken the traditional foods as preventive and curative diets is the need of the hour, even though organizations such as AYUSH (Ayurveda, Yoga, Unani, Siddha and Homeopathy) are doing their night in India in saving this knowledge. The conservation, translation, consolidation, and interpretation of this is simply not a matter of converting from one language to the other, but to understand the meaning of each word and web it into a meaningful deciphered telescope of information is fundamental for its exposition to bring it to the world's knowledge. The Indian culinary art loaded with medicinal values for health and wellness, the number of dishes, the ingredients, and the combinatorial foods are certainly the need of the hour to understand through modern science, their composition, the benefits, and the bioavailability of the nutraceuticals as the world calls it, but perhaps since the origin of many of these is from a traditional Indian holistic lifestyle approach called *Ayurveda*, it is more appropriate to term it as Ayurceuticals! Such a treatise that also gives importance to water, and the effect of the various recipes on the human body has an advantage because it is still used with the importance of process of making the product than the end product itself. These remarkable documents with a wide array of food stuffs and even the container in which they are cooked are simply Himalayan in its volume of information and the depth, and the height to which it goes perhaps cannot be summarized in this one small chapter. The classification of these traditional foods is based on grains; grasses; edible plant parts (roots, leaves, fruits, and vegetables); different spices and even different kinds of salts used in cooking; the different meats of animals, birds, and aquatic animals, and the sensorial profiles of which change using different waters in cooking such as river water, springwater, and seawater; and the varieties of milk and milk products as well as the containers in which they are cooked. The vessels spread from pure metals all the way to earthen pots and from iron to gold vessels to various mixtures of metals such as brass and bronze and other mixtures simply shows the fundamental understanding of the traditional foods in which they are cooked and the chemistry of metals, one has to remember that those times the population did not have the modern high-performance liquid chromatography (HPLC) but had the human trials and acceptance over generations dominating the knowledge transfers. Therefore, this chapter rightfully addresses the various deep insights into the traditional foods and their values for health and wellness as a humble attempt for an evidence-based approach covering several aspects of how food was viewed in India in the long history of tradition and culture, the ancient knowledge of the taste theory of food, the sensorial and chemical profiles of foods and how they are related to health through traditional wisdom. The aspect of documentation for it being transmitted without errors from generation to generation, the embodiment of food technology in traditional food systems and the scale of operation beyond one's imagination, the harmonization of nature with a combination of foods and such a knowledge was transmitted from one generation to another many a times without IPR (Intellectual Property Rights) and trademarks (!) and such management of traditional foods beyond 5000 years baffles anybody of the enormous amount of knowledge that the Indian subcontinent had and has and it is time that it is not enough that it finds a place in a chapter in a book but perhaps libraries can be created of this knowledge and such knowledge should be reverberating by support of science to convince the rest of the world that it is this knowledge of food which is fuel for the body by not just calories but thousands of molecules going into body every day and making it functional both physically and mentally to ensure a healthy and quality life. It is this knowledge that the authors have tried to deliver through the menu using this article and perhaps those who are more interested in their own country of the traditional foods they have. We are very positive that such databases around the globe would probably simply is unlimited and more to learn from each other in this world of hunger and diseases. Perhaps here is the tool that is neglected over time and today it is coming back in a very different demand from the pharmacy angle, from the functional food angle, from the nutraceutical angle, from a diet angle and also from a healthy living angle not that by following it the longevity will be doubled but will deliver that *quality longevity* which is most important as the human cells get older and older with aging. Perhaps this concept of food plus human being is the least understood of all the sciences that we know today (Achaya 1994; Patwardhan et al. 2004; Prakash 2003, 2008; Raghunathasuri 2012; Swaminathan 2012).

24.2 FOOD, TRADITION, AND CULTURE

The Indian Food Science and Technology has a great history of at least 5000 years. Information about historical aspects of foods is available in a fragmentary way in the basic texts of *Ayurveda* such as *Charaka Samhita*, *Sushruta Samhita*, and *Bhavaprakasa* and also the later texts of *Ayurveda*.

The scattered information required a systematic treatment and presentation in a succinct way so that people at large could understand the nature and properties of each and every natural ingredient and the foodstuffs prepared utilizing the natural food materials (Achaya 1994).

Thus, there arose a necessity to systematize food science based on strictly scientific principles, keeping in view of the impact of the different food ingredients and the foodstuffs prepared out of them on the human system so that the rest of the world also understands this system.

The human body, as is well known, according to the Indian traditional knowledge of medicine, is a combination of five great elements, namely, ether, air, fire, water, and earth. Even the food ingredients and foodstuffs prepared by using them are a combination of the five great elements. Therefore, any ingredients of the food or the foodstuffs are bound to have a considerable impact on the human system at large at the physical, mental, and spiritual levels according to the attributes and properties they possess. It is often said that "you are defined by the food you eat." A famous Indian saying states that a person who knows his food seldom suffers from diseases. In addition, even the mental dispositions depend on the food one consumes. The *Taittiriyopanisat* explicitly states that "Every being is born out of food, every being is sustained by food, and ultimately dissolves into food." Such hoary statements of our sages of yore alert us that we should be very careful with regard to the food we consume, not only from the point of view of our health but also from the point of view of our mental dispositions and spiritual and other pursuits (Raghunathasuri 2012).

Thus, Food Science and Technology developed by our ancient saints has value in them to adopt for the benefit of modern man. In fact, analysis of food (as was done in the works of yore) is becoming more and more relevant nowadays, even as people in the modern scientific world indiscriminately consume various foodstuffs and even junk diets without bothering about the impacts of these junk diets on the human system. It is observed that taste alone plays the all-important role in people consuming them and not the properties that may have serious impact on their physical and mental well-being.

Ayurveda, the science of life developed in India with its holistic approach, has viewed man (the microcosm) as part and parcel of the universe (macrocosm). According to this science, man is the combination of the five great elements with his consciousness being the sixth component. It also accepts that man is the replica of the universe (*purusoyamloka-sammitah*) keeping in view the impact of the food materials and the foodstuffs on the three humors of the human being, namely, *vata*, *pitta*, and *kapha* that play a major role in the well-being of human beings. The attributes (*Dravyagunas*) of each and every food material and foodstuff that are being used in India from long ago, but also the effects that take place on persons who consume them, having different types of body constitutions (*Prakritis*). This information is vital from the point of view of our approach into holistic health. Modern science has also contributed much to this field as it has concentrated much on the nutritional value of each and every food material and foodstuff. However, modern science seems to take in the medicinal properties of these for a better benefit holistically (Patwardhan et al. 2004).

24.3 *RASA* (TASTE) THEORY OF FOOD

Food materials that are naturally produced and foodstuffs that are produced using the food nutrients have different tastes depending on the proportions of the five elements present in them as mentioned in Subsection 24.2 as per the traditional Indian knowledge base.

The doctrines of *Ayurveda* opine that the basic tastes—namely, sweet (*madhura*), sour (*amla*), salty (*lavana*), astringent (*kashaya*), pungent (*katu*), and bitter (*tikta*)—and innumerable tastes created by the admixture of these six tastes have a remarkable impact in maintaining the health of the individual. These six tastes and also the subvarieties generated by their permutation and combination in various proportions could be used by expert dieticians to restore the health of an individual who is suffering from some disease or the other (Raghunathasuri 2012).

Thus, according to the Indian traditional knowledge, food is the elixir of life as it is preventive and curative. If one does not have proper diet, any amount of medicine cannot restore one's health completely. At the same time, if one has a perfect diet, many diseases are prevented, and because proper food is consumed in the prescribed manner, it can cure many of the diseases and even has the capacity to rebuild the body's mechanism of immune system. Hence, these are maxims such as "A person who has perfect knowledge about his diet does not suffer from any disease" (*hitaballavarigerogavilla*). It is difficult to prove, but one cannot ignore this old saying, which is in all cultures globally. With this background, we can proceed to understand the traditional foods of India and their live medicinal properties. There are a number of independent works dealing with dietetics such as *Bhojanakutuhalam*, *Ksemakutuhalam*, and *Pathyapathyavivekam* apart from specific chapters dealing with food science in Ayurvedic texts such as *Charaka Samhita* and *Sushruta Samhita*. Let us integrate this knowledge with the scientific files we have for a translational approach for health and wellness (Raghunathasuri 2012).

24.4 ROLE OF CHEMICALS IN ILLNESS AND VALUE OF TRADITIONAL KNOWLEDGE

The unchecked use of chemicals and chemical compounds in many fields of human activity has led to significant adverse effects on the ecology and the human being himself or herself. The quality of human, animal, and plant life has deteriorated over hundreds of years to the extent that experts have begun to seriously question the long-term benefits of modern practices based on organic and synthetic chemical compounds. There is a growing realization that man must live in harmony with nature and that time-tested natural practices must be revived. There is considerable support for a more holistic approach to the solution of our problems of health through food and nutrition. In the field of medicine, while there is little doubt that modern research and development has enabled man to overcome many dangerous diseases, the downside is that health care is not affordable by a vast majority of the world population. Other disadvantages of the widespread use of chemicals are more and more resistant in organisms that cause infections and weakening of the body's defense and immune mechanisms. All over the globe, there is a growing body of opinion for a more holistic approach of food for medicine. The wealth of knowledge contained in ancient knowledge systems and practices is no longer considered of little consequence. There is growing interest among scholars, researchers, doctors, practitioners, and other experts in ways and means to tap into ancient knowledge and investigate their relevance with many a times evidence-based approach and reverse nutraceutics (ICMR 1989, 2010).

In this context, traditional foods and their medicinal properties is one area that merits serious research due to the organic and holistic approach toward health and treatment of ailments inherent in the ancient system such as *Ayurveda* in India. One finds that in such a system of traditional approach for many diseases, natural treatment and recovery protocols based on proper food and organic formulations would be more beneficial for health in the long run. It has been shown that mortality all over the world has dropped considerably more due to intervention of better nutrition along with modern medicines and also with a holistic approach added onto it. Traditional foods with medicinal properties have been found to be a good complement to modern medicine in providing affordable health care, while restoring the defense mechanism of the human body to fight diseases (Narayana 2010).

24.5 DOCUMENTATION

The holistic approach of food, medicine, health, wellness, exercise, and age is perhaps more important to balance in life than looking at any one of them in isolation. A classic example for this is the role of traditional foods and their distilled wisdom that should never be set aside and along with it the holistic adaptation of lifestyle itself through such wisdom of traditional knowledge in every region and country globally. This is especially so in the use of herbs and condiments in different forms of cooking and processed foods. The traditional knowledge of ethnic population regarding the biodiversity of world food resources must be preserved, documented, and used, and this knowledge must be combined with scientifically supported data that in some cases might be already available and must be used in bringing the awareness. Therefore, how does one use this knowledge from generation to generation of this treasure of informatics on nutrients to prevent or delay diseases and sustain recoveries from diseases? One of the key points in traditional and ethnic foods in India is as a result of *Ayurveda* that uses herbs, condiments, spices, and plants. Such a vast data are now available in many of the ancient libraries through writings on palm leaves and then lacquering them and ultimately to preserve them as a wonderful document. This even today perplexes chemists as to how such a technology existed in "digitization over many centuries ago in spite of cultural, social, economic and political revolutions that have happened with invasions with the supporting role of the society with dedicated scholars who were true Scientists indeed." Nowadays, consumers are looking for foods that have positive health or nutritional attributes. The power of such traditional formulations for a better health using the knowledge of these foods certainly can go beyond the science and engineering. Similarly, the awareness of dietary guidelines may promote a healthy lifestyle in both rural and urban areas that need to be clubbed to traditional knowledge and wisdom and digitization for easy retrieval. In this Council of Scientific and Industrial Research (CSIR), India's lead in Traditional Knowledge Digital Library is indeed a great global contribution and needs to be cited here without fail (Prakash and van Boekel 2010).

24.6 TECHNOLOGIES FROM TRADITIONAL FOOD KNOWLEDGE

The role of understanding crops is particularly very important in terms of food security, nutrition security, and health security. It is not only the yield in the crop management but also the nutrition enrichment, especially in the staples such as legumes, pulses, and millets, that can lead to benefits in human health in terms of holistic approach which the Indian system of medicine strongly advocates. When we keenly evaluate a typical nutrients in many of the traditional and generation-to-generation carryover of certain food preparation in practices, it is very clear that many of them have a role to play in advocating health and better nutrition from a sustainable point of view. Apart from the taste and flavor, the removal of antinutritional factors by proper cooking such as trypsin inhibitors that affect digestion and absorption of food and the knowledge of such processing techniques are very evident in hundreds of dishes such that the food is safe, healthy, and nutritious (Clydesdale 2004; Rao 2002, 2012).

We all know that food and health are two important angles and nutrition in the community at large today looks at focusing the value of such foods through functional foods, and also the sustainability of nutrition on a food-based approach is clearly emphasized on a life cycle approach. This is a true reflection of tradition that always emphasizes through a food-based approach and a natural way for getting the various molecules into the body. In this entire chain of travel of food from the farm to the gut, several processing habits are a part of many cultures globally. This is to give value to the food with better digestion and better functional roles; from thousands of generations, the wrong food is eliminated and what we have today is the distilled knowledge of those several hundreds of generations leading to the understanding of food–gene interaction at a perspective level of the organism with a sustainable consumption of food for better health and wellness.

One of the challenging task for the modern scientist is to explain to the consumer the benefits of food with a focus on nutraceuticals, nutritionals, and the good biomolecules in the food as well as why the body needs such a holistic approach of different foods is to be clearly characterized and a pathway established for such a communication. We have even addressed the crop fortification many a times in feeding the masses for bridging the nutrition gap. But it is not clear globally how much investment we have put in the traditional and ethnic knowledge dissipation to bridge the modern knowledge with the worldwide biodiversity of food resources. It is also important that such knowledge has to be combined with today's basic science methodologies and at the same time generates the translational, traditional, and ethnic knowledge base with evidence-based approach (Prakash 2008).

There are several definitions for different parts of the foods that can be combined to get medical or health benefits and even different ways of processing them, and in fact there are documentations of prevention of noncommunicable diseases, and such products with very specific targeted nutrient and dietary supplements and diets to genetically engineer them have been in the forefront. However, the market is also complicated and many a times confuses the buyer in terms of specialty foods, medical foods, fortified foods, and functional foods. However, when we look at traditional and ethnic knowledge of foods that have got special properties, the amazing combination of foods for a great amount of bioavailability of nutrients through such food factors and the overall holistic health is simply one of the rare values that are much less used. This also includes immunity. Therefore, the traditional wisdom and traditional knowledge of foods which are inclusive of traditional foods and ethnic foods have to be explored for their beneficial effects with the right combination of different foods for disease management, weight management, both physical and mental health, and heart health as also to include the molecular gastronomy for the better quality of life for the consumer. It is not just ensuring food that has better absorption and delivering the biomolecules to the body but also looking at the holistic approach with the power of nutrigenomics that is as much important as nutrition. One may not claim as personalized food at this point; however, a day will come when tradition and value and the ethnic knowledge can be very well related to an absolute personalized designer food for an individual's health just like the medicines of today, which are custom made by combinatorial methods many a times when the generic medicines do not work. This treasure has to be explored and the knowledge has to be shared for better health to the community and society keeping the individual gene–food interaction in view beyond nutrigenomics (Prakash 2003).

24.7 KNOWLEDGE OF IPR AND GLOBAL HARMONIZATION

In today's competitive world, the effort of research organizations, industries, nongovernmental organizations (NGOs), policy makers, and legal systems in the background of Trade-Related Aspects of Intellectual Property Rights (TRIPS) in the patent regime is to be innovative every moment. However incremental, it is that innovation has to be ahead of others in terms of intellectual property. This needs to be addressed without violation of copyrights, trademarks, and geographical appellations and not infringing into others patents as well as to not attempt even in the figment of imagination, patenting societal knowledge, traditional knowledge, and tribal knowledge (including use of herbs, condiments, and spices on the one side of preventive health, and medicine on the other side of trying to patent traditional foods which obviously must be avoided). If somebody is trying to patent a herb, luckily documentation such as *Sushruta Samhita* and *Charaka Samhita* in India would be invoked for herbals or on management angle to Chanakya's *Arthashastra* before even granting a patent on a global basis. These are nearly 3000 years of traditional knowledge documented very well.

How does the knowledge reach the society when one patents an innovation? After all, a patent will lead to some kind of a profit for an innovation that has taken place as a result of hard work that could be of an engineer, a doctor, a scientist, or even a farmer, and as a matter of fact, it could be even an ordinary person who has innovated a product or process.

Therefore, it is not limited to only a few people. Everybody can hope to be an innovator, provided he or she has documented the innovation and has a support of no prior art of such a process or product through writing, by photo, by video, or by evidence and witnesses. This means that no such new innovated information is available in any form or in the social practices. Example: cooking of rice cannot be patented! By the normal means such as washing the rice and putting it into the boiling water for a fixed period of time (depending on the type of rice), the rice is ready. Nobody can patent this because this information is already in the society and in use. However, if a technology is there to change the method of cooking such as use of a pressure cooker, a rice cooker, and a solar cooker, then that equipment can be patented, which results in a process patent leading to product innovation and better product. However, at the same time, one can give this knowledge free of cost for the society if the innovator decides to do it. The owner of the patent or the author of the patent has every right to give his patent without charging for it for use depending on the customer that he or she would like to be benefitted. This is a great social aspect of patenting. Perhaps this is what our ancestors did with our traditional knowledge. Very seldom it is understood and in fact it is a subject that comes into a lot of deliberations more often than climate change. This does not mean that an innovator should not make any profit from his or her innovation. The classic example is Thomas Alva Edison who had more than 1000 patents. A large amount of industries came out of it; therefore, a huge amount of employment, investments, new products, and convenient products came in, and life-saving products were in the market, and thus, the quality of life of a person in the society is improved overall. Hence, it is not just measuring a patent and IPR only by money or royalty to the innovator, but one must also see the larger picture of the benefits to the civil societies, but how does food with medicinal value can be harmonized and reach every kitchen?

If we look at the statistics of patents over the past 10 years, there has been a huge increase in patent filings globally and in India. At the Indian Patent Office in India, compared to about 4,000 patents in 2002, today we are talking about 40,000 patent applications, which means that there is a tenfold increase in the past 10 years. In fact in the year 2011, it rose by nearly 13% compared to the previous year. When we look at biotechnology, pharmacy, medical equipments, and patents compared to international and national patents, there was a large slip of about 20% down in international patents during 2007, but India went up by twofold increase. Today in the life sciences and biotechnology, there are about 8000 international patent applications globally and in India about 2000 patent applications. When we compare this to pharmacy, the international scenario is about 150,000 patents filed internationally. Therefore, it is important to realize that the portfolio must increase, and not necessarily all the patents will generate money. Some patents actually protect other patents what we call as "bubble patents," and we have also to be patient with the understanding that the innovations require a long gestation period and certainly require patient investors. Therefore, it is appropriate that it will need a place that richly deserves not only for fighting of the patents when infringed and restore a country's prestige in areas such as turmeric, basmati, wheat, and some other traditional foods; now not many battles are fought on patents but are busy filing at the same time with new and incremental innovations. The social aspects of helping the needy by ensuring that the traditional knowledge and traditional wisdom are not lost but protected as they cannot be patented obviously; however, documentation and data protection through digitization are perhaps more important.

24.8 MANAGEMENT OF TRADITIONAL KNOWLEDGE BASE

The key to any productivity is not only new ideas, new paths, and new products but the approach of "translational/innovation." The word *translational/innovation* is pebbled with fire stones which one need to walk and is yet not an easy path. To a large extent, today in the market, both small and medium sectors, and global companies, more money rolls in and rolls out by "incremental innovative technologies." The delta incremental innovation makes a huge change in the risk of taking and gaining consumer confidence. This also touches the purse of the consumer who is already wedded to a product and would certainly love the "incremental innovation in his/her products" at an affordable cost. But how does one do it? This is a very difficult question to answer in a single word. Today the market penetration through translational knowledge reverberating and echoing from the bottom of the pyramid is one of the primary roles of involving the basic sciences on the one hand and to address the major cross-cutting issues of a product on the other hand using the traditional knowledge base. It then empowers the manufacturer to achieve what the consumer needs in that innovation cycle with a total rethinking. However, the market is moving today with a paradigm shift of urban to rural, quantity to quality, production to processes, and technology to policy, and the challenges become very competitive. Hence, application of cutting-edge assessment of a "translational technology" becomes very critical from time to time to induce innovation in the process and challenge oneself in replacing one's own product before the competitor does. This requires a strategy, a global knowledge with a local flavor, quick actions, high science (not necessarily always high technology), and ultimately a firm foundation of a scientific evidence to ensure that the nature of such innovations is sustainable and all inclusive in traditional foods they have survived for hundreds of generations. This should have a clear mandate of future perspective of a challenging new market, and this is where the number games of billions for India and China in the internal market makes a difference and has the advantage. Have we capitalized it fully? Perhaps we need to do this with translational innovation models in management in the chain of traditional foods to bring out its value of health and wellness to reach out rural and urban, and the traditional knowledge and wisdom will add to this quantum jump and on edge over other regions in the health and wellness agenda of the global society.

DISCLAIMER

The authors have documented their personal knowledge and the societal knowledge as well as some of the cited knowledge in the reference list, and have used the information from several seminars, symposiums, and workshops and meetings. The reference list is not complete to reflect the 5000 years of traditional foods and *Ayurveda*. The authors also do not claim any of the information in this chapter as their own since most of it is from ancient knowledge, and acknowledge the original authors such as Charaka, Sushrutha, Bhavaprakasa, Panini, Patanjali, Raghunathsuri, and many others who are never cited in most of the literature around the world.

REFERENCES

Achaya, K.T. 1994. *Indian Food—A Historical Companion.* Mumbai, India: Oxford University Press.

Clydesdale, F. 2004. Functional foods: Opportunities and challenges. *Food Technology*, 58: 35–40.

Indian Council of Medical Research. 1989. Nutrient requirement and recommended dietary allowances for Indian. A report of the expert group of Indian Council of Medical Research. National Institute of Nutrition, Hyderabad, India.

Indian Council of Medical Research. 2010. Nutrient requirement and recommended dietary allowances for Indian. A report of the expert group of Indian Council of Medical Research. National Institute of Nutrition, Hyderabad, India.

Indian Horticultural Database 2010–2011. nhb.gov.in/area-pro/database-2011.pdf, accessed on December 16, 2012.

Integrated Child Development Services (ICDS) Scheme. http://wcd.nic.in/icds.htm.

Narayana, D.B.A. 2010. Reverse pharmacology for developing functional foods/herbal supplements: Approaches, framework and case studies. In: *Functional Food Product Development*. Smith, J. and Charter, E. eds. Oxford: Wiley-Blackwell, pp. 244–256.

National Institute of Nutrition. Report of the National Nutrition Monitoring Bureau during 1983–1996. www.ninindia.org/nnmb.htm.

Patwardhan, B., Vaidya, A.D.B., and Chorghade, M. 2004. Ayurveda and natural products drug discovery. *Current Science*, 86: 789–799.

Prakash, V. 2003. The role of nutraceuticals and quality food parameters for better nutrition and better health. *Proceedings of the IX Asian Congress of Nutrition, Nutrition Goals for Asia—Vision 2020.* Nutrition Foundation of India, New Delhi, India, pp. 591–593.

Prakash, V. 2008. Nutraceuticals and functional foods—What the future holds? *The 14th World Congress of Food Sciences and Technology. International Union of Food Science and Technology (IUFoST)*, Shanghai, People's Republic of China, September 9-14.

Prakash, V. and van Boekel, M.A.J.S. 2010. Nutraceuticals: Possible future ingredients and food safety aspects. In: *Ensuring Global Food Safety*. Boisrobert, C.E., Stjepanovic, A., Oh, S. and Lelieveld, H. eds. London: Oxford Academic Press, pp. 333–338.

Raghunathasuri. 2012. *Bhojanakuthuhalam.* [original text in Sanskrit]. Translated by I-AIM, Trivandrum, India: Śūranād Kunjan Pillai.

Rao, B.S.N. 2002. Pulses and legumes as functional foods. *NFI Bulletin*, 23(1): 1–4.
Rao, B.S.N. 2012. Millets in Indian diets. An overview. *NFI Bulletin*, 33(3): 1–7.
Swaminathan, M.S. 2012. FAO and the eight millennium development goals. In: *The State of Food Insecurity in the World*. Rome, Italy: Food and Agriculture Organization of the United Nations. FAO Home page of Millennium Development Goals. http://www.fao.org/mdg/en/.
The World Bank. 2012a. The World Bank in India. *Newsletter*, vol. 10. http:/go.worldbank.org/NKFGI3VF50.
The World Bank. 2012b. South Asia. Data, projects and research. http:/go.worldbank.org/NKFGI3VF50.

25 Impact of Personalized Nutrition on Public Health

Lynnette Ferguson, Karen Bishop, and Nishi Karunsinghe

CONTENTS

25.1 INTRODUCTION

Optimal diet may vary from one individual to the other, depending on genotype [1], and certainly the diet many of us consume today does not resemble that which our species has adapted to over time, namely, one of energy saving [2]. It has been estimated that 40% of cancers could be prevented by adopting healthy lifestyles. In personalized medicine, the challenge is to deliver the benefits of recent scientific research by delivering the right drug to the right patient at the right time. Such research includes discovering the molecular causes of disease and, in some cases, developing the ability to predict the patient's response to therapy. Likewise, in personalized nutrition, the scientific challenges are similar in that both require the sifting through of research results to identify the most clinically significant genetic markers, as well as to elucidate the effect of genetics and epigenetics on gene–diet interactions. Just as Herceptin (Roche, Basel, Switzerland) is accompanied by diagnostic tests to determine the suitability of treatment for a particular patient [3], some foods and neutraceuticals need to be accompanied by genotypic tests to ensure that recommendations are made on a personalized level.

Although technology is sufficiently sensitive to provide extensive genomic and metabolomic data within a short time frame, little research involving nutrition in the form of dietary intervention studies has been used to stratify the population with respect to benefit. For a small number of products or ingredients, sufficient information is available; personalized nutrition in a broad sense is not yet ready for the consumer [4]. However, there are specific diseases that have been well characterized [5,6], and certain foods or food groups [7] that can be personalized. The severity of disease and the ability to prevent or ameliorate its impact through nutrition and other lifestyle factors, together with acceptance of personalized nutrition by the general public, will contribute toward determining the impact on public health.

A news release dated August 8, 2013, on the University of Auckland website (www.auckland.ac.nz), was entitled "Food app to fight nation's killers." The release described a new smartphone application claimed to "empower New Zealand shoppers to make healthier food choices—reducing their risk of dying early from two of the nation's biggest killers, heart attack and stroke." In particular, the application allows scanning of the bar code of packaged foods in supermarkets to identify foods with high levels of fat, salt, and sugar. These data are then converted to nutritional advice, theoretically leading to healthier food choices. The information content is based on the nutritional value of more than 8000 packaged food products found in New Zealand supermarkets (www.foodswitch.co.nz). Nevertheless, following consumer surveys, study developers acknowledge that this approach will not alone be adequate to solve the growing risks of chronic disease in this country [8]. Successful dietary interventions would need to use tailored and personalized content, and include social support. The authors suggest that the mobile phone application may be utilized to provide this.

25.2 NUTRIGENETIC APPROACHES TO PERSONALIZING NUTRITION

There is considerable variation in individual response to diet and/or lifestyle interventions, some of which may benefit certain individuals or population subgroups more than others. To apply nutrigenetics, there is a need to determine the influence of diet and other lifestyle factors such as physical activity and air quality on genotype (both genetic and epigenetic) and the interaction of these factors on health.

Celiac disease is a good example in which genetic screening could be used to identify those who may benefit from dietary intervention. Various genetic changes are necessary but not sufficient for disease development, and these changes indicate a high probability of disease occurrence [9]. Diet (i.e., avoidance of gluten) is the only sustainable means of treating celiac disease and hence would suit a nutrigenetic approach [9].

In addition to macronutrients, the effect of micronutrients may also influence the health status differently depending on genotype, lifestyle, and environmental factors. Vitamin D is an example that can be used to demonstrate the complexity of the interactions between genotypes, which influences absorption, distribution, metabolism, and excretion, and how this can influence the enzymatic activity and interact with environmental factors [10]. Vitamin D is well known for its interaction with calcium levels and influence on bone density, and in addition, it interacts with other micronutrients such as many of the B vitamins, selenium (Se), and zinc, and acts on both the innate and adaptive immune systems [10]. In addition to vitamin D, the requirement for the micronutrient selenium has shown huge variability with demographics, lifestyle factors, health status, and genetic polymorphisms, and as such, it is discussed in detail in Subsection 25.3.

25.3 SELENIUM STUDIES

The lack of Se-sufficient soil levels to maintain healthy farm animals in New Zealand has been known for a number of decades [11,12]. Plant analyses for the whole of the North Island have shown that 20%–30% of samples have low levels of both cobalt and Se, which could affect animal health [13]. In 1980, Thomson and Robinson reviewed Se status in New Zealanders and predicted that the population could be having a borderline Se deficiency, but there could be vulnerable groups that require attention [14]. In the early 1980s, it was hypothesized that increased levels of colorectal cancers within the United States and other Western countries since 1950s could be Se deficiency related, and further the deficiencies were not just from decreased consumption of Se, but also from increased intakes of antagonizing microelements of zinc and fluoride, which may impact the effect of Se [15]. Our own analyses revealed a significant association of regional colorectal cancer incidence with soil Se levels in New Zealand [16]. We have previously discussed the molecular mechanisms affecting DNA methylation through deficiencies of Se together with zinc and folate deficiencies that could account for high colon cancer rates in New Zealand [16]. Based on the intake necessary for the maximization of plasma levels of a major Se-containing antioxidant enzyme glutathione peroxidase (GPx) activity, an upper estimated requirement of 90 μg Se per day has been calculated with a lower estimated requirement of 39 μg Se per day as the intake necessary to reach two-thirds of maximal GPx activity [17]. Such information tends to support Se supplementation strategies for better health, especially in vulnerable groups with low intakes or low circulating levels of Se. However, Se supplementation benefits have varied because of a multitude of reasons as shown in our own New Zealand studies.

In a group of 43 men, who were considered as having a high risk of prostate cancer but with negative biopsies, from urology clinics in Auckland, we observed that DNA damage has a significant inverse correlation with baseline serum Se only up to a level of 100 ng/mL [18]. The levels of GPx and another Se-containing antioxidant enzyme thioredoxin reductase (TR) activities showed a significant positive correlation with baseline serum Se level of this group with a mean serum Se level of 97.8 ± 16.6 ng/mL. When this group was supplemented with 200 μg Se per day in the form of selenized yeast for six months, their GPx activity remained unchanged, whereas their TR activity showed an 80% average increase from baseline [19]. Se supplementation on this group also showed a narrowing of variance in DNA damage assessed after a peroxide challenge, with those having a higher baseline level of DNA damage showing a reduction with supplementation and those with a lower baseline level showing an increase in DNA damage [18].

These observations were repeated in a larger cohort supplemented with 200 μg Se per day supplied as selenized yeast and tested for six months with a group of men without any recorded cancers except for skin cancers. This group carried a slightly higher mean serum Se level of 111.01 ± 1.01 ng/mL compared to the negative biopsy group. In this group, those with the lowest and medium tertiles of peroxide-induced baseline DNA damage showed a significant increase in DNA damage with Se supplementation, while those with the highest tertile showed a significant lowering of DNA damage. Similar observations were made with those having the lowest and highest tertiles of baseline DNA damage. TR activity increased significantly in all strata except for the tertile with the highest baseline peroxide-induced DNA damage [20,21]. These studies also showed that 200 μg Se per day supplementation for six months significantly reduced peroxide-induced DNA damage among those being treated for cardiovascular diseases, whereas other health groups showed no change in DNA damage.

In a three-month Se supplementation (100 μg Se per day as L-selenomethionine) study carried out with men having coronary artery disease in Otago, New Zealand, a significant increase in GPx activity was recorded. These men had plasma Se levels ranging from 105 to 107 ng/mL. The increase in GPx activity did not differ in this group between GPx1 rs1050450 single-nucleotide polymorphisms unless the baseline plasma Se level was <90 ng/mL. When the plasma Se level was <90 ng/mL, those carrying the C homozygote recorded a twofold increase in GPx activity after Se supplementation [22].

Our studies with 479 men completing a supplementation protocol of 200 μg Se per day for six months with selenized yeast have shown a similar effect of GPx rs1050450 allele C, whereas those with cardiovascular conditions generally showed an increased GPx activity after Se supplementation. Although supplementation significantly increased the average TR activity in this cohort, a significant reduction in TR activity was noted among carriers of the *rs3877899 A* allele of the *selenoprotein P* (*SEPP1*) gene polymorphism [20]. Although the average GPx activity showed no significant change with Se supplementation, a group of men having a baseline serum Se level of >140 ng/mL showed a significant decline in circulating GPx activity [23]. Se supplementation in our cohort has shown beneficial effects on those who have a history of tobacco smoking. GPx activity was generally lower among those with a tobacco smoking history compared to nonsmokers. When Se is supplemented, those with a smoking history had increase in circulating GPx activity significantly compared to never smokers [20]. Those with a tobacco smoking history have been previously recorded in our studies as carrying significantly lower levels of circulating Se [23]. This study has also shown a serum Se level corresponding to the lowest level of DNA damage at 116 ng/mL for those carrying GPx rs1050450 CC genotype to 149 ng/mL for those carrying *GPx4* gene polymorphism of rs713041 TT. We have also recorded the benefits of Se as the age of men increased with a significant increase in circulating GPx activity measured from erythrocyte lysates [20]. Se supplementation has also benefitted those with higher body mass indices (BMIs) in resolving basal DNA damage [20]. Clearly, there is an interaction between Se levels and genotype, and this may be influenced by lifestyle or environmental factors to affect health.

25.3.1 Integrating Genetic and Diet Information with Other Genomic Technologies

It has proved very challenging to predict individual responses to complex diets based solely on nutrigenetic techniques, which consider common genetic variations [24]. Genetic variants influencing nutrient absorption, distribution, and metabolism have been identified, but there are only rare examples in which individual variants have been conclusively linked to chronic disease risk, in association with a given nutrient [25]. It is clear that multiple genetic variants, in association with multiple genetic and environmental factors, influence health outcomes. We require better bioinformatics tools to provide comprehensive mathematical modeling of these complex interactions. We also lack clear biomarkers for health status with respect to a number of diseases. In addition, it is apparent that good study designs need to address consumer acceptance and privacy protection before this approach becomes commercially viable [24]. Ultimately, this will improve the ability of medical practitioners and dieticians to provide personalized dietary advice.

While there was an early assumption that gene–nutrient interactions might provide sufficient information to personalize nutrition [26], the situation now appears considerably more complex than this. In addition to genetic data, we are able to compile epigenetic, metabolomic, and gut microbiome data [27]. This provides a more complete understanding of metabolic and physiologic factors and influences on gut bacteria, as well as genetic differences among individuals. It will be important to incorporate genetics, gene expression, metabolomics, proteomics, demographic, lifestyle, disease factors, and information on gut microbes in developing accurate predictive measures of health.

Metabolomics approaches are sensitive to both the host genome effects and the gut microbiome. Nicholson et al. [28] use the term "pharmacometabolomics" to describe the use of metabolic profiling of biological fluids, tissues, and tissue extracts to predict both beneficial and adverse effects of an intervention such as diet and lifestyle. Despite these advances, we are not yet in a position to effectively integrate these biologic data to provide comprehensive phenotypic profiles. More comprehensive human trials will be essential to effectively develop personalized dietary recommendations.

25.4 THE GUT MICROFLORA ADD ANOTHER DIMENSION

The gastrointestinal tract maintains a fragile balance between health and disease. This balance is maintained via a number of factors, including the interaction between the microbes inhabiting the gastrointestinal tract and our diet [2]. Enormous diversity exists within the gastrointestinal microbiota and species composition varies from person to person, yet remains reasonably stable over time [29].

In a study carried out on mice raised conventionally versus those raised in a sterile environment by Backhed et al. [30], the conventionally raised mice had a body fat content 40% higher than those raised in the sterile environment. It is believed that the gastrointestinal microbes increase the capillary density in gastrointestinal mucosa and therefore improve the efficiency of absorption of monosaccharides derived from otherwise nondigestible polysaccharides [30]. Studies by Backhed et al. [30] and Ley et al. [31] show that diet does influence the composition of microbiota, which in turn has an impact on metabolism. Following on from this, pro- and prebiotics have been developed for use as functional foods in an effort to improve the gastrointestinal microbial balance by introducing "desirable" bacteria and providing food components that will promote the growth of "desirable" bacteria, respectively [29].

A great deal more research needs to be performed in order for us to understand the interaction between the human and microbial genomes, as well as diet [2]. Only once this is understood, we can begin to personalize the approach, thus enabling us to eat for both "bugs" and humans.

25.5 PUBLIC ACCEPTANCE OF PERSONALIZED NUTRITION

Foods and food products have become increasingly personalized during the late twentieth century. Whether foods personalized through nutrigenomic technologies significantly enter the marketplace depends on numerous hurdles being overcome. These include ethical, legal, and social issues [32].

The question of public acceptance was critically evaluated by Fallaize et al. [33]. Given the high current and predicted rates of noncommunicable diseases internationally, and their relationship with diet and lifestyle, these authors consider it imperative that current recommendations are better informed—and more rigorously followed. Indeed, it has been predicted that up to 80% of chronic disease might be preventable through modifications in diet and lifestyle, thereby significantly reducing global morbidity. The human genome project has led to clarification of the interactions between diet and genes, thereby providing potential for individualizing diets according to phenotypic and genotypic data. This would lead to dietary interventions moving from population-based guidance toward potentially more efficacious "personalized nutrition." While there is still some general public skepticism, individuals with early symptoms of disease appear more willing to consider personalized nutrition. Importantly, this group of individuals has a greater motivation to change their behavior. Fallaize et al. (2013) consider that there needs to be more work on the efficacy and methods of implementing personalized nutrition [33].

People throughout the world are questioning the quality and safety of the foods they consume. In addition, producers, manufacturers, and suppliers are increasingly investing in educating the public with respect to the health benefits of consuming their products. However, we know that no single diet or food component will benefit everyone. But will the general public accept testing to identify dietary items that are most likely to benefit them? In a study carried out on 452 volunteers in Germany, the authors found that 45% of the volunteers would like to obtain genetic profiling and the related nutritional advice [5]. Furthermore, 40% of the study participants said that they would be willing to purchase functional food products [5].

Does personalized genetic information reduce high health risk food choices, that is, does this knowledge result in behavior modification? Does personalized risk prediction promote behavioral change? Just as poor lipid profiles are an early motivator to change behavior before harm is done, so too could the knowledge of the presence of genetic markers act to initiate behavioral change. However, where behavioral change can bring about changes in lipid profile, no such feedback can be provided when making a behavioral change in response to knowledge of a high-risk genetic marker.

Heart disease, obesity, adult-onset diabetes, and cancer are largely attributed to diet, inactivity, and smoking status [6,34]. However, genotype clearly plays a role as evidenced by the clustering of these chronic diseases in families. Even if people are screened, adapting to change is difficult—what are the stumbling blocks? It is essential that communication is clear. In the past, communication has provided one message taking a "one-size-fits-all" approach. If one considers that certain recommendations do not provide a benefit for the entire population, as is the case with a high polyunsaturated fatty acid (PUFA) diet, it is important to provide relevant messages to each genotypic group (e.g., different advice should be provided to the *APOA1* AA and GA vs. GG genotypes) [7].

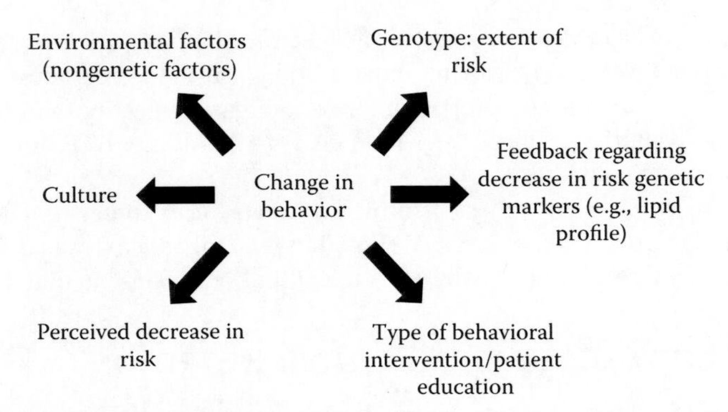

FIGURE 25.1 Personalized nutrition: Factors likely to influence the modification of behavior.

Genotype can influence a person's ability or willingness to modify his or her behavior [35], and this ability is also influenced by culture, social knowledge, and public consultation and engagement (Figure 25.1) [36]. Nutrigenetic screening provides information that could help predict personalized risk that may motivate healthy individuals to adopt beneficial eating patterns and levels of physical activity, to adhere to prescribed risk-reducing chronic medication, and/or to go for screening that is related to health conditions [6]. All of these behavioral changes would help reduce risk by enhancing compliance to risk reduction behaviors. Information from genetic screening is highly personalized, and it is believed that this characteristic may increase its motivational potential to effect change [6]. However, public acceptance of nutrigenomic testing is linked to personalized nutrition.

When considering public acceptance of genetic testing to personalize nutrition, there are a number of factors to consider. These include, but are not limited to, acceptance of functional foods, the effect of testing on perceived risk, the influence of existing levels of motivation on willingness to modify behavior [37], the influence of carrier status on motivation [37,38], belief or trust in results [39], the influence of existing health problems on attitude to testing [39], potential for stigmatization, and perceived impact on privacy [39].

25.5.1 Functional Foods

When determining public acceptance of personalized nutrition, we can learn a few lessons from functional foods. Clearly, personalized nutrition is somewhat more complex, but the marketing of functional foods can certainly help to inform us with regard to the promotion and acceptance of personalized nutrition.

Functional foods are marketed as having health benefits without stratifying the population. Menrad (2003) found that public acceptance of functional foods depended on the familiarity of the consumer with the health benefits of the particular ingredients [40]. For example, consumers were more familiar with the health benefits of calcium rather than flavonoids, and this familiarity was necessary but not sufficient to ensure market success [40]. It was also found that consumers were not prepared to change their eating patterns for functional foods [40], but this may not hold true for personalized nutrition as the extent of the health benefit might be perceived to be greater. Looking at the functional food industry in Europe, it is believed that a number of products have failed simply because of high-price premium rather than the products themselves [40]. It is also important that functional foods deliver an added value but otherwise deliver the same attributes as other successful foods with respect to taste, texture, convenience, and packaging size [40,41]. Not only is it necessary to have an excellent product with a health benefit, but it is also important to communicate this carefully as attitudes can determine whether information is accepted or rejected [42].

25.5.2 Possible Harm

Concerns have been raised regarding the possible harm caused by misinformation and misinterpretation of genetic advice provided to apply personalized nutrition [43]. San-Cristobal et al. [43] feel that care needs to be taken to prevent unnecessary anxiety that may occur when people interpret their genotypic data with respect to their likelihood of developing a disease. In addition, even if counseled and the information is properly understood, knowing one's risk of a particular disease may increase anxiety levels in the short term, but generally these levels returned to baseline over the long term [44].

25.5.3 Effect of Testing and Carrier Status on Perceived Risk and Motivation

Perceived risk needs to be considered in both carriers and noncarriers of a risk allele. San-Cristobal et al. [43] raised a concern that people who had been told that they were noncarriers of risk alleles for a particular disease might become overconfident and not follow the basic guidelines for good health. By contrast, carriers of risk alleles might become fatalistic and reduce their efforts to modify behavior, or their motivation levels might increase in response to a treatment designed for their genotype [38].

Surprisingly, it was found that there was no increased perceived risk between carriers and noncarriers of risk alleles (for hereditary breast or ovarian cancer or colorectal cancer) at 12 months after testing [44]. With regard to behavioral change, and after all, this is where we hope personalized nutrition will make an impact, it was found that knowledge of one's disease risk status generally brought about an increase in screening and prophylactic surgery or chemotherapy during the first 12 months following genetic testing, but this varied depending on the disease and intervention [44]. Although Heshka et al. [44] expected perceived risk to be higher in carriers versus noncarriers of a risk allele for a particular disease, this was not the case 12 months after genetic testing. This leads one to question whether the genetic counseling received was having the desired outcome with respect to educating the patient, as perceived risk influences behavior modification.

25.5.4 Influence of Existing Levels of Motivation on Willingness to Modify Behavior

In a small study carried out by Markowitz et al. [37], it was found that levels of motivation varied among participants, and this impacted on people's willingness to modify their behavior [38]. The authors concluded that counseling should be personalized based on the existing levels of motivation to modify behavior as well as carrier status [37]. Knowledge of risk status may motivate greater commitment to behavioral change in those already highly motivated, whereas those with low motivation may not feel inspired to initiate change based on high-risk status and some may feel fatalistic [37,38].

25.5.5 Influence of Existing Health Problems on Attitude to Testing

Both Stewart-Knox et al. [45] and Pin and Gutteling [46] found that positive attitudes to uptake of genetic testing, in order to personalize nutrition, were based on the presence of existing health problems [45] or perceived benefit [46].

25.5.6 Potential for Stigmatization

Depending on the culture and societal values, people may be concerned about being stigmatized if their genetic tests indicate that they are carrying risk alleles for various diseases. In a small study focused on Arab-American women and hereditary breast cancer, it was found that there was stigma and shame associated with a cancer diagnosis and that the discovery of cancer in the family might negatively influence their daughter's marriage prospects [47].

25.5.7 Privacy and Trust

It is believed that trust in service providers and regulators of genetic testing will influence the acceptance of genetic testing for personalized medicine [48]. In a study by Markowitz et al. [37], some participants expressed concern regarding privacy of genetic results and the effect this may have on health-care insurance. Perceived risks are related to data security in the form of commercial entities selling information or not providing sufficient security to prevent hacking of databases, rather than actual personalized nutrition [39]. In some cases, people preferred an online service if it permitted anonymity, as well as a service that was provided by health professionals working in the public sector (i.e., it was anticipated that such professionals would not be motivated by personal gain) [39]. Although online services were deemed to offer greater privacy than face-to-face contact with a health-care provider, online providers were regarded as lacking credibility, whereas a service via a health-care provider was seen as being credible and trusted [39].

25.6 CONCLUDING COMMENTS

Personalized nutrition could be implemented as a public health strategy for the early detection of individuals susceptible to specific diseases and therefore increase disease prevention through diets tailored to individuals or groups. The purpose of implementing personalized nutrition is to empower people to be involved in their own health care by modifying their behavior toward a healthier lifestyle, which is individual specific. Environment and genotype both contribute to the development of numerous chronic diseases (e.g., diabetes, heart disease, asthma, various cancers, and depression), and hence risk stratification rather than prediction is feasible [49]. Although we are not at the point of providing definitive genotypic data with respect to disease certainty for all diseases, personalized genomics can be used to provide another line of evidence when health professionals are interacting with patients [50] and would help to empower the patient if counseled correctly. However, it is also possible that in those people with no increased risk for a disease, a sense of overconfidence might arise and the health guidelines ignored [43].

REFERENCES

1. Madden, J. et al. 2011. The impact of common gene variants on the response of biomarkers of cardiovascular disease (CVD) risk to increased fish oil fatty acids intakes. *Annual Review of Nutrition*, 31: 203–234.
2. Constantin, N. and W. Wahli. 2013. Nutrigenomic foods. *Nutrafoods*, 12(1): 3–12.
3. Hamburg, M.A. and F.S. Collins. 2010. The path to personalized medicine. *The New England Journal of Medicine*, 363(4): 301–304.
4. Ferguson, L.R., 2013. Nutrigenetics and nutrigenomics: Importance for functional foods and personalized nutrition. In *Nutrigenomics and Nutrigenetics in Functional Foods and Personalized Nutrition*. L.R. Ferguson, ed. Boca Raton, FL: CRC Press, pp. 2–23.
5. Roosen, J. et al. 2008. Consumer demand for personalized nutrition and functional food. *International Journal for Vitamin and Nutrition Research*, 78(6): 269–274.
6. McBride, C.M. et al. 2010. Future health applications of genomics priorities for communication, behavioral, and social sciences research. *American Journal of Preventive Medicine*, 38(5): 556–565.
7. Ordovas, J.M. 2004. The quest for cardiovascular health in the genomic era: Nutrigenetics and plasma lipoproteins. *Proceedings of the Nutrition Society*, 63(1): 145–152.
8. Gorton, D. et al. 2011. Consumer views on the potential use of mobile phones for the delivery of weight-loss interventions. *Journal of Human Nutrition and Dietetics*, 24(6): 616–619.
9. Fenech, M. et al. 2011. Nutrigenetics and nutrigenomics: Viewpoints on the current status and applications in nutrition research and practice. *Journal of Nutrigenetics and Nutrigenomics*, 4(2): 69–89.
10. van Ommen, B. et al. 2010. The micronutrient genomics project: A community-driven knowledge base for micronutrient research. *Genes and Nutrition*, 5(4): 285–296.
11. Wichtel, J.J. 1998. A review of selenium deficiency in grazing ruminants. Part 2: Towards a more rational approach to diagnosis and prevention. *New Zealand Veterinary Journal*, 46(2): 54–58.

12. Wichtel, J.J. 1998. A review of selenium deficiency in grazing ruminants. Part 1: New roles for selenium in ruminant metabolism. *New Zealand Veterinary Journal*, 46(2): 47–52.
13. Ledgard, S.F. et al. 1991. Getting it right with the soil and pasture—Cost effective use of fertiliser. *Proceedings of the New Zealand Grassland Association* 53: 175–179.
14. Thomson, C.D. and M.F. Robinson. 1980. Selenium in human health and disease with emphasis on those aspects peculiar to New Zealand. *American Journal of Clinical Nutrition*, 33(2): 303–323.
15. Nelson, R.L. 1984. Is the changing pattern of colorectal cancer caused by selenium deficiency? *Diseases of the Colon & Rectum*, 27(7): 459–461.
16. Ferguson, L.R. et al. 2004. Dietary cancer and prevention using antimutagens. *Toxicology*, 198(1–3): 147–159.
17. Duffield, A.J. et al. 1999. An estimation of selenium requirements for New Zealanders. *American Journal of Clinical Nutrition*, 70(5): 896–903.
18. Karunasinghe, N. et al. 2004. DNA stability and serum selenium levels in a high-risk group for prostate cancer. *Cancer Epidemiology, Biomarkers & Prevention*, 13(3): 391–397.
19. Karunasinghe, N. et al. 2006. Hemolysate thioredoxin reductase and glutathione peroxidase activities correlate with serum selenium in a group of New Zealand men at high prostate cancer risk. *Journal of Nutrition*, 136(8): 2232–2235.
20. Ferguson, L.R. et al. 2012. Understanding heterogeneity in supplementation effects of selenium in men: A study of stratification variables and human genetics in a prospective sample from New Zealand. *Current Pharmacogenomics and Personalized Medicine*, 10(3): 204–216.
21. Karunasinghe, N. et al. 2013. Effects of supplementation with selenium, as selenized yeast, in a healthy male population from New Zealand. *Nutrition and Cancer*, 65(3): 355–366.
22. Miller, J.C. et al. 2012. Influence of the glutathione peroxidase 1 Pro200Leu polymorphism on the response of glutathione peroxidase activity to selenium supplementation: A randomized controlled trial. *American Journal of Clinical Nutrition*, 96(4): 923–931.
23. Karunasinghe, N. et al. 2012. Serum selenium and single-nucleotide polymorphisms in genes for selenoproteins: Relationship to markers of oxidative stress in men from Auckland, New Zealand. *Genes and Nutrition*, 7(2): 179–190.
24. De Roos, B. 2013. Personalised nutrition: Ready for practice? *Proceedings of the Nutrition Society*, 72(1): 48–52.
25. Hesketh, J. 2013. Personalised nutrition: How far has nutrigenomics progressed [quest]. *European Journal of Clinical Nutrition*, 67(5): 430–435.
26. Ferguson, L.R. and J. Kaput. 2004. Nutrigenomics and the New Zealand food industry. *Food New Zealand*, 4(2): 29–36.
27. Lampe, J.W. et al. 2013. Inter-individual differences in response to dietary intervention: Integrating omics platforms towards personalised dietary recommendations. *Proceedings of the Nutrition Society*, 72(2): 207–218.
28. Nicholson, J.K. et al. 2012. Longitudinal pharmacometabonomics for predicting patient responses to therapy: Drug metabolism, toxicity and efficacy. *Expert Opinion on Drug Metabolism & Toxicology*, 8(2): 135–139.
29. Fava, F. et al. 2006. The gut microbiota and lipid metabolism: Implications for human health and coronary heart disease. *Current Medicinal Chemistry*, 13(25): 3005–3021.
30. Backhed, F. et al. 2004. The gut microbiota as an environmental factor that regulates fat storage. *Proceedings of the National Academy of Sciences of the United States of America*, 101(44): 15718–15723.
31. Ley, R.E. et al. 2005. Obesity alters gut microbial ecology. *Proceedings of the National Academy of Sciences of the United States of America*, 102(31): 11070–11075.
32. Ghosh, D. 2010. Personalised food: How personal is it? *Genes and Nutrition*, 5(1): 51–53.
33. Fallaize, R. et al. 2013. An insight into the public acceptance of nutrigenomic-based personalised nutrition. *Nutrition Research Reviews*, 26(1): 39–48.
34. Strong, K. et al. 2006. Preventing chronic disease: A priority for global health. *International Journal of Epidemiology*, 35(2): 492–494.
35. Dockray, G.J. 2009. Cholecystokinin and gut–brain signalling. *Regulatory Peptides*, 155(1–3): 6–10.
36. Haga, S.B. and H.F. Willard. 2006. Defining the spectrum of genome policy. *Nature Reviews Genetics*, 7(12): 966–972.
37. Markowitz, S.M. et al. 2011. Perceived impact of diabetes genetic risk testing among patients at high phenotypic risk for type 2 diabetes. *Diabetes Care*, 34(3): 568–573.
38. Marteau, T.M. and C.Y. Lerman. 2001. Genetic risk and behavioural change. *British Medical Journal*, 322(7293): 1056–1059.

39. Stewart-Knox, B. et al. 2013. Factors influencing European consumer uptake of personalised nutrition. Results of a qualitative analysis. *Appetite*, 66: 67–74.
40. Menrad, K. 2003. Market and marketing of functional food in Europe. *Journal of Food Engineering*, 56(2/3): 181–188.
41. Urala, N. and L. Lähteenmäki. 2007. Consumers' changing attitudes towards functional foods. *Food Quality and Preference*, 18(1): 1–12.
42. Wilcock, A. et al. 2004. Consumer attitudes, knowledge and behaviour: A review of food safety issues. *Trends in Food Science & Technology*, 15(2): 56–66.
43. San-Cristobal, R. et al. 2013. Future challenges and present ethical considerations in the use of personalized nutrition based on genetic advice. *Journal of the Academy of Nutrition and Dietetics*, 113(11): 1447–1454
44. Heshka, J.T. et al. 2008. A systematic review of perceived risks, psychological and behavioral impacts of genetic testing. *Genetics in Medicine*, 10(1): 19–32.
45. Stewart-Knox, B.J. et al. 2009. Attitudes toward genetic testing and personalised nutrition in a representative sample of European consumers. *British Journal of Nutrition*, 101(7): 982–989.
46. Pin, R.R. and J.M. Gutteling. 2009. The development of public perception research in the genomics field: An empirical analysis of the literature in the Field. *Science Communication*, 31(1): 57–83.
47. Mellon, S. et al. 2013. Knowledge, attitudes, and beliefs of Arab-American women regarding inherited cancer risk. *Journal of Genetic Counseling*, 22(2): 268–276.
48. Castle, D. and N.A. Ries. 2007. Ethical, legal and social issues in nutrigenomics: The challenges of regulating service delivery and building health professional capacity. *Mutation Research—Fundamental and Molecular Mechanisms of Mutagenesis*, 622(1/2): 138–143.
49. Feero, W.G. et al. 2010. Genomic medicine—An updated primer. *The New England Journal of Medicine*, 362(21): 2001–2011.
50. Patel, C.J. et al. 2013. Whole genome sequencing in support of wellness and health maintenance. *Genome Medicine*, 5(6): 58.

26 Effects of Ginger on Metabolic Syndrome

A Review of Evidence

Srinivas Nammi, Yu-Ting Sun, and Dennis Chang

CONTENTS

26.1 INTRODUCTION

The use of plants for the treatment of various ailments has been documented since antiquity. People rely on these plants not only for their nutritional values but also for maintenance of health and protection from various pathological threats to the body.

Ginger (*Zingiber officinale*) belongs to the Zingiberaceae family, which also includes cardamom (*Elettaria cardamomum*) and turmeric (*Curcuma longa*). It originates in Southeast Asia and has become widespread across numerous ecological zones. Historically, ginger has been highly regarded for both its medicinal and its mercantile value. Ginger is commonly known as a condiment or spice used for its aroma and flavor in culinary arts; however, it is also widely used as herbal medicine to treat ailments in China, India, and Europe for centuries (Ali et al. 2008). In Ayurveda and Unani-Tibb medicine, ginger is predominantly used for digestive imbalance and gastric upset, such as nausea (Butt and Sultan 2011), whereas in traditional Chinese medicine (TCM), in addition to digestive complaints, ginger is also used for common cold as well as pain symptoms (Bensky 2004).

Over the past few decades, tremendous efforts have been made in the discovery of pharmaceutical medicine from herbal sources for the management of chronic illnesses. Ginger has a long history of medicinal use. An extensive preclinical research has demonstrated that ginger possesses a broad range of pharmacological properties such as antiemetic, anti-inflammatory, antioxidant, antihyperglycemic, and antihyperlipidemic effects. Clinical research into ginger has been predominately conducted on its antiemetic effects in pregnancy or motion sickness and anti-inflammatory effects for pain management in osteoarthritis. There is a general lack of clinical evidence for ginger to be used outside of these conditions (White 2007).

The metabolic syndrome is a group of risk factors that include high blood sugar (hyperglycemia), abnormal blood fat (dyslipidemia), high blood pressure (hypertension), and obesity (Alberti et al. 2005). There is preponderance of laboratory-based research supporting the application of ginger for the metabolic syndrome even though the relevant clinical evidence is generally lacking. This chapter aims to provide an overview of the clinical and preclinical evidence regarding ginger, emphasizing its effect on the metabolic syndrome.

26.2 METABOLIC SYNDROME

The metabolic syndrome represents a group of risk factors that increase the chance of diabetes and cardiovascular diseases (CVDs). Furthermore, individuals suffering from metabolic syndrome also appear to be more susceptible to other conditions including polycystic ovarian syndrome, fatty liver, cholesterol gallstones, asthma, sleep disturbance, and some forms of cancer (Grundy et al. 2004). In 1988, Gerald Reaven, a noted endocrinologist, propounded that these risk factors were commonly clustered together and designated as *Syndrome X*. He postulated that insulin resistance was the underlying cause of Syndrome X resulting in the commonly referenced term "insulin resistance syndrome" (IRS) (Reaven 1988). However, as insulin resistance is not the only or primary condition associated with this cluster of risk factors, the National Cholesterol Education Program's Adult Treatment Panel III (ATP-III) uses the term "metabolic syndrome" to avoid possible confusion (Grundy et al. 2004).

Research suggests that the pathogenesis of the metabolic syndrome potentially correlates to obesity and disorders of adipose tissue, insulin resistance, or a constellation of independent factors. These independent factors include genetic alternation, insulin secretory dysfunction, lipoprotein metabolism, and blood pressure regulation. Other factors such as proinflammatory states, aging, and hormonal changes are also considered to contribute to the metabolic syndrome (Grundy et al. 2004).

Insulin resistance is a common condition in people with metabolic syndrome and is considered as the secondary cause of the metabolic syndrome. This is a condition in which cells fail to respond to insulin, which leads to impaired glucose tolerance resulting in hyperglycemia (Ferrannini et al. 1991; Reaven 1988). Hyperglycemia contributes to the high prevalence of type 2 diabetes and the development of CVDs. While insulin resistance may occur within subjects who are not obese, numerous studies have shown a distinct association between obesity and insulin resistance (Grundy et al. 2004).

Currently, there are no viable pharmaceutical options for the management of the metabolic syndrome. Pharmaceuticals used for this purpose primarily act on individual components of the disease and are usually accompanied with numerous side effects (Nammi et al. 2004). Ginger is highly regarded for the promotion of health due to its rich phytochemistry (Butt and Sultan 2011). Preliminary studies have demonstrated the potential benefits of ginger on hyperglycemia and lipid profiles. This may represent a suitable alternative to the current symptomatic management of the metabolic syndrome.

26.3 OVERVIEW OF GINGER AND ITS CHEMICAL COMPOSITION

The portion of ginger that is typically consumed is the rhizome (underground stem), often referred to as ginger root. However, this is a misnomer because it is the horizontal stem of ginger plant that sends out the roots not the root itself (Butt and Sultan 2011). Ginger has a distinctive pungent flavor and contains a large number of compounds including carbohydrates, minerals, vitamins, enzymes as well as other phytochemicals. Gingerols, which is the pungent fraction of ginger, are one of the active constituents from lipophilic extracts and can be converted to other nonvolatile phenolic compounds such as shogaols, zingerone, and paradol (Ali et al. 2008; Butt and Sultan 2011). Despite common and widespread use of ginger, variations in chemical composition in regard to quantity and types of active ingredients are dependent on the place of origin and the rhizome form (i.e., fresh or dry) and therefore result in variation in their therapeutic effects (Schwertner and Rios 2007; Schwertner et al. 2006). For example, fresh ginger possesses a smaller amount of zingerone and shogaols in comparison with the dried or extracted ginger products. Of the numerous compounds reported in ginger, this chapter summarizes the major components that have been implicated in the pharmacological activities of ginger.

The major chemical constituents of ginger rhizome are classified into (1) essential volatile oils and (2) nonvolatile pungent compounds (Govindarajan 1982a, 1982b). The distinct aromatic odor and taste of ginger mainly derives from its volatile oil (1%–3%). Over 50% volatile oil components have been characterized, which mainly consist of various terpenoids including at least 40 different monoterpenoids (β-phellandrene, [+]-camphene, cineole, geraniol, neral, curcumene, citral, terpineol, borneol); sesquiterpenoids (α-zingeberene [35%], curcumene [18%], and α-farnesene [10%], with lesser amounts of β-bisabolene, β-sesquiphellandrene, zingiberol); and a smaller percentage of diterpenes such as galanolactone (Figure 26.1). Some of the oil components are converted into lesser odor-defining compounds during the drying process (Gong et al. 2004). However, the literature clearly shows that the composition of the ginger oil is highly variable based on the factors such

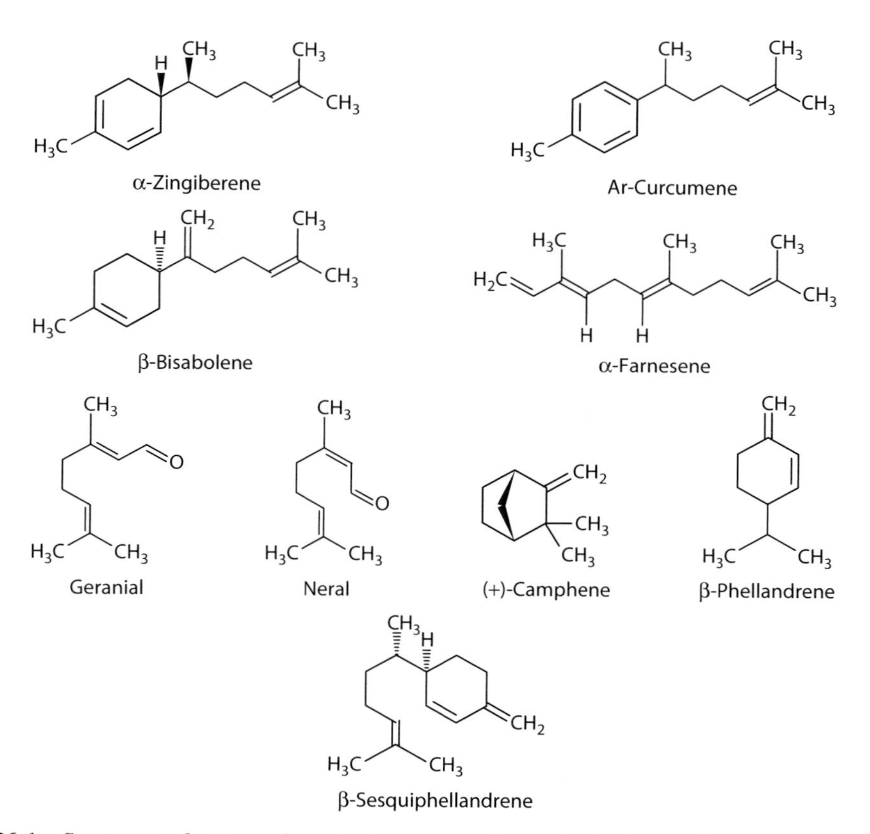

FIGURE 26.1 Structures of some major volatile compounds present in ginger rhizome (monoterpenes and sesquiterpenes).

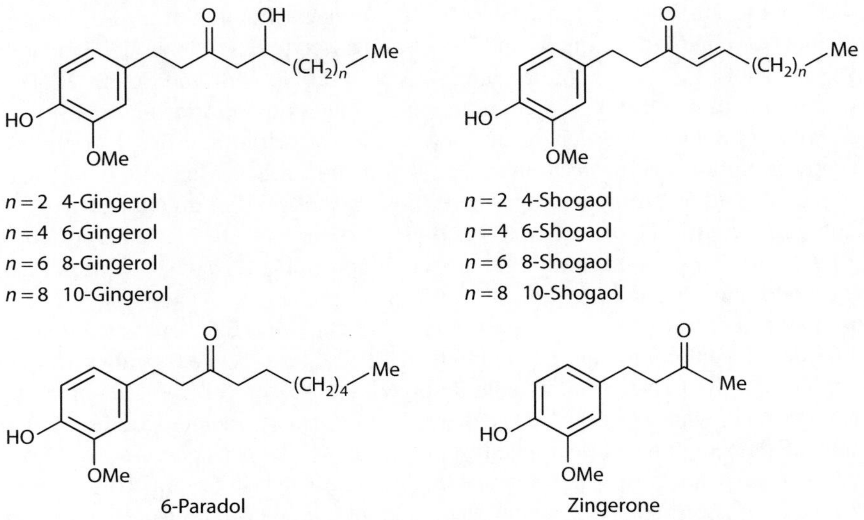

$n = 2$ 4-Gingerol
$n = 4$ 6-Gingerol
$n = 6$ 8-Gingerol
$n = 8$ 10-Gingerol

$n = 2$ 4-Shogaol
$n = 4$ 6-Shogaol
$n = 6$ 8-Shogaol
$n = 8$ 10-Shogaol

6-Paradol

Zingerone

FIGURE 26.2 Structures of some major nonvolatile pungent compounds present in ginger rhizome.

as its geographical origin, fresh or dried rhizome used, drying process and the temperature, and the analytical methodology used (Vernin and Parkanyi 2005).

The pungency of fresh ginger is primarily due to the nonvolatile compounds, which are a homologous series of phenols differing in their length of unbranched alkyl chains (Zhan et al. 2008). These include, as shown in Figure 26.2, gingerols (4-, 6-, 7-, 8-, and 10-gingerols), of which 6-gingerol is most abundant; shogaols (4-, 6-, 8-, 10-, and 12-shogaols), the dehydrated forms of gingerols of which 6-shogaol is most abundant; paradols (6-, 7-, 8-, 9-, 10-, 11-, and 13-paradols), which are 5-deoxygingerols; and zingerones, which are formed by degradation of gingerols and shogaols (Gong et al. 2004; Jiang et al. 2005; Zhan et al. 2008). Moreover, the conversion of gingerols to shogaols is found to be both heat and pH dependent (Bhattarai et al. 2007; Jolad et al. 2004; Wohlmuth et al. 2005). Thus, gingerols are the major nonvolatile constituents of fresh ginger, whereas shogaols are the major gingerol dehydrated products in dry ginger. Among the nonvolatile compounds, gingerols and shogaols are identified as the major ginger-derived pungent bioactive constituents contributing to the medical applications of ginger (Dedov et al. 2002; Jolad et al. 2004).

26.4 PHARMACOLOGICAL EFFECTS ASSOCIATED WITH METABOLIC SYNDROME

26.4.1 Antihyperglycemic Effects of Ginger

26.4.1.1 *In Vivo* Observations in Animal Studies

A number of *in vivo* studies have examined the efficacy of ginger in controlling hyperglycemia in the chemical- or diet-induced experimental animal models of metabolic syndrome and diabetes mellitus. Investigation of the active principles and mechanisms of action have also been undertaken. Type 1 or insulin-dependent diabetes mellitus, characterized by the lack of insulin production due to complete destruction of the β cells of pancreas, can only be controlled by insulin therapy. Thus, optimal functioning of pancreatic β cells is essential for the regulation of glucose homeostasis and its impairment leads to the development of diabetes (Kaul et al. 2012). Acute administration of an ethanolic extract of ginger (50–800 mg/kg; po) caused a concentration-dependent hypoglycemic effect in normoglycemic and streptozotocin (STZ)-induced type 1 diabetic rats (Ojewole 2006). The effect peaked after four hours, with ginger producing a 16%–53% reduction in blood glucose at the doses studied. Another study showed that administering a single dose of fresh juice of ginger (4 mL/kg; po) for six weeks prevented 5-hydroxytryptamine (5-HT)-induced acute hyperglycemia suggesting a 5-HT receptor's antagonistic effect of ginger (Akhani et al. 2004). In the oral glucose tolerance test, ginger treatment caused a significant decrease and increase in the area under the curves of blood glucose and insulin, respectively, in STZ-induced diabetic rats, suggesting an insulinotropic effect of ginger (Akhani et al. 2004). Studies have also shown that both ethanolic (200 mg/kg for 20 days; po) and aqueous (500 mg/kg for seven weeks; ip) extracts of ginger significantly lowered serum glucose and increased the insulin levels in STZ-induced diabetic rats, indicating that both polar and nonpolar compounds present in ginger possess antihyperglycemic effects (Al-Amin et al. 2006; Bhandari et al. 2005). A more recent study also confirmed a dose-dependent decrease in plasma glucose after chronic administration of an aqueous extract (100, 300, and 500 mg/kg for 30 days; po) of ginger in STZ-induced diabetic rats (Abdulrazaq et al. 2012). In the same study, ginger (500 mg/kg) significantly decreased kidney glycogen and increased liver and skeletal muscle glycogen in STZ-induced diabetic rats. Activities of glucokinase, phosphofructokinase, and pyruvate kinase were significantly increased with ginger treatment (Abdulrazaq et al. 2012).

Type 2 or non-insulin-dependent diabetes mellitus (NIDDM) is a chronic metabolic disorder caused by insulin resistance and/or relative insulin deficiency. It is also associated with low physical activity and high energy intake (Ganti et al. 1999; Roberts et al. 2013). In a preclinical study, the 6-gingerol-rich methanolic extract but not 6-gingerol-poor ethyl acetate extract of ginger (200 mg/kg for seven weeks) was effective in reducing elevated blood glucose levels, insulin levels, and body weight in fructose-induced type 2 diabetic rats (Kadnur and Goyal 2005). Studies in a combined high-fat diet (HFD) and STZ-induced type 2 diabetic rats also showed that freeze-dried ginger (0.5% and 2% for four weeks) resulted in a better glucose tolerance and significant insulinotropic effects (Islam and Choi 2008). Furthermore, the pungent active principle of 6-gingerol (100 mg/kg for two weeks) caused a significant decrease in fasting blood glucose and improved glucose tolerance in genetically programmed db/db type 2 diabetic mice (Singh et al. 2009). Studies from our laboratory on HFD-fed rats have demonstrated a protective effect of ginger in the development of various parameters of metabolic syndrome, a condition that predisposes to a high risk of type 2 diabetes (Nammi et al. 2009). A 6-gingerol-rich ethanolic extract of ginger (100, 200, and 400 mg/kg for six weeks; po) significantly reduced the marked rise of serum glucose, insulin, and homeostatic model assessment–insulin resistance index (HOMA–IR) induced by HFD in rats (Nammi et al. 2009). Oral administration of ginger powder (200 mg/kg for four weeks) significantly alleviated the signs of metabolic syndrome, including decrease of blood glucose and total lipid, and increase in total antioxidant level in nicotinamide and low-dose STZ-induced type 2 diabetic rats (Madkor et al. 2011). An *in vivo* mechanistic study conducted on rats showed that chronic treatment as opposed to acute treatment with ginger (50 mg% for eight weeks; po) significantly enhanced the activities of pancreatic lipase, amylase, trypsin, and chymotrypsin (Platel and Srinivasan 2000). In arsenic-induced type 2 diabetic rats, 6-gingerol treatment reduced blood glucose, increased plasma insulin levels, and protected pancreatic β cells from arsenite-induced impaired insulin signaling and oxidative damage (Chakraborty et al. 2012).

The effects of ginger on blood glucose in normal animals were inconsistent. The juice (Sharma and Shukla 1977), the whole powder (Ahmed and Sharma 1997), and the ethanolic extracts (Mascolo et al. 1989; Ojewole 2006) of ginger showed a low-to-moderate but not significant blood glucose-lowering effect in normal animals. However, acute treatment of 1.9% gingerol-containing ethanolic ginger extract (25, 50, and 100 mg/kg; po) failed to show any effect on blood glucose in normal rats after three hours post dose (Weidner and Sigwart 2000). In consistent with the later findings, chronic treatment with fresh juice of ginger (4 mL/kg for six weeks; po) did not alter blood insulin. By contrast, an elevation of blood glucose in normal rats has been observed with ginger powder (Singhal and Joshi 1983).

26.4.1.2 *In Vitro* Observations in Cell Culture Studies

Although gingerols are the major biologically active pungent compounds in ginger, direct evidence for the action mechanisms of gingerols in improving glucose homeostasis has not been reported until recently. In an *in vitro* study, 6-gingerol increased the protein and gene expressions of glucose transporter type 4 (GLUT4), insulin receptor substrate (IRS)-1, IRS-2, phosphatidylinositol 3-kinase (PI3K), Akt, and peroxisome proliferator-activated receptor (PPAR)-r signaling molecules in arsenite-intoxicated cultured β cells and hepatocytes (Chakraborty et al. 2012). In addition, ginger extract increased insulin release in insulin 1 (INS-1) insulin-secreting cells and this effect became more prominent in the presence of exogenous serotonin (Heimes et al. 2009). *In vivo* glucose tolerance test further confirmed that this ginger extract also increased plasma insulin with a concomitant decrease in blood glucose. The underlying mechanism of this action may, in part, be due to the interaction of gingerols and shogaols with serotonin receptors (Abdel-Aziz et al. 2006).

Evidence from *in vitro* studies demonstrated that ginger extract and its pungent gingerol principles promote glucose clearance in insulin-responsive peripheral tissues to maintain glucose homeostasis in the blood. Ginger extract and its active constituent, 6-gingerol, were found to enhance adipocyte differentiation and insulin-sensitive glucose uptake in 3T3-L1 adipocytes, indicating that ginger enhanced insulin sensitivity (Noipha et al. 2010; Sekiya et al. 2004). In a recent study, the efficacy and mechanisms of antidiabetic potential of ginger and its bioactive compounds have been investigated in a number of *in vitro* studies using cell culture and cell-free systems. The 6-gingerol- and 6-shogaol-rich ethyl acetate extract of ginger at concentrations ranging between 5 and 50 μg/mL enhanced glucose uptake and GLUT4 receptor expression in L6 mouse myotube cells. The extract also suppressed low-density lipoprotein (LDL) oxidation, inhibited protein glycation, and reduced lipid content in 3T3-L1 adipocytes (Rani et al. 2011, 2012). In a more recent study, the ginger extract rich in 6- and 8-gingerols stimulated glucose uptake significantly in cultured rat L6 skeletal muscle cells and attributed to increased translocation of GLUT4 glucose transporter to the cell membrane, together with increased GLUT4 protein expression (Li et al. 2012b).

The key enzymes controlling carbohydrate metabolism in the gut are salivary and pancreatic α-amylase and intestinal α-glucosidase. Thus, these enzymes are important targets to reduce postprandial hyperglycemia and in the treatment of type 2 diabetes (Tundis et al. 2010). *In vitro* enzyme inhibition study showed that hexane, ethyl acetate, methanol, 70% methanol–water, and aqueous extracts of ginger were effective in inhibiting these enzymes in cell-free assays (Rani et al. 2011). The ethyl acetate extracts showed the highest α-glucosidase and α-amylase inhibitory effects and correlated with the gingerol and shogaol contents in these extracts (Rani et al. 2011). However, an aqueous extract of ginger showed an inhibitory effect only on α-glucosidase but not on α-amylase and attributed to the low phenolic contents present in the extract (Ranilla et al. 2010).

26.4.2 Antihyperlipidemic Effects of Ginger

It is well established that insulin resistance in the peripheral tissues is tightly associated with elevated circulating lipids and tissue lipid accumulation. The resulting atherogenic dyslipidemia, characterized by increased levels of triglycerides, total cholesterol, and LDL cholesterol, and accompanied by the low levels of high-density lipoprotein (HDL) cholesterol, is an important biochemical abnormality of diabetes mellitus that predisposes to CVDs (McGarry 2002). Furthermore, the LDL particles in diabetic patients tend to be smaller and denser, and therefore more atherogenic. Hypertension also frequently accompanies central obesity, insulin resistance, and dyslipidemia, and the cluster of risk factors termed "metabolic syndrome" (Raz et al. 2005).

A number of studies have demonstrated that ginger possessed prominent lipid-lowering effects and subsequently increased insulin sensitivity. In laboratory experiments, chronic treatment with the ethanolic extract (200 mg/kg for 10 weeks; po) of ginger to cholesterol-fed hyperlipidemic rabbits significantly reduced serum triglycerides, serum lipoproteins, and phospholipids. Ginger also reduced the severity of aortic atherosclerosis (Bhandari et al. 1998). In another study, chronic administration of air dried ginger powder (0.1 g/kg for 75 days; po) to cholesterol-fed hyperlipidemic rabbits significantly reduced the development of atherosclerosis in the aorta and coronary arteries. This effect was associated with decreased lipid peroxidation and enhanced the fibrinolytic activity and prostaglandin inhibition with ginger, but blood lipids were not affected (Verma et al. 2004). Studies have also shown that both ethanolic (200 mg/kg for 20 days; po) and aqueous (500 mg/kg for seven weeks; ip) extracts of ginger significantly lowered serum triglyceride and total cholesterol, and increased HDL cholesterol levels in STZ-induced diabetic rats (Al-Amin et al. 2006; Bhandari et al. 2005). Studies have also shown that chronic treatment with both the 6-gingerol-rich methanolic extract and the 6-gingerol-poor ethyl acetate extract of ginger (200 mg/kg for seven weeks) was effective in reducing fructose-induced elevated blood lipids and body weight in type 2 diabetic rats (Kadnur and Goyal 2005). In parallel studies, chronic treatment of aqueous ginger extract (1% and 3% for eight weeks; in diet) to HFD-fed mice showed antiobesity effects with marked reduction in body weight and *in vitro* pancreatic lipase activity and increased fecal excretion of cholesterol, suggesting that ginger may block the absorption of cholesterol in the gut (Han et al. 2005). Similarly, administration of both the methanolic and ethanolic ginger extracts (250 mg/kg for eight weeks; po) improved insulin sensitivity in gold thioglucose-induced obese mice (Goyal and Kadnur 2006).

In our laboratory, administering a 6-gingerol-rich ethanolic extract of ginger (100, 200, and 400 mg/kg for six weeks; po) to HFD-fed rats significantly protected from rise in body weight, serum triglycerides, total cholesterol, LDL cholesterol, free fatty acids, and phospholipid caused by HFD (Nammi et al. 2009). Ginger (400 mg/kg) also effectively reduced triglyceride and cholesterol levels in the liver. Mechanistic studies have further shown that ginger attains its hypolipidemic effect by promoting uptake and catabolism of circulating LDL through upregulation of LDL receptor gene and protein expression in the liver and by repressing cholesterol biosynthesis through downregulation of 3-hydroxy-3-methylglutaryl-CoA (HMG-CoA) reductase protein expression in the liver (Nammi et al. 2010). Studies in atherosclerotic apolipoprotein E-deficient mice have shown that chronic treatment of ginger (25 and 250 μg/day for 10 weeks; po) was effective in reducing plasma triglycerides, total and LDL cholesterol, and the level of aortic atherosclerotic lesion by inhibiting LDL oxidation, and inhibited the rate of cholesterol biosynthesis (Fuhrman et al. 2000). This observation supported an earlier observation that an active component of ginger (50, 100, and 200 mg/kg; po) when administered to triton-induced hypercholesterolemic mice significantly lowered serum cholesterol and inhibited cholesterol biosynthesis in the liver (Tanabe et al. 1993).

In addition, an aqueous extract of ginger (50 and 500 mg/kg for four weeks; po and ip) reduced the serum levels of cholesterol, thromboxane B_2, and prostaglandin E_3 but not triglycerides in normal rats (Thomson et al. 2002). Although fresh juice (4 mL/kg for six weeks; po) of ginger showed no change in plasma lipids in normal rats (Akhani et al. 2004), dietary feeding of normal rats with 0.5% w/w ginger significantly reduced serum total cholesterol and HDL cholesterol but not serum triglyceride levels (Ahmed and Sharma 1997). Earlier studies in rats demonstrated that ginger induces bile acid synthesis through increased hepatic cholesterol-7α-hydroxylase activity (a rate-limiting enzyme of bile acid biosynthesis), which stimulates the conversion of hepatic cholesterol to bile acids, an important pathway for cholesterol elimination from the body (Srinivasan and Sambaiah 1991). Additionally, supplementing ginger powder (0.5 g per rat per day for 12 weeks; po) in a cholesterol-enriched diet improved lipid metabolism by decreasing the expression of retinoid-binding protein (RBP) and fatty acid-binding protein (FABP) genes in the liver and adipose tissue of rats (Matsuda et al. 2009). Furthermore, administration of aqueous infusion of ginger (100, 200, and 400 mg/kg for four weeks; po) to cholesterol-fed rats significantly decreased all the lipid profile parameters and improved the risk ratio of total cholesterol to HDL cholesterol (Elrokh El et al. 2010).

26.4.3 Protective Effect on Diabetic Complications

Diabetes affects carbohydrate, fat, and protein metabolism, which consequently leads to severe complications of the vital organs. Chronic hyperglycemia leads to retinopathy, neuropathy, and nephropathy, and is more fatal than the primary disease (Brownlee 2001; Campos 2012). Therefore, efforts are always at reducing the secondary complications.

26.4.3.1 Protection against Diabetic Retinopathy

The activity of aldose reductase, the enzyme that catalyses the formation of fructose from glucose in the sorbitol (polyol) pathway, is in greater concentrations in the non-insulin-sensitive cells such as the lens, the peripheral nerves, and the glomerulus of people with diabetes. Additionally, as sorbitol does not diffuse easily through cell membranes, the accumulated sorbitol causes osmotic damage and leads to diabetic retinopathy and cataract (Pollreisz and Schmidt-Erfurth 2010; Ramana 2011). Two compounds, 2-(4-hydroxy-3-methoxyphenyl) ethanol and 2-(4-hydroxy-3-methoxyphenyl) ethanoic acid, were identified to be effective even preventing galactose-induced cataractogenesis in rats, indicating that the *in vitro* observations extended to the animal models (Kato et al. 2006). In addition, these compounds of ginger inhibited human recombinant aldose reductase, suppressed the accumulation of sorbitol in human erythrocytes, and reduced the accumulation of galactitol in the lens (Kato et al. 2006).

Chronic hyperglycemia leads to nonenzymatic glycosylation of proteins and lipoproteins, thereby leading to the formation of advanced glycation end products (AGEs). The AGEs formed alter the cell physiology, generate the reactive oxygen intermediates, and cause diabetic microvascular and macrovascular complications (Brownlee 2001; Zarina et al. 2000). An *in vitro* test showed that an aqueous extract of ginger at 0.1 and 1.0 mg/mL reduced chemical-derived AGEs (Saraswat et al. 2009). Administration of aqueous extract of ginger (0.5% or 3% in the diet for two months) showed antiglycating activity in STZ-induced diabetic rats and inhibited the formation of fructose-mediated AGEs of eye lens-soluble proteins *in vitro* together with delayed progression and onset of cataract (Saraswat et al. 2009, 2010).

26.4.3.2 Protection against Diabetic Nephropathy

The hyperglycemia-induced kidney damage is a major secondary complication in diabetes and follows a well-outlined clinical course from microalbuminuria to proteinuria to azotema to the final renal failure (Satirapoj 2012; Sun et al. 2013). Preclinical studies have shown that chronic administration of ginger extract (500 mg/kg for seven weeks; ip) to STZ-induced diabetic rats significantly decreased blood glucose and protein clearance levels. The histological observations clearly revealed that ginger effectively attenuated the progression of structural nephropathy in diabetic rats. Ginger significantly reduced the capsular space shrinkage, glomerular hypertrophy and diffusion, glomerular and microvascular eosinophilic precipitation, and cytoplasm fragmentation and retraction confirming the renoprotective effects of ginger (Al-Qattan et al. 2008). In a renal ischemia/reperfusion model, ginger supplementation (5%) resulted in high antioxidant capacity and experienced less kidney damage due to oxidative stress induced by ischemia/reperfusion (Uz et al. 2009). In another study, chronic treatment with ethanolic extract of ginger (100 and 200 mg/kg for 30 days; po) significantly restored the levels of uric acid and antioxidant enzymes—superoxide dismutase (SOD), ascorbic acid, and glutathione—and the activities of xanthine oxidase and glutathione *S*-transferase in alcohol-induced renal damage in rats. Histopathological examination revealed restoration of alcohol-induced tubular congestion and degeneration confirming renoprotection with ginger (Ramudu et al. 2011a; Shanmugam et al. 2010). The same group further investigated the renal cytosolic and mitochondrial enzymatic activities. Ginger treatment (200 mg/kg for 30 days) increased the activities of glucose 6-phosphate dehydrogenase (G6PD), succinate dehydrogenase (SDH), malate dehydrogenase (MDH), and glutamate dehydrogenase (GDH) with concomitant decrease in lactate dehydrogenase in diabetic rats (Ramudu et al. 2011b).

26.4.3.3 Protection against Diabetic Neuropathy

Scientific studies have proved that chronic hyperglycemia, insulin resistance, oxidative stress, AGEs, inflammatory cytokines, and microvascular and macrovascular disease contribute toward decrease in cognitive function, increase in neurodegeneration, and the risk of developing dementia (Campos 2012). Recent studies have indicated that administering ethanolic extract of ginger to STZ-induced diabetic rats increased SOD, catalase (CAT), glutathione peroxidase (GPx), glutathione reductase (GR), reduced glutathione (GSH), and concomitantly decreased malondialdehyde (MDA) in the cerebral cortex, the cerebellum, the hippocampus, and the hypothalamus, suggesting that ginger exhibited neuroprotective effect by accelerating brain antioxidant defense mechanisms and downregulating the MDA (Shanmugam et al. 2011).

26.4.4 Anti-Inflammatory Effects of Ginger

Recent studies have indicated that chronic low-grade inflammation plays a key role in the pathogenesis of diabetes and that certain proinflammatory cytokines and transcription factors (nuclear factor kappa-light-chain-enhancer of activated B cells [NF-κB]) are involved in the development of diabetic complications (Kahn et al. 2006). Ginger is reported to be effective in preventing inflammation by modulating the key proinflammatory genes and transcription factors responsible for destruction of β cells in type 1 diabetes and by mitigating the inflammatory reactions that contribute to the diabetic complications in type 2 diabetes. Preclinical studies with gingerol in cultured cell lines (Aggarwal and Shishodia 2004) and animal (aged) studies with another component of ginger, zingerone (Kim et al. 2010), have also shown that these suppress the activation of NF-κB and the proinflammatory enzymes such as cyclooxygenase (COX)-2 and inducible nitric oxide synthase (iNOS), which upregulate through NF-κB activation and IκB kinase/mitogen-activated protein (MAP) kinases (IKK/MAPK) signaling pathway. Zingerone was also found to suppress the inflammatory response of adipose tissue in obesity by suppressing the inflammatory action of macrophages and release of monocyte chemotactic protein-1 (MCP-1) (Woo et al. 2007). Investigations in our laboratory revealed that an ethanolic extract of ginger (400 mg/kg for six weeks; po) suppressed the gene and protein expression of hepatic inflammatory markers, tumor necrosis factor (TNF)-α, and interleukin (IL)-6, and decreased the NF-κB activity (Li et al. 2012a). Thus, these studies also suggest that ginger has the potential to prevent obesity and obesity-linked metabolic effects.

26.4.5 Antioxidant Potential of Ginger

Oxidative stress resulting from an imbalance of increased free radical generation and decreased antioxidants is implicated in the pathogenesis of many diseases including metabolic disorders (Yang et al. 2011). Chronic hyperglycemia is known to increase the generation of free radicals due to autoxidation of glucose, nonenzymatic glycosylation of body proteins, and stimulation of polyol pathway. This leads to depletion of antioxidant defense systems and also progressively leads to the development of diabetic complications (De Bandeira et al. 2013; Nawroth et al. 1999). Numerous studies have shown that ginger and its constituents gingerol, zingerone, and shogaol possess free radical-scavenging properties against 2,2-diphenyl-1-picrylhydrazyl (DPPH), 2,2′-azino-bis(3-ethylbenzthiazoline-6-sulphonic acid) (ABTS), and hydroxyl, superoxide, nitric oxide, and peroxyl radicals *in vitro* (Aeschbach et al. 1994; El-Ghorab et al. 2010; Krishnakantha and Lokesh 1993; Reddy and Lokesh 1992). Ginger was reported to decrease age-related oxidative stress markers and suppress oxidative consequences in ethanol-induced hepatotoxicity in rats (Mallikarjuna et al. 2008; Topic et al. 2002). Ginger was reported to suppress 12-*O*-tetradecanoylphorbol-13-acetate (TPA)-induced oxidative stress in differentiated HL-60 and AS52 cells (Kim et al. 2002). Several reports indicate that ginger suppresses lipid peroxidation and protects the levels of GSH *in vivo* (Ahmed et al. 2000a, 2000b, 2008; El-Sharaky et al. 2009; Shobana and Naidu 2000). Recent studies have also shown that administering ginger to diabetic rats increased the levels of antioxidants

GSH and enzymes SOD, CAT, GPx, and GR, and concomitantly reduced the elevated levels of malondialdehyde in the liver (Shanmugam et al. 2011), brain (Shanmugam et al. 2011), and kidney (Afshari et al. 2007) tissues compared to concurrent diabetic controls. A group of studies further showed that the ginger constituents 6-gingerol and 6-shogaol inhibit nitric oxide and peroxynitrite production and reduce iNOS in lipopolysaccharide (LPS)-induced oxidative damage (Ippoushi et al. 2003, 2005; Koh et al. 2009). These findings conclusively show that ginger exerts a protective effect in diabetes by decreasing oxidative stress and promoting antioxidants in tissues.

26.4.6 Anti-Infertility Effect of Ginger

Diabetes mellitus is adversely linked with fertility in both males and females (Downs et al. 2010). Compared with people free of diabetes, the onset of menarche is late but menopause is earlier in women with diabetes. Diabetic women are also more prone to polycystic ovarian syndrome, the most common cause of female infertility, and also have a greater tendency for abortions (Cardozo et al. 2011). In diabetic males, spermatogenesis is impaired and the sperms are aberrant. This affects the quality and quantity of the sperms and fertility (La Vignera et al. 2012; Sermondade et al. 2012). Studies have shown that chronic administration of methanolic (100 and 200 mg/kg for 65 days; po), aqueous (150 and 300 mg/kg for 65 days; po), or 90% ethanolic (250 and 500 mg/kg for 65 days; po) extracts of ginger increased the fertility index, the weight of sexual organs, the level of serum testosterone, the semen quality and quantity, the sperm count, and the motility in alloxan-induced diabetic rats. Histopathological examination of the testes showed that ginger caused reduction in diabetes-induced degenerative changes of spermatogenic cells, diffuse edema, and incomplete arrest of spermatogenesis (Shalaby and Hamowieh 2010). On the contrary, an *in vitro* study with ginger methanolic extract (0.1%–0.6%) showed a dose-dependent decrease in sperm motility and concomitant decrease in grading and changes in sperm morphology (Jorsaraei et al. 2008). A recent study has shown that chronic ginger treatment (40 mg/kg; po) alleviated aluminum chloride-induced reproductive toxicity in male rats (Moselhy et al. 2012).

26.5 SAFETY AND TOXICITY PROFILE OF GINGER

Ginger is used as a food additive and is generally considered safe. Its effects for easing nausea and vomiting are documented in major pharmacopeias and have been recognized by the Food and Drug Administration authorities in a number of countries. Toxicological studies showed a broad safety range for ginger usage. While the acute toxicity (median lethal dose [LD50]) values of the major pungent compounds in ginger, 6-gingerol and 6-shogaol, were 250 and 687 mg/kg, respectively, the LD50 value of ethanolic ginger extract after intraperitoneal administration was found to be 1.55 g/kg in mice (Ojewole 2006; Suekawa et al. 1984). However, the acute LD50 after oral administration of methanolic and aqueous ginger extract in mice was 10.25 and 11.75 g/kg, respectively (Shalaby and Hamowieh 2010).

The teratogenic effects of ginger were studied in pregnant rats. Oral administration of an ethanolic ginger extract (1000 mg/kg) was well tolerated by pregnant rats and exerted no adverse effects on the mothers or in the developing fetuses (Weidner and Sigwart 2001). This result is in contrast to an earlier observation that administration of ginger tea to pregnant rats resulted in loss of embryos and heavier surviving fetuses (Wilkinson 2000). Furthermore, ginger extract (0.5–10 g/kg; ip) administered to mice showed no clastogenic effects compared to ginger oil, which produced some chromosomal irregularities (Mukhopadhyay and Mukherjee 2000).

Observational studies in humans suggest no evidence of teratogenicity from ginger treatments for nausea and vomiting in early pregnancy (Jewell and Young 2003). The results were confirmed in a similar multicentered, double-blind, randomized clinical trial, showing that the administration of ginger (1–1.5 g/day) during pregnancy did not show an increased risk of major malformations or other birth defects (Borrelli et al. 2005; Portnoi et al. 2003).

In a recent report, the administration of ginger powder (500, 1000, or 2000 mg/kg; po) for 35 days did not cause any overall mortality or abnormality of the general condition or hematological parameters in either male or female rats (Rong et al. 2009). All blood biochemical parameters remained unchanged with exception of the serum lactate dehydrogenase level and a slight but significant decrease in absolute and relative weight of the testes in the highest dose-treated male rats (Rong et al. 2009). However, treating diabetic rats with ginger extract for 65 days enhanced the fertility index and sexual organ weight (Shalaby and Hamowieh 2010). In sum, these data indicate that ginger consumption appears to be very safe with very limited side effects.

26.6 CLINICAL STUDIES OF GINGER

Despite the strong preclinical evidence of ginger on lipid lowering and glycemic control, the use of ginger as a potential management option for the metabolic syndrome has not been widely explored in clinical settings. Studies to date have been generally small in sample size and poor in methodology, and the evidence obtained from these clinical trials is somewhat controversial. This may be due to the variations in sample population, differences in preparation of ginger products, and the amount and type of active constituents present within the ginger products.

A placebo-controlled clinical trial conducted by Alizadeh-Navaei et al. (2008) assessed the efficacy of 3 g of ginger powder per day over a period of 45 days in 95 hyperlipidemia subjects. The results showed that both treatment and placebo groups had a significant reduction in triglyceride, cholesterol, LDL, and very-LDL (VLDL) between the baseline and the posttreatment. However, the mean changes of triglycerides, cholesterol, and HDL in the ginger group were significantly higher than those in the placebo group. These results were in contrast to those of Atashak et al. (2010) who assessed a 10-week resistance training program in addition to administration of ginger on C-reactive protein (CRP) levels, obesity markers, and lipid profiles in 32 obese men. There were no changes on lipid profiles and insulin resistance in participants who received the ginger treatment. Reductions in CRP levels were significant in both resistance training groups (with ginger and placebo) and the ginger-only group (Atashak et al. 2010).

Another placebo-controlled trial conducted by Bordia et al. (1997) investigated the effects of ginger in 20 patients with coronary artery disease (CAD) who received 4 g of ginger powder daily over three months. The study found no effects of gingeron adenosin diphosphate and epinephrine-induced platelet aggregation compared to the placebo. However, in a separated group of 10 patients, a single dose of 10 g ginger treatment caused a significant reduction in platelet aggregation induced by adenosin diphosphate and epinephrine. This result was corroborated by a study conducted by Verma and Bordia (2001) where the fibrinolytic activities of ginger were investigated in 30 healthy adults with high fat consumption. The participants randomly received 50 g of fat with and without the administration of a single 5 g dose of ginger powder. The results demonstrated that the ginger treatment significantly enhanced the fibrinolytic effects.

Clinical studies were also conducted with ginger as part of complex herbal formula in humans with obesity. A pilot study conducted by Roberts et al. (2007) investigated a dietary supplement containing rhubarb, ginger, astragulus, red sage, and turmeric for the reduction of food intake. The study found that the formula had no significant effects on weight reduction. In another study, which assessed the effects of panchratna juice (amla, tulsi, ginger, mint, and turmeric) for the management of type 2 diabetes (Iyer et al. 2010), no significant effects were found on glycemic status, anthropometric profile, or lipid profiles in patients after 45 days of fresh juice or 90 days of processed juice.

A pharmacokinetic study of the active constituents of ginger including 6-, 8-, and 10-gingerols and 6-shogaol in healthy adults reported that free 10- and 6-gingerols were detected in plasma from participants who received 2 g of ginger extracts with peak concentrations of 9.5 ± 2.2 and 13.6 ± 6.9 ng/mL, respectively, at one hour after administration. The plasma concentrations of 6-, 8-, and 10-gingerol and 6-shagaol conjugates were also detected and peaked at one hour. The half-life of the analytes was one to three hours in plasma (Yu et al. 2011).

26.7 CONCLUSION

Scientific evidence from laboratory-based research exists to support the ginger's effects on metabolic syndrome. However, the results from clinical studies appear to be weak and inconsistent. The clinical trials conducted were generally poor in methodology. Randomized controlled trials with rigorous design are warranted to evaluate the effectiveness and safety of ginger on metabolic syndrome.

REFERENCES

Abdel-Aziz, H., Windeck, T., Ploch, M., and Verspohl, E.J. 2006. Mode of action of gingerols and shogaols on 5-HT3 receptors: Binding studies, cation uptake by the receptor channel and contraction of isolated guinea-pig ileum. *European Journal of Pharmacology*, 530: 136–143.

Abdulrazaq, N.B., Cho, M.M., Win, N.N., Zaman, R., and Rahman, M.T. 2012. Beneficial effects of ginger (*Zingiber officinale*) on carbohydrate metabolism in streptozotocin-induced diabetic rats. *The British Journal of Nutrition*, 108: 1194–1201.

Aeschbach, R., Loliger, J., Scott, B.C., Murcia, A., Butler, J., Halliwell, B., and Aruoma, O.I. 1994. Antioxidant actions of thymol, carvacrol, 6-gingerol, zingerone and hydroxytyrosol. *Food and Chemical Toxicology*, 32: 31–36.

Afshari, A.T., Shirpoor, A., Farshid, A., Saadatian, R., Rasmi, Y., Saboory, E., Ilkhanizadeh, B., and Allameh, A. 2007. The effect of ginger on diabetic nephropathy, plasma antioxidant capacity and lipid peroxidation in rats. *Food Chemistry*, 101: 148–153.

Aggarwal, B.B. and Shishodia, S. 2004. Suppression of the nuclear factor-kappaB activation pathway by spice-derived phytochemicals: Reasoning for seasoning. *Annals of the New York Academy of Sciences*, 1030: 434–441.

Ahmed, R.S., Seth, V., and Banerjee, B.D. 2000a. Influence of dietary ginger (*Zingiber officinales* Rosc.) on antioxidant defense system in rat: Comparison with ascorbic acid. *Indian Journal of Experimental Biology*, 38: 604–606.

Ahmed, R.S., Seth, V., Pasha, S.T., and Banerjee, B.D. 2000b. Influence of dietary ginger (*Zingiber officinales* Rosc.) on oxidative stress induced by malathion in rats. *Food and Chemical Toxicology*, 38, 443–450.

Ahmed, R.S. and Sharma, S.B. 1997. Biochemical studies on combined effects of garlic (*Allium sativum* Linn.) and ginger (*Zingiber officinale* Rosc.) in albino rats. *Indian Journal of Experimental Biology*, 35: 841–843.

Ahmed, R.S., Suke, S.G., Seth, V., Chakraborti, A., Tripathi, A.K., and Banerjee, B.D. 2008. Protective effects of dietary ginger (*Zingiber officinales* Rosc.) on lindane-induced oxidative stress in rats. *Phytotherapy Research*, 22: 902–906.

Akhani, S.P., Vishwakarma, S.L., and Goyal, R.K. 2004. Anti-diabetic activity of *Zingiber officinale* in streptozotocin-induced type I diabetic rats. *Journal of Pharmacy and Pharmacology*, 56: 101–105.

Al-Amin, Z.M., Thomson, M., Al-Qattan, K.K., Peltonen-Shalaby, R., and Ali, M. 2006. Anti-diabetic and hypolipidaemic properties of ginger (*Zingiber officinale*) in streptozotocin-induced diabetic rats. *The British Journal of Nutrition*, 96: 660–666.

Alberti, K., Zimmet, P., and Shaw, J. 2005. The metabolic syndrome—A new worldwide definition. *Lancet*, 366: 1059.

Ali, B.H., Blunden, G., Tanira, M.O., and Nemmar, A. 2008. Some phytochemical, pharmacological and toxicological properties of ginger: A review of recent research. *Food and Chemical Toxicology*, 46: 409–420.

Alizadeh-Navaei, R., Roozbeh, F., Saravi, M., Pouramir, M., Jalali, F., and Moghadamnia, A.A. 2008. Investigation of the effect of ginger on the lipid levels. *Saudi Medical Journal*, 29: 1280–1284.

Al-Qattan, K., Thomson, M., and Ali, M. 2008. Garlic (*Allium sativum*) and ginger (*Zingiber officinale*) attenuate structural nephropathy progression in streptozotocin-induced diabetic rats. *European e-Journal of Clinical Nutrition and Metabolism*, 3: e62–e71.

Atashak, S., Peeri, M., and Jafari, A. 2010. Effects of 10 week resistance training and ginger consumption on C-reactive protein and some cardiovascular risk factors in obese men. *Physiology and Pharmacology*, 14: 318–328.

Bensky, D. 2004. *Chinese Herbal Medicine: Materia Medica*. Seattle, WA: Eastland Press.

Bhandari, U., Kanojia, R., and Pillai, K.K. 2005. Effect of ethanolic extract of *Zingiber officinale* on dyslipidaemia in diabetic rats. *Journal of Ethnopharmacology*, 97: 227–230.

Bhandari, U., Sharma, J.N., and Zafar, R. 1998. The protective action of ethanolic ginger (*Zingiber officinale*) extract in cholesterol fed rabbits. *Journal of Ethnopharmacology*, 61, 167–171.

Bhattarai, S., Tran, V.H., and Duke, C.C. 2007. Stability of [6]-gingerol and [6]-shogaol in simulated gastric and intestinal fluids. *Journal of Pharmaceutical Biomedical Analysis*, 45: 648–653.

Bordia, A., Verma, S., and Srivastava, K. 1997. Effect of ginger (*Zingiber officinale* Rosc.) and fenugreek (*Trigonella foenumgraecum* L.) on blood lipids, blood sugar and platelet aggregation in patients with coronary artery disease. *Prostaglandins, Leukotrienes, and Essential Fatty Acids*, 56: 379–384.

Borrelli, F., Capasso, R., Aviello, G., Pittler, M.H., and Izzo, A.A. 2005. Effectiveness and safety of ginger in the treatment of pregnancy-induced nausea and vomiting. *Obstetrics and Gynecology*, 105: 849–856.

Brownlee, M. 2001. Biochemistry and molecular cell biology of diabetic complications. *Nature*, 414: 813–820.

Butt, M.S. and Sultan, M.T. 2011. Ginger and its health claims: Molecular aspects. *Critical Reviews in Food Science and Nutrition*, 51: 383–393.

Campos, C. 2012. Chronic hyperglycemia and glucose toxicity: Pathology and clinical sequelae. *Postgraduate Medicine*, 124: 90–97.

Cardozo, E., Pavone, M.E., and Hirshfeld-Cytron, J.E. 2011. Metabolic syndrome and oocyte quality. *Trends in Endocrinology and Metabolism*, 22: 103–109.

Chakraborty, D., Mukherjee, A., Sikdar, S., Paul, A., Ghosh, S., and Khuda-Bukhsh, A.R. 2012. [6]-Gingerol isolated from ginger attenuates sodium arsenite induced oxidative stress and plays a corrective role in improving insulin signaling in mice. *Toxicology Letters*, 210: 34–43.

De Bandeira, M.S., Da Fonseca, L.J.S., Da Guedes, S.G., Rabelo, L.A., Goulart, M.O., and Vasconcelos, S.M.L. 2013. Oxidative stress as an underlying contributor in the development of chronic complications in diabetes mellitus. *International Journal of Molecular Science*, 14: 3265–3284.

Dedov, V.N., Tran, V.H., Duke, C.C., Connor, M., Christie, M.J., Mandadi, S., and Roufogalis, B.D. 2002. Gingerols: A novel class of vanilloid receptor (VR1) agonists. *The British Journal of Pharmacology*, 137: 793–798.

Downs, J.S., Arslanian, S., De Bruin, W.B., Copeland, V.C., Doswell, W., Herman, W., Lain, K. et al. 2010. Implications of type 2 diabetes on adolescent reproductive health risk: An expert model. *Diabetes Educator*, 36: 911–919.

El-Ghorab, A.H., Nauman, M., Anjum, F.M., Hussain, S., and Nadeem, M. 2010. A comparative study on chemical composition and antioxidant activity of ginger (*Zingiber officinale*) and cumin (*Cuminum cyminum*). *Journal of Agricultural and Food Chemistry*, 58: 8231–8237.

Elrokh El, S.M., Yassin, N.A., El-Shenawy, S.M., and Ibrahim, B.M. 2010. Antihypercholesterolaemic effect of ginger rhizome (*Zingiber officinale*) in rats. *Inflammopharmacology*, 18: 309–315.

El-Sharaky, A.S., Newairy, A.A., Kamel, M.A., and Eweda, S.M. 2009. Protective effect of ginger extract against bromobenzene-induced hepatotoxicity in male rats. *Food and Chemical Toxicology*, 47: 1584–1590.

Ferrannini, E., Haffner, S., Mitchell, B., and Stern, M. 1991. Hyperinsulinaemia: The key feature of a cardiovascular and metabolic syndrome. *Diabetologia*, 34: 416–422.

Fuhrman, B., Rosenblat, M., Hayek, T., Coleman, R., and Aviram, M. 2000. Ginger extract consumption reduces plasma cholesterol, inhibits LDL oxidation and attenuates development of athrosclerosis in atherosclerotic, apolipoprotein E-deficient mice. *Journal of Nutrition*, 130: 1124–1131.

Ganti, S.S., Nammi, S., and Lodagala, D.S. 1999. Molecular mechanisms of insulin action. *Proceedings of the Indian National Science Academy*, B65(5): 245–256.

Gong, F., Fung, Y.S., and Liang, Y.Z. 2004. Determination of volatile components in ginger using gas chromatography-mass spectrometry with resolution improved by data processing techniques. *Journal of Agricultural and Food Chemistry*, 52: 6378–6383.

Govindarajan, V.S. 1982a. Ginger—Chemistry, technology, and quality evaluation: Part 1. *Critical Reviews in Food Science and Nutrition*, 17: 1–96.

Govindarajan, V.S. 1982b. Ginger—Chemistry, technology, and quality evaluation: Part 2. *Critical Reviews in Food Science and Nutrition*, 17: 189–258.

Goyal, R.K. and Kadnur, S.V. 2006. Beneficial effects of *Zingiber officinale* on goldthioglucose induced obesity. *Fitoterapia*, 77: 160–163.

Grundy, S.M., Brewer Jr, H.B., Cleeman, J.I., Smith Jr, S.C., and Lenfant, C. 2004. Definition of metabolic syndrome report of the National Heart, Lung, and Blood Institute/American Heart Association Conference on scientific issues related to definition. *Circulation*, 109: 433–438.

Han, L.K., Gong, X.J., Kawano, S., Saito, M., Kimura, Y., and Okuda, H. 2005. Antiobesity actions of *Zingiber officinale* Roscoe. *Yakugaku Zasshi*, 125: 213–217.

Heimes, K., Feistel, B., and Verspohl, E.J. 2009. Impact of the 5-HT3 receptor channel system for insulin secretion and interaction of ginger extracts. *European Journal of Pharmacology*, 624: 58–65.

Ippoushi, K., Azuma, K., Ito, H., Horie, H., and Higashio, H. 2003. [6]-Gingerol inhibits nitric oxide synthesis in activated J774.1 mouse macrophages and prevents peroxynitrite-induced oxidation and nitration reactions. *Life Sciences*, 73: 3427–3437.

Ippoushi, K., Ito, H., Horie, H., and Azuma, K. 2005. Mechanism of inhibition of peroxynitrite-induced oxidation and nitration by [6]-gingerol. *Planta Medica*, 71: 563–566.

Islam, M.S. and Choi, H. 2008. Comparative effects of dietary ginger (*Zingiber officinale*) and garlic (*Allium sativum*) investigated in a type 2 diabetes model of rats. *Journal of Medicinal Food*, 11: 152–159.

Iyer, U., Desai, P., and Venugopal, S. 2010. Impact of panchratna juice in the management of diabetes mellitus: Fresh vs. processed product. *International Journal of Green Pharmacy*, 4: 122.

Jewell, D. and Young, G. 2003. Interventions for nausea and vomiting in early pregnancy. *Cochrane Database of Systematic Reviews*, 9: CD000145.

Jiang, H., Solyom, A.M., Timmermann, B.N., and Gang, D.R. 2005. Characterization of gingerol-related compounds in ginger rhizome (*Zingiber officinale* Rosc.) by high-performance liquid chromatography/electrospray ionization mass spectrometry. *Rapid Communications in Mass Spectrometry*, 19: 2957–2964.

Jolad, S.D., Lantz, R.C., Solyom, A.M., Chen, G.J., Bates, R.B., and Timmermann, B.N. 2004. Fresh organically grown ginger (*Zingiber officinale*): Composition and effects on LPS-induced PGE2 production. *Phytochemistry*, 65: 1937–1954.

Jorsaraei, S.G., Yousefnia, Y.R., Zainalzadeh, M., Moghadamnia, A.A., Beiky, A.A., and Damavandi, M.R. 2008. The effects of methanolic extracts of ginger (*Zingiber officinale*) on human sperm parameters; an in vitro study. *Pakistan Journal of Biological Sciences*, 11: 1723–1727.

Kadnur, S.V. and Goyal, R.K. 2005. Beneficial effects of *Zingiber officinale* Roscoe on fructose induced hyperlipidemia and hyperinsulinemia in rats. *Indian Journal of Experimental Biology*, 43: 1161–1164.

Kahn, S.E., Hull, R.L., and Utzschneider, K.M. 2006. Mechanisms linking obesity to insulin resistance and type 2 diabetes. *Nature*, 444: 840–846.

Kato, A., Higuchi, Y., Goto, H., Kizu, H., Okamoto, T., Asano, N., Hollinshead, J., Nash, R.J., and Adachi, I. 2006. Inhibitory effects of *Zingiber officinale* Roscoe derived components on aldose reductase activity in vitro and in vivo. *Journal of Agricultural and Food Chemistry*, 54: 6640–6644.

Kaul, K., Tarr, J.M., Ahmad, S.I., Kohner, E.M., and Chibber, R. 2012. Introduction to diabetes mellitus. *Advances in Experimental Medicine and Biology*, 771: 1–11.

Kim, H.W., Murakami, A., Nakamura, Y., and Ohigashi, H. 2002. Screening of edible Japanese plants for suppressive effects on phorbol ester-induced superoxide generation in differentiated HL-60 cells and AS52 cells. *Cancer Letters*, 176: 7–16.

Kim, M.K., Chung, S.W., Kim, D.H., Kim, J.M., Lee, E.K., Kim, J.Y., Ha, Y.M. et al. 2010. Modulation of age-related NF-kappaB activation by dietary zingerone via MAPK pathway. *Experimental Gerontology*, 45: 419–426.

Koh, E.M., Kim, H.J., Kim, S., Choi, W.H., Choi, Y.H., Ryu, S.Y., Kim, Y.S., Koh, W.S., and Park, S.Y. 2009. Modulation of macrophage functions by compounds isolated from *Zingiber officinale*. *Planta Medica*, 75: 148–151.

Krishnakantha, T.P. and Lokesh, B.R. 1993. Scavenging of superoxide anions by spice principles. *Indian Journal of Biochemistry and Biophysics*, 30: 133–134.

La Vignera, S., Condorelli, R., Vicari, E., D'Agata, R., and Calogero, A.E. 2012. Diabetes mellitus and sperm parameters. *Journal of Andrology*, 33: 145–153.

Li, X.H., McGrath, K.C., Nammi, S., Heather, A.K., and Roufogalis, B.D. 2012a. Attenuation of liver proinflammatory responses by *Zingiber officinale* via inhibition of NF-kappa B activation in high-fat diet-fed rats. *Basic & Clinical Pharmacology & Toxicology*, 110: 238–244.

Li, Y., Tran, V.H., Duke, C.C., and Roufogalis, B.D. 2012b. Gingerols of *Zingiber officinale* enhance glucose uptake by increasing cell surface GLUT4 in cultured L6 myotubes. *Planta Medica*, 78: 1549–1555.

Madkor, H.R., Mansour, S.W., and Ramadan, G. 2011. Modulatory effects of garlic, ginger, turmeric and their mixture on hyperglycaemia, dyslipidaemia and oxidative stress in streptozotocin-nicotinamide diabetic rats. *The British Journal of Nutrition*, 105: 1210–1217.

Mallikarjuna, K., Sahitya Chetan, P., Sathyavelu Reddy, K., and Rajendra, W. 2008. Ethanol toxicity: Rehabilitation of hepatic antioxidant defense system with dietary ginger. *Fitoterapia*, 79: 174–178.

Mascolo, N., Jain, R., Jain, S.C., and Capasso, F. 1989. Ethnopharmacologic investigation of ginger (*Zingiber officinale*). *Journal of Ethnopharmacology*, 27: 129–140.

Matsuda, A., Wang, Z., Takahashi, S., Tokuda, T., Miura, N., and Hasegawa, J. 2009. Upregulation of mRNA of retinoid binding protein and fatty acid binding protein by cholesterol enriched-diet and effect of ginger on lipid metabolism. *Life Sciences*, 84: 903–907.

McGarry, J.D. 2002. Banting lecture 2001: Dysregulation of fatty acid metabolism in the etiology of type 2 diabetes. *Diabetes*, 51: 7–18.

Moselhy, W.A., Helmy, N.A., Abdel-Halim, B.R., Nabil, T.M., and Abdel-Hamid, M.I. 2012. Role of ginger against the reproductive toxicity of aluminium chloride in albino male rats. *Reproduction in Domestic Animals*, 47: 335–343.

Mukhopadhyay, M.J. and Mukherjee, A. 2000. Clastogenic effect of ginger rhizome in mice. *Phytotherapy Research*, 14: 555–557.

Nammi, S., Kim, M.S., Gavande, N.S., Li, G.Q., and Roufogalis, B.D. 2010. Regulation of low-density lipoprotein receptor and 3-hydroxy-3-methylglutaryl coenzyme A reductase expression by *Zingiber officinale* in the liver of high-fat diet-fed rats. *Basic & Clinical Pharmacology & Toxicology*, 106: 389–395.

Nammi, S., Koka, S., Chinnala, K.M., and Boini, K.M. 2004. Obesity: An overview on its current perspectives and treatment options. *Nutrition Journal*, 3: 3.

Nammi, S., Sreemantula, S., and Roufogalis, B.D. 2009. Protective effects of ethanolic extract of *Zingiber officinale* rhizome on the development of metabolic syndrome in high-fat diet-fed rats. *Basic & Clinical Pharmacology & Toxicology*, 104: 366–373.

Nawroth, P.P., Bierhaus, A., Vogel, G.E., Hofmann, M.A., Zumbach, M., Wahl, P., and Ziegler, R. 1999. Non-enzymatic glycation and oxidative stress in chronic illnesses and diabetes mellitus. *Medizinische Klinik*, 94: 29–38.

Noipha, K., Ratanachaiyavong, S., and Ninla-Aesong, P. 2010. Enhancement of glucose transport by selected plant foods in muscle cell line L6. *Diabetes Research and Clinical Practice*, 89: e22–e26.

Ojewole, J.A. 2006. Analgesic, antiinflammatory and hypoglycaemic effects of ethanol extract of *Zingiber officinale* (Roscoe) rhizomes (Zingiberaceae) in mice and rats. *Phytotheraphy Research*, 20: 764–772.

Platel, K. and Srinivasan, K. 2000. Influence of dietary spices and their active principles on pancreatic digestive enzymes in albino rats. *Nahrung*, 44: 42–46.

Pollreisz, A. and Schmidt-Erfurth, U. 2010. Diabetic cataract—Pathogenesis, epidemiology and treatment. *Journal of Ophthalmology*, 2010: 608751.

Portnoi, G., Chng, L.A., Karimi-Tabesh, L., Koren, G., Tan, M.P., and Einarson, A. 2003. Prospective comparative study of the safety and effectiveness of ginger for the treatment of nausea and vomiting in pregnancy. *American Journal of Obstetrics and Gynecology*, 189: 1374–1377.

Ramana, K.V. 2011. Aldose reductase: New insights for an old enzyme. *Biomolecular Concepts*, 2: 103–114.

Ramudu, S.K., Korivi, M., Kesireddy, N., Chen, C.Y., Kuo, C.H., and Kesireddy, S.R. 2011a. Ginger feeding protects against renal oxidative damage caused by alcohol consumption in rats. *Journal of Renal Nutrition*, 21: 263–270.

Ramudu, S.K., Korivi, M., Kesireddy, N., Lee, L.C., Cheng, I.S., Kuo, C.H., and Kesireddy, S.R. 2011b. Nephro-protective effects of a ginger extract on cytosolic and mitochondrial enzymes against streptozotocin (STZ)-induced diabetic complications in rats. *Chinese Journal of Physiology*, 54: 79–86.

Rani, M.P., Krishna, M.S., Padmakumari, K.P., Raghu, K.G., and Sundaresan, A. 2012. *Zingiber officinale* extract exhibits antidiabetic potential via modulating glucose uptake, protein glycation and inhibiting adipocyte differentiation: An in vitro study. *Journal of the Science of Food and Agriculture*, 92: 1948–1955.

Rani, M.P., Padmakumari, K.P., Sankarikutty, B., Cherian, O.L., Nisha, V.M., and Raghu, K.G. 2011. Inhibitory potential of ginger extracts against enzymes linked to type 2 diabetes, inflammation and induced oxidative stress. *International Journal of Food Sciences and Nutrition*, 62: 106–110.

Ranilla, L.G., Kwon, Y.I., Apostolidis, E., and Shetty, K. 2010. Phenolic compounds, antioxidant activity and in vitro inhibitory potential against key enzymes relevant for hyperglycemia and hypertension of commonly used medicinal plants, herbs and spices in Latin America. *Bioresource Technology*, 101: 4676–4689.

Raz, I., Eldor, R., Cernea, S., and Shafrir, E. 2005. Diabetes: Insulin resistance and derangements in lipid metabolism. Cure through intervention in fat transport and storage. *Diabetes/Metabolism Research and Reviews*, 21: 3–14.

Reaven, G.M. 1988. Role of insulin resistance in human disease. *Diabetes*, 37: 1595–1607.

Reddy, A.C. and Lokesh, B.R. 1992. Studies on spice principles as antioxidants in the inhibition of lipid peroxidation of rat liver microsomes. *Molecular and Cellular Biochemistry*, 111: 117–124.

Roberts, A.T., Martin, C.K., Liu, Z., Amen, R.J., Woltering, E.A., Rood, J.C., Caruso, M.K., Yu, Y., Xie, H., and Greenway, F.L. 2007. The safety and efficacy of a dietary herbal supplement and gallic acid for weight loss. *Journal of Medicinal Food*, 10: 184–188.

Roberts, C.K., Hevener, A.L., and Barnard, R.J. 2013. Metabolic syndrome and insulin resistance: Underlying causes and modification by exercise training. *Comprehensive Physiology*, 3: 1–58.

Rong, X., Peng, G., Suzuki, T., Yang, Q., Yamahara, J., and Li, Y. 2009. A 35-day gavage safety assessment of ginger in rats. *Regulatory Toxicology and Pharmacology*, 54: 118–123.

Saraswat, M., Reddy, P.Y., Muthenna, P., and Reddy, G.B. 2009. Prevention of non-enzymic glycation of proteins by dietary agents: Prospects for alleviating diabetic complications. *The British Journal of Nutrition*, 101: 1714–1721.

Saraswat, M., Suryanarayana, P., Reddy, P.Y., Patil, M.A., Balakrishna, N., and Reddy, G.B. 2010. Antiglycating potential of *Zingiber officinalis* and delay of diabetic cataract in rats. *Molecular Vision*, 16: 1525–1537.

Satirapoj, B. 2012. Nephropathy in diabetes. *Advances in Experimental Medicine and Biology*, 771: 107–122.

Schwertner, H.A. and Rios, D.C. 2007. High-performance liquid chromatographic analysis of 6-gingerol, 8-gingerol, 10-gingerol, and 6-shogaol in ginger-containing dietary supplements, spices, teas, and beverages. *Journal of Chromatography B: Analytical Technologies in the Biomedical and Life Sciences*, 856: 41–47.

Schwertner, H.A., Rios, D.C., and Pascoe, J.E. 2006. Variation in concentration and labeling of ginger root dietary supplements. *Obstetrics and Gynecology*, 107: 1337–1343.

Sekiya, K., Ohtani, A., and Kusano, S. 2004. Enhancement of insulin sensitivity in adipocytes by ginger. *BioFactors*, 22: 153–156.

Sermondade, N., Faure, C., Fezeu, L., Levy, R., and Czernichow, S. 2012. Obesity and increased risk for oligozoospermia and azoospermia. *Archives of Internal Medicine*, 172: 440442.

Shalaby, M.A. and Hamowieh, A.R. 2010. Safety and efficacy of *Zingiber officinale* roots on fertility of male diabetic rats. *Food and Chemical Toxicology*, 48: 2920–2924.

Shanmugam, K.R., Mallikarjuna, K., Kesireddy, N., and Sathyavelu Reddy, K. 2011. Neuroprotective effect of ginger on anti-oxidant enzymes in streptozotocin-induced diabetic rats. *Food and Chemical Toxicology*, 49: 893–897.

Shanmugam, K.R., Ramakrishna, C.H., Mallikarjuna, K., and Reddy, K.S. 2010. Protective effect of ginger against alcohol-induced renal damage and antioxidant enzymes in male albino rats. *Indian Journal of Experimental Biology*, 48: 143–149.

Sharma, M. and S. Shukla. 1977. Hypoglycemic effect of ginger. *The Journal of Research in Indian Yoga and Homoeopathy*, 12: 127–130.

Shobana, S. and Naidu, K.A. 2000. Antioxidant activity of selected Indian spices. *Prostaglandins, Leukotrienes, and Essential Fatty Acids*, 62: 107–110.

Singh, A.B., Akanksha, N. Singh, Maurya, R., and Srivastava, A.K. 2009. Anti-hyperglycaemic, lipid lowering and antioxidant properties of [6]-gingerol in db/db mice. *International Journal of Medicine and Medical Sciences*, 1: 536–544.

Singhal, P.C. and Joshi, L.D. 1983. Glycaemic and cholesterolemic role of ginger and till. *Journal of Scientific Research in Plants & Medicine*, 4: 32–34.

Srinivasan, K. and Sambaiah, K. 1991. The effect of spices on cholesterol 7 alpha-hydroxylase activity and on serum and hepatic cholesterol levels in the rat. *International Journal for Vitamin and Nutrition Research*, 61: 364–369.

Suekawa, M., Ishige, A., Yuasa, K., Sudo, K., Aburada, M., and Hosoya, E. 1984. Pharmacological studies on ginger. I. Pharmacological actions of pungent constituents, (6)-gingerol and (6)-shogaol. *Journal of Pharmacobio-Dynamics*, 7: 836–848.

Sun, Y.M., Su, Y., Li, J., and Wang, L.F. 2013. Recent advances in understanding the biochemical and molecular mechanism of diabetic nephropathy. *Biochemical and Biophysical Research Communications*, 433: 359–361.

Tanabe, M., Chen, Y.D., Saito, K., and Kano, Y. 1993. Cholesterol biosynthesis inhibitory component from *Zingiber officinale* Roscoe. *Chemical & Pharmaceutical Bulletin (Tokyo)*, 41: 710–713.

Thomson, M., Al-Qattan, K.K., Al-Sawan, S.M., Alnaqeeb, M.A., Khan, I., and Ali, M. 2002. The use of ginger (*Zingiber officinale* Rosc.) as a potential anti-inflammatory and antithrombotic agent. *Prostaglandins, Leukotrienes, and Essential Fatty Acids*, 67: 475–478.

Topic, B., Tani, E., Tsiakitzis, K., Kourounakis, P.N., Dere, E., Hasenohrl, R.U., Hacker, R., Mattern, C.M., and Huston, J.P. 2002. Enhanced maze performance and reduced oxidative stress by combined extracts of *Zingiber officinale* and *Ginkgo biloba* in the aged rat. *Neurobiology of Aging*, 23: 135–143.

Tundis, R., Loizzo, M.R., and Menichini, F. 2010. Natural products as alpha-amylase and alpha-glucosidase inhibitors and their hypoglycaemic potential in the treatment of diabetes: An update. *Mini Reviews in Medicinal Chemistry*, 10: 315–331.

Uz, E., Karatas, O.F., Mete, E., Bayrak, R., Bayrak, O., Atmaca, A.F., Atis, O., Yildirim, M.E. and Akcay, A. 2009. The effect of dietary ginger (*Zingiber officinals* Rosc.) on renal ischemia/reperfusion injury in rat kidneys. *Renal Failure*, 31: 251–260.

Verma, S. and Bordia, A. 2001. Ginger, fat and fibrinolysis. *Indian Journal of Medical Sciences*, 55: 83.

Verma, S.K., Singh, M., Jain, P., and Bordia, A. 2004. Protective effect of ginger, *Zingiber officinale* Rosc. on experimental atherosclerosis in rabbits. *Indian Journal of Experimental Biology*, 42: 736–738.

Vernin, G. and Parkanyi, C. 2005. Chemistry of ginger. In: *Ginger: The Genus Zingiber (Medicinal and Aromatic Plants—Industrial Profiles)*. Ravindran, P.N. and Nirmal Babu, K. eds., Boca Raton, FL: CRC Press, pp. 87–180.

Weidner, M.S. and Sigwart, K. 2000. The safety of a ginger extract in the rat. *Journal of Ethnopharmacology*, 73: 513–520.

Weidner, M.S. and Sigwart, K. 2001. Investigation of the teratogenic potential of a *Zingiber officinale* extract in the rat. *Reproductive Toxicology*, 15, 75–80.

White, B. 2007. Ginger: An overview. *American Family Physician*, 75: 1689–1691.

Wilkinson, J.M. 2000. Effect of ginger tea on the fetal development of Sprague–Dawley rats. *Reproductive Toxicology*, 14: 507–512.

Wohlmuth, H., Leach, D.N., Smith, M.K., and Myers, S.P. 2005. Gingerol content of diploid and tetraploid clones of ginger (*Zingiber officinale* Roscoe). *Journal of Agricultural and Food Chemistry*, 53: 5772–5778.

Woo, H.M., Kang, J.H., Kawada, T., Yoo, H., Sung, M.K., and Yu, R. 2007. Active spice-derived components can inhibit inflammatory responses of adipose tissue in obesity by suppressing inflammatory actions of macrophages and release of monocyte chemoattractant protein-1 from adipocytes. *Life Sciences*, 80: 926–931.

Yang, H., Jin, X., Kei Lam, C.W., and Yan, S.K. 2011. Oxidative stress and diabetes mellitus. *Clinical Chemistry and Laboratory Medicine*, 49: 1773–1782.

Yu, Y., Zick, S., Li, X., Zou, P., Wright, B., and Sun, D. 2011. Examination of the pharmacokinetics of active ingredients of ginger in humans. *The AAPS Journal*, 13: 417–426.

Zarina, S., Zhao, H.R., and Abraham, E.C. 2000. Advanced glycation end products in human senile and diabetic cataractous lenses. *Molecular and Cellular Biochemistry*, 210: 29–34.

Zhan, K., Wang, C., Xu, K., and Yin, H. 2008. Analysis of volatile and non-volatile compositions in ginger oleoresin by gas chromatography-mass spectrometry. *Se Pu Chinese Journal of Chromatography*, 26: 692–696.

27 Local Food Futures and Healthy Communities
Role of Sustainable Practices

Sumita Ghosh

CONTENTS

27.1 INTRODUCTION

"Local food" links to a short food supply chain where production and distribution of healthy, nutritious, and fresh foods are geographically located in close proximity to consumers (Gaynor 2006; Daniels et al. 2008; Ghosh 2012). Within a food system, local food could provide energy and carbon benefits by reducing the distance of food travel and the needs for refrigeration and storage (Halweil 2002; Garnett 2007); initiate self-sufficient local economies through farmers' markets and local shops (Halweil 2002; McGill University 2013); facilitate stronger social networks and community building (Winklerprins 2002; Gaynor 2006); and improve physical and mental health through stress relief, psychological wellness, and opportunities for adopting an active lifestyle (Van Den Berg and Custers 2011; Canadian Horticultural Therapy Association 2013). Therefore, local food production practices could be intrinsically connected to sustainability. However, all local food ventures may not be always synonymous to sustainable local food unless a project demonstrates an adequate sustainability performance either individually or collectively encompassing social, economic, environmental, and health domains. For example, if a customer's trip to purchase organic vegetables is more than 7.4 km, the CO_2 emissions associated with the trip is larger than the doorstep food delivery system (Coley et al. 2009). Environmental contribution of buying the food locally within a 20 km radius is higher than the food sourced from organic farms (BBC News 2005). A survey in the United States reveals diverse understandings of consumers on geographic boundary of local food, varies up to 100 miles of distance, and is shaped by consumer's "... diverse connotations depending on place, culture and lifestyle" (Hartman Group 2008). Spatial dimensions of local food "... itself involves the social construction of scale" (Hinrichs 2003). There is no one definition of local food as it needs to be defined holistically from a transdisciplinary context.

Importance of local food is increasingly visible and is realized in communities' daily life. A local food movement from grassroot up to a global scale is already happening. Havana, Cuba, is a leading example of growing food locally on any available urban spaces such as balconies, backyards, and empty lots in achieving self-reliance in the event of a severe food shortage (Organic Vision 2011). Buchmann's (2009) study supports that Cuban home gardens act as community's important land resource for providing householders' food demand and socioecological resilience. In 1999, allotment gardens in London produced 232,000 tons of fruits and vegetables supplying 18% of the population's daily intake as per the World Health Organization (WHO) recommendations (Garnett 1999). Spatial hierarchy of spaces for growing local food could include different typologies of spaces: urban farms, home gardens, community and allotment gardens, school gardens, green roofs, street gardens, and others (Ghosh 2013). These spaces differ in their ownership patterns, participation and capacity building, production potential, quality and amount of produce, locational aspects, health contributions, and therapeutic supports (Ghosh 2010). As an alternative food production system, open land areas locked in millions of private and public outdoor spaces if put to productive uses could significantly improve global food security and public health (Ghosh 2010). Vegetables could be easily grown in these smaller spaces within urban-built environments close to home. Vegetables intake as an important part of our daily diet could reduce the occurrence of cardiovascular diseases, micronutrient deficiencies, obesity, diabetes, and cerebrovascular disease, and lower the risks of some cancers to 30%–40% (WHO 1990 and WCRF 1997 as quoted in Pederson and Robertson [2001]).

Local food and public health are intrinsically linked. Current research is increasingly focusing on understanding the connections of local food, public health, and urban planning in creating healthy communities and sustainable cities. "Healthy by Design—A planners' guide to environments for active living" developed by the National Heart Foundation (Victoria Division) in Australia aims to create supportive environments for physical activity (SEPA) (National Heart Foundation 2013). However, this is still a less explored area of research, and local food-efficient designs of current and future built environments are absolutely essential. In this chapter, a review is conducted to investigate nutritional, health, well-being, and therapeutic benefits; sustainability potential; and communities' attitudes and perceptions associated with local food production. Mainly growing vegetables in comparatively smaller growing spaces such as community, allotment, home, and school gardens located within the existing built environments are considered. This chapter explores the barriers to change and analyzes the significance, performance, and synergies of "local food" from sustainability and public health perspectives. Sustainability attributes of local food production important for delivering meaningful outcomes for building healthy communities for healthy future are systematically collated to provide a synthesis and recommendations and to comprehend potential future research directions.

27.2 LOCAL FOOD AND PUBLIC HEALTH—A REVIEW

27.2.1 Nutritional and Health Impacts of Local Food

The WHO (2003) recommends a daily intake of five servings of vegetables as an important part of our daily diet for adequate nutrition. A collaborative promotion initiative by the WHO (2003) and the Food and Agriculture Organization of the United Nations (FAO) for improving fruit and vegetable intake promotes small-scale food production for better access to fresh food and vegetables. The New Zealand's national nutritional survey identified the major difficulties in eating vegetables as important parts of the diets: do not always have at home (16%), cost (15%), takes too long to prepare (15%), and not enough time (15%) (Ministry of Health 1999). Locally grown vegetables produced in the backyards and community and allotment gardens are easily available, accessible, and fresh, and could reduce the needs for refrigeration and storage.

Home and community or allotment gardens could play important roles in generating health awareness, improving food access, and preventing diseases such as obesity. A pilot study was conducted on the fourth- to sixth-grade children to understand the intervention of "The Delicious and Nutritious Garden" in their home food environments in Southeastern Minnesota (Heim et al. 2011). A parental reporting method through pre–postsurveys was used. This study reported that almost 99% of the children enjoyed the food-growing activities and have actively engaged themselves in planting and growing fruits and vegetables in their gardens. The participation in growing vegetables promoted positive attitudes in children to eat daily fresh fruits and vegetables grown by them in their own gardens (Heim et al. 2011).

Vitamin A deficiency is a major health problem for children in low-income households and countries. This could lead to vision problems such as night blindness and other respiratory diseases. Eating vitamin A-rich fruits and vegetables such as papaya, cantaloupe, apricots, carrots, butternuts, and dark leafy vegetables can be effectively sourced from vegetables and fruits grown in home gardens with lesser costs. Helen Keller International a reputed non profit organization from New York, USA had conducted a pilot study with 1000 poor households in Bangladesh to analyze nutritional improvements in dietary intake as a result of growing vitamin A-rich foods in low-cost vegetable gardens and providing nutrition education (Talukder et al. 2000). This large-scale home gardening program had increased food access, availability, and consumption of vitamin A-rich fruits and vegetables produced in household's daily diets (Talukder et al. 2000). The Medical Research Council (MRC) and the Agricultural Research Council collaboratively promoted β-carotene-rich vegetables and fruits, butternuts, carrots, spinach, and orange-fleshed sweet potato grown in home-based gardens in rural community of KwaZulu-Natal in South Africa to improve the nutritional status of children (Faber et al. 2002). Nutrition education was provided at a household level, and households with project gardens were compared to households without project gardens. Evidence shows a significant increase in vitamin A intake in children from households with project gardens (Faber et al. 2002). As additional butternuts were produced and were available at local shops, the children from households without project gardens had also improved their vitamin A intake. This research exemplifies a significant achievement that local food production can improve children's health and could positively impact on the health of an entire population (Faber et al. 2002).

Schools and universities could effectively participate and contribute to generating sustainable food systems through staff's and students' participation in institutional vegetables gardens (Figure 27.1). Carlsson and Williams (2008) examined school gardens as one of the three avenues (school food gardens, farm-to-school programs, and school procurement policies) to contribute to sustainable food systems. This study indicates that smaller school gardens could act as demonstration gardens while larger food gardens could supply vegetables for food preparation to school cafeteria, thus creating a self-sufficient closed-loop local sustainable food system (Carlsson and Williams 2008). Teachers' perceptions in Californian schools show that gardens in schools can improve children's nutritional awareness and healthy eating habits (Graham and Zidenberg-Cherr 2005). Edible Campus Gardens in the University of Utah allows gardening over four seasons to cultivate crops in raised beds throughout the year and is supported by the Office of Sustainability. The produces including vegetables such as tomatoes, beans, peppers, and cucumbers are sold in a student-run market stall and also supplied for campus dinner services (The University of Utah 2013). In 2010, a campus at McGill University, Canada, harvested 28 different types of vegetables and produced a total of produce of one ton from 275 containers and two gardens located within the university campus (McGill University 2013).

A study on low-income Latino immigrant families' participation using intervention of weekly gardening activities in community gardens, cooking, nutritional workshops, and social networking events for families can help to prevent childhood obesity through increased consumption of fruits and vegetables and easy access to nutritious food (Castro et al. 2013). Research on household participation in the community garden and related intake of fruits and vegetables among 766 adults in Flint, Michigan, shows an impressive result (Alaimo et al. 2008). An adult household member

FIGURE 27.1 New sustainability garden growing vegetables at Pymble Ladies College, Sydney, Australia. (Courtesy of Sumita Ghosh.)

who participated in a community garden consumed fruits and vegetables 1.4 times per day more than those who did not participate. This household member who had participated in community gardening was 3.5 times more likely to consume fruits and vegetables and could consume fruits and vegetables at least five times daily (Alaimo et al. 2008).

An objective measurement and comparison of benefits of allotment gardening on health, well-being, and physical activity of older and younger age groups in the Netherlands has been conducted. From the same locality, 12 allotment sites with 121 allotment gardeners and a control group consisting of 63 members without any allotment gardens were surveyed (Van Den Berg et al. 2010). The measures included five self-reported health attributes (such as perception of general health, acute health complaints, physical barriers, chronic illnesses, and general practice [GP] consultation), four self-reported well-being variables (such as stress, life satisfaction, loneliness, and social networks), and one self-reported level of physical activity during the summer (Van Den Berg et al. 2010). After adjusting for socioeconomic variables (such as education and income), access, and stress exposure levels of the allotment gardeners, the study found that both younger and older age groups reported increased physical activity during summer compared with members of the control group, who were living as neighbors. Allotment gardeners in the 62 years and older category scored significantly or marginally better in all health and well-being measures than nongardeners in the same age group. Interestingly, there was almost no difference between health and well-being measures of young age groups of allotment gardeners and control group of nongardeners. Allotment gardening could help health and well-being of older age groups significantly and could promote an active lifestyle. Allotment gardening for a period of 30 minutes can help to achieve 22% reduction in elevated salivary cortisol levels compared to 11% reduction in the same measured in a control group doing a passive indoor reading task for the same duration of time (Van Den Berg and Custers 2011).

27.2.2 Making Choices for Local Food

Making a choice for eating local food would depend on the awareness as an individual or a group, backgrounds, growing environments, lifestyle patterns, and age groups. Robinson-O'Brien and Neumark-Sztainer's (2009) survey identified a total of 2516 adolescents or high-school students as supporters of alternative food practices from 31 schools in Minnesota. A survey was conducted with the adolescent supporters of alternative food practices considering four categories: locally grown, organic, nongenetically engineered, and nonprocessed food. The survey measures included the attitudes toward alternative food production practices, personal health, dietary quality, fast-food use, and demographics (Robinson-O'Brien and Neumark-Sztainer 2009). Results demonstrated the percentages of supporters in the four categories are 20.9% for locally grown, 23.2% for organic, 34.1% for nongenetically engineered, and 29.8% for nonprocessed. Another cross-sectional study has been conducted on a sample of 1201 university students consisting of 95% of the sample below 33 years of age, 53% female and students who provide low, moderate, or high considerations to alternative food production practices (Pelletier et al. 2013). This survey examined the consumptions of fruits, vegetables, and fast food to identify their dietary behavioral patterns. This study confirmed that the cohort of young adults who provide importance to alternative food production practices consumes 1.3 times more servings of fruits and vegetables, more dietary fiber, and less sugar and fat than the participants who place lower importance to the same. These young adults can maintain healthy dietary quality or behavior and would make choices to source foods from an alternative production network including local food and more likely to understand better social, environmental, and health consequences of eating food produced through alternative and sustainable practices (Pelletier et al. 2013). Harmon and Maretzki's (2006) research reported that although adolescents may have supportive attitudes toward local and organic food, they are not sure how to implement this into healthy eating behaviors in practice.

27.2.3 Therapeutic Values of Horticultural Practices

Through the process of growing local food, valuable therapeutic benefits are generated among the participants. *Horticultural therapy* includes a formal process of mental and physical rehabilitation through continued associations of a patient in gardening with plants and is guided by a professional therapist to achieve specific clinical goals essential for improvement of patient's health (Canadian Horticultural Therapy Association 2013). Horticultural therapy is comparable with physical therapy, occupational therapy, music therapy, and animal therapy (Mizuno-Matsumoto et al. 2008). Growing local food could assist as a part of horticulture therapy in different kinds of diseases such as dementia and cerebrovascular diseases. Improvements in brain functions of cerebrovascular diseases were investigated in five case studies using horticulture therapy (Mizuno-Matsumoto et al. 2008). This research demonstrates that patient's capabilities to observe and to assist in the whole process of growing vegetation from seed to plants, clear representation of one's efforts, and possibilities of sharing the achievements with other people could create spontaneous rehabilitations in the patients (Mizuno-Matsumoto et al. 2008). Local food production could cultivate important health benefits for communities in addition to supplying nutritious and fresh fruits and vegetables. "Therapeutic Horticulture" differs from horticultural therapy as the former is an informal process and is targeted to "purposeful use of plants and plant-related activities to promote health and wellness for an individual or group" (Canadian Horticultural Therapy Association 2013). Horticultural therapy and therapeutic horticulture could assist in mental and social well-being in community and allotment gardens or in individual home gardens.

27.2.4 Community Perceptions of Local Food Production

Growing traditional and culturally appropriate food connects communities and strengthens their cultural identity. Productive home gardens in Greek and Vietnamese communities in Marrickville in Sydney reflect aspirations for traditional cuisine, connections of food growing to homeland practices, and cultural continuity (Graham and Connell 2006). Food production in home and community gardens generates social networks of exchange (Winklerprins 2002) and biotic interactions with nature (Ghosh 2010). To be able to grow and eat culturally, appropriate food and improved food access were positively perceived by the community gardeners. A number of health and nutritional benefits were available to the gardeners, and food gardening practices facilitated community cohesion and better mental health conditions in community gardens of Toronto in Canada (Wakefield et al. 2007).

The Northern Rivers Food Links (NRFL) established by seven Northern Rivers councils in New South Wales, Australia, conducted a random telephone survey with 500 grocery buyers. Out of a total of 11 different factors and performance on a 5-point importance scale, freshness (4.82—first), quality of produce (4.72—second), and purchase locally grown produce (4.20—fourth) establish the prospects of local food (Parker 2010). Out of total survey participants, 88% eat fresh fruit or vegetables daily and 56% believe that better quality is available if one decides to buy local produce. Importantly, 81% grew their own vegetables, 34% grew fruits, and 27% cultivated herbs or spices (Parker 2010).

Aboriginal cultural identity is linked to land and their culture is intertwined with active participation in natural resource management, traditional food production, gathering and hunting, and social networks of exchange. The Northern Territory Food Gardens in the Remote Indigenous Gardens (RIG) network have altogether 41 indigenous gardens consisting of 24 community gardens, 13 school gardens, and four pilot gardens. These gardens produce vegetables such as sweet potato, tomato, pumpkin, cucumber, and cassava, and fruits such water melon, banana, pawpaw, and mango (Northern Territory Food Gardens 2013). The RIG's (2013) network in Australia aims to build community resilience, better food access, food security, and improved health of remote indigenous communities through local food projects.

27.2.5 Measuring Sustainability and Supply Potential of Local Food

Limited research has been done to understand to what extent locally produced food and associated smaller-scale production practices could support recommended healthy dietary intake.

The Australian Government Department of Health and Ageing (2013) has published the "healthy eating" recommendations that include average daily five servings of vegetables consisting of 2.5 cups of vegetable from three different groups of vegetables: starchy, deep green and legumes, and other vegetables. The vegetable requirements in daily diet could vary depending on the total dietary recommendations and the age groups. Out of 2150 kcal Australian recommended daily average food energy intake per person (Haug et al. 2007), 12%–20% of total recommended daily energy intake should be sourced from the vegetables (Ghosh 2011a). Considering daily vegetable servings and intake distributions in the three groups of vegetables, the energy value of daily vegetable demand for an average person is estimated to vary between 425 and 255 kcal (Ghosh 2011a).

Research on productive land areas required to grow household food demand at a local scale indicates that a home garden of a size of 97 m^2 (7.6 × 12.8 m) could produce almost the annual vegetable demand for two people in the United States (North Carolina State University 1996). Over a period of five months in a total 123 containers comprising 30 m^2, a total of 176.9 kg (5.9 kg/m^2) of vegetables were harvested from the edible gardens in the campus of McGill University (2013). This value of vegetable production capacity per square meter was very similar to the value of vegetable production capacity per square meter of home garden area (5.94 kg/m^2) calculated by Ghosh and Head (2009) and Ghosh et al. (2008) in New Zealand and Australia. Suburban residential blocks with plot areas ranging between 500 and 1000 m^2 could produce nearly most of the household vegetable demand and surplus fruits (Francis 2010).

A local food energy model (Ghosh 2012, 2014; Ghosh et al. 2008) using geographic information systems (GISs) and ecological footprinting method (Wackerangel and Rees 1996) is applied to determine local food production potential of different density residential developments in Australia and New Zealand at different local spatial scales (Ghosh and Head 2009; Ghosh 2011a, 2012, 2014). Total annual nutritional energy equivalent of "vegetable demand" for the neighborhood as a share of annual dietary requirement is calculated. The potential "supply" of vegetables from the onsite

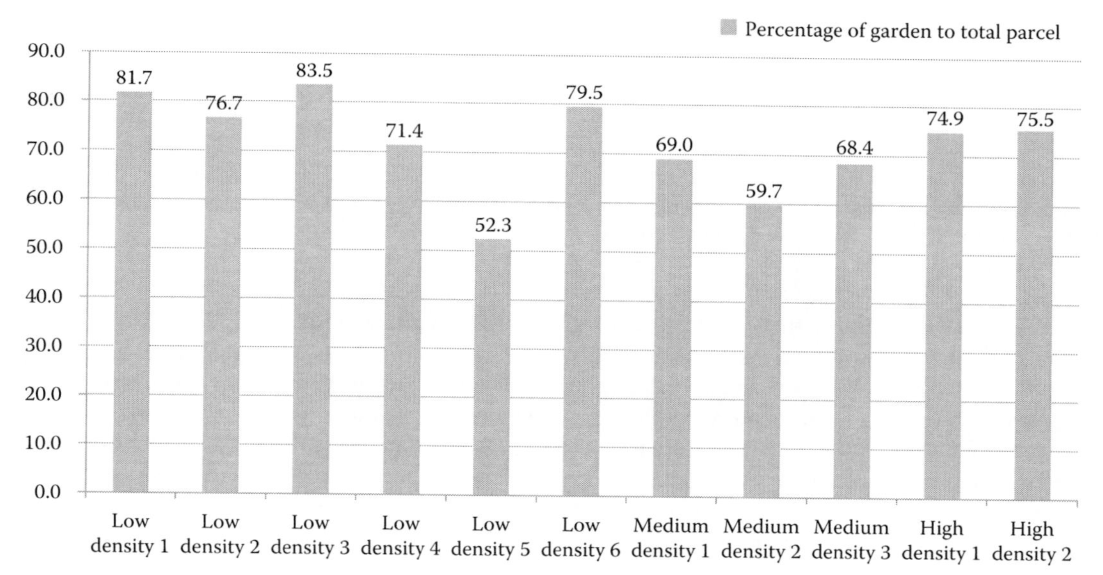

FIGURE 27.2 Percentages of home garden spaces to total parcel areas in 11 case studies. (Adapted from Ghosh, S. Morphologies and spatial distributions of domestic gardens in residential developments. *Institute of Australian Geographers (IAG) Conference*, Wollongong, Australia, July 3–6, 2011b.)

available productive land areas and the "deficit" that needs to be supplied from offsite or "surplus" produced on site in energy units are estimated per household and on per capita basis using this model (Ghosh 2011a, 2012, 2014). Using the local food model, the available onsite area for vegetable production is estimated to be equal to 33,227 m^2 in the home gardens of a residential low-density neighborhood with 309 dwellings (Ghosh 2011a). Assuming 330 kcal is the vegetable intake required per capita per day and vegetable production capacity of 1847 kcal/m^2 of home garden, on-site productive land areas in this neighborhood could supply a total of 55% of all the annual average vegetable nutritional energy "demand" of the residents (Ghosh 2011a). In another study, local food potential of other productive spaces such as street verges and parks and including a limited share of home gardens were estimated for a mixed use development with residential, retail, and commercial office uses. This mixed development with a total site area of 51 hectares and on-site productive land of 9.8 hectares could supply 62% of the total vegetable demand of the resident community (Ghosh 2012). Garden sizes influence garden composition, and a study on the five cities in the United Kingdom has demonstrated that larger gardens are more likely to contain more features such as vegetable patches, uncultivated land, and trees (Loram et al. 2008). Percentages of home garden spaces to total parcel or plot areas (low, medium, and high density) in 11 case studies in New Zealand and Australia show variations within the same density patterns as well as between different densities (Figure 27.2).

27.2.6 Improving Health in Cities and Role of Food-Efficient Design and Planning

Glouberman et al. (2006) identified three different approaches: the Urban Health, the Healthy Cities, and the Health in Cities to building and improving health in cities. The Urban Health approach has three main focuses: (1) developing understanding on diseases and at-risk populations, (2) spatial distribution of health problems and sanitary environment, and (3) its impacts and delivery of services. The Healthy Cities approach explores the influences of city environments on the health of residents and diversity and interconnectedness of the multiple urban living factors on health (Glouberman et al. 2006). The Health in Cities approach identifies cities and health as complex adaptive systems. This approach recognizes the diverse requirements of multiple groups of people with varying health needs. There are competing interests between groups, and all these factors together have an overall impact on health of a population (Glouberman et al. 2006). The WHO's (2013) Healthy Cities project aims to reintegrate health and planning and to establish urban planning as a determinant of health. It is a global health movement and a long-term international development initiative.

> A healthy city is one that is continually creating and improving the physical and social environments and expanding the community resources that enable people to mutually support each other in performing all the functions of life and developing to their maximum potential. (Health Promotion Glossary 1998 as quoted in WHO [2013])

Food-efficient design and planning and provisions for local food production could make vital contributions toward building a healthy city. Duany (2011) identifies four models of local food production linked to human settlements: agricultural retention, urban agriculture, agricultural urbanism, and agrarian urbanism to growing food within human environments. Agricultural retention connects to conservation of existing food-producing spaces, whereas urban agriculture refers to producing food on any or all available spaces within built environments such as community gardens, roof gardens, and other vacant lots. Agricultural urbanism focuses on settlements that are dependent on farming practices, whereas agrarian urbanism is a complex network in which a sustainable food system, society, and food-efficient design and planning of settlements are holistically integrated (Duany 2011). "Agrarian urbanism" concept could be applied across urban to natural six zones of "the Transect." The conceptual framework of the Transect considers a cross section of human habitats along a rural and urban continuum, contains six systematic urban to natural zones, and applies sustainable design principles in practice for creating responsive environments (Bohl and Plater-Zyberk 2006). Conservation subdivision approach assists in protection of more natural areas and food production spaces through design and provides harmonious solutions (Arendt 2010). This approach recognizes the Transect as an important and comprehensive geographic analysis of all human environments. Arendt (2010) argues that using a conservation subdivision approach, food-efficient designs could be maximized in new or retrofitted in existing urban developments.

Using a case study of Story County, Iowa, a "typology of continuous productive landscapes" was identified (Grimms 2009). Six typologies of local food production sites are private residence garden, community or allotment garden, food boulevard, institutional gardens (including religious and educational gardens), neighborhood farm, and urban farm that could be designed and planned for food efficiency (Grimm 2009). Research has shown that retrofitting a contemporary low-density residential neighborhood with 5% reduction in the existing ground footprints of houses, but without any changes in density, could further enhance the on-site local food growing potential by an additional 56%, reaching surplus on-site vegetable production (Ghosh 2012). This study reinforces the importance of design and planning in creating food-efficient built environments.

Increasingly, importance of developing appropriate food policies at state and local levels is visible in different parts of the world. Identifying long-term gardening sites in 1998 in Berkeley, California, Seattle's 1994 Comprehensive Plan's emphasis on setting up community gardens (Wekerle 2002), community gardens policies of the Auckland City Council (2014) in New Zealand, the City of Sydney (2012) in Australia, and the Toronto Food Policy Council (2013), Toronto, Canada's agenda to develop a healthy food system highlights the importance of creating sustainable local food system for health and well-being of communities.

27.3 BARRIERS TO LOCAL FOOD PRODUCTION

Soil contamination and air pollution absorption are significant barriers to local food production within urban and suburban built environments. Foods grown on contaminated soil could contain lead (Pb), cadmium (Cd), arsenic (As), solvents, pesticides, and petroleum hydrocarbons as the main contaminants that could impact on human health as a result of growing food on the contaminated sites (Minnesota Institute of Sustainable Agriculture 2010; Crozier et al. 2013) in community or allotment gardens, empty lots, street verges, and home gardens. Compost applications in vegetables such as green bean, lettuce, and carrot in soils of five urban gardens showed higher concentrations of cadmium or lead creating significant risks to human health (Murray et al. 2011). Food grown close to roads or traffic could absorb air pollutants such as concentrations of heavy metals. As witnessed, vegetables (e.g., tomatoes, green beans, carrots, potatoes, and leafy vegetables) grown in high traffic areas in the inner city of Berlin in Germany have absorbed air pollutants (Saeumel et al. 2012). The concentrations of pollutants in vegetables depend on the specific locations of the sites of production. Reducing food production in the inner city in high traffic locations, testing soils, and understanding the background of the soils in urban gardens prior to setup should be able to address any health-risk issues appropriately. Raised bed provides a solution for safe local food production in urban gardens (Minnesota Institute of Sustainable Agriculture 2010). A number of guidelines on how to grow safer food in urban areas are very useful for minimizing these associated health risks from growing on the contaminated sites (Minnesota Institute of Sustainable Agriculture 2010; US Environmental Protection Agency 2011).

There are other organizational and behavioral barriers that could impact the future of local food production in urban areas. Growing food within built environments needs to be recognized as an important part of the overall sustainable local food system by the community and decision makers. Lack of community awareness on positive implications for growing food, motivations to changing sedentary lifestyle patterns, utilization of underutilized land areas, inadequate provisions for ongoing financial supports, incentives and training for gardening practices, and absence of appropriate local food planning policies for implementation could impact on the uptake of local food production. However, the worldwide movement to growing food locally demonstrates that the urban and suburban patterns are changing and integrating local food production within the fabrics of developments.

27.4 SYNTHESIS AND DISCUSSIONS

A review in this chapter has demonstrated that within a complex web of multiple issues linked to food access, food security, safe food production, land ownership, soil contamination, socioeconomic status, motivations, and other concerns, it is possible to build community resilience and to improve health of communities through local food production. An active engagement in growing food and an increase in intake of fruits and vegetables could improve vitamin A deficiency and could reduce night blindness and obesity. This could progress physical and mental health of children, young and older adults, and overall population considerably. Educational institutions could play pioneering roles in educating future generations on healthy eating habits (Briggs et al 2003) and social, environmental, and economic benefits of growing fruits and vegetables locally. McGill University's award-winning campus gardens showcase the potential of schools and universities in creating a sustainable food system. Behavioral change toward healthy eating is articulated very well in Robinson-O'Brien and Neumark-Sztainer's (2009) study. This study highlights how positive attitudes can be developed in young adults to give importance to alternative food production practices and opting for healthy consumption choices. Traditional cuisine and herbs grown in migrant home gardens strengthen their cultural identity and connections to memories of distant homeland. The NRFL presents an important organizational collaboration developed. Local food system in that region performed very well in providing nutritious fresh produce and in inspiring communities to grow their own fruits and vegetables. Local food production in RIG provides a unique way to recreate their connections to land and reestablish cultural identity in new urban, suburban, and remote settings, and has positive implications for resource management, better health, and social connectedness. The local food energy model is able to provide an objective measure of nutritional equivalent of on-site local food production potential for different patterns of developments. Local food production as formal therapeutic horticulture and informal horticultural therapy could generate important health benefits for specific groups of people and for all communities.

Sustainable food-efficient design and planning of cities and suburbs play an important role in promoting local food production and in building healthy communities (New South Wales Government 2009). Research indicates that the availability of on-site productive land areas within urban environments is influenced by the urban morphological processes, peoples' lifestyle preferences, cultural backgrounds, household participation, motivation factors, and knowledge and awareness (Gaynor 2006; Graham and Connell 2006; Ghosh and Head 2009; Ghosh 2010, 2011a). The concept of agrarian urbanism, conscious design and planning of human environments along "the Transect" (Duany 2011), has a greater meaning for sustainability, planning, community's food access and food security, open-space conservation, qualities and aesthetics, and establishment of social networks of exchange. Population characteristics, collective community performance, new local economies, formulation of meaningful and safer methods of cultivation practices, and promotion of sustainable behavior for local food production in communities will be essential and should be understood within a broader context. All these foundations will enable transdisciplinary connections across science, nutrition, alternative medicine, architecture, planning, urban design, horticulture, geography, and social science. Future research should explore the "feasibility and effectiveness of an educational program" that links to developing understanding on the overall food system and alternative food production practices (Robinson-O'Brien and Neumark-Sztainer 2009). Appropriate intervention methods that could assist in improving nutritional intake should be explored in future research for improving practices of growing local food. Related organizational and governance structures, incentives, regulatory aspects, and policy directions should also be explored in detail.

27.5 CONCLUSIONS

As a network of productive landscapes, distributions of local food production spaces from urban to natural areas should be spatially integrated with other green infrastructure elements of the cities and suburbs. These small food production spaces can provide populations' nutritional demand to a great extent, create social networks and engagement, and can positively contribute toward improving public health and in building healthy cities of today and tomorrow.

ACKNOWLEDGMENTS

The author acknowledges the support of those who provided help and expresses her thanks to the anonymous referees for their valuable comments.

REFERENCES

Alaimo, K., Packnett, E., Miles, R.A., and Kruger, D.J. 2008. Fruit and vegetable intake among urban community gardeners. *Journal of Nutrition Education and Behavior*, 40(2): 94–101.

Arendt, R. 2010. *Envisioning Better Communities: Seeing More Options, Making Wiser Choices*. Washington, DC: American Planning Association.

Auckland City Council. 2014. Auckland City Council Community Garden Policy. http://communitygarden.org.au/2005/10/10/auckland-com-garden-policy/ (accessed on April 19, 2014).

BBC News. 2005. Local food "greener than organic." *BBC News*, March 2. http://news.bbc.co.uk/2/hi/science/nature/4312591.stm (accessed on January 12, 2013).

Bohl, C.C. and Plater-Zyberk, E. 2006. Building community across the rural-to-urban transect. *Places*, 18(1): 4–17.

Briggs, M., Safaii, S., and Beall, D.L. 2003. Position of the American Dietetic Association, Society for Nutrition Education, and American School Food Service Association—Nutrition services: An essential component of comprehensive school health programs. *Journal of the American Dietetic Association*, 103(4): 505–514.

Buchmann, C. 2009. Cuban home gardens and their role in socio-ecological resilience. *Human Ecology*, 37: 705–721.

Canadian Horticultural Therapy Association. 2013. About horticultural therapy and therapeutic horticulture. http://www.chta.ca/about_ht.htm (accessed on March 12, 2013).

Carlsson, L. and Williams, P.L. 2008. New approaches to the health promoting school: Participation in sustainable food systems. *Journal of Hunger and Environmental Nutrition*, 3(4): 400–417.

Castro, D.C., Samuels, M., and Harman, A.E. 2013. Growing healthy kids: A community garden-based obesity prevention program. *American Journal of Preventive Medicine*, 4(3 Suppl 3): S193–S199.

City of Sydney. 2012. Community Gardens Policy. http://www.cityofsydney.nsw.gov.au/Residents/ParksAndLeisure/CommunityGardens/CommunityGardensPolicy.asp (accessed on May 29, 2012).

Coley, D., Howard, M., and Winter, M. 2009. Local food, food miles and carbon emissions: A comparison of farm shop and mass distribution approaches. *Food Policy*, 34: 150–155.

Crozier, C.R., Polizzotto, M., and Bradley, L. 2013. Soil facts: Minimizing risks of soil contaminants in urban gardens. AGW 439-78. North Carolina Cooperative Extension Service, Raleigh, NC. http://www.soil.ncsu.edu/publications/Soilfacts/AG-439-78_Urban_Soil_Contaminants.pdf (accessed on June 19, 2013).

Daniels, P., Bradshaw, M., Shaw, D., and Sidaway, J. 2008. *An Introduction to Human Geography*. 3rd edition. Harlow: Pearson Prentice Hall.

Department of Health and Ageing. 2013. Eat for health: Australian Dietary Guidelines Summary, Australian Government. http://www.eatforhealth.gov.au/sites/default/files/files/the_guidelines/n55a_australian_dietary_guidelines_summary_131014.pdf (accessed on April 20, 2014).

Duany, A. 2011. *Garden Cities: Theory & Practice of Agrarian Urbanism*. London: The Prince's Foundation for the Built Environment.

Faber, M., Venter, S.L., Spinnler, A.J., and Benade, Â. 2002. Increased vitamin A intake in children aged 2–5 years through targeted home-gardens in a rural South African community. *Public Health Nutrition*, 5(1): 11–16.

Francis, R. 2010. Harvesting the suburbs and small-space gardens: Micro-Eden series #2. Permaculture College Australia. http://permaculture.com.au/harvesting-the-suburbs-and-small-space-gardens-micro-eden-series-2/ (accessed on April 19, 2014).

Garnett, T. 1999. City harvest: The feasibility of growing more food in London. Sustain: The alliance for better food and farming. A National Food Alliance Publication. http://www.sustainweb.org/publications/?id=134 (accessed on April 19, 2014).

Garnett, T. 2007. Food refrigeration: What is the contribution to greenhouse gas emissions and how might emissions be reduced. *Paper presented at the Food Climate Research Network,* Centre for Environmental Strategy, University of Surrey.

Garnett, T. 2011. Where are the best opportunities for reducing greenhouse gas emissions in the food system (including the food chain)? *Food Policy*, 36: 23–32.

Gaynor, A. 2006. *Harvest of the Suburbs: An Environmental History of Growing Food in Australian Cities*. Crawley, WA: University of Western Australia Press.

Ghosh, S. 2010. Sustainability potential of suburban gardens: Review and new directions. *Australasian Journal of Environmental Management*, 17: 49–59.

Ghosh, S. 2011a. Growing healthy local food: Sustainability potential and household participation in home gardens. *The 5th State of Australian Cities Conference (SOAC)*, Melbourne, Australia, November 29–December 2.

Ghosh, S. 2011b. Morphologies and spatial distributions of domestic gardens in residential developments. *Institute of Australian Geographers (IAG) Conference*, Wollongong, Australia, July 3–6.

Ghosh, S. 2012. Designing food efficient urban forms. *The 10th International Urban Planning and Environment Association Symposium (UPE 10): Next City, Planning for a New Energy and Climate Future*. University of Sydney, Sydney, Australia, July 24–27.

Ghosh, S. 2013. Local food production in action: Exploring its potential for improving food security. *The 19th International Conference of the Society for Human Ecology Jointly with 4th International Conference on Sustainability Science in Asia,* Australian National University, Canberra, Australia, February 5–8.

Ghosh S. 2014. Measuring sustainability performance of local food production in home gardens. *Local Environment: The International Journal of Justice and Sustainability*, 19(1): 33–55.

Ghosh, S. and Head, L. 2009. Retrofitting suburban garden: Morphologies and some elements of sustainability potential of two Australian residential suburbs compared. *Australian Geographer*, 40(3): 319–346.

Ghosh, S., Vale, R.J.D., and Vale, B.A. 2008. Local food production in home gardens: Measuring onsite sustainability potential of residential development. *International Journal of Environment and Sustainable Development*, 7(4): 430–451.

Glouberman, S., Gemar, M., Campsie, P., Miller, G., Armstrong, J., Newman, C., Siotis, A., and Groff, P. 2006. A framework for improving health in cities: A discussion paper. *Journal of Urban Health: Bulletin of the New York Academy of Medicine*, 83(2): 325–338.

Graham, H. and Zidenberg-Cherr, S. 2005. California teachers perceive school gardens as an effective nutritional tool to promote healthful eating habits. *Journal of the American Dietetic Association*, 105: 1797–1800.

Graham, S. and Connell, J. 2006. Nurturing relationships: The gardens of Greek and Vietnamese migrants in Marrickville, Sydney. *Australian Geographer*, 37(3): 375–393.

Grimms, J. 2009. Food urbanism—A sustainable design option for sustainable communities. *Landscape Architecture and Environmental Studies,* Iowa State University, Ames, IA. http://johnsonlinn-localfood.webs.com/Planning%20Resources/Food%20Urbanism_Grimm.pdf (accessed on April 19, 2014).

Halweil, B. 2002. Home grown: The case for local food in a global market. Worldwatch paper 163. WorldWatch Institute, Washington, DC.

Harmon, A. and Maretzki, A. 2006. A survey of food system knowledge, attitudes, and experiences among high school students. *Journal Hunger and Environmental Nutrition*, 1: 59–82.

Hartman Group. 2008. Consumer understanding of buying local. http://www.hartman-group.com/hartbeat/2008-02-27 (accessed on March 20, 2011).

Haug, A., Brand-Miller, J.C., Christoperson, O.A., McArthur, J., Fayet, F., and Truswell, S. 2007. A food "lifeboat": Food nutrition considerations in the event of a pandemic or other catastrophe. *Medical Journal of Australia*, 187(11/12): 674–676.

Heim, S., Bauer, K.W., Stang J., and Ireland, M. 2011. Can a community-based intervention improve the home food environment? Parental perspectives of the influence of the delicious and nutritious garden. *Journal of Nutrition Education and Behavior*, 43(2): 130–134.

Hinrichs, C.C. 2003. The practice and politics of food system localization. *Journal of Rural Studies,* 19: 33–45.

Loram, A., Warren. P.H., and Gaston, K.J. 2008. Urban domestic gardens (XIV): The characteristics of gardens in five cities. *Environmental Management*, 42(3): 361–376.

McGill University. 2013. Making the edible campus. http://www.mcgill.ca/mchg/sites/mcgill.ca.mchg/files/MakingtheEdibleCampus.pdf (accessed on March 27, 2013).

Ministry of Health. 1999. NZ Food: NZ people. Key results of the 1997 National Nutrition Survey. Wellington, New Zealand.

Minnesota Institute of Sustainable Agriculture. 2010. Urban gardens and soil contaminants—A gardener's guide to healthy soil. http://www.misa.umn.edu/prod/groups/cfans/@pub/@cfans/@misa/documents/asset/cfans_asset_287228.pdf (accessed on March 29, 2013).

Mizuno-Matsumoto, Y., Kobashi, S., Hata, Y., Ishikawa, O., and Asano, F. 2008. Horticultural therapy has beneficial effects on brain functions in cerebrovascular diseases. *International Journal of Intelligent Computing in Medical Sciences and Image Processing (IC-MED)*, 2(3): 169–182.

Murray, H., Pinchin, T.A., and Macfie, S.M.2011. Compost application affects metal uptake in plants grown in urban garden soils and potential human health risk. *Journal of Soils and Sediments*, 11(5): 815–829.

National Heart Foundation of Australia. 2013. Healthy by design—A planners' guide to environments for active living. http://www.heartfoundation.org.au/SiteCollectionDocuments/Healthy-by-Design.pdf (accessed on March 20, 2013).

New South Wales Government. 2009. Healthy urban development checklist: A guide for health services when commenting on development policies, plans and proposals. Sydney, Australia. http://www0.health.nsw.gov.au/pubs/2010/pdf/hud_checklist.pdf (accessed on April 19, 2014).

North Carolina State University. 1996. Home Vegetable Gardening. http://www.ces.ncsu.edu/depts/hort/hil/ag-06.html (accessed on April 19, 2014).

Northern Territory Food Gardens. 2013. NT gardens. Northern Territory Food Gardens. http://ntgardens.org/ (accessed on April 2, 2013).

Organic Vision. 2011. Planet Cuba: Complex, complicated but sustainable. http://72.18.132.73/~organicv/2011/03/planet-cuba-complex-complicated-but-sustainable/ (accessed on November 8, 2012).

Parker, J. 2010. Consumer awareness and behaviour survey relating to production, distribution and consumption of local foods. Northern Rivers Food Links Report. New South Wales, Australia.

Pederson, R.M. and Robertson, A. 2001. Food policies are essential for healthy cities. *Urban Agriculture Magazine*, 3: 9–10.

Pelletier, J.E., Laska, M.N., and Neumark-Sztainer, D. 2013. Positive attitudes toward organic, local, and sustainable foods are associated with higher dietary quality among young adults. *Journal of the Academy of Nutrition and Dietetics*, 113(1): 127–132.

Remote Indigenous Gardens' Network. 2013. The remote indigenous gardens network—RIG network. http://www.remoteindigenousgardens.net/ (accessed on November 8, 2012).

Robinson-O'Brien, R. and Neumark-Sztainer, D. 2009. Characteristics and dietary patterns of adolescents who value eating locally grown, organic, nongenetically engineered, and nonprocessed food. *Journal of Nutrition Education and Behavior*, 41(1): 11–19.

Saeumel, I., Kotsyuk, I., Hoelscher, M., Lenkereit, C., Weber, F., and Kowarik, I. 2012. How healthy is urban horticulture in high traffic areas? Trace metal concentrations in vegetable crops from plantings within inner city neighbourhoods in Berlin, Germany. *Environmental Pollution*, 165: 124–132.

Talukder, A., Kiess, L., Huq, N., de Pee, S., Darnton-Hill. I., and Bloem, M.W. 2000. Increasing the production and consumption of vitamin A-rich fruits and vegetables: Lessons learned in taking the Bangladesh homestead gardening programme to a national scale. *Food Nutrition Bulletin*, 21(2): 165–172.

The University of Utah. 2013. Edible Campus Gardens. http://sustainability.utah.edu/operations/food/edible-gardens.php (accessed on January 29, 2013).

Toronto Food Policy Council. 2013. Food policy. Toronto, ON. http://www.toronto.ca/health/tfpc/ (accessed on January 12, 2013).

US Environmental Protection Agency. 2011. Reusing potentially contaminated landscapes: Growing gardens in urban soils. EPA/542/F-10/011. http://www.cluin.org/download/misc/urban_gardening_fact_sheet.pdf (accessed on April 19, 2014).

Van Den Berg, A.E. and Custers, M.H.G. 2011. Gardening promotes neuroendocrine and affective restoration from stress. *Journal of Health Psychology*, 16: 3–11.

Van Den Berg, A.E., Winsum-Westra, M.V., Vries, S.D., and Dillen, S.M.E.V. 2010. Allotment gardening and health: A comparative survey among allotment gardeners and their neighbors without an allotment. *Environmental Health*, 9: 1–12.

Wackerangel, M. and Rees, W. 1996. *Our Ecological Footprint: Reducing Human Impact on Earth.* Gabriola Island, BC: New Society Publishers.

Wakefield, S., Yeudall, F., Taron, C., Reynold, J., and Skinner, A. 2007. Growing urban health: Community gardening in South-East Toronto. *Health Promotion International*, 22(2): 92–101.

Wekerle, G.R. 2002. Toronto's official plan from the perspective of community gardening and urban agriculture. City Farmer, Toronto, ON. http://www.cityfarmer.org/torontoplan.html (accessed on April 4, 2011).

Winklerprins, A.M.G.A. 2002. House-lot gardens in Santarém, Pará, Brazil: Linking rural with urban. *Urban Ecosystems*, 6: 43–65.

World Health Organization. 2003. *Fruit and Vegetable Promotion Initiative*. A meeting report/25-27/08/03. http://www.sochinut.cl/pdf/fruit_and_vegetable_report.pdf (accessed on March 20, 2013).

World Health Organization. 2013. Types of healthy settings. http://www.who.int/healthy_settings/types/cities/en/ (accessed on March 27, 2013).

28 Oriental Traditional Philosophy and Food Function

Young Rok Seo and Yeo Jin Kim

CONTENTS

28.1 INTRODUCTION

Different food preparation and consumption have arisen due to the geographic, social, and economic separation among societies, and those patterns comprise food cultures when adaptation to surrounding environmental conditions has been successful. Food culture is considered central part of human's life. Ancient people cooked with the ingredients that could be gathered or raised in their own region in accordance with their cultures. The traditional Oriental vegetarian food culture has descended from Eastern philosophy and ancient Asian thought. Buddhism, Hinduism, and Taoism are representative of traditional Eastern philosophy and religion. Traditionally, the majority have not allowed the use of animal products as food ingredients due to the ideology of *ahimsa* that prohibits the taking of life. Accordingly, various vegetal diet cultures were developed with a foundation in religious ritual activity.

Western dietary life developed in the context of nomadic life, so the amount of meat consumption was relatively higher than in agricultural civilization. The core of the Western diet is meat with bread made from wheat. In many modern Western countries, the meat-based diets, excess of food available per capita, and government subsidies for production of certain ingredients have led to the excessive intake of saturated fats and high-calorie foods. It is worth noting the correlation between the amount of meat intake per person and the rate of death due to colorectal cancer (Armstrong and Doll 1975; Rose et al. 1986; Santarelli et al. 2008). Carcinogens such as heterocyclic amine (HCA) are formed when meat is cooked at a high temperature leading to an increase in colorectal cancer (Norat et al. 2002). Not only HCA but also *N*-nitroso compounds in meat are reported to be carcinogenic based on many animal studies (Lijinsky 1999). Because of the high risk of cancer due to the high consumption of fat, the US Department of Health and Human Services (2005) has suggested in its diet guidelines that the public choose product low in fats and limit intake of fats saturated and/or *trans* fatty acid.

In contrast to the Western society, Asian countries were more agrarian, so their livestock were valuable assets as agricultural draft animals than as a source of meat. That is why rice became the staple food with side dishes composed of vegetables such as wild greens rather than meats. In addition, they chose to salt or ferment in-season ingredients to keep them fresh for a longer period of time. Consequently, a larger proportion of the Oriental diet has traditionally been given over to vegetables than the Western diet. The Eastern diet low in fat and cholesterol has been passed down over the generations. Fruits and vegetables are rich in phytochemicals that act as antioxidants. It is reported that vegetarian foods not only supply basic nutrition but also decrease the risk of chronic diseases including cancer and diabetes because they contain abundant plant compounds with anti-cancer or anti-inflammatory properties. Thus, because of the traditional vegetal diet, Asians tend to have a lower risk of cardiovascular disease or obesity than Westerners.

Recently, however, traditional Oriental food culture has been overtaken by the global spread of Western-origin multinational food processing corporations and restaurant chains. The consumption of these companies' foods has increased the incidence of diseases such as obesity, diabetes, and colorectal cancer. Though, at the same time, the rise in the standard of living in many Asian countries has brought with it a greater concern for quality of life including health and nutrition. Accordingly, the public has become interested in the results of studies on the diet management for maintaining a healthy life. As a result, it is expected that Asians will address the threat of modern diseases by returning to a traditional Eastern vegetable-based diet with a new foundation of knowledge based on modern research. It is the aim of this chapter to overview both the traditions that inform the Eastern diet and the latest research on functional food.

28.2 THE VEGETARIAN DIET IN BUDDHISM

Among the traditional systems of ideas that have originated and spread from the Eastern countries, Buddhism is a representative philosophy. Buddhism is based on mindfulness, which is called *sati* in the ancient Indian language of Pali. From the dietary perspective, the practice of *sati* involves observing carefully what one eats and drinks as well as savoring one's food. Buddhist foods passed down from ancient India have diversified to fit the environment of each geographic area to which Buddhism spread. They are known as "temple food" in the East Asian region. Temple food is spiritually based on the respect for life, of which vegetarianism is the logical consequence. Temple foods differ a bit in each region; some are vegan. In other words, in addition to meat itself, no animal products such as milk, egg, or honey are ingested. In other regions, food can be obtained from living creatures as long as the animals are not killed. In addition, temple food also plays a role in ascetic exercises that help lead to enlightenment. Buddhism teaches that the change of seasons influences human physiology, so seasonal foods are important ingredients in temple food. In addition, focusing on the taste of unprocessed ingredients, eating lightly and eating vegetarian all contribute to the effort to find harmony between the body and the mind.

Some famous Buddhist foods are shiitake mushroom rice made up of shiitake, miso made from fermented soy beans, and lotus leaf rice, which is boiled rice and cereals wrapped in a steamed lotus leaf. Shiitake mushroom is the main ingredient of shiitake mushroom rice and other vegetarian foods. The scientific name of the shiitake mushroom is *Lentinula edodes*; it mainly inhabits in East Asia, including Korea, China, and Japan. When reconstituted in water after drying, shiitake mushroom becomes chewier and has a stronger mushroom taste. Shiitake is used as an important ingredient in strict vegetarian diet, *shojin* dishes, which are strictly vegetarian in Japanese Buddhist food. Shiitake can be put in rice to accentuate its flavor. Since it has a meat-like texture, it can also be fried. Shiitake is stored dry after dehydrating in the sun, and it is used as a *dashi*, which is a basic ingredient of miso soup in Japan. In *L. edode*, there are variety of vital bioactive elements such as proteins, lipids, carbohydrates, fibers, minerals, and the antitumor polysaccharide lentinan (Sadler 2003). Like other types of mushrooms, it was reported that *L. edode* can produce a large quantity of vitamin D when it is exposed to sunlight or UV light (Lee et al. 2009). It is also known for its ability to lower the blood pressure, reduce cholesterol, and strengthen the immune system (Bobek et al. 1991). A research team in Brazil determined that a mutagen-injected mouse that was fed *L. edode* showed less frequency of micronucleated polychromatic erythrocytes (MNPCEs) (Sugui et al. 2003). According to another study, *L. edode* reduces platelet aggregation and plays an antiviral or antibacterial role because it possesses protein inhibitors (Bisen et al. 2010). In other studies,

L. edode containing bioactive substances such as lentinan was reported to inhibit cancer incidence (Nanba et al. 1987; Hasegawa et al. 1989). An active hexose-correlated compound (AHCC), which can be extracted from shiitake and is rich in α-glucan, is used as an alternative therapy for cancer patients in Japan (Hyodo et al. 2005). In an animal study, it was reported that AHCCs increase resistance to pathogens (Shah et al. 2011). *Lentinula edodes* has also shown efficacy in the liver cells. According to a 2010 study from Malaysia, animal experiments showed that *L. edode* extracts protected the hepatocytes by an antioxidant effect against hepatotoxicity induced by paracetamol (acetaminophen) (Sasidharan et al. 2010). Accordingly, it has been concluded that the intake of *L. edode*, containing vital biochemical substances, strengthens the immune system.

A variety of Buddhist foods contain beans notably; Korean soybean paste and Japanese soy miso are traditional fermented bean foods. The soybean (*Glycine max*) is recognized as a complete protein food by the US Food and Drug Administration (FDA) (Henkel 2000). A complete protein contains a certain amount of essential amino acids that must be supplied because the human body cannot synthesize them on its own. *Glycine max* is composed of 40% protein, 20% oil, 35% carbohydrates, and 5% vitamins and inorganic salts. Most of the proteins have heat stability, so it can be cooked at high temperatures without destroying the essential amino acids. It also has a variety of biologically active substances. According to a Serbian study, the phenolic substances included in *G. max* are more effective antioxidants than its proteins or oil (Malencic et al. 2007). These phenolic substances include isoflavones, saponins, and antocyanins. Isoflavones can also act as an estrogen antagonist. The common isoflavones in *G. max* are genistein and daidzein. It was reported that they can block the growth of estrogen receptor-associated cancer cells *in vitro* (Boué et al. 2009). It was also shown in an *in vitro* study that isoflavones and saponins in *G. max* prevent colorectal cancer (MacDonald et al. 2005). Anthocyanins, a component of *G. max*, could be an immune suppressor by suppressing cyclooxygenase (COX)-2 and inducible nitric oxide synthase (iNOS) mRNA, as found by an *in vitro* study of human colorectal cancer cells and an *in vivo* study of animals exposed to carcinogens (Kim et al. 2008). In sum, soybeans, which contain various phytochemicals, have antioxidant, anti-inflammatory, and anticancer properties.

The lotus plant is the symbol of Buddhism, and most of its parts, including the rhizome, stem, and leaves, are edible. In Asia, lotus is prepared in a variety of ways, such as lotus rhizome boiled in soy sauce and lotus leaf rice. The lotus (*Nelumbo nucifera*) grows in the soil submerged in ponds. In China, the extract of lotus or lotus tea is used to treat obesity. It has been reported that lotus leaf can be used in the treatment of hyperlipidemia in experiments with rodent models (Onishi et al. 1984; La Cour et al. 1995). Using mice with high-fat diet-induced obesity, the rate of digestive enzymatic activity, lipid metabolism, and thermogenesis was measured to observe the antiobesity effects of lotus leaf extract. The results of this Japanese study show the great of *N. nucifera* in drugs for obesity suppression based on the regulation of fat absorption by increasing energy expenditure and lipid degradation (Ono et al. 2006). Lotus leaf is abundant in flavonoids and alkaloids. The antioxidants Qc-3-Glc and Qc-3-Gln can be extracted from lotus leaf, and consuming them can prevent disease caused by oxidative stress (Jung et al. 2008). Lotus rhizome extracts reduce blood sugar levels in normal and diabetic rats (Mukherjee et al. 1997). Thus, *N. nucifera* might be used as an antiobesity agent and also contributes to lowering the risk of diabetes and increasing immunity.

28.3 THE HINDU DIET FROM ANCIENT INDIA

Hinduism is a tradition handed down from Central Asia. It is not based on dogma, but it is a religion developed over a long period of time associated with diverse customs and systems of thought. Brahmanism, an early form of Hinduism also known as Vedism, was transformed by absorbing indigenous beliefs into Hinduism. An important characteristic of Hinduism is veneration of cows, so eating beef is strictly forbidden. More generally, an Indian sacred text, *the Mahabharata*, says "Nonviolence is the best duty and the best teaching." In a spirit of nonviolence, Hindus practice vegetarianism, but the extent varies depending on the region. It is believed that nonvegetarian foods interfere with the maturation of the heart and soul. For these reasons, vegetarian meals should be the main in Hinduism.

Most Hindu foods are derived from the ancient Indian foods. Spices, herbs, vegetables, and fruits are the main ingredients. Lassi, curry, and dal are some of the representative foods of Hinduism. Lassi is a yogurt-based drink originating from India and Pakistan. The main ingredient is yogurt, which is made from fermented milk. It is usually drunk with fruit, honey, and spices to taste, but traditional lassi is simply a blend of yogurt with water and salt. Lassi is a coagulated dairy product generated from lactic acid fermentation by *Lactobacillus delbrueckii* subsp. *bulgaricus* and *Streptococcus salivarius* subsp. *thermophilus* in milk (Bourlioux and Pochart 1988). Lactic acid bacteria (LAB) have recently drawn attention because they are accepted as probiotics. Over a century ago, Metchnikoff, a Russian scientist, demonstrated that *L. bulgaricus* inhibits toxic substances from putrefactive bacteria in the human intestine (Metchnikoff and Mitchell 1908). More recently, several studies have been conducted on the effects of LAB and have reported that the grater the quantity of LAB is the lower quantity of pathogenic bacteria (Gilliland 1990; Perdigon et al. 2001; Hummel et al. 2007). It has also been revealed that *Lactobacillus acidophilus* and *Bifidobacterium longum* have the capacity to resist gastric juice (Charteris et al. 1998; Kailasapathy and Chin 2000). LAB can induce adjuvant activity on the colon mucosal surface by excreted peptidoglycan that is digested by lysozyme in the intestine. This might lead to antibody formation, which in turn would increase the immune response (Link-Amster et al. 1994). In addition, it was demonstrated that glycopeptides in LAB's cell wall help to enhance immunity (Bogdanov et al. 1975). An animal study showed that the production of secreted immunoglobulin A (sIgA) and the number of sIgA-producing cells are increased in the small intestine of mice orally administered LAB and fed yogurt (Perdigon et al. 1995; Bakker-Zierikzee et al. 2006). Using yogurt containing living bacteria, *L. bulgaricus* and *S. thermophilus*, the yogurt-fed mice had a higher percentage of B lymphocytes in Peyer's patch cells than milk-fed control mice (DeSimone et al. 1987; Adolfsson et al. 2004). In epidemiologic research, the intake of yogurt containing living LAB strengthened the immune system over the long time by stimulating production of cytokines from *L. bulgaricus* and *S. thermophilus*. Folates from vitamin B_{12} are also generated during yogurt fermentation. Yogurt also contains higher quantities of minerals such as calcium than milk (Meydani and Ha 2000). It was reported that consumption of probiotic yogurt containing *La. acidophilus* and *Bifidobacterium lactis* has the antioxidant property and improves fasting blood glucose in type 2 diabetic patients (Ejtahed et al. 2012). Consequently, yogurt-based lassi containing LAB enhances immunity.

Curry, mainly eaten with rice or bread, originated in ancient India. It is usually made with various vegetables and spices depending on the nation or region, and one of the common ingredients is turmeric. This plant inhabits India, Southeast Asia, and Southwest China, and mainly its rhizome is used for food. Its scientific name is *Curcuma longa* L. and it contains curcumin as a major component (Araújo and Leon 2001). Curcumin, a well-known anticancer agent, is reported to inhibit carcinogenesis and metastasis (Duvoix et al. 2005). It was demonstrated that curcumin interferes with angiogenesis, particularly neovascularization by fibroblast growth factor (Arbiser et al. 1998). When cancer was chemically induced in another animal study, carcinogenesis was lower in a curcumin-administered group than a control group. The weight in the curcumin group was also reduced in the same study (Inano et al. 1999). Moreover, it could prevent angiogenesis and metastasis of cancer cells by regulating cell adhesion molecules such as intracellular adhesion molecule (ICAM)-1, vascular cell adhesion molecule (VCAM)-1, and endothelial leukocyte adhesion molecule (ELAM)-1 (Bhandarkar and Arbiser 2007). Curcumin in *C. longa* L. affects signal transduction, especially in essential elements such as nuclear factor kappa B (NF-κB) and Akt. It was reported that curcumin regulates several molecules related to the cell cycle or inflammation downstream of NF-κB by suppressing it (Shishodia et al. 2005). Curcumin inhibits the activity of Akt that controls apoptosis via protein phosphorylation. In the study, using curcumin-treated human kidney cancer cells, the level of Akt expression and phosphorylation is definitely decreased (Woo et al. 2003). In an animal study, the anticancer effect of curcumin under dose-dependent conditions was found in the colon, the stomach, and the mouth (Maheshwari et al. 2006). More recently, it has been shown that curcumin can inhibit insulin-induced colon cancer cell proliferation *in vitro* through a methyl ethyl ketone (MEK)-mediated mechanism (Fenton and McCaskey 2013). In epidemiologic research, curcumin is beneficial in delaying development of type 2 diabetes in the prediabetes population (Chuengsamarn et al. 2012). In short, curcumin, one of turmeric components, provides anticancer and antidiabetic effects.

Dal, which is mainly eaten on the Indian subcontinent, is stew made with pulses. In South India, it is served with rice and vegetables in a diverse combination of pulses including soybeans, which have already been discussed, peas, and so on. It was revealed that abundant intake of pulses contributes to the lower rates of chronic diseases such as cardiovascular disease or cancers in Asia compared to Western countries (Messina 1995). In addition to the benefits of soy described in Subsection 28.2, many isoflavones and proteins that also exist in other beans have positive effects on health. 7S protein, one of soy proteins, is a polymer with α, α', and β subunits (Maruyama et al. 1998). It was shown that soybean proteins activate a process that reduces membrane cholesterol and triglycerides (Reynolds et al. 2006). When human liver cells were treated with the α' chain, it was found that ingestion and degradation of cholesterol increased and more low-density lipoprotein (LDL) receptor RNAs were expressed (Duranti et al. 2004). Another study demonstrated that genistein, an isoflavone in pulses, can help treat type 2 diabetes (Gilbert and Liu 2013). In particular, it stimulates proliferation of β cells in the pancreas for an antidiabetic effect. In *in vitro* and *in vivo* studies of genistein treatment, pancreatic β-cell division is activated (Fu et al. 2010). In an epidemiologic study, bean components reduced the rate of diabetes (Liu et al. 2011). In short, intake of diverse combination of pulses contributes to improve chronic disease.

28.4 TAOIST FOODS BASED ON YIN–YANG AND THE FIVE ELEMENTS

Taoism was developed by Lao Tzu in China, and it spread widely throughout East Asian countries. It is a Chinese representative philosophy that tends to emphasize living in harmony. This idea includes the idea of Yin–Yang and the Five Elements. Even before Lao Tzu, the idea of Yin–Yang and the Five Elements existed in traditional Asian ideology and shaped the establishment of traditional Chinese medicine, Taoism, and other systems of thought in China. Yin–Yang is a concept from Oriental philosophy that consists of two opposing forces, Yin and Yang, and the harmony between them. Every object or situation must be in an interdependent relationship with each other, and the separation of Yin and Yang is necessary for their relationship. However, the separation is not absolute but relative, so the two cannot exist independently from each other. The Five Elements are the first principles of all things: wood, fire, soil, metal, and water. The universe consists of the movements of these five auras, in accordance with Taoism philosophy. Maintaining the harmonious relationship among the elements influences the universe. When Ying–Yang and the Five Elements stay in harmony, it is called the balanced state. In Oriental and Chinese medicines, a Yin–Yang balanced diet improves one's constitution or cures diseases according to differences of seasonal climate, gender, and body parts. However, since these traditional claims have not yet been fully proven scientifically, they need to be reinterpreted cautiously in view of modern molecular dietetics.

From a medical and physiological perspective, when Yin–Yang and the Five Elements are in a state of equilibrium, homeostasis is maintained in the human body by negative feedback. This concept has been explicated as self-harmonized Yin–Yang. This idea is recognized that modern scientific evidences are insufficient, but they can be considered positive ways from the perspective of maintaining equilibrium. When Yang is exceeded, Yin is added to maintain the balance point and vice versa. The paired Yin–Yang at different levels is common in the human body from the systemic level, such as the body temperature and blood pressure, to the molecular level. One set of cellular proteins, the kinases, is said to have Yang properties and counterbalance the Yin of phosphatase (Tan 1993; Schwoerer et al. 2008). Additional Yin–Yang pairs include lipoproteins, such as cholesteryl ester, and vitamin E, which are pro-oxidants and an antioxidant, respectively (Aviram

1999). In addition, it was reported that the proteins p53 and c-*Myc*, which play a role in regulating cell growth and proliferation, are regulated through independent ubiquitylation by several E3 ubiquitin ligases and their regulator ARF (alternate reading frame). According to this study, p53–c-*Myc* networks for the control of cell growth maintain cellular homeostasis in an interdependence and fluctuating balance between Yin and Yang. ARF and p53 act as the Yin force to suppress neoplasia provided by the Yang of c-*Myc* from undergoing uncontrolled cell growth (Dai et al. 2006). The Oriental ideology that the human body is composed of a balance of energy, Yin and Yang, can be applied to the diet. According to the principle that every creature is associated with the Yin–Yang and the Five Elements philosophy, each part of the human body also contains these principles.

In traditional Chinese medicines, the five colors and tastes, which are based on the Yin–Yang and the Five Elements theory, are closely related to the organs of the human body. The five elements are represented by green, red, yellow, white, and black. Chromatic foods, which are foods with a variety of colors, are drawing interests as a healthy diet. Beginning in 1988, with the supports from the National Cancer Research Center, the "5 A Day" campaign was conducted in California. It suggests eating at least five types of fruits or vegetables per day, and in 2003, the World Health Organization (Wholey et al. 2003) recommended eating at least 400 g of vegetables a day. Eating a variety of colorful fruits and vegetables ensure the intake of color-specific phytochemicals such as chlorophyll, lycopene, carotenoid, flavonoid, and allicin. These phytochemicals act as an antioxidant or antiaging factor so that they improve the immune system and promote general health. The correlation between color and health, according to the principles of Yin–Yang and the Five Elements, has been applied in traditional Chinese medicine and common folk remedies through the ages. However, since these principles have not been fully explained by modern medicine and dietetics due to a lack of scientific data, we should approach them cautiously. In addition, plenty of exceptions to these principles should be considered.

28.5 DISCUSSION

General fat, an essential component of the body, plays a vital role in physiological functioning and as an energy source. However, modern high-fat, high-calorie diets have been associated with various chronic diseases. It has been reported that the expression of lipid metabolism genes is repressed by consumption of high-fat diet in the liver and white adipose tissue of rats (Tovar et al. 2011). The expression of lipogenin, which is involved in lipogenesis and the differentiation of fat cells, is increased by high-fat diet-induced obesity (Li et al. 2002). The consumption of excessive saturated fat and cholesterol may contribute to increases in blood pressure by plaques on the vessel walls, which increase the risk of cardiovascular diseases (Stamler et al. 1986; Siri-Tarino et al. 2010). It has also been demonstrated that even a single high-fat meal may affect cardiovascular function in healthy adults (Jakulj et al. 2007).

The vegetal diets based on traditional Oriental philosophy and religion ensure the intake of a variety of phytochemicals and prevent several diseases including cancer, diabetes, and obesity. It is reported that bioactive phytochemicals from various fruits and vegetables can lower the risk of stroke (Larsson et al. 2013). Vegetable-derived compounds have been found to afford cancer protection as protective agents (MacDonald and Wagner 2012; Lampe 2009). Most phytochemicals act as antioxidants. The phytochemicals that have an anticancer effect are curcumin, AHCC, lentinan, and saponin. Genistein has an antidiabetic effect and the phytochemicals in *N. nucifera* act as antiobesity agents. In addition, some anthocyanin and dairy bacteria stimulate the immune system.

The consumption of low-fat, low-calorie vegetal foods as encouraged by traditional Oriental philosophies reduces the risk of chronic illness. However, a vegetal diet is not the only means to maintaining health. Other factors contribute to the preservation of health, for example, calorie restriction, which is the reduction of caloric intake without malnutrition. It is reported that calorie restriction reduces oxidative stress, which in turn reduces protein and DNA damage, and potentially extends the life span (Sohal and Weindruch 1996). During calorie restriction, blood pressure is decreased by the decline in insulin concentration and sympathetic nervous activity (Velthuis-te Wierik et al. 1994).

Low-calorie intake with adequate nutrient consumption could improve health and life expectancy by reducing the risk of chronic diseases (Fontana and Klein 2007).

The other important factor in maintaining general health is lifestyle behaviors, including exercise, sleeping patterns, smoking and drinking, and meditation. A great number of studies suggested that physical activity contributes to a protective effect on cardiovascular risk and other chronic disease (Adamska et al. 2012; Zoeller 2009; Vogel et al. 2009; Lin et al. 2010). Additional factors other than food which also promote the quality of life and health are too hard to enumerate here and vary according to the individual and context. Even so, reducing the intake of common injurious substances is very useful in maintenance of physical, mental, and social well-being.

The Eastern philosophies on food are nonscientific and remain largely unexplored by modern science, and thus scientific study of the diets prescribed by the Eastern religions is worth pursuing. Specifically, studies are needed to determine the physiological effects of functional foods of the Oriental traditions in conjunction with various factors. In addition, it is necessary to investigate the points of view of the detractors of these religions. Where Eastern traditional thought has accumulated a helpful understanding of health and nutrition over the centuries, although expressed in different language than that of modern science, it will be found to agree with scientific research in the future.

ACKNOWLEDGMENT

We are grateful to Han Sol Bae, Dongguk University, Seoul, for carefully modifying the chapter and giving us helpful advice.

REFERENCES

Adamska, A., M. Przegaliński, I. Rutkowska, A. Nikołajuk, M. Karczewska-Kupczewska, M. Górska, and M. Strączkowski. 2012. Relationship between regular aerobic physical exercise and glucose and lipid oxidation in obese subjects—A preliminary report. *Polish Annals of Medicine*, 19(2): 117–121. doi:10.1016/j.poamed.2012.05.001.

Adolfsson, O., S.N. Meydani, and R.M. Russell. 2004. Yogurt and gut function. *American Journal of Clinical Nutrition*, 80(2): 245–256.

Araújo, C.A.C. and L.L. Leon. 2001. Biological activities of *Curcuma longa* L. *Memórias do Instituto Oswaldo Cruz*, 96: 723–728.

Arbiser, J.L., N. Klauber, R. Rohan, R. van Leeuwen, M.T. Huang, C. Fisher, E. Flynn, and H.R. Byers. 1998. Curcumin is an in vivo inhibitor of angiogenesis. *Molecular Medicine*, 4(6): 376–383.

Armstrong, B. and R. Doll. 1975. Environmental factors and cancer incidence and mortality in different countries, with special reference to dietary practices. *International Journal of Cancer*, 15(4): 617–631.

Aviram, M. 1999. Macrophage foam cell formation during early atherogenesis is determined by the balance between pro-oxidants and anti-oxidants in arterial cells and blood lipoproteins. *Antioxidants and Redox Signaling*, 1(4): 585–594.

Bakker-Zierikzee, A.M., E.A.F. van Tol, H. Kroes, M.S. Alles, F.J. Kok, and J.G. Bindels. 2006. Faecal SIgA secretion in infants fed on pre- or probiotic infant formula. *Pediatric Allergy and Immunology*, 17(2): 134–140. doi:10.1111/j.1399-3038.2005.00370.x.

Bhandarkar, S.S. and J.L. Arbiser. 2007. Curcumin as an inhibitor of angiogenesis. *Advances in Experimental Medicine and Biology*, 595: 185–195. doi:10.1007/978-0-387-46401-5_7.

Bisen, P.S., R.K. Baghel, B.S. Sanodiya, G.S. Thakur, and G.B. Prasad. 2010. *Lentinus edodes*: A macrofungus with pharmacological activities. *Current Medicinal Chemistry*, 17(22): 2419–2430.

Bobek, P., E. Ginter, L. Kuniak, J. Babala, M. Jurcovicova, L. Ozdin, and J. Cerven. 1991. Effect of mushroom *Pleurotus ostreatus* and isolated fungal polysaccharide on serum and liver lipids in Syrian hamsters with hyperlipoproteinemia. *Nutrition*, 7(2): 105–108.

Bogdanov, I.G., P.G. Dalev, A.I. Gurevich, M.N. Kolosov, V.P. Malekova, L.A. Plemyannikova, and I.B. Sorokina. 1975. Antitumour glycopeptides from *Lactobacillus bulgaricus* cell wall. *FEBS Letters*, 57(3): 259–261.

Boué, S.M., S.L. Tilghman, S. Elliott, M.C. Zimmerman, K.Y. Williams, F. Payton-Stewart, A.P. Miraflor et al. 2009. Identification of the potent phytoestrogen glycinol in elicited soybean (*Glycine max*). *Endocrinology*, 150(5): 2446–2453. doi:10.1210/en.2008-1235.

Bourlioux, P. and P. Pochart. 1988. Nutritional and health properties of yogurt. *World Review of Nutrition and Dietetics*, 56: 217–258.

Charteris,W.P., P.M. Kelly, L. Morelli, and J.K. Collins. 1998. Development and application of an in vitro methodology to determine the transit tolerance of potentially probiotic *Lactobacillus* and *Bifidobacterium* species in the upper human gastrointestinal tract. *Journal of Applied Microbiology*, 84(5): 759–768. doi:10.1046/j.1365-2672.1998.00407.x.

Chuengsamarn, S., S. Rattanamongkolgul, R. Luechapudiporn, C. Phisalaphong, and S. Jirawatnotai. 2012. Curcumin extract for prevention of type 2 diabetes. *Diabetes Care*, 35(11): 2121–2127. doi:10.2337/dc12-0116.

Dai, M.S., Y. Jin, J.R. Gallegos, and H. Lu. 2006. Balance of Yin and Yang: Ubiquitylation-mediated regulation of p53 and c-Myc. *Neoplasia*, 8(8): 630–644. doi:10.1593/neo.06334.

De Simone, C., R. Vesely, R. Negri, B. Bianchi Salvadori, S. Zanzoglu, A. Cilli, and L. Lucci. 1987. Enhancement of immune response of murine Peyer's patches by a diet supplemented with yogurt. *Immunopharmacology and Immunotoxicology*, 9(1): 87–100. doi:10.3109/08923978709035203.

Duranti, M., M.R. Lovati, V. Dani, A. Barbiroli, A. Scarafoni, S. Castiglioni, C. Ponzone, and P. Morazzoni. 2004. The alpha' subunit from soybean 7S globulin lowers plasma lipids and upregulates liver beta-VLDL receptors in rats fed a hypercholesterolemic diet. *Journal of Nutrition*, 134(6): 1334–1339.

Duvoix, A., R. Blasius, S. Delhalle, M. Schnekenburger, F. Morceau, E. Henry, M. Dicato, and M. Diederich. 2005. Chemopreventive and therapeutic effects of curcumin. *Cancer Letters*, 223(2): 181–190. doi:10.1016/j.canlet.2004.09.041.

Ejtahed, H.S., J. Mohtadi-Nia, A. Homayouni-Rad, M. Niafar, M. Asghari-Jafarabadi, and V. Mofid. 2012. Probiotic yogurt improves antioxidant status in type 2 diabetic patients. Nutrition 28(5): 539–543. doi:10.1016/j.nut.2011.08.013.

Fenton, J.I. and S.J. McCaskey. 2013. Curcumin and docosahexaenoic acid block insulin-induced colon carcinoma cell proliferation. *Prostaglandins, Leukotrienes and Essential Fatty Acids*, 88(3): 219–226. doi:10.1016/j.plefa.2012.11.010.

Fontana, L. and S. Klein. 2007. Aging, adiposity, and calorie restriction. *Journal of the American Medical Association*. 297(9): 986–994. doi:10.1001/jama.297.9.986.

Fu, Z., W. Zhang, W. Zhen, H. Lum, J. Nadler, J. Bassaganya-Riera, Z. Jia, Y. Wang, H. Misra, and D. Liu. 2010. Genistein induces pancreatic beta-cell proliferation through activation of multiple signaling pathways and prevents insulin-deficient diabetes in mice. *Endocrinology*, 151(7): 3026–3037. doi:10.1210/en.2009-1294.

Gilbert, E.R. and D. Liu. 2013. Anti-diabetic functions of soy isoflavone genistein: Mechanisms underlying its effects on pancreatic [small beta]-cell function. *Food & Function*, 4(2): 200–212.

Gilliland, S.E. 1990. Health and nutritional benefits from lactic acid bacteria. *FEMS Microbiology Letters*, 87(1/2): 175–188. doi:10.1016/0378-1097(90)90705-U.

Hasegawa, J., M. Hosokawa, F. Okada, and H. Kobayashi. 1989. Inhibition of mitomycin C-induced sister-chromatid exchanges in mouse bone marrow cells by the immunopotentiators Krestin and Lentinan. *Mutation Research*, 226(1): 9–12.

Henkel, J. 2000. Soy. Health claims for soy protein, questions about other components. *FDA Consumer*, 34(3): 13–15, 18–20.

Hummel, A.S., C. Hertel, W.H. Holzapfel, and C.M. Franz. 2007. Antibiotic resistances of starter and probiotic strains of lactic acid bacteria. *Applied and Environmental Microbiology*, 73(3): 730–739. doi:10.1128/AEM.02105-06.

Hyodo, I., N. Amano, K. Eguchi, M. Narabayashi, J. Imanishi, M. Hirai, T. Nakano, and S. Takashima. 2005. Nationwide survey on complementary and alternative medicine in cancer patients in Japan. *Journal of Clinical Oncology*, 23(12): 2645–2654. doi:10.1200/jco.2005.04.126.

Inano, H., M. Onoda, N. Inafuku, M. Kubota, Y. Kamada, T. Osawa, H. Kobayashi, and K. Wakabayashi. 1999. Chemoprevention by curcumin during the promotion stage of tumorigenesis of mammary gland in rats irradiated with gamma-rays. *Carcinogenesis*, 20(6): 1011–1018.

Jakulj, F., K. Zernicke, S.L. Bacon, L.E. van Wielingen, B.L. Key, S.G. West, and T.S. Campbell. 2007. A high-fat meal increases cardiovascular reactivity to psychological stress in healthy young adults. *Journal of Nutrition*, 137(4): 935–939.

Jung, H.A., Y.J. Jung, N.Y. Yoon, M. Jeong da, H.J. Bae, D.W. Kim, D.H. Na, and J.S. Choi. 2008. Inhibitory effects of *Nelumbo nucifera* leaves on rat lens aldose reductase, advanced glycation endproducts formation, and oxidative stress. *Food and Chemical Toxicology*, 46(12): 3818–3826. doi:10.1016/j.fct.2008.10.004.

Kailasapathy, K. and J. Chin. 2000. Survival and therapeutic potential of probiotic organisms with reference to *Lactobacillus acidophilus* and *Bifidobacterium* spp. *Immunology and Cell Biology*, 78(1): 80–88.

Kim, J.-M., J.-S. Kim, H. Yoo, M.-G. Choung, and M.-K. Sung. 2008. Effects of black soybean [*Glycine max* (L.) Merr.] seed coats and its anthocyanidins on colonic inflammation and cell proliferation in vitro and in vivo. *Journal of Agricultural and Food Chemistry*, 56(18): 8427–8433. doi:10.1021/jf801342p.

La Cour, B., P. Molgaard, and Z. Yi. 1995. Traditional Chinese medicine in treatment of hyperlipidaemia. *Journal of Ethnopharmacology*, 46(2): 125–129.

Lampe, J.W. 2009. Interindividual differences in response to plant-based diets: Implications for cancer risk. *The American Journal of Clinical Nutrition*, 89(5): 1553S–1557S. doi:10.3945/ajcn.2009.26736D.

Larsson, S.C., J. Virtamo, and A. Wolk. 2013. Total and specific fruit and vegetable consumption and risk of stroke: A prospective study. *Atherosclerosis*, 227(1): 147–152. doi:10.1016/j.atherosclerosis.2012.12.022.

Lee, G.S., H.S. Byun, K.H. Yoon, J.S. Lee, K.C. Choi, and E.B. Jeung. 2009. Dietary calcium and vitamin D2 supplementation with enhanced *Lentinula edodes* improves osteoporosis-like symptoms and induces duodenal and renal active calcium transport gene expression in mice. *European Journal of Nutrition*, 48(2): 75–83. doi:10.1007/s00394-008-0763-2.

Li, J., X. Yu, W. Pan, and R.H. Unger. 2002. Gene expression profile of rat adipose tissue at the onset of high-fat-diet obesity. *American Journal of Physiology: Endocrinology Metabolism*, 282(6): E1334–E1341. doi:10.1152/ajpendo.00516.2001.

Lijinsky, W. 1999. N-Nitroso compounds in the diet. Mutation Research, 443(1/2): 129–138.

Lin, H.-H., Y.-F. Tsai, P.-J. Lin, and P.-K. Tsay. 2010. Effects of a therapeutic lifestyle-change programme on cardiac risk factors after coronary artery bypass graft. *Journal of Clinical Nursing*, 19(1/2): 60–68. doi:10.1111/j.1365-2702.2009.02980.x.

Link-Amster, H., F. Rochat, K.Y. Saudan, O. Mignot, and J.M. Aeschlimann. 1994. Modulation of a specific humoral immune response and changes in intestinal flora mediated through fermented milk intake. *FEMS Immunology and Medical Microbiology*, 10(1): 55–63. doi:http://dx.doi.org/.

Liu, Z.-M., Y.-M. Chen, and S.C. Ho. 2011. Effects of soy intake on glycemic control: A meta-analysis of randomized controlled trials. *The American Journal of Clinical Nutrition*, 93(5): 1092–1101. doi:10.3945/ajcn.110.007187.

MacDonald, R.S., J. Guo, J. Copeland, J.D. Browning, D. Sleper, G.E. Rottinghaus, and M.A. Berhow. 2005. Environmental influences on isoflavones and saponins in soybeans and their role in colon cancer. *The Journal of Nutrition*, 135(5): 1239–1242.

MacDonald, R.S. and K. Wagner. 2012. Influence of dietary phytochemicals and microbiota on colon cancer risk. *Journal of Agricultural and Food Chemistry*, 60(27): 6728–6735. doi:10.1021/jf204230r.

Maheshwari, R.K., A.K. Singh, J. Gaddipati, and R.C. Srimal. 2006. Multiple biological activities of curcumin: A short review. *Life Sciences*, 78(18): 2081–2087. doi:10.1016/j.lfs.2005.12.007.

Malencic, D., M. Popovic, and J. Miladinovic. 2007. Phenolic content and antioxidant properties of soybean (*Glycine max* (L.) Merr.) seeds. *Molecules*, 12(3): 576–581.

Maruyama, N., T. Katsube, Y. Wada, M.H. Oh, A.P. Barba De La Rosa, E. Okuda, S. Nakagawa, and S. Utsumi. 1998. The roles of the N-linked glycans and extension regions of soybean β-conglycinin in folding, assembly and structural features. *European Journal of Biochemistry*, 258(2): 854–862. doi:10.1046/j.1432-1327.1998.2580854.x.

Messina, M. 1995. Modern applications for an ancient bean: Soybeans and the prevention and treatment of chronic disease. *Journal of Nutrition*, 125(3 Suppl): 567S–569S.

Metchnikoff, E. and P.C. Mitchell. 1908. *The Prolongation of Life: Optimistic Studies*. New York: G.P. Putnam's Sons.

Meydani, S.N. and W.-K. Ha. 2000. Immunologic effects of yogurt. *The American Journal of Clinical Nutrition*, 71(4): 861–872.

Mukherjee, P.K., K. Saha, M. Pal, and B.P. Saha. 1997. Effect of *Nelumbo nucifera* rhizome extract on blood sugar level in rats. *Journal of Ethnopharmacology*, 58(3): 207–213.

Nanba, H., K. Mori, T. Toyomasu, and H. Kuroda. 1987. Antitumor action of shiitake (*Lentinus edodes*) fruit bodies orally administered to mice. *Chemical & Pharmaceutical Bulletin (Tokyo)*, 35(6): 2453–2458.

Norat, T., A. Lukanova, P. Ferrari, and E. Riboli. 2002. Meat consumption and colorectal cancer risk: dose-response meta-analysis of epidemiological studies. *International Journal of Cancer*, 98(2): 241–256.

Onishi, E., K. Yamada, T. Yamada, K. Kaji, H. Inoue, Y. Seyama, and S. Yamashita. 1984. Comparative effects of crude drugs on serum lipids. *Chemical & Pharmaceutical Bulletin (Tokyo)*, 32(2): 646–650.

Ono, Y., E. Hattori, Y. Fukaya, S. Imai, and Y. Ohizumi. 2006. Anti-obesity effect of *Nelumbo nucifera* leaves extract in mice and rats. *Journal of Ethnopharmacology*, 106(2): 238–244. doi:10.1016/j.jep.2005.12.036.

Perdigon, G., S. Alvarez, M. Rachid, G. Aguero, and N. Gobbato. 1995. Immune system stimulation by probiotics. *Journal of Dairy Science*, 78(7): 1597–1606. doi:10.3168/jds.S0022-0302(95)76784-4.

Perdigon, G., R. Fuller, and R. Raya. 2001. Lactic acid bacteria and their effect on the immune system. *Current Issues in Intestinal Microbiology*, 2(1): 27–42.

Reynolds, K., A. Chin, K.A. Lees, A. Nguyen, D. Bujnowski, and J. He. 2006. A meta-analysis of the effect of soy protein supplementation on serum lipids. *The American Journal of Cardiology*, 98(5): 633–640. doi:10.1016/j.amjcard.2006.03.042.

Rose, D.P., A.P. Boyar, and E.L. Wynder. 1986. International comparisons of mortality rates for cancer of the breast, ovary, prostate, and colon, and per capita food consumption. *Cancer*, 58(11): 2363–2371.

Sadler, M. 2003. Nutritional properties of edible fungi. *Nutrition Bulletin*, 28(3): 305–308. doi:10.1046/j.1467-3010.2003.00354.x.

Santarelli, R.L., F. Pierre, and D.E. Corpet. 2008. Processed meat and colorectal cancer: A review of epidemiologic and experimental evidence. *Nutrition and Cancer*, 60(2): 131–144. doi:10.1080/01635580701684872.

Sasidharan, S., S. Aravindran, L.Y. Latha, R. Vijenthi, D. Saravanan, and S. Amutha. 2010. In vitro antioxidant activity and hepatoprotective effects of *Lentinula edodes* against paracetamol-induced hepatotoxicity. *Molecules*, 15(6): 4478–4489. doi:10.3390/molecules15064478.

Schwoerer, A.P., C. Neuber, A. Schmechel, I. Melnychenko, G. Mearini, P. Boknik, U. Kirchhefer et al. 2008. Mechanical unloading of the rat heart involves marked changes in the protein kinase-phosphatase balance. *Journal of Molecular Cell Cardiology*, 45(6): 846–852. doi:10.1016/j.yjmcc.2008.09.003.

Shah, S.K., P.A. Walker, S.D. Moore-Olufemi, A. Sundaresan, A.D. Kulkarni, and R.J. Andrassy. 2011. An evidence-based review of a *Lentinula edodes* mushroom extract as complementary therapy in the surgical oncology patient. *Journal of Parenteral and Enteral Nutrition*, 35(4): 449–458. doi:10.1177/0148607110380684.

Shishodia, S., G. Sethi, and B.B. Aggarwal. 2005. Curcumin: Getting back to the roots. *Annals of the New York Academy of Sciences*, 1056: 206–217. doi:10.1196/annals.1352.010.

Siri-Tarino, P.W., Q. Sun, F.B. Hu, and R.M. Krauss. 2010. Meta-analysis of prospective cohort studies evaluating the association of saturated fat with cardiovascular disease. *The American Journal of Clinical Nutrition*, 91(3): 535–546. doi:10.3945/ajcn.2009.27725.

Sohal, R.S. and R. Weindruch. 1996. Oxidative stress, caloric restriction, and aging. *Science*, 273 (5271): 59–63.

Stamler, J., D. Wentworth, and J.D. Neaton. 1986. Is relationship between serum cholesterol and risk of premature death from coronary heart disease continuous and graded? Findings in 356,222 primary screenees of the Multiple Risk Factor Intervention Trial (MRFIT). *Journal of the American Medical Association*, 256(20): 2823–2828.

Sugui, M.M., P.L. Alves de Lima, R.D. Delmanto, A.F. da Eira, D.M.F. Salvadori, and L.R. Ribeiro. 2003. Antimutagenic effect of *Lentinula edodes* (BERK.) Pegler mushroom and possible variation among lineages. *Food and Chemical Toxicology*, 41(4): 555–560. doi:http://dx.doi.org/10.1016/S0278-6915(02)00306-X.

Tan, Y.H. 1993. Yin and Yang of phosphorylation in cytokine signaling. *Science*, 262(5132): 376–377.

Tovar, A.R., A. Díaz-Villaseñor, N. Cruz-Salazar, G. Ordáz, O. Granados, B. Palacios-González, C. Tovar-Palacio, P. López, and N. Torres. 2011. Dietary type and amount of fat modulate lipid metabolism gene expression in liver and in adipose tissue in high-fat diet-fed rats. *Archives of Medical Research*, 42(6): 540–553. doi:10.1016/j.arcmed.2011.10.004.

U.S. Department of Health and Human Services, U.S. Department of Agriculture, and U.S. Dietary Guidelines Advisory Committee. 2005. *Dietary guidelines for Americans, 2005*. 6th edn., HHS publication. Washington, DC: G.P.O.

Velthuis-te Wierik, E.J., H. van den Berg, G. Schaafsma, H.F. Hendriks, and A. Brouwer. 1994. Energy restriction, a useful intervention to retard human ageing? Results of a feasibility study. *European Journal of Clinical Nutrition*, 48(2): 138–148.

Vogel, T., P.H. Brechat, P.M. Leprêtre, G. Kaltenbach, M. Berthel, and J. Lonsdorfer. 2009. Health benefits of physical activity in older patients: A review. *International Journal of Clinical Practice*, 63(2): 303–320. doi:10.1111/j.1742-1241.2008.01957.x.

Wholey, M.H., M.H. Wholey, W.A. Tan, G. Eles, C. Jarmolowski, and S. Cho. 2003. A comparison of balloon-mounted and self-expanding stents in the carotid arteries: Immediate and long-term results of more than 500 patients. *Journal of Endovascular Therapy*, 10(2): 171–181. doi:10.1583/1545-1550(2003)010<0171:ACOBAS>2.0.CO;2.

Woo, J.H., Y.H. Kim, Y.J. Choi, D.G. Kim, K.S. Lee, J.H. Bae, D.S. Min et al. 2003. Molecular mechanisms of curcumin-induced cytotoxicity: Induction of apoptosis through generation of reactive oxygen species, down-regulation of Bcl-XL and IAP, the release of cytochrome c and inhibition of Akt. *Carcinogenesis*, 24(7): 1199–1208. doi:10.1093/carcin/bgg082.

Zoeller, R.F. 2009. Physical activity and fitness in African Americans: Implications for cardiovascular health. *American Journal of Lifestyle Medicine*, 3(3): 188–194. doi:10.1177/1559827609331915.

29 Ayurnutrigenomics
Traditional Knowledge-Inspired Approach toward Personalized Nutrition

Parikshit Debnath, Subhadip Banerjee, and Pratip Kumar Debnath

CONTENTS

Tell me what you eat, and I shall tell you what you are.

Jean Anthelme Brillat-Savarin (1783–1833)

29.1 INTRODUCTION

Food is the stepping stone to nutrition, which finds its primordial causality in nature since the creation of life started on the Earth. Ludwig Andreas Feuerbach (1804–1872), a German philosopher and anthropologist, truly commented that what is occurring to science currently and food is intermingled with life and soul have been known to man through experience. Thus, food and nutrition have been connected to human ever since, which naturally interact with their body, mind, and society as a whole.

Food components have the ability to interfere with molecular mechanisms causal to an organism's physiome, which has incited a revolution in thinking about what we eat (Mutch et al. 2005a). Pharmacogenomics and nutrigenomics are mostly intersecting concepts (Ghosh et al. 2007); surprisingly, it is food that we are more exposed to in comparison with drugs. Ayurveda is an ancient science of life practiced for thousands of years evolved around the concept of preventive and personalized medicine by maintaining a balance of the three entities called *Tridoshas*, namely, *vata*, *pitta*, and *kapha* (Mukherjee et al. 2012). Ayurveda takes a holistic approach toward medicine that integrates mind, body, and soul, and toward several other stratifications (Figure 29.1). Interestingly, we find that Ayurveda merges food (*Pathya* or *Ahara*) and drug (*Aushadh*) inside the concept of therapeutics to maintain harmonization of the *Doshas* or physiological factors according to individualistic variability or *Prakriti* and other environmental factors (Debnath et al. 2010). An herb that is consumed as food or spices such as *haridra* or *haldi* (*Curcuma longa*) is also indicated in different ailments from gastrointestinal tract (GIT) disorder to cancer (Datta and Debnath 2000). Thus, nutrition takes a central stage in Ayurvedic therapy, which is personalized according to individual constitution (*Prakriti*).

Nutrition research started back in 1785, when elementary metabolic and respiratory processes were discovered (Carpenter 2003; Ordovas and Mooser 2004). However, this personalized approach toward nutrition research has evolved recently to understand the mechanisms of individualized nutritional responses and nutrigenomics (Godard and Ozdemir 2008). As a matter of fact, nutrition impacts people predominantly in both health and disease together with prevention and treatment of some widespread multifactorial chronic diseases. Nutrigenomics assures its relevance in public

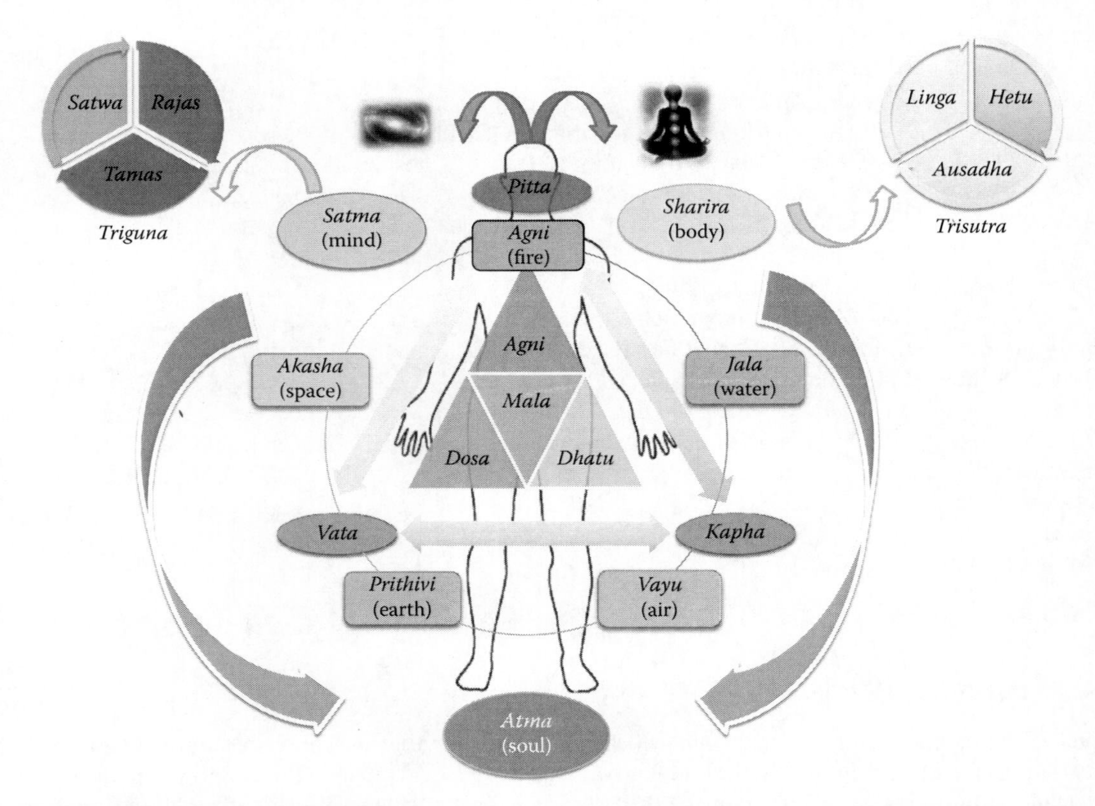

FIGURE 29.1 Basic principles of Ayurveda that integrates the body, the mind, and the soul connecting to nature.

health and nutritional interference guided by human genomic variation. Moreover, the scope of analysis in nutrigenomics research is broad and genome wide, which may recognize new biological mechanisms governing host response to food (Ozdemir et al. 2009). While the heritability estimates vary across studies, there is evidence supporting an appreciable genetic contribution to host response to food, dietary habits, and food preferences (Breen et al. 2006; Teucher et al. 2007).

The authors coin the term "Ayurnutrigenomics," which is a systematic integration of nutritional practices according to Ayurveda in relation to the *Prakriti* of an individual that amalgamates the information from genomics, proteomics, and metabolomics projected to provide solid evidence-based scientific foundation for the advancement of personalized nutrigenomic dietetics. The search for nutritionally and pharmacologically active food supplements or herbs suitable to one's genetic constitution and environmental factors can be greatly advanced by the combination of various "omics" approaches with a range of bioactivity assays in biological systems toward futuristic development. In a harmonizing development, the use of metabolome-refined foods or nutraceuticals with other biochemical components in combination may prove to be very useful as personalized holistic food substances for a variety of traditionally inspired human health-care and preventive applications. Chapter 29 details on various salient features of Ayurveda and cutting-edge nutrigenomic applications in an inclusive endeavor to explain the novel concept of Ayurnutrigenomics.

29.2 AYURVEDA SYSTEMS BIOLOGY AND DISEASE ETIOPATHOGENESIS

Ayurvedic principles are developed on the basis of innumerable clinical observations (time tested). The scientific rationality of these principles is based not merely on ancient texts but because they can demonstrate results as can be found in *Astanga Hridaya Uttaratantra* in verse 40:81 (Valiathan 2006). A primary reason of this demonstrable reproducibility is its development over a long period of what we call translational research today. Understanding of the systems biology of a disease or a disease complex at the genomics level gives us strategic advantage to identify target and also related challenge to find out ways that may help us to mitigate, treat, or manage the disease. The Ayurvedic internal medicine known as *Kayachikitsa* mainly consists of etiopathogenesis, diagnosis, and treatment of the diseases (Roy and Debnath 1998). Genome-wide association studies (GWAS) in which variations in its entirety for many diseased and healthy subjects could be compared to identify regions in the genome that had sufficiently different frequency and could be associated with the disease (Frazer et al. 2009). We find that Ayurvedic drugs and diseases have been classified according to phenotypic classifications (symptomatic complex) correlated with the concept of genomic concept of *Prakriti* (*vata*, *pitta*, *kapha*) (Sethi et al. 2011). The diseases or phenotypes should be properly classified and research should be initiated to know the molecular pathways or network associated with the diseases. This will help us to know the pharmacology of the herbs which interferes with the disease complex that positively modulates the system (body and mind) against the disease (Debnath et al. 2006). This translation of disease and treatment philosophy in terms of recent findings of omics level of research may discover novel targets or strategy to develop drugs and combinations of drugs, most importantly dietetic considerations against diseases.

29.3 FOOD AND LIFESTYLE IN *DINACHARYA* AND *RITUCHARYA*

Ayurveda provides a guiding principle for attainment and preservation of perfect health in its *Swastha Vritta* (Ayurvedic preventive health), *Dinacharya* (daily regimens), and *Ritucharya* (seasonal regimens). Specific food/dietary schedules for different times of the day and different seasons, according to one's age and, most importantly, to suit one's individual constitution or *Prakriti*, have been instructed (Datar 1922; Debnath 1993a, 1993b, 1995, 2002; Shastri 1985). Natural physical urges such as hunger, thirst, sleep, micturition, and defecation are not to be suppressed. Sexual urges are to be controlled following the regimens. Along with nutrition measures for personal hygiene such as use of medicated gargles, oil massages, and Yoga are prescribed (Datar 1922; Debnath 1975; Satyavati 1982). Ayurveda suggests avoidance of negative emotions such as anger, fear, greed, vanity, jealously, and malice along with excessive attachment to anything for psychological cleansing. Positive attitude, *Sadvritta* (adherence to a strict code of moral principles and conduct throughout life), and measures of relaxation such as meditation, prayer, and group activities are very important for maintenance of mental health (Datar 1922; Sastry 2005; Shastri 2003; Tripathi 1992). In prevention and management of chronic diseases such as cardiovascular disorders, diabetes, hypertension, arthritic disorders, stress, and cancers, nondrug measures as supplements to medications and surgical principles found in Ayurveda are being advocated routinely in the present era by health-care providers of developed nations. The ultimate goal of *Ayurveda* (the science of living) is to help mankind to live a healthy long life, mainly to achieve the well-recognized fourfold purpose of human life encompassing seasonal consideration of dietetics. In classical Ayurvedic texts, seasons of the year and diurnality besides dietetic regimens are considered (C.Su.-6) (Sharma and Dash 2001).

29.4 THE CONCEPT OF FOOD AND DIETETICS (*AHARA* AND *PATHYA*) IN AYURVEDA

Nature has provided diet for human beings to achieve healthy long life. Ancient classics emphasized on *Ahara* (diet) to promote health and prevention of diseases. The Vedas, which are considered to be the first written records of Indian literature, contain innumerable dietetic references. Rig Veda enumerates *Purusha* (man) governed by *Atma* (immortal soul) which is manifested by *Anna* (diet) (R.V.-10/90/2) (Kashyap and Sadagopan 1998), while different diseases are treated with herbs as medicine (Kashyap and Sadagopan 1998; Singh et al. 2003). The Rig Veda also gives valuable information of about 67 medicinal plants (Heyn 1990; Kashyap and Sadagopan 1998). Yajur Veda provides information about 81 plants (Samant et al. 2002), whereas Atharva Veda mentions 293 medicinal herbs (Acharya 2002; Samant et al. 2002).

In *Taittiriyopanishad*, the *Anna* has been regarded as "Brahma" (mythologically the creator of the universe) as all animates are produced from *Anna*, after production life is maintained by *Anna* and at the end assimilates in the *Anna* (Tait.Up.Bh.V.-2/1) (Nathani 2013). The *Bhagavad Gita* acknowledged diet as a source for creation of life (Bhag.G.-3/14.1). In addition, to achieve success in Yoga, appropriate diet along with other activities and regimen of life is in fact addressed upon (Bhag.G.-6/17) (Prabhupada 2010).

The *Ahara* as a causative factor, in the context of the origin of *Purusha* (man) and his disease, carries the historical value of dietetics according to *Charaka Samhita* (C.Su.-25/31) (Sharma and Dash 2001). However, *Sushruta Samhita* comprehensively narrated dietetics concerning the applicability and significance of diet in human life establishing the historical importance of it.

Ahara has been defined in Ayurveda as anything that is consumed by mouth into the alimentary canal, which after proper digestion transformed into the tissue elements and do the functions such as promotion of growth, recovery due to loss, protection from diseases for survival is termed to be *Ahara* (Shastri 1985). *Charaka* proclaims that *Ahara* maintains the balance of *Doshas* and *Dhatus* by promoting healthiness and disease avoidance (C.Su.-25/33) (Sharma and Dash 2001). *Ahara* restores the vigor, provides strength, sustains body, and increases the lifetime, bliss, memory, *Ojas*, and digestive capability (S.Ci.-24/68) (Shastri 2003).

Dietary consideration in terms of wholesome and unwholesome is an important component of Ayurvedic therapeutics, which eventually leads to happiness or misery. Sometimes, dietary management in itself is a complete treatment. Ayurvedic dietetics is concerned primarily with the energetic of food as a means of balancing the biological humors (*Dosha*). As opposed to the present-day approach, Ayurvedic nutrition not only deals with the detailed nutritional aspect of food but also takes into account the manner and food we intake, the nature of foodstuff, *Agnibala* (enzymatic activity of digestive metabolism), cooking process, blending, time of year, surrounding and settings, and so on.

29.5 THE CONCEPT OF AYURGENOMICS

Ayurgenomics presents a personalized approach in predictive, preventive, and curative aspects of stratified medicine with molecular variability that intersects the mind and the body (Patwardhan and Bodekar 2008; Sethi et al. 2011). It embodies the study of interindividual variability due to genetic variability in humans (Frazer et al. 2009) for assessing susceptibility, establishing diagnosis, and assessing prognosis mainly on the basis of constitution type of the individual persons called *Prakriti* (Mukherjee and Prasher 2011; Prasher et al. 2008). Selection of a suitable dietary, therapeutic, and lifestyle regime is made on the basis of clinical assessment of the individual keeping one's *Prakriti* in mind (Patwardhan and Mashelkar 2009; Patwardhan et al. 2005). *Prakriti* is a corollary of the comparative proportion of three entities (*Tridoshas*): *vata* (V), *pitta* (P), and *kapha* (K). This is not only genetically determined (*Shukra Shonita*), but also influenced by the environment (*Mahabhuta Vikara*), chiefly by maternal diet and lifestyle (*Matur Ahara Vihara*), and age of the parents (*Kala Garbhashaya*). The ethnicity (*Jati*), familial characteristics (*Satmya*) as well as place of origin of an individual (*Desha*) are also described to influence the development of *Prakriti* besides the aforementioned individual specific factors. Metabolic variability has been correlated with CYP2C19 genetic variability and HLA gene polymorphism to elucidate the concept of pharmacogenomics with the *Prakriti* types is interesting (Ghodke et al. 2011; Patwardhan et al. 2005). However, knowledge regarding population-wide unevenness athwart Indian populations endows with a major thrust to promote the quest for understanding the variability in healthy individuals (Indian Genome Variation Consortium 2008). Transcriptional profiles of pooled RNA from *Vata*, *Pitta*, and *Kapha* revealed differences in core biological processes between the *Prakriti* groups that overlapped with the biochemical pathways and biochemical profiles signify the existence of genetic variation and their cellular manifestation as mentioned in Ayurvedic text. Molecular connections about variation in *EGLN1* gene with high-altitude adaptation (Aggarwal et al. 2010; Bigham et al. 2010; Peng et al. 2011; Storz 2010; Xu et al. 2011) and its association with asthma (Huerta-Yepez et al. 2011) substantiate the hypothesis. This EGLN1 genetic variation has been extrapolated in hypoxia-inducing factor (HIF) variations that can be correlated at the cellular function level in diverse diseases such as cancer, asthma, chronic obstructive pulmonary disease, ischemia, and stroke (Eltzschig and Carmeliet 2011; Huerta-Yepez et al. 2011; Semenza 2000; Smith et al. 2008). We also come across indications in which such genetic involvement may be associated with cardiovascular function (Schubert et al. 2009; Shiomi et al. 1996), systemic inflammation (Das 2004; Haensel et al. 2008), gut microbiota axis (Arumugam et al. 2011; Kau et al. 2011), and anthropometry (Manning and Bundred 2001) and related diseases that open up a new age of Ayurgenomics-guided therapy based on Ayurvedic herbal products. It can very easily solve the disease pattern and treatment efficacy variations among the populations seen from personal clinical experiences. As stressed by the authors in Subsection 29.2, integration of the knowledge of *Nidan* (prognosis) with phenotypic ensembles and the upcoming genetic association can elucidate disease pathways; hence, food regimen and nutraceutical discovery will be accentuated in the lines of Ayurveda systems biology.

29.6 AYURNUTRIGENOMICS: NUTRIGENOMICS IN AYURVEDA

Ayurveda emphasizes on *Prakriti* or body constitution in consideration of food intake. Three extreme human phenotypes—*Vata* (V), *Pitta* (P), and *Kapha* (K)—form seven types of *Prakriti* among human beings with contrasting phenotypic differences. This *Tridosha* theory has been deciphered as Ayurgenomics at the genomic expression level (Prasher et al. 2008). They have found contrasting differences with respect to the biochemical and hematological parameters at the genome-wide expression level. The Ayurvedic concept essentially integrates personalized food and drugs together with this *Prakriti* concept. Every individual should take a diet suitable to his or her predominant constitutional *doshas* to balance them in different seasons. Table 29.1 shows the different types of foods.

TABLE 29.1
Food Prescribed for Different *Prakriti* of People

	Dietary guidelines according to *Prakriti*					
	Vata		*Pitta*		*Kapha*	
	Use	Avoid	Use	Avoid	Use	Avoid
Rasa (taste)	Sweet, sour, salty	Bitter, pungent, astringent	Sweet, bitter, astringent	Sour, salty, pungent	Bitter, pungent, astringent	Sweet, sour, salty
Cereals (grains)	Rice, wheat, oats, ragi	Corn, maize, white bread	Rice, wheat, oats, barley	Corn, millet, ragi	Rice, corn, oats (dry), barley	White rice, wheat, flour, white bread
Pulses (legumes)	Green gram, split red gram, split black gram, soybean	Bengal gram, peas red kidney beans, horse gram	Green gram, split red gram, split black gram, soybean	Red kidney beans/horse gram	Green gram, split red gram, split black gram, peas, red kidney beans, horse gram	Soy products
Vegetables	Carrots, beetroot, sweet potato, green beans, radish, onion, sweet corn, capsicum	Cabbage, cauliflower, broccoli, turnip, green beans, leafy greens, ladyfinger	Cabbage, cauliflower, broccoli, turnip, green beans, leafy greens, ladyfinger, peas, potato, cucumber, sweet potato	Brinjal, radish, onion/garlic, carrots, capsicum, spinach, mushroom	Cabbage family, leafy greens, bitter gourd, beans, brinjal, radish, onion, capsicum, spinach, ladyfinger, lettuce, mushrooms	
Fruits	Banana, berries, grapes, citrus fruits, mango, sweet melons, papaya, pineapple, peaches, plum	Dry fruit, raw apple, pears, sour melons, pomegranate	Apple, avocado, orange, mosambi, guava, mango, papaya, sweet pineapple, pomegranate, melons	Berries, banana, grapes, lemon, orange (sour), peaches, pineapple (sour), plums, mango, raisins	Apple, pears, berries, cherries, dry fig, prunes, peaches, guava	Banana, avocado, grapes, lemons, melons, oranges, papaya (ripe)

Spices	Coriander, curry leaves, fenugreek, turmeric mustard, cumin, carom seed, ginger, garlic, mint, asafetida, cinnamon, cardamom		Coriander, cardamom, cinnamon, fennel, turmeric, fresh ginger	Dry ginger, chillies, mustard	Black pepper, chillies, ginger, turmeric, cumin, carom seed, fennel, mint, coriander, cinnamon, asafetida	Salt
Nuts	Almond, cashew, apricot, peanuts		Coconut			Cashew, walnut, almond, peanut, coconut
Oils	Groundnut, sesame, sunflower		Sunflower, olive, soy, groundnut, coconut	Sesame, safflower, corn oil	No fried foods	
Dairy products	Milk, butter, ghee, cheese, yogurt		Unsalted butter, ghee, milk, sweet yogurt (nonfat)		No dairy product, goat's milk occasionally, nonfat buttermilk	Yogurt
Animal foods	Beef, chicken, turkey, egg, seafood, oily fish		Chicken, turkey, egg white, shrimp, oily fish	Beef, lamb, pork, seafood	Chicken, turkey, eggs, fish (freshwater)	Beef, lamb, pork (all dark meat)

However, this warrants for a need of epidemiological research on nutrition concepts of Ayurveda. Research on *Prakriti* with nutrition is going to emerge as a major field that may be termed as "Ayurnutrigenomics." As we develop toward the omics age of science, we find notable parallelism in thoughts with our ancient scientific lineage including Ayurveda. At that time, they also felt the necessity of correlating nutrition with the differences in biological phenotypes, which is the expression of individual genome or variome. We find the concept of Ayurgenomics and then Ayurnutrigenomics back at the time of Ayurveda, which is quite novel and contemporary even today. The National Institutes of Health (NIH) defines genomics as the study of all of a person's genes, including interactions of those genes with each other and with the person's environment (Norheim et al. 2012). Nutrigenomics extends from the study of the genome-wide influence of nutrition to the ensuing time-dependent response in transcriptomics, proteomics, and metabolomics to express the phenotype of a biological system (Muller and Kersten 2003; Afman and Muller 2006). The influence of genetic variation on pharmacokinetics and nutrikinetics (i.e., absorption, metabolism, elimination, or biological effects of nutrients on human body) has also traditionally been incorporated in the concept of nutrigenomics to optimize nutrition according to the subject's genotype. The Ayurvedic concept of nutrition can be explained or translated in the same direction. However, molecular nutrition research is broader than nutrigenomics, as it includes the effect of nutrients and foods/food components on whole body physiology and health status at the systems biology level. Nutrigenomics also includes precise determination of molecular mechanisms essential to human health and disease advocating an enormous prospective for promoting health and lowering mortality and morbidity. Sophisticated molecular techniques based on the different *omics* (genomics, epigenomics, transcriptomics, proteomics, and metabolomics) may help us in this direction to develop better understanding toward Ayurvedic principles on nutrition and genomics. The nutritional epidemiology approach discussed later, where we study the role of nutrition in causes and prevention of disease to guarantee precision of health recommendations, is the need of the day to develop evidence-based quantification (Margetts and Nelson 1997; Sempos et al. 1999). Ayurveda is evidence-based science, but it has not been updated in thousands of years when genetic evolution and environmental changes have occurred. Therefore, research should be guided following Ayurvedic understanding to develop evidences to find their justification in the present times. Ayurvedic principles have also given the concept of adaptability termed as *Satmya*. Satmya is explained by a concept that even if a food habit is harmful considering a person's genetic constitution, due to climatic (*Ritu Satmya*), geographical (*Desha Satmya*), disease (*Roga Satmya*), regular habit (*Oak Satmya*), and sociocultural (*Jati Satmya*) factors that food habit may get habituated to his or her nutriome. Nutriome is referred to as the entire food habit of a person interacting with his or her genome or physiome. A simple example is rice rich in carbohydrate diet among Indians, yet they are prone to diabetes. Different clinical principles have been mentioned in Ayurveda to advise preventive and personalized treatment to a human being such as *Prakriti* (individual constitution), *Dosha Dushya* (disease state and localization), and *Satma* (habituation factor) to determine food, medicine, or lifestyle modification to a person (Figure 29.2).

Thus, we find that Ayurnutrigenomics presents a huge scope of development toward an understanding of nutrigenomics and molecular nutrition research. The fundamental recommendations can be very useful in framing health recommendation and personalized food design. In the subsequent Subsections 29.8 and 29.9, we described many such recommendations and directions toward a rational approach in modern Ayurnutrigenomic research.

29.7 CLASSIFICATION OF FOOD

Acharya Charaka has classified the food articles in different ways: according to source—animal origin and plant origin—and according to effect—wholesome and unwholesome. Food may be drinkables, eatables, chewable, and so on (C.Su.-25/36) (Sharma and Dash 2001) according to the way of intake as also stated by *Bhavaprakasha* (B.P.Pu.Kh.-5/144) (Mishra 1993) and *Sharangadhara* (Sh.Pu.Kh.-6/2) (Sharma 1988). Acharya Kashyapa distinguished food on its

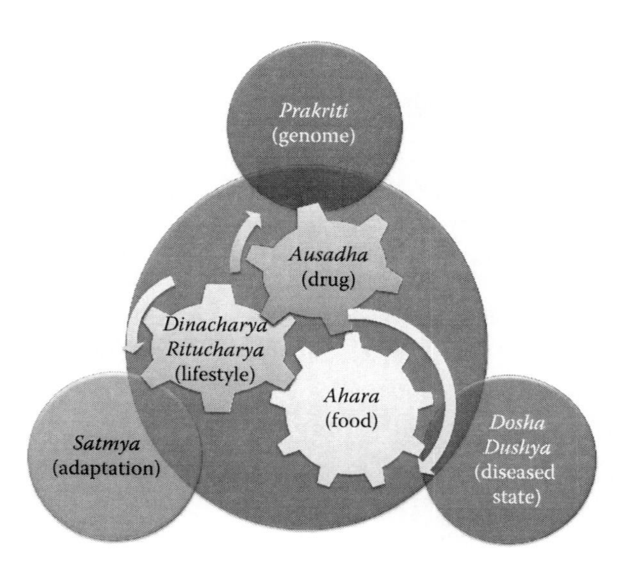

FIGURE 29.2 Clinical principles followed in Ayurveda to achieve personalized medicine and nutrition.

PanchaMahabhuta composition, namely, *Akasheeya*, *Vayavya*, *Agneya*, *Apya*, and *Parthiva* (containing the five elements—space, air, fire, water, and earth) (K.S. Khi.-4/8) (Tiwari 1976). In other context, Charaka has classified food articles into 12 groups (C.Su.-27/6-7) (Sharma and Dash 2001) and Acharya Sushruta classified them into 21 groups (S.Su.-45) (Shastri 2003). According to properties, food is of 20 types and of innumerable varieties due to abundance of substances, their combinations, and preparations.

The human body is endowed with strength, complexion, and growth, and continues up till the full life span, being supported by three *Upastambhas* (supporting factors) of life: *Ahara* (diet), *Nidra* (sleep), and *Brahmacharya* (celibacy). Among these three regulating factors, diet is an essential factor for maintenance of healthy life (C.Su.-11/35). Charaka also mentioned that *Anna* (food) is the best sustainer of life (C.Su.-25/40) (Datar 1922). Sushruta stated *Ahara* determines the origin of beings and forms a chief source of bodily strength, complexion, and *Ojas* (S.Su.-1/36) (Shastri 2003). Citing the importance of *Pathya* (*Ahara*), Acharya Lolimbaraja says that, if wholesome diet is given in a planned way, there is no need for separate medical treatment, and if unwholesome diet is being permitted, there is no benefit of any medication (Vaidyajeevanam-1/10) (Tripathi 1978).

29.7.1 Six *Rasas* (Taste-Based Classification) in Diet

Ayurveda has recommended that all the six tastes should be in every diet to enhance the *bala* or immunity. Six *Rasas* are *Madhura Rasa* (sweet taste), *Amla Rasa* (sour taste), *Lavana Rasa* (salty taste), *Katu Rasa* (pungent taste), *Tikta Rasa* (bitter taste), and *Kashaya Rasa* (astringent taste). These six *Rasas* directly influence the *Tridoshas* and also influence the nutrition and transformation of bodily tissues (Datar 1922; Dwarkanath 2010; Srinivasulu 2010).

29.7.2 Climate and Dietetics

Vata-alleviating diet is more appropriate to cold, dry, windy climates, including the high desert or high plains regions. *Pitta*-alleviating diet is suitable for hot climates including the Southern United States and the lower desert of the Southwest. *Kapha*-alleviating diet is more congenial to damp and cold climates, including the Midwest, most of the East and Northeast, and the Northwest (C.Su.26/88; A.S.Su.7/232–234; K.S.Ka.7/41–55) (Datar 1922; Tiwari 1976; Tripathi 1992).

29.8 FOOD–DRUG SYNERGISM

Ayurveda precisely distinguishes food to be the most important part of Ayurvedic therapy. It can observe that the concept of food and drug merges in Ayurveda. It classifies and states various dietetic principles that add stratified personalization considering very crucial variables. Together with *Prakriti* constitution of the body, various dietary principles and other factors such as *matra* (quantity), *kala* (time or season), *kriya* (mode of preparation), *bhumi* (habitat or climate), *deha* (constitution of person), and *desha* (body humor and environment) also play a significant role in the acceptability of wholesome diet. The quantity of diet depends on the power of digestion and metabolism (C.Su.-5/3). There is no fixed quantity in which different food articles are to be taken. One-third of the stomach should be filled with solid food, one-third should be filled with liquids, and one-third should be left empty (C.Vi.-2/3).

29.8.1 *Ashta Ahara Vidhi Visheshayatan* (the Eight Factors of Diet)

It is impossible to derive the total benefit out of food, simply on the basis of the quantity of intake, without considering the eight factors: *Prakriti* (nature of food articles), *Karana* (method of their processing), *Samyoga* (combination), *Rashi* (quantity), *Desha* (habitat), *Kala* (time in the form of day, night, or seasons and state of individual), *Upayogasamstha* (rules governing the food intake), and *Upayokta* (health of the individual who takes it).

29.8.2 *Dvadasha Ashana Pravichara* (12 Dietary Considerations)

Sushruta has described 12 dietary considerations and mentioned that which type of food is suitable for which individual (S.U.-64/56): *Shita*, *Ushna*, *Snigdha*, *Ruksa*, *Shuska*, *Drava*, single diet a day, twice daily, *Ausadhiyukta Ahara*, *Mitahara*, *Shamana*, and *Vrittiprayojaka Ahara*.

29.8.3 *Viruddhahara* (Dietetic Incompatibility)

Incompatible diets are responsible for various disorders in human beings. Charaka has described 18 factors responsible for dietetic incompatibility (C.Su.-26/86–87): *Desha* (climate), *Vidhi* (rules of eating), *Kala* (season), *Agni* (digestive power), *Matra* (quantity), *Satmya* (accustom), *Doshas* (*tridosha*, i.e., *Vata*, *Pitta*, *Kapha*), *Samskara* (mode of processing), *Ahara virya* (potency of food), *Kostha* (bowel habits), *Avastha* (state of health), *Krama* (order of food intake), *Parihara* (restriction), *Upachara* (prescription), *Paka* (cooking), *Sanyoga* (combination), *Hridya* (palatability), and *Sampad* (richness of quality).

29.9 FOOD INTERACTIONS

29.9.1 Food–Food Interactions

Ayurvedic principles state that fish should be avoided along with milk, which may cause constipation. Meat of domestic, marshy, and aquatic animals negatively interacts with honey, sesame seeds, sugar candy, milk, radish, lotus stalk, germinated grains, and seeds of *masha* (*Phaseolus radiatus*) (Datar 1922). The vegetable *pushkara* (*Nelumbo nucifera*) and meat of dove fried in mustard oil with honey and milk may lead to obstruction of *srotas* (circulatory system in Ayurveda) and diseases of the head and neck (Srikanthamurthy 2010). Hot honey, or intake of honey in fever, may lead to a range of illnesses. Diarrhea may be caused due to intake of rice prepared with milk and sugar or with sour gruel (Tripathi 2005). Intake of honey and ghee; honey, rain water, and ghee in equal quantities; and honey with hot water or lotus seeds may lead to various adverse effects. Unwholesome combinations such as sour foods with milk, hot foods after taking pork, cold foods after taking ghee, and honey and ghee in equal quantity may lead to various adverse effects such as blindness, ascites, sterility, eruptions, insanity, fainting, intoxication, tympanitis, skin diseases, sprue, edema, dyspepsia, fever, and rhinitis (Tripathi 2007).

29.9.2 Food–Drug Interactions

Pippali (*Piper longum*) combined with fish fat, *kakmachi* (*Solanum nigrum*) with honey or roasted meat, seeds of *pushkara* (*N. nucifera*) with honey, *bhallataka* (*Semecarpus anacardium*) with hot water, and *kampillaka* (*Mallotus philippinensis*) with buttermilk should not be taken; these combinations may lead to various harmful effects. Milk should not be taken after intake of radish, garlic, and bitter gourd (*Momordica charantia*). Leaves of *Ocimum sanctum* may cause various types of skin diseases. Leaves of asafetida (*Ferula narthex*) or ripe fruit of *lakoocha* (*Artocarpus lakoocha*) should not be taken with honey and milk; it may cause loss of strength and complexion and may lead to oligospermia, sterility, and many other diseases, even death (Datar 1922). Consuming *bhallataka* seeds with hot water and *kampillaka* with buttermilk are also unwholesome and may cause various ailments in the body. *Shilajatu* is advised not to be taken with *kakmachi* and meat of pigeon. Fruits of *lakoocha* and decoction of *masha* in combination have been forbidden. Paste of sesame seeds with leaf of black cumin (*Nigella sativa* Linn.) may cause diarrhea (Srikanthamurthy 2008). Formulations prepared from mercury and mercurial compounds must not be consumed with pumpkin fruit, banana, bitter gourd, sour gruel, pigeon flesh, and so on; they may interact and lead to an increase in body temperature and formation of blisters and rashes (Kulkarni 1998).

29.9.3 Food–Disease Interactions

The concept of interaction between two antagonistic principles is regarded as the worst for the production of skin disorders. Interaction between foods, drugs, or foods and drugs in various levels can cause various disorders affecting all the body tissues ranging from systemic to localized manifestations, even death (Tripathi 2007). Studies report that piperine from *marich* (black pepper, *Piper nigrum* Linn.) and *pippali* (long pepper, *P. longum* Linn.) augmented the concentration–time curve of phenytoin, propranolol, and theophylline in healthy volunteers and plasma bioavailability of rifampicin in patients with pulmonary tuberculosis (Baravkar et al. 2009).

29.10 IMMUNONUTRIENTS IN TRADITIONAL FOODS

The *Rasayana* therapy is a comprehensive regimen that encompasses the concept of "rejuvenation" of mental and physical health (Datta and Debnath 2001; Debnath et al. 1998, 1999). Ayurveda entails that *Rasayana* refers to acquisition, movement, or circulation of nutrition needed to the body tissue (known as *Dhatus* in Ayurveda). The *Rasayana* therapy has classifications as per various approaches that are primarily of promotive value and are essentially meant to rejuvenate the body and mind to impart longevity against aging and immunity against disease. The herbs and foods mentioned under this are perceived as immunomodulators, adaptogenic, antiaging, antistress, and memory enhancers.

It has been mentioned by *Charaka Samhita* in the very first chapter of *Chikitsa Sthana*. However, other texts such as *Sushruta Samhita* and *Astanga Hridaya* by Bagbhatahas mentioned it briefly with few additions or changes. Dalhan's commentary on *Sushruta Samhita* (Ci. 27,1–2) poses a very rational classification, namely, *Ajasrika* (nutrition), *Kamya* (vigor and vitality), and *Naimittika* (disease specific), on the basis of the scope of use:

1. *Kamya Rasayana* (promoter of general health)
   a. *Pranakamya* (longevity), namely, antiaging and adaptogenic
   b. *Medhakamya* (promoter of intellect), namely, antistress, memory enhancer, Alzheimer's disease, or Parkinson's disease
   c. *Srikamya* (promotion of luster and complexion), namely, antiaging and antistress
2. *Naimittika Rasayana* (promoter of specific vitality in specific diseases)

However, based on the contents, Rasayana has been divided into three types:

1. *Ausadha Rasayana* (drug)
2. *Ajasstrika Rasayana* (nutrition)
3. *Acara Rasayana* (conduct)

Rasayana is attained through direct enrichment of the nutritional flow (referred to as *Posaka Rasa*). Foods such as *Shatavari* (*Asparagus racemosus*), milk, and ghee helps in this process. The second approach is to improve the metabolic process (referred to as *Agnivyapara*) which increases the anabolic effect, thereby improving the overall health of the human body. *Bhallataka* is an example that acts at the metabolomic level to promote health. Another way of attaining rejuvenation is to boost the circulation by promoting the competent flow of nutrients through the channels (known as *srotas*) of the body. This may help in better bioavailability of the nutrients all over the body improving health and desired benefits. Herbs and foods such as *Guggulu* work on the similar way of exhibiting hypolipidemic and anti-atherosclerotic activity. *Rasayana* herbs such as *Centella asiatica, Pueraria tuberosa*, and *O. sanctum* have been proven for antistress or antianxiety activity in clinical or experimental setup (Bhattacharya et al. 2008; Jana et al. 2010; Pramanik et al. 2011).

There is general idea that *Rasayana* is geriatrics, but it is actually science of nutrition that encompasses application in all the ages to augment our vitality. *Charaka Samhita* clearly mentioned that nutrition is the primary aim of *Rasayana* and other effects such as antiaging, adaptogenic, and memory enhancer are the secondary functions or attributes associated with the main aim. It is important to emphasize that *Rasayana* is not just a drug therapy that encompasses nutrition (*Ajasrika*) and good conduct (*Acara rasayana*). Medicated ghee, milk, and foods have been mentioned under this *Ajasrika rasayana*. Different *Rasayana* herbs and food supplements are used nowadays. Classical food supplements such as *Chyawanprash* and *Brahmi Rasayana* are very quite effective at all age groups (Sur et al. 2004). However, *Sharangadhara Samhita* has augmented further personalization or specificity of the *Rasayana* herbs according to age group of a person (Sharma 1988).

At present, we observe that many herbal preparations alter immune function and display an array of immunomodulatory effects. In various *in vitro* and *in vivo* studies, herbal medicines have been reported to modulate cytokine secretion, histamine release, immunoglobulin secretion, class switching, cellular coreceptor expression, lymphocyte expression, phagocytosis, and so on (Plaeger 2003). Ayurveda gives a separate class of immunomodulatory botanicals named *Rasayanas*. Several botanicals from these texts have been studied for their immunomodulatory properties and have the potential to provide new scaffolds for safer, synergistic, cocktail immunodrugs (Rege et al. 1999).

The root *Glycyrrhiza glabra*, commonly known as liquorice, has been used since ancient times in Indian, Chinese, Egyptian, Greek, and Roman medicine. It is prominent in Ayurveda as *Rasayana* with cytoprotective and demulcent effects and is a popular home remedy for minor throat infections. Biologically active substances in liquorice roots include glycyrrhizic acid (GL) and its aglycone (GA), phenolic compounds, oligosaccharides and polysaccharides, lipids, sterine, and so on. Recently, GL and GA have been shown to exert a hepatoprotective effect via modulation of immune-mediated hepatocyte toxicity, nuclear factor kappa B (NF-κB), and interleukin (IL)-10, which explains how their administration resulted in a downregulation of inflammation in the liver (Abe et al. 2003; Yoshikawa et al. 1997). GL has been reported to increase the resistance to *Candida albicans* and herpes simplex virus-1 infection in animal models (Sekizawa et al. 2001; Utsunomiya et al. 2000). Many researchers have suggested that the effects on the production of interferon (IFN) and Th2 cytokines might be one of the mechanisms involved in the anti-infective process. Recently, glycyrrhizin has been found to be active in inhibiting replication of the severe acute respiratory syndrome (SARS)-associated virus (FFM-1 and FFM-2) and also H5N1 influenza A virus-infected cells (Cinatl et al. 2003; Michaelis et al. 2011). GL is also reported to have modulatory effects on the complement system. Reports indicate that GL blocks C5 or a more distal stage of the complement cascade, suggesting that it might have a role in preventing tissue injury not only in chronic hepatitis but also in autoimmune and inflammatory diseases (Fujisawa et al. 2000). Many researchers have reported that the response-modifying activity in flavonoids and chalcones isolated from the root extract and even food supplement that modulates the Bcl-2/Bax family of apoptotic regulatory factors has been suggested as a probable mechanism for its reported cytoprotective activity against MCF-7 breast cancer cells (Hu et al. 2009; Jo et al. 2004; Ju et al. 2008). These activities include antioxidant, chemopreventive, and antimicrobial activities (Barfod et al. 2002). Chemical modification of GL and GA has been tried, and a significant improvement in anti-inflammatory, antiallergic, and antiulcer activities was observed (Baltina 2003). These observations indicate immune-modulating and biological response modifier activities associated with GL (Abe et al. 2003).

Withania somnifera (L.) Dunal (Solanaceae), commonly known as *Ashwagandha*, Indian ginseng or winter cherry, is one of the most esteemed medicinal plants used in Indian Ayurveda for over 3000 years (Gupta and Rana 2007; Singh et al. 2011). It is used as herbal medicine in various forms (decoctions, infusions, ointments, powder, and syrup) in different parts of the world (Kumar et al. 2007) for all age groups of patients without any side effects even during pregnancy (Gupta and Rana 2007; Sharma et al. 1985). The extracts as well as different isolated bioactive constituents of *W. somnifera* have been reported to possess adaptogenic, anticancer, anticonvulsant, immunomodulatory, antioxidative, and neurological effects. The plant is also considered efficacious in the treatment of arthritis, geriatric, behavioral, and stress-related problems (Dhuley 2001; Mishra et al. 2000). Sitoindosides and acylsterylglucosides in *Ashwagandha* are antistress agents. Active principles of *Ashwagandha*, for instance, the sitoindosides VII–X and withaferin A, have been shown to have a significant antistress activity against acute models of experimental stress (Bhattacharya et al. 1997). Many of its constituents support immunomodulatory actions (Agarwal et al. 1999). High-performance liquid chromatography–tandem mass spectroscopy (HPLC–MS/MS) study shows the favorable pharmacokinetic property of aqueous extract of *W. somnifera* taking withaferin A and withanolide A as metabolic markers in animal model (Patil 2013). Recently, *Ashwagandha* has been used as an adjunct therapy in tuberculosis management and patented in the United States as a vaccine adjuvant (Debnath et al. 2012; Patwardhan 2013).

Another Ayurvedic *Rasayana* known for immunomodulatory and cytoprotective activities is *Tinospora cordifolia* (TC). Quaternary alkaloids and biotherapeutic diterpene glucosides of TC (syringin, cordiol, cordioside, and coriofolioside) showed an immunopotentiating activity (Kapil and Sharma 1997). Research work has been conducted on berberine, jatrorrhizine, tinosporaside, and columbin, which shows a possible mechanism of immunomodulatory activity as an activation of macrophages, leading to increases in granulocyte–macrophage colony-stimulating factor (GM-CSF), leukocytosis, and improved neutrophil function. TC also inhibits C3 convertase of the classical complement pathway (Thatte et al. 1994). Research on polysaccharide (α-D-glucan) derived from TC shows the activation of nuclear killer (NK) cells, complement system, and Th1 pathway cytokines, coupled with low nitric oxide synthesis (Nair et al. 2004). The antitumor activity of TC extract was comparable and better than doxorubicin treatment on HeLa cells. The reported hepatoprotective activity of TC against carbon tetrachloride-induced liver damage proves its metabolic rejuvenation effects (Diwanay et al. 2004).

29.11 TRADITIONAL FOOD PRACTICES AS ADJUNCT THERAPY

Chronic diseases such as diabetes, obesity, and cancer are caused by defects in multiple genes and pathways. Thus, it is not surprising that the current one-target-one-compound approach in drug discovery and development has failed to deliver as many efficacious medicines as expected in the postgenomic era. The research tools that provide extensive data include omics tools, such as genomics, transcriptomics, proteomics, interactomics, metabolomics, localizomics, and phenomics (Choi 2010). It is interesting to observe that Ayurveda merges the concept of pharmagenomics and nutrigenomics that is drug (*Aushadh*) and diet (*Ahara*) according to *Prakriti*. It has been shown experimentally and accepted traditionally that the combination of certain drugs could be more potent than the simple sum expected from the action of two individual drugs (Adhikari et al. 2011, 2012; Bhattacharya et al. 2010; Lehár et al. 2007; Maity et al. 2012). There should be a unique treatment that fits best the molecular profile of his or her disease and his or her lifestyle habits. Translational systems biology (Borisy et al. 2003; Vodovotz and An 2007) is termed as the design of models that are created within a primary focus on rapid translational application in areas such as *in silico* clinical trials, patient diagnostics, rational drug design, and long-term rehabilitative care (Vodovotz et al. 2008, 2009). *Chyawanprash*, a multiherbal food supplement, has been clinically proven to exhibit immunomodulatory effects. Ayurveda advocates a unique treatment for every patient from a pool of drugs combined with food as an adjunct therapy that potentially fits many different patients according to their constitution (*Prakriti*) and cause of disease. Drug discovery using reverse pharmacology and bedside-to-bench approach from Ayurvedic herbs has been proclaimed as "goldmine" in case of chronic inflammatory diseases such as cancer (Aggarwal et al. 2011). *Naimittika Rasayana* has been referred to a special class of *Rasayana* drugs, elaborated to be used against specific diseases as an adjunctive therapy to others drugs (Su. Ch.-27/1) (Table 29.2).

29.11.1 *Anupana* or Vehicle of Drug Therapy

Anupana or vehicle of drug therapy is a crucial concept. In medicine, the word *Anupana* might be in use for vehicle, adjuvant, or carrier in the course of action such as drug absorption, drug companion, and drug interaction (Saraswat et al. 2012). *Anupana* is a substance that is taken along with or after *Aushadh* or *Aahar*. It enhances the action of *Dravya* that is administered in the body. Though the

TABLE 29.2
***Naimittika Rasayana* Drugs in Various Disease Conditions**

Disease	Drugs	Scientific Name
Eye diseases	*Triphala*	*Emblica officinalis*
		Terminalia chebula
		Terminalia belerica
Heart diseases	*Salaparni*	*Desmodium gangeticum*
Skin diseases	*Tuvaraka*	*Hydnocarpus laurifolia*
	Gandhaka Rasayana	
Grahani (lower GIT diseases)	*Pippali*	*Piper longum*
Tuberculosis	*Rasona*	*Allium sativum*
	Pippali	*P. longum*
	Shilajit	*Asphaltum*
Anemia	*Lauha*	Iron preparations
Bronchial asthma	*Agastya Rasayana*	*T. chebula*
Neuromuscular diseases	*Rasona*	*A. sativum*
	Guggulu	*Commiphora mukul*
	Bala	*Sida cordifolia*
Diabetes	*Shilajit*	*Asphaltum*
	Amalaki	*E. officinalis*
	Haridra	*Curcuma longa*
Obesity and lipid disorders	*Guggulu*	*C. mukul*
	Haritaki	*T. chebula*
Hypertension	*Rasona*	*A. sativum*
	Medhya Rasayana	
Hypotension	*Kupilu*	*Strychnos nux-vomica*
Allergic diseases	*Haridra*	*C. longa*

Anupana is administered along with *Aushadh* to improve the taste and to mask the bad odor of the *Dravya*, it is mainly given for carrying the essential substance to the target place.

Ayurvedic classics vividly deals with *Anupanas* regarding the descriptions, properties, mode of action, criteria for selection, specific actions, and so on. Further, if the prescribed *Anupana* is not available, then based on the habitat, suitable *Anupanas* are to be considered and evaluated from the list of *Asavas* (medicated alcoholic preparations) and other *Peya* (medicated drinks). Apart from this, methods of selection of *Anupana* based on *Roga* (disease), *Rogi* (patient), *Doshic* prevalence, *Prakriti*, *Kalpana* (method of preparation), *Rasa* (taste), *Kala* (season and time), *Vayah* (age), and such factors are incorporated (Datar 1922; Sastry 2005; Shastri 2003; Tripathi 1992).

The mode of action of *Anupana* is depicted by *Sharangadhara* with the illustration of oil spreading swiftly when dropped on the water surface. Here, the oil drop is compared to the *Pradhana Aushadha* (specific medicine), whereas water simulates the role of *Anupana* (Sharma 1988). Yogaratnakara has prescribed diversified routes of administration of *Anupana*. To refer to a few, *Virechana* (medicated purgation) as *Anupana* in *Udara* (ascites), *Nidra* (sleep) in *Ajirna* (indigestion) and *Aruchi* (anorexia), *Nasya* (nasal application of medicated drugs) in *Urdhvajatruroga* (diseases of head and neck), *Raktamokshana* (bloodletting) in *Vidradhi* (abscess), and many more. Further, the author says that *Anupana* influences on the prime drugs' strength and effectiveness (Shastri 2003). The treatise *Rasa Tarangini* brilliantly described the mode of action of *Anupana* as that it disintegrates the *Paramanus* (molecules) of the *Yoga* (combinatorial drugs) and therefore aids in carrying it swiftly in the body (Sharma 2004).

Sushruta describes the two types of *Anupana*: *Aushadhopyogi* (useful for medicine)—that is *Anupana* in the context of *Aushadha sevana* (intake of medicine)—and *Aharopyogi Anupana* (useful for diet)—that is *Anupana* in the context of *Ahara sevana* (intake of diet). *Aharopyogi Anupana* is divided into three categories: (1) *Adipana*, the *Anupana* to be taken prior to the meal; (2) *Madhyapana*, the *Anupana* to be taken during or in between the meals; and (3) *Antapana*, the *Anupana* to be taken after the meals (C.Su.-46/420) (Shastri 2003).

29.12 AYURGENOMICS WAY FORWARD: HERBAL METABOLOMICS IN RATIONAL NUTRACEUTICAL DESIGN

Prevention and development of diseases to augment the repairing processes is a crucial area of nutrition research to cure already fully developed diseases (Van Ommen et al. 2010). Correlating Ayurvedic terminology, we find genetic (*Prakriti*) and environmental factors (*Mahabhuta Vikara*) together with diet (*Ahara*) and lifestyle (*Dinacharya*), particularly overnutrition (*Satmya*) and sedentary behavior (*eksthanasna*) that are interrelated to promote the progression and pathogenesis (*nidana*) of these polygenic diet-related diseases (Phillips 2013). Novel advanced methods for mass measurements of genes, transcripts, proteins, and metabolites are united with advanced imaging, epidemiology, clinical interventions with diverse threats, and ultimately bioinformatics to amalgamate all information in systems biology (Norheim et al. 2012). The Subsections 29.2 through 29.6 have explained Ayurvedic principles of diet correlating to various factors such as genomics (*Prakriti*) and other environmental factors. Research utilizing these advanced methods from bedside to bench only can augment better understanding toward Ayurnutrigenomics. Though powerful analytical platforms are accessible to analyze genes, proteins, and metabolites, their inclusive and combined use to appreciate the interferences of nutritional factors on metabolism is quite few (Griffin et al. 2004; Hirai et al. 2004; Mutch et al. 2005b). Studies utilizing these analytical platforms (i.e., genomics, proteomics, metabolomics) offer precise information unfolding a given phenotype, whereas their integration endow with the best possible resources to unfold the effects of a biological dispute on an organism at the level of integrative metabolism) (Nicholson et al. 2004; Zeisel et al. 2001). Undoubtedly, these methods are crucial to be incorporated in standard routine investigations, but we face various challenges due to lack of standardized statistical methods and public databases as their practice is still seldom (Arab 2004; Cahill and Nordhoff 2003; Corthals et al. 2000; Mendes 2002; Rabilloud 2002; Rose et al. 2004).

Nutrigenomics maps the influence of dietary molecules on the genome to correlate the consequential phenotypical divergence in the cellular response such as metabolic pathways and homeostasis of the biological systems that may also be further controlled by genetic interactions (Muller and Kersten 2003; Ordovas and Mooser 2004). Traditionally, plants that are both food and a source of medicine embody a great public and medical significance worldwide as foundation of nutraceutical development in personalized food design as utilized in novel lead compounds for drug development (Shyur and Yang 2008). Metabolomics is an promising and rapidly evolving science and technology system of broad experimental analysis of metabolite profiles, either as a targeted subset of related chemicals or more globally, for diverse applications in diagnosis, toxicology, disease development and animal disease models, genetic modification of specific organisms, drug discovery and development, and phytomedicines (Kell 2006; Lindon et al. 2007; Ulrich-Merzenich et al. 2007). Metabolomics is also a crucial module of the systems biology approach. In nutrigenomics, it can be used for the measurement of metabolite profiles, behaviors, and responses toward the body milieu, disease, and counteractive changes due to food or medication of a given tissue or biological fluid (Dunn and Ellis 2005; Sumner et al. 2003). Metabolite analysis or metabolite profiling is done utilizing a meticulous set of systematic technique(s) such as gas chromatography–mass spectrometry (GC–MS) and liquid chromatography–MS (LC–MS), together with an estimate of quantity. Analytical techniques from thin-layer chromatography (TLC) to high-end technologies such as Fourier transform infrared spectroscopy (FT-IR), Raman spectroscopy, and nuclear magnetic resonance (NMR) are also included in the metabolite analysis arsenal. GC–MS has been considerably used in metabolite profiling of human body fluids or plant extracts (Horning and Horning 1971; Sandberg et al. 1965; Sauter et al. 1991) and also for efficient quality control (Zeng et al. 2007). The Human Metabolome Database (HMDB) is currently the largest and most inclusive database offering spectral, physicochemical, clinical, biochemical, genomic, and metabolism information for a library of >2500 known human metabolites (Wishart et al. 2007). The diversity of plant metabolites evolved through the unremitting interaction with intimidating environments, together with distinctive species and agronomic differences, and these metabolites generally present a definite phenotypic expression correlated to their biochemical structures (Schauer and Fernie 2006). A spectrum of pharmacological or nutraceutical efficacy experienced traditionally arises only as a synergistic action of multiple ingredients in a single plant or from a multiple component herbal supplement recently reported by traditional Chinese medicine (TCM) (Williamson 2001). Wang et al. (2005) connects metabolomics to this effort that can provide the required associations between the complex chemical mixtures used in TCM and molecular pharmacology. Herbal metabolite signatures in gene and/or protein expression profiles can also be of eminent significance in nutraceutical standardization, such as their use in "biological fingerprinting" of medicinal plant extracts (i.e., bioactivity spectra of phytoextracts or phytocompounds vs. their medicinal efficacy in test animal or human systems) (Wang et al. 2006). Metabolomics approaches using GC–MS, LC–MS, or 2D NMR are effective tools for quality control of clinically active food products as utilized for medicinal plants or herbs and medicinal products (Yang et al. 2006; Ye et al. 2007). Moreover, current metabolomics research can be used for comparison of small-metabolite signatures to discern health and disease states (Yang et al. 2006). Focused or targeted metabolomics platforms, for example, lipid profiling or lipidomics, may be applied in the corrective nutraceuticals and therapeutic personalized food development process in lipid-related metabolic disorders and inflammatory disease states as used in modern drug development (Meer et al. 2007; Morris and Watkins 2005). The use of genetically modified mouse models together with LC–MS-based metabolomics can be a useful tool for mechanistic studies of genotype-dependent food or drug metabolism (Chen et al. 2007).

Ayurnutrigenomics is a commune where food and drug intersect to project their effects according to the genetic constitution (*Prakriti*) of a person at the systems biology level. It is very evident that the techniques discussed in Subsections 29.6 and 29.12 can be of immense importance in Ayurnutrigenomics research way forward. From nutri-epidemiology to food quality control efforts, metabolomics will surely find inclusive applications in standard research methodology for obvious reasons. Traditionally inspired approach will augment distinctive research and development effort toward safe and personalized foods. The outcomes will favor not only better health and food habit recommendations but also smart Ayurnutrigenomic-inspired foods.

ACKNOWLEDGMENTS

We thank Dr. Swati Debnath and Sushruta Ghosh for their valuable input and technical support.

REFERENCES

Abe, M., Akbar, S.K.F., Hasebe, A. et al. 2003. Glycyrrhizin enhances interleukin-10 production by liver dendritic cells in mice with hepatitis. *Journal of Gastroenterology*, 38(10): 962–967.

Acharya, S.S. 2002. *Atharvaveda Samhita* [Text with Hindi Translation]. 5th edition. Haridwar, India: Bramhavarchas Shantikunj.

Adhikari, A., Sen, K., Bhattacharya, K. et al. 2011. A study on the evaluation of efficacy and safety of a multiherbal preparation (Andromet) in erectile dysfunction (ED): A randomized placebo controlled trials. *International Journal of Research in Ayurveda and Pharmacy*, 2(5): 1607–1616.

Adhikari, A., Sen, K., Biswas, S. et al. 2012. Study to evaluate the efficacy & safety of a polyherbal preparation in various common dermatological conditions. *The Antiseptic*, 109(1): 27–29.

Afman, L. and Muller, M. 2006. Nutrigenomics: From molecular nutrition to prevention of disease. *Journal of the American Dietetic Association*, 106: 569–576.

Agarwal, R., Diwanay, S., Patki, P. et al. 1999. Studies on immunomodulatory activity of *Withania somnifera* (Ashwagandha) extracts in experimental immune inflammation. *Journal of Ethnopharmacology*, 67(1): 27–35.

Aggarwal, B.B., Prasad, S., Reuter, S. et al. 2011. Identification of novel anti-inflammatory agents from Ayurvedic medicine for prevention of chronic diseases: "Reverse pharmacology" and "bedside to bench" approach. *Current Drug Targets*, 12(11): 1595–1653.

Aggarwal, S., Negi, S., Jha, P. et al. 2010. EGLN1 involvement in high-altitude adaptation revealed through genetic analysis of extreme constitution types defined in Ayurveda. *Proceedings of the National Academy Sciences of the United States of America*, 107(44): 18961–18966.

Arab, L. 2004. Individualized nutritional recommendations: Do we have the measurements needed to assess risk and make dietary recommendations? *Proceedings of the Nutritional Society*, 63: 167–172.

Arumugam, M., Raes, J., and Pelletier, E. 2011. Enterotypes of the human gut microbiome. *Nature*, 473(7346): 174–180.

Baltina, L.A. 2003. Chemical modification of glycyrrhizic acid as a route to new bioactive compounds for medicine. *Current Medicinal Chemistry*, 10: 155–171.

Baravkar, A.A., Sawant, S.D., Bhise, S.B. et al. 2009. Herb–drug interactions. *Indian Journal of Pharmacy Practice*, 2: 25–30.

Barfod, L., Kemp, K., Hansen, M. et al. 2002. Chalcones from Chinese liquorice inhibit proliferation of T cells and production of cytokines. *International Immunopharmacology*, 2: 545–555.

Bhattacharya, D., Sur, T.K., Jana, U. et al. 2008. Controlled programmed trial of *Ocimum sanctum* leaf on generalized anxiety disorders. *Nepal Medical College Journal*, 10(3): 176–179.

Bhattacharya, S., Pal, D., Ghosal, S. et al. 2010. Effects of adjunct therapy of a proprietary herbo-chromium supplement in type 2 diabetes: A randomized clinical trial. *International Journal of Diabetes in Developing Countries*, 30(3): 153–161.

Bhattacharya, S.K., Satyan, K.S., and Ghosal, S. 1997. Antioxidant activity of glycowithanolides from *Withania somnifera*. *Indian Journal of Experimental Biology*, 35(3): 236–239.

Bigham, A., Bauchet, M., Pinto, D. et al. 2010. Identifying signatures of natural selection in Tibetan and Andean populations using dense genome scan data. *PLoS Genetics*, 6(9): e100116. doi:10.1371/journal.pgen.1001116.

Borisy, A.A., Elliott, P.J., Hurst, N.W. et al. 2003. Systematic discovery of multicomponent therapeutics. *Proceedings of the National Academy of Sciences of the United States of America*, 100(13): 7977–7982.

Breen, F.M., Plomin, R., and Wardle, J. 2006. Heritability of food preferences in young children. *Physiology and Behavior*, 88: 443–447.

Cahill, D.J. and Nordhoff, E. 2003. Protein arrays and their role in proteomics. *Advances in Biochemical Engineering and Biotechnology*, 83: 177–187.

Carpenter, K.J. 2003. A short history of nutritional science: Part 1 (1785–1885). *Journal of Nutritional*, 133: 638–645.

Chen, C., Gonzalez, F.J., and Idle, J.R. 2007. LC–MS-based metabolomics in drug metabolism. *Drug Metabolism Reviews*, 39: 581–597.

Choi, S. 2010. Systems biology approaches: Solving new puzzles in a symphonic manner. In: *Systems Biology for Signaling Networks*, ed. S. Choi. New York: Springer, pp. 3–11.

Cinatl, J., Morgenstern, B., Bauer, G. et al. 2003. Glycyrrhizin, an active component of liquorice roots, and replication of SARS-associated coronavirus. *Lancet*, 361: 2045–2046.

Corthals, G.L., Wasinger, V.C., Hochstrasser, D.F. et al. 2000. The dynamic range of protein expression: A challenge for proteomic research. *Electrophoresis*, 21: 1104–1115.

Das, U.N. 2004. Metabolic syndrome X: An inflammatory condition? *Current Hypertension Reports*, 6: 66–73.

Datar, V.K. 1922. *The Charaka Samhita by Agnivesha with Ayurveda Dipika Commentary of Chakrapani Datta*. Mumbai, India: Nirnaya-Sagar Press.

Datta, G.K. and Debnath, P.K. 2000. Ancient use of spices. In: *Recent Trends in Species and Medicinal Plants Research*, ed. A.K. De. New Delhi, India: Associate Publishing, pp. 27–30.

Datta, G.K. and Debnath, P.K. 2001. Stress adaptation in Ayurveda by immunomodulatory *Rasayana. Proceedings of the National Seminar on Rasayana*. March 8–10, 1999, Central Council for Research in Ayurvedic Sciences, New Delhi, India, pp. 60–75.

Debnath, P.K. 1975. Massage as therapy in relation to rheumatic diseases. *Rheumatism*, 6: 1–10.

Debnath, P.K. 1993a. Ayurveda: La Scienza Della Vita L'antica risposta per man mantenere in salute I'uomo. *Boll Ospedale di Varese*, 22: 838–844.

Debnath, P.K. 1993b. Ayurveda—The ancient science of Indian medicine. *Paper presented at the Faculty of Medicine*, University of Insubria, Varese, Italy.

Debnath, P.K. 1995. Ayurveda the science of life—The ancient answer for the maintenance of mental health. *Paper presented at the University of Pavia*, Pavia, Italy.

Debnath, P.K. 1999. Stress adaptation. *Paper presented at the National Seminar on* Rasayan. March 8–10, Central Council for Research in Ayurvedic Science, New Delhi, India.

Debnath, P.K. 2002. Emerging global change of occidental outlook in Ayurveda for improvement in quality of life: Prospects and perspectives. *Paper presented at the 89th session of the Indian Science Congress Association (Medical & Veterinary)*. January 3–8, Lucknow, India.

Debnath, P.K., Chattopadhyay, J., Ghosal, D. et al. 1998. Immunomodulatory role of Ayurvedic *Rasayan* for quality of life. *Proceedings of the International Conference on Stress Adaptation*. January 10–12, Bose Institute, Kolkata, India, p. 38.

Debnath, P.K., Chattopadhyay, J., Mitra, A. et al. 2012. Adjunct therapy of Ayurvedic medicine with anti-tubercular drugs on the therapeutic management of pulmonary tuberculosis. *Journal of Ayurveda and Integrative Medicine*, 3: 141–149.

Debnath, P.K., Datta, G.K., Jana, U. et al. 2006. Ayurvediya Mano-Daihika Vikara (psychosomatic diseases) on modern concept strategies. *Proceedings of the National Academy of Ayurveda*. March 30–31, Government of India, New Delhi, India, pp. 110–117.

Debnath, P.K., Mitra, A., Hazra, J. et al. 2010. Evidence based medicine—A clinical experience on Ayurveda medicine. In: *Recent Advances in Herbal Drug Research and Therapy*, eds. A. Roy and K. Gulati. New Delhi, India: IK International Publishing House Pvt. Ltd., pp. 49–73.

Dhuley, J.N. 1998. Effect of *Ashwagandha* on lipid peroxidation in stress-induced animals. *Journal of Ethnopharmacology*, 60(2): 173–178.

Diwanay, S., Gautam, M., Patwardhan, B. et al. 2004. Cytoprotection and immunomodulation in cancer therapy. *Current Medicinal Chemistry—Anti-Cancer Agents*, 4: 479–490.

Dunn, W.B. and Ellis, D.I. 2005. Metabolomics: Current analytical platforms and methodologies. *Trends in Analytical Chemistry*, 24: 285–294.

Dwarkanath, C. 2010. *Digestion and Metabolism in Ayurveda*. Varanasi, India: Chowkhamba Krishnadas Academy.

Eltzschig, H.K. and Carmeliet, P. 2011. Hypoxia and inflammation. *The New England Journal of Medicine*, 364(7): 656–665.

Frazer, K.A., Murray, S.S., Schork, N.J. et al. 2009. Human genetic variation and its contribution to complex traits. *Nature Reviews Genetics*, 10(4): 241–251.

Fujisawa, Y., Sakamoto, M., Matsushita, M. et al. 2000. Glycyrrhizin inhibits the lytic pathway of complement—Possible mechanism of its anti-inflammatory effect on liver cells in viral hepatitis. *Microbiology and Immunology*, 44: 799–804.

Gautam, M., Diwanay, S.S., Gairola, S., Shinde, Y.S., Jadhav, S.S., and Patwardhan, B.K. 2004. Immune response modulation to DPT vaccine by aqueous extract of Withania somnifera in experimental system. *International Immunopharmacology*, 4(6): 841–849.

Ghodke, Y., Joshi, K., and Patwardhan, B. 2011. Traditional medicine to modern pharmacogenomics: Ayurveda *Prakriti* type and CYP2C19 gene polymorphism associated with the metabolic variability. *Evidence-Based Complementary and Alternative Medicine*, 2011: 1–5. doi:10.1093/ecam/nep206.

Ghosh, D., Skinner, M.A., and Laing, W.A. 2007. Pharmacogenomics and nutrigenomics: Synergies and differences. *European Journal of Clinical Nutrition*, 61: 567–574.

Godard, B. and Ozdemir, V. 2008. Nutrigenomics and personalized diet: From molecule to intervention and nutri-ethics. *OMICS: A Journal of Integrative Biology*, 12: 227–228.

Griffin, J. L., Bonney, S.A., Mann, C. et al. 2004. An integrated reverse functional genomic and metabolic approach to understanding orotic acid-induced fatty liver. *Physiological Genomics*, 17: 140–149.

Gulati, K., Roy, A., Debnath, P.K. et al. 2002. Immunomodulatory Indian medicinal plants. *Journal of Natural Remedies*, 2(2): 121–131.

Gupta, G.L. and Rana, A.C. 2007. *Withania somnifera* (*Ashwagandha*): A review. *Pharmacognosy Review*, 1: 129–136.

Gupta, N., Sharma, P., Santosh Kumar, R.J. et al. 2012. Functional characterization and differential expression studies of squalene synthase from *Withania somnifera. Molecular Biology Reports*, 39(9): 8803–8812.

Haensel, A., Mills, P.J., Nelesen, R.A. et al. 2008. The relationship between heart rate variability and inflammatory markers in cardiovascular diseases. *Psychoneuroendocrinology*, 33: 1305–1312.

Heyn, B. 1990. *Ayurveda: The Ancient Indian Art of Natural Medicine & Life Extension*. Rochester, VT: Healing Arts Press.

Hirai, M.Y., Yano, M., Goodenowe, D.B. et al. 2004. Integration of transcriptomics and metabolomics for under-standing of global responses to nutritional stresses in *Arabidopsis thaliana. Proceedings of the National Academy of Sciences of the United States of America*, 101: 10205–10210.

Horning, E.C. and Horning, M.G. 1971. Metabolic profiles: Gas-phase methods for analysis of metabolites. *Clinical Chemistry*, 17: 802–809.

Hu, C., Liu, H., Du, J. et al. 2009. Estrogenic activities of extracts of Chinese licorice (*Glycyrrhiza uralensis*) root in MCF-7 breast cancer cells. *Journal of Steroid Biochemistry and Molecular Biology*, 113: 209–216.

Huerta-Yepez, S., Baay-Guzman, G.J., Bebenek, I.G. et al. 2011. Hypoxia inducible factor promotes murine allergic airway inflammation and is increased in asthma and rhinitis. *Allergy*, 66(7): 909–918.

Indian Genome Variation Consortium. 2008. Genetic landscape of the people of India: A canvas for disease gene exploration. *Journal of Genetics*, 87(1): 3–20.

Jana, U., Sur, T.K., Maity, L.N. et al. 2010. A clinical study on the management of generalized anxiety disorder with *Centella asiatica. Nepal Medical College Journal*, 12(1): 8–11.

Jo, E.H., Hong, H.D., Ahn, N.C. et al. 2004. Modulations of the Bcl-2/Bax family were involved in the chemopreventive effects of licorice root (*Glycyrrhiza uralensis* Fisch) in MCF-7 human breast cancer cell. *Journal of Agricultural and Food Chemistry*, 52: 1715–1719.

Ju, Y.H., Doerge, D.R., and Helferich, W.G. 2008. A dietary supplement for female sexual dysfunction, Avlimil, stimulates the growth of estrogen-dependent breast tumors (MCF-7) implanted in ovariectomized athymic nude mice. *Food and Chemical Toxicology*, 46(1): 310–320.

Kapil, A. and Sharma, S. 1997. Immunopotentiating compounds from *Tinospora cordifolia. Journal of Ethnopharmacology*, 58: 89–95.

Kashyap, R.L. and Sadagopan, S. 1998. *Rig Veda Mantra Samhita*. Bangalore, India: Sri Aurobindo Kapali Sastry Institute of Vedic Culture.

Kau, A.L., Ahern, P.P., Grin, N.W. et al. 2011. Human nutrition, the gut microbiome and the immune system. *Nature*, 474(7351): 327–336.

Kell, D.B. 2006. Systems biology, metabolic modelling and metabolomics in drug discovery and development. *Drug Discovery Today*, 11: 1085–1092.

Kulkarni, D.A. 1998. *Vagbhattachariya's Rasa Ratna Samucchaya*. New Delhi, India: Meharchand Lachhmandas Publications.

Kumar, A., Kaul, M.K., Bhan, M.K., Khanna, P.K., and Suri, K.A. 2007. Morphological and chemical variation in 25 collections of the Indian medicinal plant, *Withania somnifera* (L.) Dunal (Solanaceae). *Genetic Resources and Crop Evolution*, 54: 655–660.

Lehár, J., Zimmermann, G.R., Krueger, A.S. et al. 2007. Chemical combination effects predict connectivity in biological systems. *Molecular Systems Biology*, 3: 80. doi:10.1038/msb4100116.

Lindon, J.C., Holmes, E., and Nicholson, J.K. 2007. Metabonomics in pharmaceutical R&D. *FEBS Journal*, 274: 1140–1151.

Maity, Y., Adhikari, A., Bhattacharya, K. et al. 2012. A study on evaluation of antidepressant effect of Imipramine adjunct with Aswagandha and Brahmi. *Nepal Medical College Journal*, 13(4): 250–253.

Manning, J.T. and Bundred, P.E. 2001. The ratio of second to fourth digit length and age at first myocardial infarction in men: A link with testosterone? *British Journal of Cardiology*, 8(12): 720–723.

Margetts, B. and Nelson, M. 1997. *Design Concepts in Nutritional Epidemiology*. New York: Oxford University Press.

Mendes, P. 2002. Emerging bioinformatics for the metabolome. *Briefings in Bioinformatics*, 3: 134–145.

Michaelis, M., Geiler, J., Naczk, P. et al. 2011. Glycyrrhizin exerts antioxidative effects in H5N1 influenza a virus-infected cells and inhibits virus replication and pro-inflammatory gene expression. *PLoS One*, 6(5): e19705. doi:10.1371/journal.pone.0019705.

Mishra, B.S. 1993. *Bhava Prakasa of Bhavamishra with Vidyotini Hindi Commentary*. 8th edition. Varanasi, India: Chaukhambha Sanskrit Series Office.

Mishra, L.C., Singh, B.B., and Dagenais, S. 2000. Scientific basis for the therapeutic use of *Withania somnifera* (*Ashwagandha*): A review. *Alternative Medicine Review*, 5(4): 334–346.

Morris, M. and Watkins, S.M. 2005. Focused metabolomic profiling in the drug development process: Advance from lipid profiling. *Current Opinion in Chemical Biology*, 9: 407–412.

Mukherjee, M. and Prasher, B. 2011. Ayurgenomics: A new approach in personalized and preventive medicine. *Science and Culture*, 77(1/2): 10–17.

Mukherjee, P.K., Nema, N.K., Venkatesh, P. et al. 2012. Changing scenario for promotion and development of Ayurveda—Way forward. *Journal of Ethnopharmacology*, 143: 424–434.

Muller, M. and Kersten, S. 2003. Nutrigenomics: Goals and strategies. *Nature Reviews Genetics*, 4: 315–322.

Mutch, D.M., Grigorov, M., Berger, A. et al. 2005. An integrative metabolism approach identifies stearoyl-CoA desaturase as a target for an arachidonate-enriched diet. *FASEB Journal*, 19(6): 599–601.

Mutch, D.M., Wahli, W., and Williamson, G. 2005. Nutrigenomics and nutrigenetics: The emerging faces of nutrition. *FASEB Journal*, 19: 1602–1616.

Nair, P.K., Rodriguez, S., Ramachandran, R. et al. 2004. Immune-stimulating properties of a novel polysaccharide from the medicinal plant *Tinospora cordifolia*. *International Immunopharmacology*, 4: 1645–1659.

Nathani, N. 2013. An appraisal of the concept of diet and dietetics in Ayurveda. *Asian Journal of Modern and Ayurvedic Medical Science*, 2(1): 1–9.

Nicholson, J. K., Holmes, E., Lindon, J.C. et al. 2004. The challenges of modeling mammalian biocomplexity. *Nature Biotechnology*, 22: 1268–1274.

Norheim, F., Gjelstad, I.M.F., and Hjorth, M. 2012. Molecular nutrition research—The modern way of performing nutritional science. *Nutrients*, 4: 1898–1944.

Ordovas, J.M. and Mooser, V. 2004. Nutrigenomics and nutrigenetics. *Current Opinion in Lipidology*, 15: 101–108.

Ozdemir, V., Motulsky, A.G., Kolker, E. et al. 2009. Genome–environment interactions and prospective technology assessment: Evolution from pharmacogenomics to nutrigenomics and ecogenomics. *OMICS: A Journal of Integrative Biology*, 13(1): 1–6.

Patil, D., Gautam, M., Mishra, S. et al. 2013. Determination of withaferin A and withanolide A in mice plasma using high-performance liquid chromatography-tandem mass spectrometry: Application to pharmacokinetics after oral administration of *Withania somnifera* aqueous extract. *Journal of Pharmaceutical and Biomedical Analysis*, 80: 203–212. doi:10.1016/j.jpba.2013.03.001.

Patwardhan, B. and Bodeker, G. 2008. Ayurvedic genomics: Establishing a genetic basis for mind-body typologies. *Journal of Alternative and Complement Medicine*, 14: 571–576.

Patwardhan, B., Joshi, K., and Arvind, C. 2005. Classification of human population based on HLA gene polymorphism and the concept of *Prakriti* in Ayurveda. *Journal of Alternative and Complementary Medicine*, 11(2): 349–353.

Patwardhan, B. and Mashelkar, R.A. 2009. Traditional medicine-inspired approaches to drug discovery: Can Ayurveda show the way forward? *Drug Discovery Today*, 14(15/16): 804–811.

Peng, Y., Yang, Z., Zhang, H. et al. 2011. Genetic variations in Tibetan populations and high-altitude adaptation at the Himalayas. *Molecular Biology and Evolution*, 28(2): 1075–1081.

Phillips, C.M. 2013. Nutrigenetics and metabolic disease: Current status and implications for personalized nutrition. *Nutrients*, 5: 32–57.

Plaeger, S.F. 2003. Clinical immunology and traditional herbal medicines. *Clinical and Diagnostic Laboratory Immunology*, 10: 337–338.

Prabhupada, A.C.B.S. 2010. *Bhagavad-Gita as It Is*. 2nd edition. Mumbai, India: The Bhaktivedanta Book Trust.

Pramanik, S.S., Sur, T.K., Debnath, P.K. et al. 2011. Effect of *Pueraria tuberosa* on cold immobilization stress induced changes in plasma corticosterone and brain monoamines. *Journal of Natural Remedies*, 11(1): 69–75.

Prasher, B., Negi, S., Aggarwal, S. et al. 2008. Whole genome expression and biochemical correlates of extreme constitutional types defined in Ayurveda. *Journal of Translational Medicine*, 6(48): 1–12.

Rabilloud, T. 2002. Two-dimensional gel electrophoresis in proteomics: Old, old fashioned, but it still climbs up the mountains. *Proteomics*, 2: 3–10.

Rege, N.N., Thatte, U.M., Dahanukar, S.A. et al. 1999. Adaptogenic properties of six *Rasayana* herbs in Ayurvedic medicine. *Phytotherapy Research*, 13: 275–291.

Rose, K., Bougueleret, L., Baussant, T. et al. 2004. Industrial-scale proteomics: From liters of plasma to chemically synthesized proteins. *Proteomics*, 4: 2125–2150.

Roy, A. and Debnath, P.K. 1998. *Rogavinischaya (Bengali Version) by Kaviraj Jamini Bhushan Roy*. Kolkata, India: Jaminibhushan Rashtriya Ayurved Medical College & Hospital.

Samant, S.S., Dhar, U., and Palni, L.M.S. 2002. *Himalayan Medicinal Plants: Potential and Prospects*. Almora, India: Gyanodaya Prakashan.

Sandberg, D.H., Sjoevall, J., Sjoevall, K. et al. 1965. Measurement of human serum and bile acids by gas–liquid chromatography. *Journal of Lipid Research*, 6: 182–192.

Saraswat, V.K., Balram, Nathani, S., Jaiswal, M.L., and Kotecha, M. 2012. Significance of Anupana in diet and drug. *International Journal of Ayurvedic and Herbal Medicine* 2(4): 628–635.

Sastry, H.S. 2005. *Vaghabhata: Astanga Hridaya, with Sarvanga Sundara Commentary by Arunadatta & Ayurvedarasayana of Hemadri*. 9th edition. Varanasi, India: Chaukhamba Orientalia.

Satyavati, G.V. 1982. Some traditional medical systems and practices of global importance. *Indian Journal of Medicine Research*, 76: 1–26.

Sauter, H., Lauer, M., and Fritsch H. 1991. Metabolic profiling of plant: A new diagnostic technique. In: *Synthesis and Chemistry of Agrochemicals II*, eds. D.R. Baker, W.K. Moberg, and J.G. Fenyes. Washington, DC: American Chemical Society Press, pp. 288–299.

Schauer, N. and Fernie, A.R. 2006. Plant metabolomics: Towards biological function and mechanism. *Trends in Plant Science*, 11: 508–516.

Schubert, C., Lambertz, M., Nelesen, R.A. et al. 2009. Effects of stress on heart rate complexity—A comparison between short-term and chronic stress. *Biological Psychology*, 80(3): 325–332.

Sekizawa, T., Yanagi, K., Itoyama, Y. et al. 2001. Glycyrrhizin increases survival of mice with herpes simplex encephalitis. *Acta Virologica*, 45: 51–54.

Semenza, G.L. 2000. HIF-1 and human disease: One highly involved factor. *Genes and Development*, 14(16): 1983–1991.

Sempos, C.T., Liu, K., and Ernst, N.D. 1999. Food and nutrient exposures: What to consider when evaluating epidemiologic evidence. *American Journal of Clinical Nutrition*, 69: 1330S–1338S.

Sethi, T.P., Prashe, R B., and Mukherjee, M. 2011. Ayurgenomics: A new way of threading molecular variability for stratified medicine. *ACS Chemical Biology*, 6: 875–880.

Sharma, P. 1988. *Sharangadhara Samhita, Subodhini Hindi Commentary*. 7th edition. Varanasi, India: Chaukh Amarbharati Prakashana.

Sharma, P.V. 2004. *Charaka Samhita*, vol. I, Chaukhamba Orientalia Publication, Varanasi, India.

Sharma, R.K. and Dash, B. 2001. *Agnivesa's Charaka Samhita, Text with English Translation & Critical Exposition Based on Cakrapani Datta's Ayurvedadipika*. 2nd edition. Varanasi, India: Chaukhambha Sanskrit Series Office.

Sharma, S., Dahanukar, S., and Karandikar, S.M., 1985. Effect of long-term administration of the roots of Ashwagandha and Shatvari in rat. *Indian Drugs*, 29: 133–139.

Shastri, A. 2003. *Sushruta: Sushruta Samhita with Ayurveda Tattva Sandipika Hindi Commentary*. 14th edition. Varanasi, India: Chaukhambha Sanskrit Sansthan.

Shastri, R.D. 1985. *Swasthavritta Samucchaya*. 11th edition. Varanasi, India: Kamlavasa, Assi.

Shiomi, T., Guilleminault, C., Sasanabe, R. et al. 1996. Augmented very low frequency component of heart rate variability during obstructive sleep apnea. *Sleep*, 19: 370–377.

Shyur, L.F. and Yang, N.S. 2008. Metabolomics for phytomedicine research and drug development. *Current Opinion* in Chemical *Biology*, 12: 66–71.

Singh, J., Bagchi, G.D., and Khanuja, S.P.S. 2003. Manufacturing and quality control of Ayurvedic and Herbal preparations. In *GMP for Botanicals*, eds. R. Verpoorte, and P.K. Mukherjee, New Delhi, India: Business Horizons Ltd., pp. 201–230.

Singh, N., Bhalla, M., de Jager, P. et al. 2011. An overview on Ashwagandha: A Rasayana (Rejuvenator) of Ayurveda. *African Journal of Traditional, Complementary and Alternative Medicines*, 8(5): 208–213.

Smith, T.G., Robbins, P.A., and Ratclie, P.J. 2008. The human side of hypoxia-inducible factor. *British Journal of Haematology*, 141(3): 325–334.

Srikanthamurthy, K.R. 2008. *Sushruta's Sushruta Samhita*. Varanasi, India: Chaukhamba Orientalia.

Srikanthamurthy, K.R. 2010. *Vagbhat's Astanga Hridayam*. Varanasi, India: Chowkhamba Krishnadas Academy.

Srinivasulu, M. 2010. *Concept of Ama in Ayurveda*. Varanasi, India: Chaukhambha Sanskrit Series Office.

Storz, J.F. 2010. Genes for high altitudes. *Science*, 329(5987): 40–41.

Sumner, L.W., Mendes, P., and Dixon, R. 2003. Plant metabolomics: Large-scale phytochemistry in the functional genomics era. *Phytochemistry*, 62: 817–836.

Sur, T.K., Pandit, S., Mukherjee, R. et al. 2004. Effect of *Sonachandi Chyawanprash* and *Chyawanprash Plus*—Two herbal formulations on immunomodulation. *Nepal Medical Collage Journal*, 6(2): 126–128.

Teucher, B., Skinner, J., Skidmore, P.M. et al. 2007. Dietary patterns and heritability of food choice in a UK female twin cohort. *Twin Research and Human Genetics*, 10: 734–748.

Thatte, U.M., Rao, S.G., Dahanukar, S.A. et al. 1994. Tinospora cordifolia induces colony-stimulating activity in serum. *Journal of Postgraduate Medicine*, 40: 202–203.

Tiwari, P.V. 1976. *Kashyapa Samhita*. 2nd edition. Varanasi, India: Chaukhambha Vishvabharati Prakashan.

Tripathi, B. 2007. *Acharya Charaks' Charak Samhita*. Varanasi, India: Chaukhamba Surbharati Prakashan.

Tripathi, I. 1978. *Vaidya Jeevanam by Lolimbaraj with Vidyotini Hindi commentary*. Varanasi, India: Chaukhambha Orientalia.

Tripathi, R.B. 2005. *Vagbhat's Astanga Samgraha*. New Delhi, India: Chaukhamba Sanskrit Pratishthan.

Tripathi, R.D. 1992. *Ashtanga Samgraha, Soroj Hindi Commentary*. 2nd edition. New Delhi, India: Chaukhambha Sanskrit Sansthan.

Ulrich-Merzenich, G., Zeitler, H., Jobst, D. et al. 2007. Application of the "omic" technologies in phytomedicine. *Phytomedicine*, 14: 70–82.

Utsunomiya, T., Kobayashi, M., Ito, M. et al. 2000. Glycyrrhizin improves the resistance of MAIDS mice to opportunistic infection of *Candida albicans* through the modulation of MAIDS-associated type 2 T cell responses. *Clinical Immunology*, 95: 145–155.

Valiathan, M.S. 2006. *Towards Ayurvedic Biology: A Decadal Vision Document*. Bangalore, India: Indian Academy of Sciences.

Van der Meer-van Kraaij, C., Kramer, E., Jonker-Termont, D. et al. 2005. Differential gene expression in rat colon by dietary heme and calcium. *Carcinogenesis*, 26: 73–79.

Van Meer, G., Leeflang, B.R., Liebisch, G. et al. 2007. The European lipidomics initiative: Enabling technologies. *Methods in Enzymology*, 432: 213–232.

Van Ommen, B., Bouwman, J., Dragsted, L.O. et al. 2010. Challenges of molecular nutrition research: The nutritional phenotype database to store, share and evaluate nutritional systems biology studies. *Genes and Nutrition*, 5: 189–203.

Vodovotz, Y. and An, G. 2009. Systems biology and inflammation. In: *Systems Biology in Drug Discovery and Development: Methods and Protocols*, ed. Q. Yan. Totowa, NJ: Springer Science & Business Media, pp. 181–201.

Vodovotz, Y., Constantine, G., Rubin, J. et al. 2009. Mechanistic simulations of inflammation: Current state and future prospects. *Mathematical Biosciences*, 217: 1–10.

Vodovotz, Y., Csete, M., Bartels, J. et al. 2008. Translational systems biology of inflammation. *PLoS Computational Biology*, 4: 1–6.

Wang, C.Y., Chiao, M.T., Yen, P.J. et al. 2006. Modulatory effects of *Echinacea purpurea* extracts on human dendritic cells: A cell and gene-based study. *Genomics*, 88: 9801–9808.

Wang, M., Lamers, R.J., Korthout, H.A. et al. 2005. Metabolomics in the context of systems biology: Bridging traditional Chinese medicine and molecular pharmacology. *Phytotherapy Research*, 19: 173–182.

Williamson, E.M. 2001. Synergy and other interactions in phytomedicines. *Phytomedicine*, 8: 401–409.

Wishart, D.S., Tzur, D., Knox, C. et al. 2007. HMDB: The Human Metabolome Database. *Nucleic Acids Research*, 35: D521–D526.

Xu, S., Li, S., Yang, Y. et al. 2011. A genome-wide search for signals of high-altitude adaptation in Tibetans. *Molecular Biology and Evolution*, 28(2): 1003–1011.

Yang, N.S., Shyur, L.F., Chen, C.H. et al. 2004. Medicinal herb extract and a single-compound drug confer similar complex pharmacogenomic activities in MCF-7 cells. *Journal of Biomedical Science*, 11: 418–422.

Yang, S.Y., Kim, H.K., Lefeber, A.W.M. et al. 2006. Application of two-dimensional nuclear magnetic resonance spectroscopy to quality control of ginseng commercial products. *Planta Medica*, 72: 364–369.

Ye, M., Liu, S.H., Jiang, Z. et al. 2007. Liquid chromatography/mass spectrometry analysis of PHY906, a Chinese medicine formulation for cancer therapy. *Rapid Communications in Mass Spectrometry*, 21: 3593–3607.

Yoshikawa, M., Matsui, Y., Kawamoto, H. et al. 1997. Effects of glycyrrhizin on immune-mediated cytotoxicity. *Journal of Gastroenterology and Hepatology*, 12: 243–248.

Zeisel, S.H., Allen, L.H., Coburn, S.P. et al. 2001. Nutrition: A reservoir for integrative science. *Journal of Nutrition*, 131: 1319–1321.

Zeng, Z.D., Liang, Y.Z., Chau, F.T. et al. 2007. Mass spectral profiling: An effective tool for quality control of herbal medicines. *Analytica Chimica Acta*, 604: 89–98.

Index

Note: Locators "*f*" and "*t*" denote figures and tables in the text

D

E

F

G

H